Lecture Notes on Data Engineering and Communications Technologies 207

Series Editor

Fatos Xhafa, *Technical University of Catalonia, Barcelona, Spain*

The aim of the book series is to present cutting edge engineering approaches to data technologies and communications. It will publish latest advances on the engineering task of building and deploying distributed, scalable and reliable data infrastructures and communication systems.

The series will have a prominent applied focus on data technologies and communications with aim to promote the bridging from fundamental research on data science and networking to data engineering and communications that lead to industry products, business knowledge and standardisation.

Indexed by SCOPUS, INSPEC, EI Compendex.

All books published in the series are submitted for consideration in Web of Science.

Bing-Yuan Cao · Shu-Feng Wang ·
Seyed Hadi Nasseri · Yu-Bin Zhong
Editors

Intelligent Systems and Computing

 Springer

Editors
Bing-Yuan Cao
School of Mathematics and Information
Sciences
Guangzhou University
Guangzhou, Guangdong, China

Tsinghua University–IEEE Journal FIE
Editorial Department
Peking, China

Seyed Hadi Nasseri
Department of Applied Mathematics
University of Mazandaran
Bābolsar, Iran

Shu-Feng Wang
Pearl River Delta Regional Logistics
Research Center
Guangdong Baiyun University
Guangzhou, China

Yu-Bin Zhong
School of Mathematics and Information
Sciences
Guangzhou University
Guangzhou, Guangdong, China

ISSN 2367-4512   ISSN 2367-4520 (electronic)
Lecture Notes on Data Engineering and Communications Technologies
ISBN 978-981-97-2890-9   ISBN 978-981-97-2891-6 (eBook)
https://doi.org/10.1007/978-981-97-2891-6

This Springer imprint is published by the registered company Springer Nature Singapore Pte Ltd.
The registered company address is: 152 Beach Road, #21-01/04 Gateway East, Singapore 189721, Singapore

If disposing of this product, please recycle the paper.

Congratulations

Abstract. When this meeting was held during the world COVID-19 epidemic, people increasingly felt the importance of science and knowledge and the value of scientists. The author, on behalf of the Iranian Operations Research Society, warmly congratulates the convening of this conference.

Keywords: COVID-19; four sessions and one celebration; congratulates

Ladies and gentlemen:

Good morning! Today we have four meetings as follows, the 10th International Conference on Fuzzy Information and Engineering, the 2nd International Conference of Operations Research and Management, the 6th Annual Meeting of Guangdong Provincial Operations Research Society, the 3rd Annual Meeting of Guangdong, Hong Kong and Macau Operations Research Society, and the 10th Anniversary Celebration of the Guangdong Operations Research Society which are held with the efforts of dear colleagues, institutions, universities, and scientists in China.

I would like to express my warm congratulations on the "four sessions and one celebration!"

As we know, in recent years, the world's conditions have progressed with the spread of the Corona epidemic virus in such a way that it can be said, except for the scientific approach and the help of scientists, the survival of mankind was in danger.

Therefore, the importance of science and knowledge and the value of scientists became more and more felt. Such events will undoubtedly play a basic and key role in the present and in building a better future.

On behalf of the Iranian Operations Research Association, I would like to express my gratitude to all the planners, secretaries of the scientific and executive committees, associations, institutions, and universities that cooperate and support, and especially to my dear, valuable, and hard-working friend and colleague Professor Cao. I am hoping to hold an in-person event with the health and happiness of all friends and colleagues.

Sincerely yours

*Elected as President of the Iranian Operations Research Society in May 2023.

December 2022

Seyed Hadi Nasseri
Vice-president of Iranian
Operations Research Society,
Head of Research Center
of Optimization and Logistics

Preface

This international conference was held online on December 28, 2022, jointly organized by associations, including the 10th International Fuzzy Information and Engineering Association (10th ICFIE preparation), the 2nd International Operations Research and Management Association (2nd ICORG preparation), the 6th Guangdong Provincial Operations Research Society (6th GDORS), the 3rd China Guangdong–Hong Kong–Macao Operations Research Society (3rd CGHMORS), and the 10th Anniversary Celebration for the Founding of the Guangdong Provincial Operations Research Society (CTAGDORS) (also known as the "Four Meetings and One Celebration"). Representatives attended the meeting from countries and regions such as China, Canada, France, the UK, Iran, Pakistan, Malaysia, and Hong Kong.

The 12th Academic Annual Meeting of the Fuzzy Information and Engineering Branch of the Chinese Operations Research Society (FIEBORSC) was held online on September 17, 2022, with representatives and research institutions attending the meeting from 83 higher education institutions and from all over the country.

More than 500 representatives attended the two conferences online. In order to document this great historical event, a collection of Lecture Notes on Data Engineering and Communications Technologies papers was published by the world-renowned publisher Springer Nature.

The opening ceremony of the "Four Meetings and One Celebration" International Conference was presided over by Associate Professor Wang Peihua, and the President of the Conference, Professor Cao Bingyuan delivered a welcoming speech. Professor Witold Pedrycz, a Canadian academician and member of the ICFIE Association, delivered a congratulatory message. Due to the epidemic, Zhang Jingzhong, Honorary Chairman of the Guangdong Provincial Operations Research Association and Academician of the Chinese Academy of Sciences, presented a written speech a with written congratulation sent by Professor Reza Tavakkoli Moghaddam and Professor S. Hadi, the Chairman and Vice Chairman of the Iranian Operations Research Society, the pioneer and pioneer of fuzzy mathematics in China, Professor Wang Peizhuang from Beijing Normal University, and President Zhang Qiu, the local chairman.

The key speech for the conference was given by Professor Witold Pedrycz, a Canadian academician, Professor Li Zhongfei from Southern University of Science and Technology, and Professor Hao Zhifeng, the President of Shantou University. Following came their report topics

Federated Learning: A Perspective of Granular Computing (Witold Pedrycz)
Prediction of China Mutual Fund Returns Based on Machine Learning (Li Zhongfei)
On the New Thoughts of Metaverse and Operational Optimization (Hao Zhifeng).

The celebration was presided over by Fan Suohai, Vice Chairman and Secretary General of Guangdong Provincial Institute of Operations Research, and Professor Cao Bingyuan, Chairman gave a work report. Professor Wang Shufeng, Vice Chairman and

President of the Logistics Branch, and Hu Boping, Chairman of the Disabled Maker Research Branch, respectively, spoke on behalf of the two branches. Representative of Chairman Fang Bin, a French overseas Chinese entrepreneur, and Dr. Shang Benin from the UK; Chairman Li Xiaofeng of Shenzhen Non-Rabbit Health Technology Co., Ltd. delivered congratulatory speeches on behalf of the business community. During the ceremony, the main achievements and outstanding contributions from Zhang Jingzhong and Cao Bingyuan, the meritorious members of the Guangdong Provincial Institute of Operations Research were showcased with a plaque for the meritorious members awarded. The meeting awarded the prize certificate for the first Science and Technology Excellent Paper by the Guangdong Provincial Institute of Operations Research, separately to Associate Professor Hu Yaohua from Shenzhen University and Professor Yang Xiaopeng from Hanshan Normal University, respectively. FIEBORSC is chaired by Professor Zhong Yubin, Vice Chairman and Secretary General of the Fuzzy Information and Engineering Branch of the China Operations Research Society. General Liu Zengliang, Chairman of the Society, delivered the annual work report of the Society. The key paper presenters of the conference are renowned experts in the field of fuzzy information and engineering in China, pioneers and pioneers of fuzzy mathematics in China, and Professor Wang Peizhuang from Beijing Normal University. His report title is Factor Space for 40 Years; the title of General/Professor Liu Zengliang's report from Beijing National Defense University is Theory and Application of Factor Neural Networks. The conference speakers also included Professor Xu Zeshui from Sichuan University; Professor Yuan Xuehai from Dalian University of Technology; Professor He Qing, Institute of Computing, Chinese Academy of Sciences; Professor Guo Sicong from Liaoning University of Engineering and Technology, and Professor Yu Fusheng from Beijing Normal University; Professor Zhong Yubin from Guangzhou University; Professor Chen Shuishui from Jimei University; and Professor Li Taifu from Chongqing University of Science and Technology. 62 academic papers were exchanged in writing, including the theory and application of fuzzy mathematics, fuzzy information, and engineering, reflecting some new phased research achievements of the society in the theory and application of fuzzy information and engineering. Experts and scholars attending the meeting unanimously believe that the 12th FIEBORSC Academic Annual Conference will play an important role in promoting research on fuzzy mathematics and system theory and applications in China.

This book includes 32 high-quality research papers from the two sessions. Covering topics in the fields of certainty, stochastic uncertainty, and fuzzy uncertainty, especially the accessible language communication achievements of Chen Yongwen and others, the world's first; Professor Wang Shufeng's digital logistics was first proposed in China; Dr. Zuo Peijun's key to prevention and treatment of COVID-19 is forward-looking and highly praised. Finally, strong support for world-renowned publishing houses Springer Natural and Tsinghua University IEEE International Magazine Fuzzy Information and Engineering; The unremitting efforts of the International Fourth Conference and a Celebration, the FIEBORSC Organizing Committee, and the strong assistance from the domestic emerging enterprises in their original era, we sincerely appreciate the high importance attached by Guangzhou University. In particular, all the organizers of the two sessions and all the presenting delegates were fearless of the danger of COVID-19 positive, and their selfless dedication and strong support for the conference will be forever in history.

Bing-Yuan Cao
Yu-Bin Zhong

Organization

Presidium of the Joint International Assembly

Chairman

Cao Bingyuan   Tsinghua University–IEEE, China
  (Editor-in-Chief of FIAE)

Members

Academician Witold Pedrycz, Canada
Academician Zhang Jingzhong, China
Wang Peizhuang, USA
Tavakkoli Moghaddam, Iran
Hadi Nasser, Iran
Hao Zhifeng, China

General Assembly I: The 10th International Conference on Fuzzy Information and Engineering (ICFIE'2022).
The 2nd International Conference of Operations Research and Management (ICORG'2022).
The 6th Annual Meeting of Guangdong Provincial Operations Research Society.
The 3rd Annual Meeting of China, Guangdong, Hong Kong and Macau Operations Research Society.
Also the 10th Anniversary Celebration of Guangdong Operational Research Society (GDORS).
December 28, 2022
https://www.csaeep.com/#/meeting/
http://icodm.org/en/index.php

Sponsor

The International Fuzzy Information and Engineering Association (Preparation)
Guangdong, Hong Kong, and Macau Operations Research Society (Hong Kong)
Yuan Age (Hainan) Digital Technology Co., Ltd. (China)

Organizer

Guangdong Provincial Operations Research Association, China

Co-organizers

International Center of Optimization and Decision Making, Iran
Iranian Operations Research Society, Iran
China Science and Education Press, Hong Kong
China Education Research Foundation, Hong Kong
China Harmony Foundation, Hong Kong
International Institute of General Systems Studies, China Branch
Fuzzy Information and Engineering, Tsinghua University, CINA-IEEE
Research Center of Optimization and Logistics, Iran

Chairman

Cao Bingyuan, China

Honorary Chairman

Academician Zhang Jingzhong, China

Chairman of the Procedure Committee

Hadi Nasseri, Iran

Local Chairman

Zhang Qiu, China

Members

Yang Xiaopeng, China
Khizar Hayat, Pakistan
Li Zhongfei, China
Fang Bin, France
Peng Jigen, China
Shang Bening, UK
Fan Suohai, China
Abdul Samad Shibghatullah, Malaysia

General Assembly II: The 12th Academic Annual Meeting of the Fuzzy Information and Engineering Branch of the Operations Research Society of China, September 17, 2022.

Chairman

Liu Zengliang, China

Honorary Chairman

Wang Peizhuang, USA

Executive Chairman

Zhong Yubin, China

Members

Xu Zeshui, China
Yu Fusheng, China
Yuan Xuehai, China
Chen Shuili, China
He Qing, China
Guo Sicong, China
Amirhossein Nafe, Iran

Sponsor

Fuzzy Information and Engineering Branch of the Operations Research Society of China.

Organizer

Guangzhou University

Contents

Annual Meeting of Fuzzy Information and Engineering

Variable Universe Fuzzy Control Based on Adaptive Error Integral
for Uncertain Nonlinear Systems with Time-Delay 3
  Yinhua Guo, Jiayin Wang, Fusheng Yu, and Cunqin Shi

Optimization of Ideal Reflector for Radio Telescope Based on Improved
Composite Algorithm 28
  Yizhi Guo, Zhixuan Li, Anyi Yao, Yiqi Liang, and Yubin Zhong

Linear Programming Subject to Max-Product Fuzzy Relation Inequalities
with Discrete Variables 37
  Xu Fu, Chang-xin Zhu, and Zejian Qin

Global Optimization for the Concave-Concave Multiplicative
Programming with Coefficient 49
  ChangXin Zhu and YongBin OuYang

Fish Swarm Algorithm Based Reflecting Surface Adjustment Strategy
for Radio Telescopes 64
  Jin-pei Chen, Yu-cheng Wu, Xi-yue Liu, and Guo-dong He

Study on Furnace Temperature Curve Based on Heat Conduction Model 73
  Yaping Chai, Bingying Song, Congrun Zhang, and Haichang Luo

Research on Intelligent Adjustment Technology of Active Reflector
of Radio Telescope 96
  Wangwei Zhong, Weitong Chen, Jialian Li, Hongbin Lin, and Yubin Zhong

Dynamic Credit Line Decisions for MSMEs 108
  Li Lin, Zheyu Gong, and Yubin Zhong

Research on the Ordering and Transportation of Raw Materials
for Production Enterprises 119
  Jiachong Zheng, Yifeng Zhang, Yuhang Duan, Yang Mao, and Jiexia Yang

Research on Cantonese Cultural and Creative Product Design Strategy
Based on Fuzzy Kano Model 129
  Xingyi Zhong

Establishment and Application of Catalyst Optimization Model Based
on Genetic Algorithm 140
    Shuyi Wang, Weize Zhang, Ziqi Zhong, Yongtao Li, and Zhongyuan Peng

Application and Optimization of Shape Adjustment Design Based
on "FAST" Active Reflector Model 151
    Hongwei Tang, Xianglong Li, and Wanying Wu

Research and Practice of Internet+ Innovation and Entrepreneurship
Ability Cultivation Model Based on Cooperative Learning 163
    Yubin Zhong, Weitong Chen, Jialian Li, and Hongbin Lin

Research on the Optimization Model of Ethanol Coupling to C4 Olefins
Based on Regression Analysis 177
    Ying Xie, Qi Li, Wenya Zhu, Qiwen Wu, Jun Wan, and Hong Mai

An Empirical Study on Eco-Tourism Network Integral Problem 189
    Jiasheng Wu, Yongru Cen, Ziqing Li, Weiyu Liu, Lanxi Bai,
    and Shuitian Wu

Study on Iterative Desert Crossing Based on Game Theory 201
    Jiekai Cao, Wenzhu Wang, Shengqi Deng, and Qingping He

Joint International Conference for "Four Meetings and a Celebration"

Research on Hamacher Operations for q-rung Orthopair Fuzzy Information 219
    Wen Sheng Du

Healthcare Referral and Coordination in a Two-Tier Service System
via Government Subsidy Scheme 232
    Caimin Wei, Zhiyuan Tong, Zongbao Zou, and Zhongping Li

An Indirect Solution to WPLTS Group Decision Making Problem 251
    Guo-Cheng Zhu and Jian Xu

My Humble Opinion on Digital Logistics Theory 268
    Shufeng Wang

A Dual Referral Optimization Model for Medical Clusters Based
on Queuing Theory and Cooperative Game Incentives 279
    Zhiyuan Tong, Yulin Nie, Miaoxia Zhuang, Sitong Liang, Ning Liu,
    and Caimin Wei

Digitally Empowers Yiwen to Promote Barrier-Free Language
Communication   292
    Chen Yong-wen, Chen Ying-bing, Peng Bin, and Cao Bing-yuan

Research on the Recommendation Method of Postgraduate Supervisor
Based on Natural Intelligence Information Integration   300
    Xiao He, Tai-Fu Li, Yu-Yan Li, Shao-Lin Zhang, Fu-Hong Qing,
    and Qiao Zeng

The Qudratic Programming with Max-Min Fuzzy Relation Equations
Constraint   319
    Xue-Gang Zhou and YongBin OuYang

Feature Selection Algorithm for Multi-label Classification Based on Graph
Operations   335
    Qianyao Tang, Fuyi Wei, Zhihong Liu, Hang Zhang, Ying Guo,
    Peiwei Su, and Dongxin Li

A Comprehensive Evaluation Method for Tourism Value Co-creation
Based on Fuzzy Analysis   343
    Wu Nan, Hu ChuXiong, and Wang Xiaoyu

U-type Tube Vibration Based Fuel Density Portable Testing Instrument
Research   358
    Fuhong Qin, Yuyan Li, Taifu Li, Xianguo Wang, Hongchao Zhao,
    and Qiang He

The Research on System Structure Weights of Feedback, Scale, Inclusion
and Their Utility Functions   373
    Liting Zeng, Jianbo Guan, and Yunshi Fong

Screening for Late-Onset Fetal Growth Restriction in Antepartum Fetal
Monitoring Using Deep Forest and SHAP   383
    Jianhong Huo, Guohua Li, Chongwen Li, Xia Li, Guiqing Liu,
    Qinqun Chen, Jialu Li, Yuexing Hao, and Hang Wei

The Solution Closest to a Given Vector in the System of Fuzzy Relation
Inequalities   395
    Miaoxia Chen, Abdul Samad Shibghatullah, and Xiaopeng Yang

Fuzzy Analysis Model of Financial System -- Application in Credit Risk
of Commercial Banks   405
    Yin Tian-hui and Cao Bing-yuan

Mutation and Prediction of COVID-19 422
  Pei-Jun Zuo, Long-Long Zuo, Zhi-Hong Li, and Li-Ping Li

Author Index 435

About the Editors

Bing-Yuan Cao is a professor at Guangzhou University. Doctoral (postdoctoral) supervisor; Lingnan Second Class Chair Professor at Foshan University of Science and Technology; President of the Fuzzy Information and Engineering (FIE) Branch of the China Operations Research Society, the Guangdong Provincial Operations Research Society, and the Guangdong–Hong Kong–Macao Operations Research Society; Chairman of the International FIE Association (soon to be established) and lifelong editor in chief of the FIE magazine at Tsinghua University. In 1987, he was the first to propose "Fuzzy Geometric Programming" and published over 190 papers (many in Top and Q1 journals). He was also the editor in chief of over ten Springer books, received two International FIE Society Achievement Awards, three Provincial Science and Technology Awards, one Springer Most Popular Textbook by Readers, and ranked 5th in the 2021 Optimization Top 100 by Elsevier SciVal.

Prof. Shu-Feng Wang is Director of the Pearl River Delta Regional Logistics Research Center, Guangdong Baiyun University, Professor of Logistics Economics, and Master Supervisor. His main research interests include transportation economics, logistics economics, digital economics, and international trade. In the past five years, he has focused on the evolution and system integration of digital logistics theory, the efficiency of regional logistics resource allocation, the systematic reconstruction of the industrial value chain in the Guangdong–Hong Kong–Macao Greater Bay Area, the generation mechanism of systematic enterprise operation capability and its cultivation research.

He has presided over and completed more than 20 scientific research projects including provincial and ministerial projects and published five academic works and teaching materials such as "Regional Logistics Theory and Empirical Research," "Transportation Management," "Logistics System Planning and Design Theory and Method," and more than 50 academic papers.

Seyed Hadi Nasseri is Associate Professor, University of Mazandaran, Iran. He is Regional Editor of Fuzzy Information and Engineering, Managing Editor of Caspian Mathematical Science magazine, and Editor of International Journal of Operations Research and Its Applications. He has published more than 112 papers and more than 124 conference papers. He used to be Guest Editor of Journal of Applied Mathematics and Special Issue of Operations Research, Contributing Editor of Iranian Journal of Management Sciences in China, Vice President of Iranian Operational Research Society, Secretary General of Iran Fuzzy System Society, Chairman of the International Center for Optimization and Decision-making, the first place in the entrance examination of applied mathematics master.

Prof. Yu-Bin Zhong received his bachelor's degree and master's degree from Beijing Normal University and currently works as Professor at the School of Mathematics and Information Science, Guangzhou University, China. His main research direction is fuzzy system and artificial intelligence, operations research and control, systems engineering and intelligent algorithm, hyperalgebraic structure, and mathematical theory of knowledge representation. He has published more than 30 papers in core journals, including more than 20 in SCI and EI, and presided over and researched more than ten teaching and scientific research projects.

Annual Meeting of Fuzzy Information and Engineering

Variable Universe Fuzzy Control Based on Adaptive Error Integral for Uncertain Nonlinear Systems with Time-Delay

Yinhua Guo[1], Jiayin Wang[1(✉)], Fusheng Yu[1,2], and Cunqin Shi[2]

[1] School of Mathematical Science, Beijing Normal University, Beijing, China
202021130106@mail.bnu.edu.cn, {wjy,yufusheng}@bnu.edu.cn
[2] School of Mathematics and Statistics, Longdong University, Qingyang, China

Abstract. This article deals with variable universe fuzzy control (VUFC) based on adaptive error integral for uncertain nonlinear systems with time delay. A novel fuzzy control strategy is firstly addressed for the system to guarantee asymptotic stability. The strategy concludes in three modules: first, introduced an extended state observer to observe the extended state variables and estimate the unknown total disturbance and the unmodeled dynamics of the controlled plant; second, employ a tracking differentiator to track the dynamic characteristics of the input signal as quickly as possible in the strategy; and thirdly, develop a variable universe fuzzy control for the external-loop system with the observer and the differentiator. Compared with the existing results, the proposed strategy relaxes the restrictions on the conventional domain by the self-regulation of the tracking error domain during the operation of the system. Furthermore, to enhance the control accuracy of the controller, we define an extended adaptive error integral which can not only remain superior to the traditional error integral but also overcome the inferior. In addition, the boundedness of all signals of the presented systems is analyzed and proved by the Lyapunov theory. In the end, several simulation examples and some comparisons are presented to highlight the novelty and feasibility of the proposed method.

Keywords: Variable Universe Fuzzy Controller · Extended State Observer · Tracking Differentiator · Extraction-Expansion Factor · Extended Adaptive Error Integral

1 Introduction

Due to its complexity and diversity, the phenomenon of lag (time delay) always exists in the motion law of the systems. Therefore, the research work on time delay has attracted extensive attention from scholars and experts in the field of control, including system stability analysis, Smith predictor, internal model control, adaptive control, fault-tolerant control, fuzzy control, etc. [1–8]. Among the methods, due to the strong robustness, the influence of disturbance and parameter variation on the control effect is greatly weakened, and fuzzy control is especially suitable for the control of nonlinear, time-varying,

© The Author(s), under exclusive license to Springer Nature Singapore Pte Ltd. 2024
B.-Y. Cao et al. (Eds.): ICFIE 2022, LNDECT 207, pp. 3–27, 2024.
https://doi.org/10.1007/978-981-97-2891-6_1

and pure lag systems, which has been widely studied by scholars. But the performance of the early fuzzy controller is not high enough for the time delay system. A fuzzy robust tracking control controller for uncertain nonlinear time-delay systems is proposed, which ensures ideal tracking performance in the sense that all closed-loop signals are consistent and ultimately bounded [9]. A continuous-time fuzzy compensation control method is proposed to eliminate the uncertainty of the system with time delay [10]. A robust fuzzy predictive control scheme with interval delay is proposed, and experimental results show that the scheme has high control accuracy [11].

As we all know, the control rules are the core of the fuzzy controller, whether it's correct or not directly affects the performance of the controller, and the number of its is also an important factor to weigh the performance of the controller. However, the design of a fuzzy controller is highly dependent on expert experience, and the performance of the control system will be greatly reduced when the existing experience is deviated or even wrong [12]. In addition, the accuracy of traditional fuzzy control is limited by fuzzy division. If the number of fuzzy partitions is increased to improve the control accuracy, the steady state error will be increased and the reliability of the controller will be reduced. To overcome the shortcomings of the conventional fuzzy control mentioned above, Li proposed variable universe fuzzy control (VUFC) and pointed out that the discourse universe of fuzzy variables can be extract or expand under the condition of reducing or increasing systematic errors [13]. To some extent, it solves the problem of over-reliance on expert experience, and can precisely strengthen the control of certain target areas. The ability of local target densification control makes the variable universe fuzzy method widely studied by scholars, and successfully applied in robot control [14], neural control engineering [15], etc. However, the input of variable universe fuzzy controller usually uses the difference (system error) between the input signal and the output value and the difference (system error change rate) between the derivative of the input signal and the output value. When the input signal and output value contain higher noise, the error and error change rate will interact with each other. If the extraction-expansion factor of the input variable of VUFC is simply chosen as a function of error or error change rate, the input universe will be adjusted repeatedly and the convergence speed of the system will be reduced. In the current related studies, the control methods of contract-expansion factors are not uniform: [16–21] mathematical function, integral regulation, multi-population genetic algorithm, fuzzy neural network, and backpropagation algorithm are respectively used to describe and optimize the gain of contract-expansion factors. Observer-based variable theory fuzzy controller [22, 23] is proposed for chaotic systems and unknown dead zone nonlinear systems respectively. These control strategies improve the control accuracy of the system to a certain extent, but the influence between error and error change rate has not been completely solved, especially when the input signal contains higher noise, the problem of extracting the input differential signal has not been solved. In addition, the study of VUAF for uncertain nonlinear systems with time delay is very little.

In this paper, a novel variable universe fuzzy controller is proposed for uncertain nonlinear systems with time delay. That is, based on introducing the extended state observer and tracking differentiator [24], an extended adaptive error integral is designed to reduce the influence of approximation error and time delay on tracking error. This is because both the extended state observer and the tracking differentiator are constructed using the error approximation principle. The innovations of this paper are as follows:

- A nonlinear extended state observer driven by error is introduced to observe the extended state variables and estimate the unknown total disturbance and the unmodeled part of the controlled system. This observer does not depend on the exact model object and is more practical in engineering.
- An error-driven tracking differentiator is introduced to track the dynamic characteristics of the input signal as fast as possible to obtain approximate external differential signals and smooth signal.
- A variable universe fuzzy control method based on extended adaptive error integration is proposed to ensure the stability of the system. The method is suitable for complex time-delay systems, and an extended error integral is defined by using the advantages of error integral, stretching factor, and variable domain, which greatly improves the control accuracy of the proposed controller.
- The superiority of the controller is verified by simulation and comparison tests.

2   Methodology Development

2.1   Model of Uncertain Nonlinear Systems with Time Delay

The equation of uncertain nonlinear system with time delay [25] is described as:

$$\begin{cases} \dot{x} = \hat{f}(t - \tau, d, x, u) \\ y = \overline{g}(x, u, t) \end{cases} \tag{1}$$

where τ, y, d, x and u are time delay, input, state variable, external disturbance and output of system respectively. $\hat{f}(x, u, t - \tau, d)$ is the state with respect to the related variables, $\overline{g}(x, u, t)$ is output functions with respect to the related variables.

The control objective of this paper is to find a fuzzy tracking controller such that the state of the system (1) follows a given stable reference model while keeping all closed-loop signals ultimately bounded all the time.

2.2   Estimation of the System

In uncertain nonlinear systems, a key challenge is the estimation state. As a special case of uncertain nonlinear systems, the state estimation of uncertain nonlinear systems with time delay is also very important. In this study, the system state can be expanded from the original n-dimensional state $x_0 = [x_1, x_2, \cdots, x_n]$ to an $n+1$-dimensional vector $x = [x_1, x_2, \cdots, x_n, x_{n+1}]$. The basic principle of the ESO algorithm is shown in a) and has been described in detail in previous studies [24, 26, 27], and [28]. The ESO algorithm can successfully estimate the normal and perturbed system states and compensate the estimated states and total perturbations, which provides an opportunity for the efficient design of VUFC strategy.

1) Coordinate Transformation of the Original System

A *border* nonlinear time-varying dynamic system is with the following form:

$$y^{(n)}(t) = g\left(y^{(n-1)}(t), y^{(n-2)}(t), \ldots, y(t), w(t)\right) + b(x, t)u(t) \tag{2}$$

where $y(t)$ (simply y), $u(t)$ (simply u), $w(t)$ (simply w) and $b(x, t)$ (simply b) are the output, input, external disturbance and high-frequency gain of the dynamic system respectively. Let b_0 be the estimated value of b, $f(t) = g\left(y^{(n-1)}, y^{(n-2)}, \ldots, y, w\right) + (b - b_0)u$ be the so-called "total disturbance". Then, Eq. (3) is transformed into Eq. (3)

$$y^{(n)} = f\left(y^{(n-1)}, y^{(n-2)}, \ldots, y, w\right) + b_0 u \tag{3}$$

Expand $f(t)$ into a new state of system and assume that $f(t)$ is differentiable, let $\dot{f}(t) = x_{n+1}(t)$. The system in Eq. (3) can be described in a state space Eq. (4)

$$\begin{cases} \dot{x}_i = x_{i+1}(t) \quad i = 1, \ldots, n-1 \\ \dot{x}_n(t) = g\left(y^{(n-1)}, y^{(n-2)}, \ldots, y, w\right) + bu \\ y = x_1(t) \end{cases} \tag{4}$$

2) Extended State Observer

A To better estimate the state of Eq. (4), the nonlinear extended state observer (NESO) with y and u as inputs is designed as:

$$\begin{cases} e(t) = z_1(t) - y(t) \text{ or simply } e = z_1 - y \\ z_1(t + 1) = z_1(t) + h(z_2(t) - \beta_{01}e) \\ z_2(t + 1) = z_2(t) + h(z_3(t) - \beta_{02}fe_1) \\ \vdots \\ z_n(t + 1) = z_n(t) + h(z_{n+1}(t) - \beta_{0n}fe_{n-1} + b_0 u(t)) \\ z_{n+1}(t + 1) = z_{n+1}(t) + h(-\beta_{0n+1}fe_{n-1}) \\ fe_{i-1} = fal(e, \alpha_{i-1}, \delta) \\ fal = \begin{cases} e\delta^{\alpha-1} & |e| \leq \delta \\ |e|^{\alpha} sign(e) & |e| > \delta \end{cases} \end{cases} \tag{5}$$

When the observer gain β_{0i} is adjusted properly, $z_i(t)$ can respectively tracks $x_i(t)(i = 1, 2, \cdots, n+1)$. Choose the following control rate to compensate for the estimated total disturbance $f(t)$:

$$u(t) = \frac{u_0(t) - z_{n+1}(t)}{b_0} \tag{6}$$

where $u_0(t)$ will be analyzed later. Equation (6) is substituted into Eq. (3) to get Eq. (7).

$$y^{(n)}(t) = f(t) - z_{n+1}(t) + u_0(t) \approx u_0(t) \tag{7}$$

Then, Eq. (7) means that the model in Eq. (3) becomes an integral series object after estimating and compensating the total disturbance $f(t)$ through the NESO $u_0(t)$.

Because of the complexity of the actual environment, for the input signal accompanied by random noise, the differential method is used directly to obtain its differential will amplify the noise. In order to obtain the smooth input signal and its differential signal, a tracking differentiator is introduced in this paper.

2.3 Acquisition of Smooth Input Signal and Its Differential Signal

The function *fhan* is to obtain a smooth control trajectory by tracking-differentiator (TD). The algorithm of TD as is follows:

$$
\begin{cases}
fh(t) = fhan\big(v_1(t) - v_0(t), v_n(t), r(t), \hat{h}_0(t)\big) \\
v_i(t+1) = v_i(t) + \hat{h}v_{i+1}(t) \quad i = 1, 2, \ldots, n-1 \\
v_n(t+1) = v_n(t) + \hat{h}fh(t) \\
fhan : \begin{cases}
d = \hat{h}_0 r \\
d_0 = \hat{h}_0 d \\
v(t) = v_1(t) - v_0(t) + \hat{h}_0 v_n(t) \\
a_0 = \sqrt{d^2 + 8r|v(t)|} \\
a(t) = \begin{cases} v_n(t) + v(t)/\hat{h} & |v(t)| \le d_0 \\ v_n(t) + \frac{1}{2(a_0-d)sign(v(t))} & |v(t)| > d_0 \end{cases} \\
sat = \begin{cases} \frac{a(t)}{d} & |a(t)| \le d \\ sign(a) & |a(t)| > d \end{cases} \\
fh = -rsat(a(t), d)
\end{cases}
\end{cases}
\tag{8}
$$

where $v_0(t)$ is the input trajectory, $v_1(t)$ is the tracking trajectory of the input trajectory, $v_i(t)$ ($i = 1, 2, \ldots, n$) is the differential trajectory of the tracking trajectory, r is the speed factor, $\hat{h}$ is the sampling period, $\hat{h}_0$ is the filter factor, *fhan* is the fastest tracking function.

The smooth tracking signal and differential signal are further used in VUFC strategy, which is beneficial to solve the contradiction between VUFC overshoot and rapidity. The design process is as follows.

2.4 A Novel Variable Universe Adaptive Fuzzy Controller

It is well known that fuzzy control is suitable for control systems that are hard to model. And the design of a fuzzy controller depends mostly on the fuzzy inference rule base on the knowledge of field experts. In addition, the stability of fuzzy controller has been analyzed in Refs. [29, 30]; and has been widely used such as [31]. However, traditional fuzzy controller is not ideal for high-precision control situations. Therefore, variable universe adaptive fuzzy control is proposed in [32], which can effectively control nonlinear systems. The basic principle of a variable universe adaptive fuzzy controller is briefly stated in this section (Refer to [33] for details)

1) Traditional Fuzzy Controller

Fuzzy controller combines the error (e_i) between the differential signal obtained by TD and the state estimation value estimated by NESO. The $e_i(t)$ is

$$e_i = v_i - z_i \ (i = 1, \ldots, n) \tag{9}$$

According to Eq. (9), the input state error of fuzzy controller is $e^T = (e_1, e_2, \cdots, e_n)$. Let $E_i = [-I_i, I_i]$ be the universe of input variable e_i, $Y = [-Q, Q]$ be the universe of output variable u, $A_i = \{A_{ij}\}$ be a fuzzy partition on $\mathcal{X}_i$, and $\mathcal{B} = \{B_j\}$ be a fuzzy partition on $\mathcal{Y}$. Each element of A_i is considered a linguistic variable value of e_i, and each element of $\mathcal{B}$ is considered a linguistic variable value of $u(i = 1, 2, \cdots, n; j = 1, 2, \cdots, m)$. Suppose that we have m fuzzy rules, each of which is in the following form:

$$If \ e_1 \ is \ A_{j1}, \ e_2 \ is \ A_{j2}, \ldots, e_n \ is \ A_{jn}, \ then \ y_j \ is \ B_j \tag{10}$$

Let e_{ji} be the peak point of A_{ji}, and u_j be the peak point of B_j, $(i = 1, 2, \cdots, n; j = 1, 2, \cdots, m)$, [13, 33, 34]. According to [33], the fuzzy logic system (i.e., the fuzzy controller) based on Eq. (10) can be expressed as

$$y(e) = F(e) \triangleq \sum_{j=1}^{m} \prod_{i=1}^{n} A_{ij}(e_i) y_j \tag{11}$$

2) Contraction-Expansion Factors of Universes

The contraction-expansion factor of the universe of the fuzzy controller with n inputs and one output is described as follows.

A *contraction-expansion factor* on $\mathcal{X} = [-E, E]$ is a function $\xi: \mathcal{X} \to [0, 1]$, $e \mapsto \xi(e)$, which satisfies the following conditions:

i. Evenness: $\xi(e) = \xi(-e)$;
ii. Near zero: $\xi(0) = \varepsilon$, where ε is a small positive real constant;
iii. monotonicity: strictly monotonically increasing on $[0, E]$;
iv. compatibility: $|e| \leq \xi(e)E$.

For any $e \in \mathcal{X}$, $\mathcal{X}(e) \triangleq \xi(e)\mathcal{X} \triangleq [-\xi(e)E, \xi(e)E] \triangleq \{\xi(e)e'|e' \in \mathcal{X}\}$ is called a variable universe of $[-E, E]$. $\mathcal{X}$ is the initial universe of the variable universes whose situation changing is shown in Fig. 1.

Two typical contraction-expansion factors of universe $[-E, E]$ are

$$\xi(e) = (|e|/E)^\tau + \varepsilon, \ \ \tau > 0 \tag{12}$$

$$\xi(e) = 1 - \lambda exp\left(-ke^2\right), \ \lambda \in (0, 1), \ k > 0 \tag{13}$$

where ε is an arbitrary positive constant.

3) Variable Universe Adaptive Fuzzy Controller

Suppose that $\xi_i(e_i)$ and $\beta(u)$ are the contraction-expansion factors of the initial universes

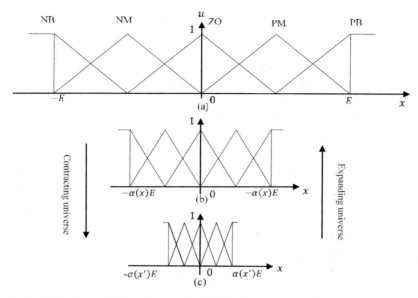

Fig. 1. Principle of a variable universe: (a) initial universe and its fuzzy portion; (b) & (c): contracting/expanding.

of $\mathcal{X}_i$ and Y respectively. By means of the conclusions in Ref. [33], a variable universe adaptive fuzzy controller based on Eq. (11) can be written as follows:

$$y(e(t+1)) = \beta(t) \sum_{j=1}^{m} \prod_{i=1}^{n} A_{ij}\left(e_i(t) \Big/ \xi_i(e_i(t))\right) y_j \tag{14}$$

where $e(t) \triangleq (e_1(t), e_2(t), \cdots, e_n(t))^T$, $\xi_i(e_i(t)) = 1 - \lambda exp\left(-ke_i(t)^2\right)$, $\beta(t) = 1 - \lambda exp\left(-ku(e(t))^2\right)$.

For convenience, we write

$$\omega(e(t)) \triangleq \sum_{j=1}^{m} \prod_{i=1}^{n} A_{ij}\left(e_i(t) \Big/ \xi_i(e_i(t))\right) y_j \tag{15}$$

Then Eq. (14) is expressed as

$$y(e(t+1)) = \beta(t)\omega(e(t)) \tag{16}$$

Compare with linear active disturbance rejective control (LADRC), the nonlinear active disturbance rejective control (NADRC) with strong noise suppression and robustness has higher control performance for multivariable and strongly coupled nonlinear systems [35]. In recent years, the LADRC and NADRC method has been widely applied and achieved a well effects [36]. But the nonlinear PD in NADRC is not the best choice. As well known, a defect of the PID controller is a stronger dependency on the model and a smaller dynamic adjustment range. To overcome the defect, in this section, we propose a novel variable universe adaptive fuzzy control which is designed by employing a variable universe adaptive fuzzy controller that does not depend on the model and improves the accuracy of control. Furthermore, to improve the control accuracy of the proposed control strategy, we will define an extended adaptive error integral in following section.

3   The Asymptotically Stable of the Variable Universe Adaptive Fuzzy Analyzed by Lyapunov Theory

3.1   Feasibility Analysis of the Proposed Method

In this section, we first set up the control objectives, and then discuss how to develop a novel variable universe adaptive fuzzy to achieve these control objectives.

According to [37, 38], if the state vector $x = (x_1, x_2, \cdots, x_n)^T = (y, \dot{y}, \cdots, y^{(n-1)})^T \in R^n$ is available measurement in the systems (1), the systems have model information accurately. The control objective is to force y to follow a given bounded reference trajectory, $v_0(t)$, under the constraint that all trajectories involved must be bounded. However, the principle of the proposed control strategy is to use the error drive to stabilize the system, that is, use the output $z_i(t)$ of the NESO to track $x_i(t)(i = 1, 2, \cdots, n + 1)$, use $v_i(t)$ to track $z_i(t)(i = 1, 2, \cdots, n)$ and $v_1(t)$ to track the input signal $v_0(t)$. In this way, the system keeps circulating the diagnostic error indirectly forces y to track given bounded reference signal, $v_0(t)$. Therefore, we only need to analyze the stability of the error feedback controller, namely just needs to analyze the stability of the variable universe fuzzy controller. Specially, we have

Control Objectives

Determine the parameter of feedback control $u(x)$ meet the following conditions:

i) In the sense, the closed-loop system must be globally stable, that is, all variables, $v(t)$, $x(t)$, $z(t)$ and $u_0(x)$, must be uniformly bounded; i.e., $|v_i(t)| \leq M_{v_i} < \infty$, $|x_i(t)| \leq M_{x_i} < \infty$, $|z_i(t)| \leq M_{z_i} < \infty$ $(i = 1, 2, \cdots, n)$ and $|u_0(x)| \leq M_u < \infty$ for all $t \geq 0$, where M_{v_i}, M_{x_i}, M_{z_i} and M_u are given in advance.

ii) Under the constraint of i), the estimated tracking error $e(t) \equiv v(t) - z(t)$, should be as small as possible.

Now, we analyze how to construct an adaptive fuzzy controller $u_a(x)$ to realize these control objectives.

First of all, let $e(t) = (v_1(t) - z_1(t), v_2(t) - z_2(t), \cdots, v_n(t) - z_n(t))^T = (e_1(t), e_2(t), \cdots, e_n(t))^T$ and $k = (k_n, k_{n-1}, \cdots, k_1)^T \in R^n$ be such that all roots of the polynomial $h(s) = s^n + k_1 s^{n-1} + \cdots + k_n$ are in the open left-half plane. Under the function g and b are known, then the control law

$$\bar{u}(t) = 1/b\left[-g(t) + v_{n+1}(t) + k^T e(t)\right] \tag{17}$$

is applied to Eq. (2) resulting in

$$e_{n+1}(t) + k_1 e_n(t) + \cdots + k_n e_1(t) = 0 \tag{18}$$

which shows that $\lim_{t \to \infty} e(t) = 0$ is a key objective of control. On account of $g(t)$ and b are unknown, the optimal control, $\bar{u}(t)$ cannot be realized. Our purpose is to design a variable universe fuzzy system extremely close to this optimal control.

Supposed that the control u is summation of an adaptive fuzzy control $u_a(x)$ and an adjustable control $u_b(x)$:

$$u = u_0(x) + u_b(x) \tag{19}$$

where $u_a(x)$ is a variable universe fuzzy system in the form of (16), and $u_b(x)$ will be discussed later in this section. If we put the Eq. (19) into the Eq. (3), we have

$$x^{(n)} = f(x) + b[u_a(x) + u_b(x)] \tag{20}$$

Now add the substructure $b\bar{u}$ to Eq. (20) and after some direct operations, the error equation governing the closed-loop system is

$$e_n(t) = -k^T e + [\bar{u}(t) - u_a(x) - u_b(x)] \tag{21}$$

or, equivalently

$$\dot{e} = E_c e + b_c \left[u^* - u_c(x) - u_s(x) \right] \tag{22}$$

where

$$E_c = \begin{bmatrix} 0 & 1 & 0 & 0 & \cdots & 0 & 0 \\ 0 & 0 & 1 & 0 & \cdots & 0 & 0 \\ \vdots & \vdots & \vdots & \vdots & \ddots & \vdots & \vdots \\ 0 & 0 & 0 & 0 & \cdots & 0 & 1 \\ -k_n & k_{n-1} & \cdots & \cdots & \cdots & \cdots & -k_1 \end{bmatrix}, \quad b_1 = \begin{bmatrix} 0 \\ \vdots \\ 0 \\ b \end{bmatrix} \tag{23}$$

Define $V_e = \frac{1}{2} e^T P e$, where P is a symmetric positive definite matrix satisfying the Lyapunov equation [39].

$$E_c^T P + P E_c = -Q \tag{24}$$

where $Q > 0$. Using Eq. (24) and the error equation Eq. (22), we have

$$\begin{aligned} \dot{V}_e &= -\frac{1}{2} e^T Q e + e^T P b_1 [\bar{u}(t) - u_a(x) - u_b(x)] \\ &\le -\frac{1}{2} e^T Q e + \left| e^T P b_1 \right| (|\bar{u}(t)| + |u_a|) - e^T P b_1 u_s \end{aligned} \tag{25}$$

Our task now is to design u_b such that $\dot{V}_e \leq 0$. To realize $\dot{V}_e \leq 0$, we need the following assumption:

Assumption: We can define a function $\overline{f}(x)$ and a constant $\tilde{b}$ such that $|f(x)| \leq \overline{f}(x)$ and $0 < \tilde{b} \leq b$.

We construct the compensator $u_b(x)$ as follows [39]:

$$u_b(x) = I_1^* sgn\left(e^T P b_1\right)\left[|u_a| + \frac{1}{\tilde{b}}\left(\overline{f}(x) + |v_{n+1}| + \left|k^T e\right|\right)\right] \tag{26}$$

Where $I_1^* = 1$ if $V_e > \overline{V}$ ($\overline{V}$ is a constant specified according to the needed), and $I_1^* = 0$, if $V_e \leq \overline{V}$. Because $b > 0$, $sgn\left(e^T P b_1\right)$ can be determined; thus, the adjustable control, u_b, of Eq. (26) can be achieved. Put Eq. (26) and Eq. (17) into Eq. (25) and considering the $I_1^* = 1$ case, we have

$$\dot{V}_e = -\frac{1}{2}e^T Q e + \left|e^T P b_1\right|\left[\frac{1}{\tilde{b}}\left(|f| + |v_{n+1}| + \left|k^T e\right|\right) + |u_a|\right.$$
$$\left. -|u_a| - \frac{1}{\tilde{b}}\left(\overline{f}(x) + |v_{n+1}| + \left|k^T e\right|\right)\right] \leq -\frac{1}{2}e^T Q e \leq 0 \tag{27}$$

For the same reason, $I_1^* = -1$, the conclusion is also valid. Therefore, using the adjustable control u_b of Eq. (26), we always have $V_e < \overline{V}$. Because $P > 0$, the boundedness of V_e heralds the boundedness of x.

Equation (26) implies that the u_b is nonzero only when the error function V_e is greater than the positive constant $\overline{V}$. Nevertheless, largely control is undesirable because it may increase the implementation cost. So, we do not choose this strategy because the u_b is usually very large. But, if the system tends to be unstable, then the compensator u_b begins to force $V_e < \overline{V}$. But the strategy must manual adjustment, which brings a lot of inconvenience to the application of the controller.

Next, we replace the $u_a(x) + u_b(x)$ by the variable universe fuzzy system (18) multiplied by a special adaptive error integral.

3.2   Variable Universe Fuzzy Controller Based on Extended Adaptive Error Integral

Many uncertain disturbance factors exist in an actual system, if a disturbance is added at a certain moment, the static error of the system will increase rapidly and the stability of the system will deteriorate. Meanwhile, the accumulation of error integral over time makes the output change of the controller increase and the system becomes sluggish. From the feasibility analysis of the variable universe adaptive fuzzy in Sect. 3.1, we see that it is necessary to add a compensator when the change of the control quantity is very large. But a compensator is added will increase the difficulty of controller design. In this section, *an extended adaptive error integral is proposed*, that is, taking the range of error integral $[-I, I]$ (Eq. (27)) and the output range of the fuzzy controller $[-P, P]$ (Eq. (28)) as Cartesian product will give universe $[-R, R]$, and name $[-R, R]$ as the range of the extended error integral. Then the range of extended error integral multiplied by the extraction-expansion factor defined on $[-R, R]$ gives the extended state error integral with variable range which named as *extended adaptive error integral*. The extended adaptive error integral not only limits the change of the control quantity, eliminates state error, and enhances the disturbance rejection of the systems, but also will not increase the difficulty of controller design.

The detailed design process of the error feedback controller of an extended adaptive error integral is divided into four steps as follows:

Step 1 *Extended contraction-expansion factor*
The error integral is calculated as:

$$\overline{\beta}(t) = \overline{\beta}\left(e^T(t)\right) = K_I \int_0^t e^T(\tau)d\tau + \overline{\beta}(0) \tag{28}$$

Let

$$[-I, I] = \{\overline{\beta}(t)|t < +\infty\} \tag{29}$$

be the range of all error integrals;
Let

$$[-P, P] = \{\omega\left(e^T(t)\right)|t < +\infty\} \tag{30}$$

is the output range of the variable universe fuzzy controller.

An *extended contraction-expansion factor* on $X = [-R, R] = [-I, I] \times [-P, P]$ is a function $\rho: X \rightarrow [0, 1], x \mapsto \rho(x)$ which satisfies the following conditions:

i. Evenness: $\rho(x) = \rho(-x)$;
ii. monotonicity: strictly monotonically increasing on $[0, R]$;
iii. compatibility: $|x| \leq \rho(x)E$.

where $[-I, I] = \{\overline{\beta}(t)|t < +\infty\}$, $[-P, P] = \{\omega(e^T(t))|t < +\infty\}$.
A practical extended contraction-expansion factor is given below:

$$\rho(t) = 1 - P_I exp\left(-\sum_{i=1}^{n} \lambda_i \rho_i^2 - \Upsilon(\omega(t))^2\right) \tag{31}$$

where $P_I(\in (0, 1))$ and $\lambda_i, \Upsilon(\geq 0)$ are parameters given in advance $(i = 1, 2, \cdots, n)$.

Step 2 *Extended adaptive variable universe*
For any $x \in X$, $X(x) \triangleq \rho(x)X \triangleq [-\rho(x)R, \rho(x)R] \triangleq \{\rho(x)x'|x' \in X\}$ is called a variable universe of $[-R, R]$. X is the initial universe of the variable universe.

Step 3 *Extended adaptive error integral*
The extended adaptive error integral $\tilde{\rho}(t)$ is defined as follows:

$$\tilde{\rho}(t) \in [-\rho(t)R, \rho(t) R] \tag{32}$$

Step 4 *Design of error feedback controller of an extended adaptive error integral*
Let $\tilde{\rho}(t) = \beta(t)$, then Eq. (16) can be expressed as

$$u_0 = \tilde{\rho}(t)\omega\left(e^T(t)\right) \tag{33}$$

Substituting Eq. (33) into Eq. (7) results in a variable universe adaptive fuzzy (34) based on an extended adaptive error integral.

$$u(t) = \frac{u_0(t) - z_{n+1}(t)}{b_0} \tag{34}$$

Next, we provide an example to show the performance of the proposed controller given by Eq. (34).

Example. Take the expression of the second-order system y as follows:

$$\ddot{y} = u(t) \tag{35}$$

The design of the proposed controller for the second-order system includes three steps; Step 1 and Step 2 are given in Sect. 4. Here, we only explain Step 3* in detail as follows.

According to Eq. (9), the state error is $e^T(t) = (e_1(t), e_2(t))$ or simply $e^T = (e_1, e_2)$, the universe of e_1 is $[-E_1, E_1] = [-100, 100]$, the universe of e_2 is $[-E_2, E_2] = [-50, 50]$, and the universe of control variable $u(t)$ is $[-U, U] = [-50, 50]$. The 49 fuzzy rules are listed in Table 5. For simplicity, the fuzzy sets NB, NS, NM, ZO, PM, PS, PB defined on $[-E_1, E_1]$ of the input variable e_1 are denoted as $A_1, A_2, ..., A_7$, respectively. Similarly, the fuzzy sets $NB, NS, NM, ZO, PM, PS, PB$ defined on $[-E_2, E_2]$ of the input variable e_2 are denoted as $B_1, B_2, ..., B_7$, respectively, and the fuzzy sets $NB, NS, NM, ZO, PM, PS, PB$ defined on $[-U, U]$ of the output variable $u(t)$ are denoted as $C_1, C_2, ..., C_7$, respectively. The membership functions of the values of variable e_1 are respectively defined by

$$A_1(e_1) = gaussmf\left(e_1, \left[\hat{a}, \hat{b}\right]\right), \quad A_2(e_1) = trimf\left(e_1, \left[\hat{a}, \hat{b}, \hat{c}\right]\right)$$

$$A_3(e_1) = trimf\left(e_1, \left[\hat{a}, \hat{b}, \hat{c}\right]\right), \quad A_4(e_1) = trimf\left(e_1, \left[\hat{a}, \hat{b}, \hat{c}\right]\right)$$

$$A_5(e_1) = trimf\left(e_1, \left[\hat{a}, \hat{b}, \hat{c}\right]\right), \quad A_6(e_1) = trimf\left(e_1, \left[\hat{a}, \hat{b}, \hat{c}\right]\right)$$

$$A_7(e_1) = gaussmf\left(e_1, \left[\hat{a}, \hat{b}\right]\right),$$

The choice of the membership functions of variables e_2 and u is identical to that of the membership functions of variable e_1. The parameters of these memberships are listed in Table 7 in the Appendix.

The error integration $\overline{\beta}(t)$ and contraction-expansion factors $\xi_1(e_1)$, and $\xi_2(e_1, e_2)$ are defined as,

$$\overline{\beta}(t) = \int_0^t (8, 8) e^T(\tau) d\tau + \beta(0) \tag{36}$$

$$\xi_1(e_1) = 1 - 0.97 exp\left(-e_1^2\right) \tag{37}$$

$$\xi_2(e_1, e_2) = 1 - 0.97 exp\left(-0.2e_1^2 - 0.8e_2^2\right) \tag{38}$$

Based on the above 49 fuzzy rules and Eq. (16), the variable universe adaptive fuzzy controller is designed as follows:

$$\omega(e_1, e_2) = U \frac{\sum_{i,j,k=1}^{7} a_i b_j u_k}{\sum_{i,j=1}^{7} a_i b_j} \tag{39}$$

where $a_i = A_i\left(\frac{e_1}{\alpha_1(e_1)}\right)$, $b_j = B_j\left(\frac{e_2}{\alpha_2(e_1,e_2)}\right)$, $(i, j = 1, 2, \cdots, 7)$.
The contraction-expansion factor $\rho(t)$ of the control quantity is defined as

$$\rho(t) = 1 - 0.9exp\left(-0.96\rho_1^2(t) + 0.03\rho_2^2(t) + 0.03\omega(e_1, e_2)\right) \tag{40}$$

According to Eq. (32) and (33), the control amount is expressed as

$$u_0(t) = \tilde{\rho}(t)\omega(e_1, e_2) \tag{41}$$

Based on Eq. (40), Eq. (41) and Eq. (7), the proposed controller for the second-order system is

$$u(t) = (u_0(t) - z_3(t))/b_0 \tag{42}$$

Let $r(t) = 1$ be the reference input signal; $x(0) = 0.5$ is the initial value of system G_1; step time $t = 2$ s, the simulation time $T = 10$ s; a disturbance is added at $T = 5$ s. The parameters of system G_1 in Table 7 in the Appendix, simulation result under the proposed controller, NADRC and VUFC, the simulation results and performance comparison are shown in Figs. 2, 3, 4 and Table 1 respectively.

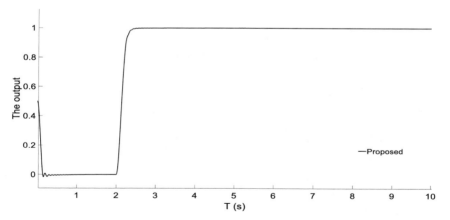

Fig. 2. Dynamic response of system G_1 under the proposed controller.

Figures 2, 3, and 4 and Table 1 show that the dynamic response of the system under the proposed controller is better than under the NADRC and VUFC. Especially, settling time is shorter, with smaller steady-state error.

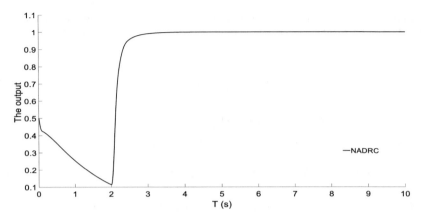

Fig. 3. Dynamic response of system G_1 under the NADRC

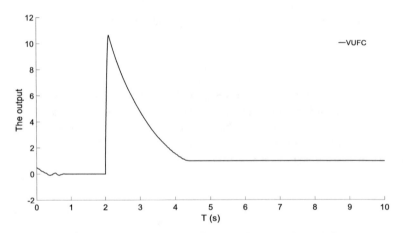

Fig. 4. Dynamic response of system G_1 under the VUFC

Table 1. Performance comparison for system: Eq. (35)

Index	Proposed	NADRC	VUFC
$IAE = \int_0^t e^2(t)dt$	0.0081	0.1785	46
Settling time (s)	0.36	3.7	0.85

So far, the design of the proposed controller is completed. Figure 5 is the framework of the proposed controller, where v_0 is the expected value, v^T is the differential signal extracted by TD, u_0 is the output of the variable universe fuzzy controller, and u is the output of the proposed controller.

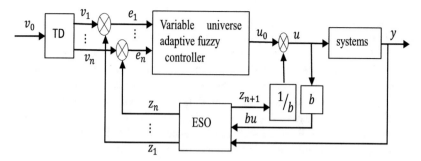

Fig. 5. Framework of the proposed controller

As shown in Fig. 5, the proposed controller consists of three parts: the TD, the variable universe fuzzy controller based on extended error integral, and the ESO. The controller is used to continuously detect the error between the differential signal generated by the TD and the state variable estimate generated by the ESO in the controlling process. According to the fuzzy control rules and contraction or expansion of the universe of e^T, the control amount u_0 is obtained. The output z_{n+1} of the ESO compensates for output u_0 of the proposed controller to form the control quality u which achieves the control effect for the controlled plant and avoids oscillation caused by a large change in the universe of e^T. The general nature of the proposed controller will be verified through different experiments in Sect. 4.

4 Simulation Analysis

In this section, the proposed controller is tested on several typical second-order systems: the uncertain system with delay-time G_2 in [40], the chaotic systems (a typical nonlinear system) G_3 in [39] and the nonlinear time-variant systems with unknown orders, G_4, in [41]. All of these systems can be generated from Eq. (1). In the following, we apply the second-order novel variable universe fuzzy designed in Sect. 3 (example) to systems: G_2 and G_3. To verify the ability of the proposed controller to control the complex system, in Sect. 4.3 designed a controller different from that in Sect. 3 (example) and used it for the complex uncertain system G_4. The parameters selected for these systems G_2–G_4 are listed in Table 7 in the Appendix.

4.1 Proposed Controller for Uncertain Systems with Time-Delay

To verify the practicability and strong robustness of the proposed controller under the combined influence of time delay and compound disturbance, consider the plant G_2 is expressed as follows:

$$G_2(s) = \frac{200}{3s^2 + 100s + 50} e^{-\tau s} \tag{43}$$

Let $r(t) = 1$ be the reference input signal, time delay variable $\tau = 0.01$, $x(0) = 0$ be the initial value of system G_2, the simulation time be $T = 5\,\text{s}$, and a disturbance $d_1 = 30\%$ be added at $T = 2\,\text{s}$ and a disturbance $d_2 = 10\%$ be added at $T = 3\,\text{s}$; the simulation results and performance comparison are shown in Fig. 7 and Table 2 respectively. The disturbance design of the system $G_2(s)$ is shown in Fig. 6.

Let $r(t) = 1$ be the reference input signal, time delay variable $\tau = 0.01$, $x(0) = 0$ be the initial value of system G_2, the simulation time be $T = 5\,\text{s}$, and a disturbance $d_1 = 30\%$ be added at $T = 2\,\text{s}$ and a disturbance $d_2 = 10\%$ be added at $T = 3\,\text{s}$; the simulation results and performance comparison are shown in Fig. 7 and Table 2 respectively. The disturbance design of the system $G_2(s)$ is shown in Fig. 6.

Let $r(t) = 1$ be the reference input signal, time delay variable $\tau = 0.01$, $x(0) = 0$ be the initial value of system G_2, the simulation time be $T = 5\,\text{s}$, and a disturbance $d_1 = 30\%$ be added at $T = 2\,\text{s}$ and a disturbance $d_2 = 10\%$ be added at $T = 3\,\text{s}$; the simulation results and performance comparison are shown in Fig. 7 and Table 2 respectively. The disturbance design of the system $G_2(s)$ is shown in Fig. 6.

Let $r(t) = 1$ be the reference input signal, time delay variable $\tau = 0.01$, $x(0) = 0$ be the initial value of system G_2, the simulation time be $T = 5\,\text{s}$, and a disturbance $d_1 = 30\%$ be added at $T = 2\,\text{s}$ and a disturbance $d_2 = 10\%$ be added at $T = 3\,\text{s}$; the simulation results and performance comparison are shown in Fig. 7 and Table 2 respectively. The disturbance design of the system $G_2(s)$ is shown in Fig. 6.

Let $r(t) = 1$ be the reference input signal, time delay variable $\tau = 0.01$, $x(0) = 0$ be the initial value of system G_2, the simulation time be $T = 5\,\text{s}$, and a disturbance $d_1 = 30\%$ be added at $T = 2\,\text{s}$ and a disturbance $d_2 = 10\%$ be added at $T = 3\,\text{s}$; the simulation results and performance comparison are shown in Fig. 7 and Table 2 respectively. The disturbance design of the system $G_2(s)$ is shown in Fig. 6.

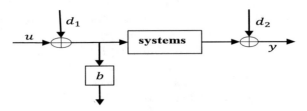

Fig. 6. Disturbance design of the system $G_2(s)$

Figure 7 and Table 2 show that the transition time and disturbance rejection performance of system G_2 under the proposed controller are better than those under NADRC-Smith, the NADRC, and VUFC.

Remark: The proposed controller has higher controlling accuracy than the controller in [41].

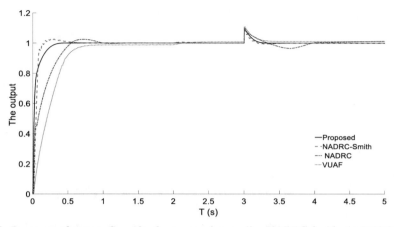

Fig. 7. Response of system G_2 under the proposed controller, NADRC-Smith, the NADRC and VUFC

Table 2. Performance comparison for system: Eq. (33)

Index	Proposed	NADRC-Smith	NADRC	VUFC
$IAE = \int_0^t e^2(t)dt$	0.039	0.13	0.018	0.128
Settling time (s)	0.38	0.41	1.210	0.750
Disturbance error at 1 s	0.01	0.02	0.011	0.243
Disturbance error at 2 s	0.27	0.58	0.980	0.38

4.2 Proposed Controller for Chaotic Systems

To verify the strongest stability of the proposed controller, we consider the duffing forced oscillation system G_3, which is a typical Chaotic system and described as follows:

$$G_3 : \begin{cases} \dot{x}_1 = x_2 \\ \dot{x}_2 = -0.1x_2 - x_1^3 + 12cost + u(t) \end{cases} \tag{44}$$

Let $r(t) = 1$ be the reference input signal, $x(0) = (2, 2)$ be the initial value of system G_3. Step time $T = 20$ s and the simulation time be $T = 100$ s; the simulation results and performance comparison are shown in Fig. 8 and Table 3 respectively.

Figure 8 and Table 3 show that under the proposed controller, the system has a better dynamic response than under the NADRC and VUFC. Especially, the proposed controller almost has steady-state errors and oscillations.

4.3 Proposed Controller for Nonlinear Time-Variant Systems with Unknown Orders

Due to the complexity of the actual system, there are a lot of uncertainties in the system, such as the order of the system. This section presents a simulation of the second-order

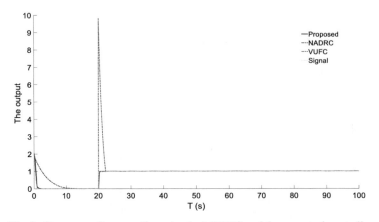

Fig. 8. Response of system G_3 under the LADRC and the proposed controller

Table 3. Performance comparison for system: Eq. (35)

Index	Proposed	NADRC	VUFC
$IAE = \int\limits_{0}^{t} e^2(t)dt$	1.3019	0.1785	31
Settling time (s)	2.17	13.51	0.85

nonlinear time-variant systems with unknown orders G_4 under the proposed controller. Observe that lower-order NADRC can handle the control problem of higher-order systems [42]. In addition, the capacity of the low-order NADRC to control high-order systems has been verified [43, 44]. The second order proposed controller will be designed and discussed for system G_4, where G_4 is

$$G_4 : \begin{cases} \dot{x}_1 = x_2 \\ \dot{x}_2 = x_3 \\ \dot{x}_3 = -3x_3 - 3x_2 - x_1^3 + (1 + 0.5sin(t)) \\ \qquad +2.1sin(t) + u(t) \end{cases} \tag{45}$$

The design of the proposed controller for the second system includes four steps; Step 1, Step 2, and Step 4 are given in Sect. 4. Here, we only explain Step 3* in detail as follows.

According to Eq. (9), the state error is $e^T(t) = (e_1(t), e_2(t))$ or simply $e^T = (e_1, e_2)$, the universe of e_1 is $[-E_1, E_1] = [-2, 2]$, the universe of e_2 is $[-E_2, E_2] = [-8, 8]$,and the universe of control variable $u(t)$ is $[-U, U] = [-3, 3]$. The 49 fuzzy rules are listed in Table 8 For simplicity, the fuzzy sets NB, NS, NM, ZO, PM, PS, PB defined on $[-E_1, E_1]$ of input variable e_1 are denoted as A_1, A_2, $\cdots$, A_7, respectively. Similarly, the fuzzy sets NB, NS, NM, ZO, PM, PS, PB defined on $[-E_2, E_2]$ of input variable e_2 are denoted as B_1, B_2, $\cdots$, B_7, respectively, and the fuzzy sets NB, NS, NM, ZO, PM, PS, PB defined on $[-U, U]$ of the output variable $u(t)$ are denoted as C_1, C_2, $\cdots$, C_7, respectively. The membership functions of the values of variable e_1 are defined by

$$A_1(e_1) = zmf\left(e_1, \left[\hat{a}, \hat{b}\right]\right), \quad A_2(e_1) = trimf\left(e_1, \left[\hat{a}, \hat{b}, \hat{c}\right]\right)$$

$$A_3(e_1) = trimf\left(e_1, \left[\hat{a}, \hat{b}, \hat{c}\right]\right), \quad A_4(e_1) = trimf\left(e_1, \left[\hat{a}, \hat{b}, \hat{c}\right]\right)$$

$$A_5(e_1) = trimf\left(e_1, \left[\hat{a}, \hat{b}, \hat{c}\right]\right), \quad A_6(e_1) = trimf\left(e_1, \left[\hat{a}, \hat{b}, \hat{c}\right]\right)$$

$$A_7(e_1) = smf\left(e_1, \left[\hat{a}, \hat{b}\right]\right),$$

The choice of the membership functions of variables e_2 and u is identical to that of the membership functions of variable e_1. The parameters of these memberships are listed in Table 9 in the Appendix. The error integration $\bar{\beta}(t)$ and contraction-expansion factors $\alpha_1(e_1)$, $\alpha_2(e_1, e_2)$ and $\gamma(\rho_1(t), \rho_2(t))$ are defined as,

$$\bar{\beta}(t) = \int_0^t (120, 30)e^T(\tau)d\tau + \beta(0) \tag{46}$$

$$\alpha_1(e_1) = 1 - 0.9exp\left(-e_1^2\right) \tag{47}$$

$$\alpha_2(e_1, e_2) = 1 - 0.9exp\left(-0.7e_1^2 - 0.2e_2^2\right) \tag{48}$$

Based on the above 49 fuzzy rules and Eq. (18), the expression of the variable universe adaptive fuzzy based on extended adaptive error integral is showed in Eq. (46):

$$\omega(e_1, e_2) = U\frac{\sum_{i,j,k=1}^7 a_i b_j u_k}{\sum_{i,j=1}^7 a_i b_j}. \tag{49}$$

where $a_i = A_i\left(\frac{e_1}{\alpha_1(e_1)}\right)$, $b_j = B_j\left(\frac{e_2}{\alpha_2(e_1, e_2)}\right)$, $(i, j = 1, 2, \cdots, 7)$.
The contraction-expansion factor $\rho(t)$ of the control quantity is defined as

$$\rho(t) = 1 - 0.9exp\left(-0.96\rho_1^2(t) + 0.03\rho_2^2(t) + 0.03\omega(e_1, e_2)\right) \tag{50}$$

Then according to Eq. (32), the control amount by expressed as

$$u_0(t) = \tilde{\rho}(t)\omega(e_1, e_2) \tag{51}$$

Based on Eq. (7), Eq. (47), and Eq. (51), the proposed controller for the second-order system is

$$u(t) = (u_0(t) - z_3(t))/b_0 \tag{52}$$

Let $r(t) = 0.5 + \sin(0.5t) + \sin(t)$ be the reference input signal; $x(0) = 0$ is the initial value of system G_4; the simulation time is $T = 25$ s; the simulation results and performance comparison are shown in Fig. 9 and Table 4 respectively.

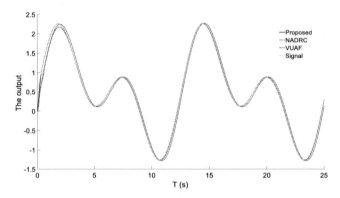

Fig. 9. Response of G_4 under NLADRC and the proposed controller.

Table 4. Performance Comparison for System: Eq. (35)

Index	Proposed	NADRC	VUFC
$IAE = \int\limits_{0}^{t} e^2(t)dt$	0.265	0.312	0.037
Settling time (s)	0.34	13.51	2.56

Figure 9 and Table 4 show that the setting time of the steady-state error has been greatly improved compared with the NADRC and VUFC.

Conclusion: The above examples explained that the proposed controller realized better control performance than NADRC and ADRC-Smith and VUFC [22]. Especially, the steady-state error is effectively weakened and the ability of disturbance rejection is distinctively enhanced.

5   Conclusions

Time delay and uncertainty are common problems in natural systems. To enhance the ability of VUFC to deal with delay and disturbance problems, a variable universe fuzzy controller based on extended adaptive error integral is developed in this paper. Firstly, the extended state observer is introduced to estimate and compensate the uncertainty and total disturbance of the system. Secondly, the tracking differentiator is introduced to obtain the smooth input signal and differential signal. Furthermore, based on the analysis and proof of the system presented by using Lyapunov theory, an extended adaptive error integral is defined. Finally, the proposed controller is applied to several typical systems. Simulation results show that the VUFC based on adaptive error integration outperforms the state-of-the-art controllers in terms of settling time, weakening overshoot, suppressing disturbance, steady-state error, and robustness. Our future research will apply the proposed control method to the minimal phase system and analyze it. In addition, the method proposed in this paper has only been verified in theory and simulation, and further experiments are needed to verify the feasibility of the control strategy proposed in this paper.

Acknowledgments. The project is supported by National Natural Science Foundation (Number: 11971065, 11571001).

Appendix

Table 5. Fuzzy Rules

e_1	e_2						
	NB	NM	NS	ZO	PS	PM	PB
NB	NB	NB	NM	NM	NS	NS	ZO
NM	NB	NM	NM	NS	NS	ZO	PS
NS	NM	NM	NS	NS	ZO	PS	PS
ZO	NM	NS	NS	ZO	PS	PS	PM
PS	NS	NS	ZO	PS	PS	PM	PM
PM	NS	ZO	PS	PS	PM	PM	PB
PB	ZO	PS	PS	PM	PM	PB	PB

See Tables 6, 7 and 8.

Table 6. The parameters of the membership functions in G_1–G_3

Variable	Parameter	membership function						
		NB	NM	NS	ZO	PS	PM	PB
e_1	$\hat{a}$	−20	−100	−60	−20	0	20	20
	$\hat{b}$	−100	−60	−30	0	30	60	100
	$\hat{c}$	0	−20	0	20	60	100	0
e_2	$\hat{a}$	7.08	−50	−30	−10	0	10	09
	$\hat{b}$	−50	−30	−15	0	15	30	50
	$\hat{c}$	0	−10	0	10	30	50	0
u	$\hat{a}$	10	−100	−60	−20	10	33	−10
	$\hat{b}$	−100	−66	−35	0	35	66	100
	$\hat{c}$	0	−33	−10	20	60	100	0

Table 7. The parameters of the membership functions in G_4

variable	parmeter	membership function						
		NB	NM	NS	ZO	PS	PM	PB
e_1	$\hat{a}$	−2	−2	−1.3	−0.7	0	0. 7	1.5
	$\hat{b}$	−1.5	−1.3	−0.7	0	0.7	1.3	2
	$\hat{c}$	0	−0.7	0	0.7	1.3	2	0
e_2	$\hat{a}$	−8	−8	−5.3	−2.7	0	2.7	5.5
	$\hat{b}$	−5.5	−5.3	−2.7	0	2.7	5.3	8
	$\hat{c}$	0	−2.7	0	2.7	5.3	8	0
u	$\hat{a}$	−3	−3	−2	−1	0	1	2
	$\hat{b}$	−2	−2	−1	0	1	2	3
	$\hat{c}$	0	−1	0	1	2	3	0

Table 8. Fuzzy rules

e_1	e_2						
	NB	NM	NS	ZO	PS	PM	PB
NB	NB	NB	NB	NB	NM	ZO	ZO
NM	NB	NB	NB	NB	NM	ZO	ZO
NS	NM	NM	NM	NM	ZO	PS	PS
ZO	NM	NM	NS	ZO	PS	PM	PM
PS	PS	PS	ZO	PM	PM	PM	PM
PM	ZO	ZO	PM	PB	PB	PB	PB
PB	ZO	ZO	PM	PB	PB	PB	PB

References

1. Garrido, J., Vázquez, F., Morilla, F.: Inverted decoupling internal model control for square stable multivariable time delay systems. J. Process. Control. **24**(11), 97–106 (2014)
2. Ding, B.C., Gao, C.B., Ping, X.B.: Dynamic output feedback robust MPC using general polyhedral state bounds for the polytopic uncertain system with bounded disturbance. Asian J. Control **18**(2), 699–708 (2016)
3. Li, Z.C., Huang, C.Z., Yan, H.C.: Stability analysis for systems with time delays via new integral inequalities. IEEE Trans. Syst. Man Cybern. Syst. **48**(12), 2495–2501 (2018)
4. Wang, L.M., Zhu, C.J., Yu, J.X., et al.: Fuzzy iterative learning control for batch processes with interval time-varying delays. Ind. Eng. Chem. Res. **56**(14), 3993–4001 (2017)
5. Yan, H.C., Yang, Q., Zhang, H., Yang, F., Zhan, X.: Distributed H state estimation for a class of filtering networks with time-varying switching topologies and packet losses. IEEE Trans. Syst. Man Cybern. Syst. **48**(12), 2047–2057 (2018)
6. Sanz, R., García, P., Albertos, P.: A generalized Smith predictor for unstable time-delay SISO systems. ISA Trans. **72**, 197–204 (2017)
7. Wang, L.M., Liu, B., Yu, J.X., et al.: Delagran-dependent-based hybrid iterative learning fault-tolerant guaranteed cost control for multi-phase batch processes. Ind. Eng. Chem. Res. **57**(8), 2932–2944 (2018)
8. Chen, C.-L., Jong, M.-J.: Fuzzy predictive control for the time-delay system. In: Second IEEE International Conference on Fuzzy Systems, vol. 1, no. 8, pp. 236–240 (1993)
9. Du, Z., Lin, T., Zhao, T.: Fuzzy robust tracking control for uncertain nonlinear time-delay system. Int. J. Comput. Commun. Control **10**(8), 52–65 (2015)
10. Du, Z.-B., Lin, T.-C., Balas, V.E.: A new approach to nonlinear tracking control based on fuzzy approximation. Int. J. Comput. Commun. Control **1**(7), 61–72 (2015)
11. Shi, H., Li, P., Cao, J., et al.: Robust fuzzy predictive control for discrete-time systems with interval time-varying delays and unknown disturbances. IEEE Trans. Fuzzy Syst. **28**(7), 1504–1516 (2020)
12. Sala, A.: On the conservativeness of fuzzy and fuzzy-polynomial control of nonlinear systems. Ann. Rev. Control **33**(1), 48–58 (2009)
13. Li, H., Miao, Z., Wang, J.: Variable universe stable adaptive fuzzy control of nonlinear system. Sci. China Ser. E Technol. Sci. **45**(3), 225–240 (2002)
14. Nurmaini, S., Chusniah: Differential drive mobile robot control using variable fuzzy universe of discourse. In: Proceedings of the International Conference on Electrical Engineering and

Computer Science (ICECOS), Palembang, IN, USA, pp. 50–55 (2017). https://doi.org/10.1109/ICECOS.2017.8167165

15. Yang, S., Deng, B., Wang, J.: Design of hidden-property-based variable universe fuzzy control for movement disorders and its efficient reconfigurable implementation. IEEE Trans. Fuzzy Syst. **27**(2), 304–318 (2019)

16. Mingxue, L., Guolai, Y., Xiaoqing, L., Guixiang, B.: Variable universe fuzzy control of adjustable hydraulic torque converter based on multipopulation genetic algorithm. IEEE Access **7**, 29236–29244 (2019). https://doi.org/10.1109/ACCESS.2019.2892181

17. Aziz, S., Wang, H., Liu, Y., Peng, J., et al.: Variable universe fuzzy logic-based hybrid LFC control with real-time implementation. IEEE Access **7**, 25535–25546 (2019). https://doi.org/10.1109/ACCESS.2019.2900047

18. Zheng, M., Yuelin, Z., Jinxin, W.: The research of course control based on variable universe fuzzy PID. In: Proceedings of the 5th International Conference on Mechatronics and Robotics Engineering (ICMRE), Rome, Italy, pp. 88–92 (2019). https://doi.org/10.1145/3314493.3314522

19. Pang, H., Liu, F., Xu, Z.: Variable universe fuzzy control for vehicle semi-active suspension system with MR damper combining fuzzy neural network and particle swarm optimization. Neurocomputing **306**, 130–140 (2018). https://doi.org/10.1016/j.neucom.2018.04.055

20. Whittington, J., Bogacz, R.: Theories of error back-propagation in the brain. Trends Cogn. Sci. **23**(3), 235–250 (2019)

21. Zhang, H., Zhang, R., He, Q., et al.: Variable universe fuzzy control of high-speed elevator horizontal vibration based on firefly algorithm and backpropagation fuzzy neural network. IEEE Access **9**(3), 57020–57032 (2021)

22. Wang, R., Liu, Y.-J., Yu, F.: Adaptive variable universe of discourse fuzzy control for a class of nonlinear systems with unknown dead zones. Int. J. Adapt. Control Sig. Process. **31**(3), 57020–57032 (2017)

23. Wang, J., Qiao, G.-D., Deng, B.: Observer-based robust adaptive variable universe fuzzy control for chaotic system. Chaos Solitons Fractals **23**(3), 1013–1032 (2005)

24. Han, J.Q.: From PID to active disturbance rejection control. IEEE Trans. Ind. Electron. **56**(3), 900–906 (2009)

25. Wu, Z.L., Li, D.H., Chen, Y.Q.: Active disturbance rejection control design based on probabilistic robustness for uncertain systems. Ind. Eng. Chem. Res. **59**(40) (2020). https://doi.org/10.1021/acs.iecr.0c03248

26. Gao, Z.: Active disturbance rejection control: a paradigm shift in feedback control system design. In: 2006 American Control Conference, Minneapolis, MN, USA. IEEE (2006). https://doi.org/10.1109/ACC.2006.1656579

27. Ran, M.P., Wang, Q., Dong, C.Y., et al.: Active disturbance rejection control for uncertain time delay systems. Automatica **112**, 108692 (2008). https://doi.org/10.1016/j.automatica.2019.108692

28. Tan, W., Fu, C.F.: Analysis of active disturbance rejection control for processes with time delay. In: Proceedings of American Control Conference (ACC), Chicago, IL, USA, pp. 3962–3967. IEEE (2015). https://doi.org/10.1109/ACC.2015.7171948

29. Lam, H.K., Leung, F.H.F.: Stability analysis of fuzzy control systems subject to uncertain grades of membership. IEEE Trans. Syst. Man Cybern. Part B (Cybern.) **35**(6), 1322–1325 (2005)

30. Cao, S.G., Rees, N.W., Feng, G.: Stability analysis of fuzzy control systems. IEEE Trans. Syst. Man Cybern. Part B (Cybern.) **26**(1), 201–204 (1996)

31. Wu, C., Liu, J., Jing, X., Li, H., et al.: Adaptive fuzzy control for nonlinear networked control systems. IEEE Trans. Syst. Man Cybern. Syst. **47**(8), 2420–2430 (2017)

32. Li, H.X.: Adaptive fuzzy controllers based on variable universe. Sci. China **42**(1), 10–20 (1999)

33. Li, H.X.: To see the success of fuzzy logic from mathematical essence of fuzzy control. Sci. China Ser. E Technol. Sci. **42**, 10–20 (1995)
34. Li, H.X.: Interpolation mechanism of fuzzy control. Sci. China Ser. E Technol. Sci. **41**(3), 312–320 (1998)
35. Zhu, E., Pang, J., Sun, N., et al.: Airship horizontal trajectory tracking control based on active disturbance rejection control. Nonlinear Dyn. **75**(4), 725–734 (2012)
36. Huang, Y., Xue, W.C.: Active disturbance rejection control: Methodology and theoretical analysis. ISA Trans. **53**(4), 963–976 (2014)
37. Shimkin, N.: Nonlinear control systems. In: Binder, M.D., Hirokawa, N., Windhorst, U. (eds.) Encyclopedia of Neuroscience, pp. 2886–2889. Springer, Heidelberg (2009). https://doi.org/10.1007/978-3-540-29678-2_4021
38. Slotine, J.E., Li, W.P.: Applied Nonlinear Control. China Machine Press (1991)
39. Wang, L.X.: Stable adaptive fuzzy control of nonlinear systems. IEEE Trans. Fuzzy Control Nonlinear Syst. **1**(2), 146–155 (1993)
40. Y.C. Liu, J. Gao, L.Y. Zhang, et al.: Smith-ADRC based Z axis impact force control for high speed wire bonding machine. In: 2018 19th International Conference on Electronic Packaging Technology, Shanghai, China, pp.1003–1008. IEEE (2018)
41. Li, X., Wei, A., Tian, S.: ADRC for nonlinear time-variant systems with unknown orders. In: 2016 35th Chinese Control Conference (CCC), Chengdu, China, pp. 9075–9080. IEEE (2016). https://doi.org/10.1109/ChiCC.2016.7554803
42. Zhao, C.Z., Li, D.G.: Control design for the SISO system with unknown order and the unknown relative degree. ISA Trans. **53**(4), 858–872 (2014)
43. Li, M.D., Li, D.H., Wang, J., et al.: Active disturbance rejection control for the fractional-order system. ISA Trans. **52**(3), 365–374 (2013)
44. Wu, Z.L., Li, D.L., Xue, Y.L., et al.: Active disturbance rejection control for fluidized bed combustor. In: 2016 16th International Conference on Control, Automation and Systems (ICCAS), Gyeongju, Korea, pp. 1286–1291. IEEE (2016). https://doi.org/10.1109/ICCAS.2016.7832479

Optimization of Ideal Reflector for Radio Telescope Based on Improved Composite Algorithm

Yizhi Guo[1], Zhixuan Li[2], Anyi Yao[1], Yiqi Liang[1], and Yubin Zhong[1(✉)]

[1] School of Mathematics and Information Science, Guangzhou University, Guangzhou, China
zhong_yb@gzhu.edu.cn
[2] School of Physics and Materials Science, Guangzhou University, Guangzhou, China

Abstract. The traditional quantum genetic algorithm lacks convergence and is easy to enter local optimal problems, so we improve the steps of population initialization, including niche technology, chaos mutation, etc. These measures can solve the problems such as too fast convergence speed to some extent. In this paper, the algorithms that go through these operations are denoted as improved composite algorithms.

It can be found that the improved composite algorithm basically reaches the optimal value when the 21st generation is around, however, the quantum genetic algorithm can reach the optimum only when the algorithm operation reaches about 35 generations. In addition, the improved composite algorithm has basically reached the optimal value of −0.9356 when iterated to about the 21st generation, while the traditional quantum genetic algorithm is still at a low level. As a result, the improved composite algorithms has quicker convergence speed and better searching ability than the traditional quantum genetic algorithm. Therefore, the improved composite algorithm is better than the traditional quantum genetic algorithm in solving the ideal reflector optimization problem based on FAST.

Keywords: Quantum genetic algorithm · Optimization of ideal reflector for telescope · Algorithm comparison and optimization

1 Introduction

With the creation of FAST, more and more people join in the optimization of radio telescope research. From the recurrence band and radio flare associated to the radio telescope to the reflector adjustment of the radio telescope itself, many scholars have done related work, and the alteration of the reflector of the radio telescope has become a hot topic in the academic field, and relevant researchers have actively launched a series of discussions and studies: For example, Li Minghui and Zhu Lichun et al. put out optimization analysis on deformation strategy of FAST instantaneous paraboloid in 2012 [1], and Huang Hao and Wang Qiming et al. proposed development strategy of active reflective surface control system of FAST based on Ethernet in 2006 [2]. In 2012, Zhu Lichun published the research on control of the 500-meter aperture round radio telescope and so on [3]. The optimized reflector of the radio telescope can soak up the signals emitted by celestial bodies from space to the greatest magnitude, so as to maximize the economic and communal good of the creation of the radio telescope. Considering various practical factors, the optimization of the ideal reflector of the radio telescope has become an important part of studying radio telescope. So it is a very important work to study the optimization of the ideal reflector of the radio telescope.

Quantum genetic algorithm is an advanced algorithm integrating genetic algorithm and quantum technology. It can make the population convergence speed faster, and also can expand the optimal search range of the whole algorithm. However, there still has no scholars at home and abroad have used the improved composite algorithm realized by the improved method used in this paper to solve the optimization problem of the ideal reflector based on FAST, so the research of this paper is valuable.

In this paper, the Sect. 1 is our introduction, the Sect. 2 is the model we use in this paper, the Sect. 3 introduces the principle of our algorithm, describes the enhancement of the quantum genetic algorithm, and the Sect. 4 gives our views on the results.

This article only discusses the case when the object observed by the radio telescope is directly above the telescope.

2 Model Building

2.1 The FAST Reflective Surface

The active reflector of FAST [4] adjusts itself with the changes of the celestial bodies which are measured by FAST so as to maximize the benefit of the feed chamber receiving celestial energy. Therefore, FAST can correspond to two states in working condition and non-working condition respectively, namely working state and reference state. Figure 1 shows the reflection surface of the FAST telescope in the baseline state and the reflection surface in the working state at a certain time:

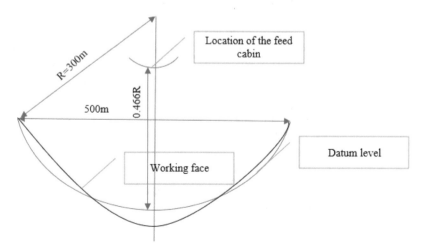

Fig. 1. Paraboloid and working face section of FAST.

Among them, the aperture of FAST is 500 m and the radius is 300 m. When the base state is changed to the working state, the telescope reflector will change its shape by adjusting the actuator expansion and other methods to obtain a greater degree of energy. During the adjustment, the working face will be squeezed so that the apex of the reflector will be shifted down or up, and when we replace the reflector, the distance between the reflector and the feed cabin will also change. We use the focal length ratio to measure this process, which is denoted as FP. In order to calculate the focal length ratio, we give a formula as follows:

$$FP = \frac{f}{R} \tag{1}$$

where, f is the focal length of the telescope at this time and R is the radius of the telescope.

2.2 The Establishment of an Ideal Reflector Model

Based on the brief introduction of FAST reflector in the previous section, we begin to build the ideal reflector model of FAST.

When the reflector changes from the reference state to the working state, the main cable node, an important component of the telescope, will be displaced along the direction of the spherical radius, which are called the radial, meridian and zonal displacement of the main cable node respectively. They are the important factors that affect the reflector variation of the telescope. Considering the partial limitations of the actuator in the process of expansion, the influence of the stress change of the main cable joints, and the emphasis on the boundary transition, the ideal reflector model requires the arc length difference between the working state reflector and the base state reflector, and the ring length difference between the working state reflector and the base state reflector to be as small as possible. At the same time, the displacement of the main cable node along the spherical radius should be as small as possible, but the smooth transition between the working state reflector and the base state reflector should be ensured, so that the working state reflector can better receive the signal sent by the celestial body, and reduce the load of each component of the telescope to a certain extent, so as to prevent unnecessary losses caused by component damage. Therefore, we can convert the problem into making the radial, meridional and zonal displacements of the main cable nodes as small as possible and keeping the curvature as consistent as possible at the intersection of the reference state reflector and the working state reflector. Thus, the following model [1] is presented:

$$\min \quad score = \frac{1}{4}L_a + \frac{1}{4}L_b + \frac{1}{4}L_c - \frac{1}{4}L_d \tag{2}$$

$$\begin{cases} L_a = R - L_{circle} \\ L_b = \int_0^{x_1} \sqrt{(\frac{x}{f})^2 + 1}\,dx - R \cdot \arcsin \frac{x_0}{R} \\ L_c = \left(\frac{x_1}{x_0} - 1\right) \times 100\% \\ L_d = -\frac{x_0}{y_0} \rightarrow \frac{x}{f} \\ Q \in [0.1 \quad 0.6] \\ FP \in \left[0.466 - Q \quad 0.466 + Q\right] \end{cases} \tag{3}$$

Score is the final evaluation value (this article uses this evaluation value as the standard to evaluate the performance of the ideal reflector). The radial displacement of the main cable, the meridional displacement, the zonal displacement and the edge curvature of the working reflecting surface are denoted as L_a, L_b, L_c and L_d respectively; R is the radius of the telescope when the telescope is in the base state; the distance between the apex of the reflector and the ground state in operation is denoted as Q; L_{circle} is the distance between the working reflector node and the center of the sphere. We make x_1 the x coordinate value corresponding to the base state node in the working state, the focal length of a radio telescope is expressed as f, x_0 is the x coordinate value of any node in the base state, y_0 is the y coordinate value of any node in the base state, and FP is the focal length ratio in the current state.

3 Algorithm Implementation

3.1 Quantum Genetic Algorithm

Quantum Genetic Algorithm. Quantum genetic algorithm [5] combines the main advantages of quantum thinking and genetic algorithm. Quantum computing can solve quite a number of problems through quantum state superposition, interference and other operations, while genetic algorithm is more classical, and its main principle is to imitate the trend and characteristics of biological evolution and produce special properties such as variation and crossover. It can achieve the optimization operation of the objective function under certain constraints. However, while the genetic algorithm has strong adaptability, it has unscientific defects such as cross mutation, which may lead to rapid decline in the convergence rate of the algorithm or local optimal solution. So quantum genetic algorithm comes into being.

Quantum Coding and Initialization. Quantum coding [6] is a kind of quantum bit coding based on genetic algorithm to extend binary coding. For example, a chromosome W contains n genes, as shown below:

$$|W\rangle = \begin{bmatrix} a_1......a_n \\ b_1......b_n \end{bmatrix} \tag{4}$$

Among them, $|a_i|^2 + |b_i|^2 = 1, i \in [1, n]$, Generally, the initial population takes a fixed value $[a_i, b_i] = [\frac{1}{\sqrt{2}}, \frac{1}{\sqrt{2}}], i \in [1, n]$.

Quantum Revolving Gate. The traditional quantum genetic algorithm updates the population iteratively by adjusting the quantum gate angle, its expression matrix [6] is shown below:

$$Y(\theta) = \begin{bmatrix} \cos\theta & -\sin\theta \\ \sin\theta & \cos\theta \end{bmatrix} \tag{5}$$

where, θ is the rotation angle of quantum gate.

3.2 Improved Composite Algorithm

Although the quantum genetic algorithm has high convergence and excellent optimization ability, it will converge prematurely and have no ability to get the best solution
sometimes. Therefore, we can adopt some measures to make the population distribution
uniform at the beginning or add some disturbance to make the population iteration jump
out of the dead end when the population iteration is about to enter the local optimal.
Next, we introduce various improvements:

Niche Technology. First of all, we discuss niche [7], which is a method to transform
the biological idea of the reproduction of the same species into a mathematical form. It
classifies each generation of the population in detail, then optimizes each category, and
then hybridizes individuals with the best quality to achieve individual iteration of the
population. Based on the optimization characteristics of niche, We use this technology
to optimize the population initialization part of the quantum genetic algorithm. Here,
we also divide it into N probability Spaces according to the references. The quantum
coding initialization proceeds as follows:

$$\begin{vmatrix} a_m \\ b_m \end{vmatrix} = \begin{vmatrix} \sqrt{\frac{i}{N}} \\ \sqrt{\frac{1-i}{N}} \end{vmatrix} \tag{6}$$

where, a_m and b_m respectively represent the corresponding quantum coding.

Circle Chaos Variation. Chaotic mapping [8] is a mapping method based on chaotic
thought. Chaotic mapping generally obtains uncertain results through some deterministic
equations, which can replace pseudo-random number generators to some extent. Circle
chaotic mapping [9] is one of them, and its formula is as follows:

$$x_{i+1} = \mathrm{mod}\left(x_i + 0.2 - \left(\frac{0.5}{2\pi}\right)\sin(2\pi x_i), 1\right) \tag{7}$$

where, x_i and x_{i+1} are the random number generated each time in this paper.

Quantum variation [6] is mainly used for the iteration of disturbed population so
that it will not converge too early and thus cannot obtain the optimal solution. We can
select individuals with predetermined probabilities and mutate certain positions in those
individuals, thus causing the population to be disturbed to a certain extent. The mutation
mode adopted here is quantum non-gate mutation, and the quantum gate matrix J is
shown as follows:

$$J = \begin{bmatrix} 0 & 1 \\ 1 & 0 \end{bmatrix} \tag{8}$$

So we use this quantum gate to update the qubit coding as follows:

$$\begin{bmatrix} 0 & 1 \\ 1 & 0 \end{bmatrix}\begin{bmatrix} a_i \\ b_i \end{bmatrix} = \begin{bmatrix} b_i \\ a_i \end{bmatrix} \tag{9}$$

where, a_i and b_i are the quantum bit encoding.

A random number ranging from 0 to 1 is generated at random for each quantum variation, which is used to determine whether qubit variation is required in this round robin. However, the random numbers generated by default by the general system pseudorandom number generator still have some rules, so in this paper, we use the circle chaotic mapping to generate each random number, thus reducing the periodicity of the pseudorandom number to a certain extent.

Quantum Total Interference Crossing. Quantum full interference crossover [10] is to shift all the individuals in the whole population in a circular manner to disrupt their original order in order to enhance the algorithm performance.

Quantum Revolving Gate Based on H_ε Gate. The quantum revolving gate adopted by traditional quantum genetic algorithm completes each population convergence through fixed rotation angle. But in fact, no matter the rotation angle is too large (which will make the algorithm skip the optimal solution contained in the angle interval), or the rotation angle is too small (which will make the population cannot converge for a long time during the algorithm operation), the probability amplitude will basically be close to 0 or 1, which keeps the population from achieving global optimization [11].

Therefore, this paper replace the original quantum revolving gate through the H_ε gate. H_ε gate [11] is a quantum gate updating strategy related to quantum evolution algorithm. It makes the probability amplitude not converge to 0 or 1, but makes it approach to $\sqrt{1-\varepsilon}$ or $\sqrt{\varepsilon}$, which can prevent the population iteration falling into local optimal and accelerate the convergence rate. Here is a pseudocode demonstration of it:

$$
\begin{aligned}
&H_\varepsilon\_algorithm(a_i, b_i, \varepsilon) \\
&if \quad |a_i|^2 \le \varepsilon \quad and \quad |b_i|^2 \ge 1 - \varepsilon \quad then \\
&\qquad \begin{bmatrix} a_i' \\ b_i' \end{bmatrix} \leftarrow \begin{bmatrix} \sqrt{\varepsilon} \\ \sqrt{1-\varepsilon} \end{bmatrix} \\
&else \\
&\qquad if \quad |a_i|^2 \ge 1 - \varepsilon \quad and \quad |b_i|^2 \le \varepsilon \quad then \\
&\qquad\qquad \begin{bmatrix} a_i' \\ b_i' \end{bmatrix} \leftarrow \begin{bmatrix} \sqrt{1-\varepsilon} \\ \sqrt{\varepsilon} \end{bmatrix} \\
&\qquad else \\
&\qquad\qquad \begin{bmatrix} a_i' \\ b_i' \end{bmatrix} \leftarrow \begin{bmatrix} \sqrt{a_i} \\ \sqrt{b_i} \end{bmatrix}
\end{aligned}
\tag{10}
$$

where, a_i and b_i are the I-th gene in chromosome, and ε is a parameter.

Algorithmic Flow. A flowchart is an important part of displaying an algorithm and also a shortcut for readers to quickly understand the method. In order to facilitate readers' understanding, we have provided a brief explanation of our method in this article. Figure 2 shows our operation flow chart:

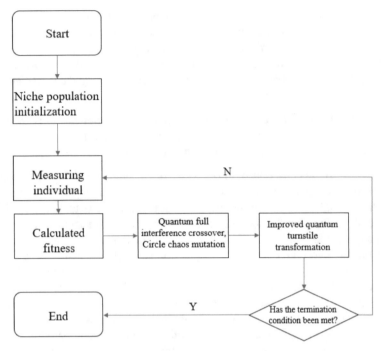

Fig. 2. Algorithm flow chart.

As can be seen, this article first uses small habitat population initialization for data preprocessing, then measures the individual of the population and calculates fitness. Then, it performs a fully connected transformation and uses Circle chaotic mutation for further operations, performing a turnstile transformation until the target is met to stop the iteration. Otherwise, it returns to the measure stage.

4 Result Analysis

Our running machine model is LAPTOP-6MSUAQBB, processor is AMD Ryzen 7 4700U with Radeon Graphics 2.00 GHz, operating system is 64-bit Windows10 Home Edition, compiler is MATLAB R2019a.

We add all the optimization methods in Sect. 3 into the traditional quantum genetic algorithm, and take the variation rate as 0.5, population size as 50, binary length of each variable as 10, iteration times as 50, ε in the improved quantum revolving gate as 0.01 [12], Through many experiments, we showed the difference between the two algorithms, and we selected the situation with the most obvious experimental effect to show, as shown in Fig. 3:

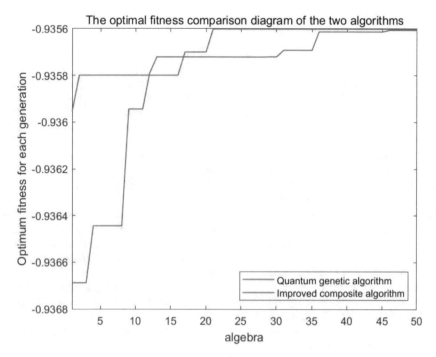

Fig. 3. Algorithm optimal fitness comparison diagram.

Here, in order to facilitate the presentation of the results, we take the opposite number of iteration process values. It can be seen that after adding a series of improvement measures, the improved composite algorithm converge around the 21st generation, but the quantum genetic algorithm only converge around the 35th generation, and the improved composite algorithm has reached the best value of −0.9356 when it iterates to about 21st generation, while the quantum genetic algorithm still stays at a low position. Therefore, our results indicate convergence rate of the improved composite algorithm is faster than that of the traditional quantum genetic algorithm, it can also reach a relatively good fitness earlier, which indicates that our series of improvements are effective. This just shows that the improved composite algorithm is better than the traditional quantum genetic algorithm in the study of the optimization of the ideal reflective surface based on FAST, and the improved composite algorithm also solves the optimization problem correctly. What's more, it also provides some reference for us to study the optimization of the ideal reflector of the telescope.

Acknowledgements. This work is supported by 2022 National Student Innovation Training Project of Ministry of Education of P.R.C (202211078154).

References

1. Li, M., Zhu Zhu, C.: Optimization analysis of FAST instantaneous paraboloid deformation strategy. J. Guizhou Univ. Nat. Sci. Ed. **2012**(06), 30–34+49 (2012)
2. Huang, H., Wang, Q., Zhang, H., Zhu, C.: Development of active reflective surface control system of FAST based on Ethernet. Comput. Eng. Appl. **05**, 97–100 (2006)
3. Zhu, L.-C.: Deformation control of the entire network of active reflecting surface of the Five-hundred-meter Aperture Spherical Radio Telescope (FAST). Res. Inf. Technol. Appl. **3**(04), 67–75 (2012)
4. Qian, H.: Theoretical and Experimental Research on Support Structure of FAST Active Reflective Surface. Harbin Institute of Technology (2007)
5. Shi, F., Wang, H., Yu, L., et al.: MATLAB Intelligent Algorithm Analysis of 30 Cases, p. 108. Beihang University Press, Beijing (2011)
6. Feng, J., Wang, Q., Wang, Y., Xu, B.: Fuzzy PID control of ultrasonic motor based on improved quantum genetic algorithm. J. Jilin Univ. (Eng. Technol. Edn.) **51**(06), 1990–1996 (2021)
7. Gai, J.: Improvement and Research of Quantum Genetic Algorithm. Bohai University (2017)
8. Chen, T., Yan, M., Hu, Y., Qin, T.: Photovoltaic power prediction of 5G base stations based on improved firefly algorithm. Smart Grid **12**(2), 43–55 (2022)
9. Song, L., Chen, W., Chen, W., Lin, Y., Sun, X.: Improvement and application of Sparrow search algorithm based on Hybrid strategy. J. Beijing Univ. Aeronaut. Astronaut. 1–16 (2022). https://doi.org/10.13700/j.bh.1001-5965.2021.0629
10. Fu, L.: Improvement of Quantum Genetic Algorithm and Its Application in Cargo Assembly Problem. Guangxi University (2015)
11. Zhang, X., Sui, G., Zheng, R., Li, Z., Yang, G.: An improved Quantum Revolving gate quantum Genetic Algorithm. Comput. Eng. **39**(04), 234–238 (2013). (in Chinese)
12. Han, K.-H., Kim, J.-H.: Quantum-inspired evolutionary algorithms with a new termination criterion, H/sub /spl epsi// gate, and two-phase scheme. IEEE Trans. Evol. Comput. **8**(2), 156–169 (2004). https://doi.org/10.1109/TEVC.2004.823467

Linear Programming Subject to Max-Product Fuzzy Relation Inequalities with Discrete Variables

Xu Fu[1], Chang-xin Zhu[2], and Zejian Qin[3](✉) ⓘ

[1] Human Resources Office,
Guangzhou Vocational and Technical University of Science and Technology,
Guangzhou 510550, Guangdong, China
[2] Department of Basic Course,
Guangzhou Vocational and Technical University of Science and Technology,
Guangzhou 510550, China
[3] Department of Applied Mathematics, Guangdong University of Finance,
Guangzhou 510521, China
735692885@qq.com

Abstract. In this paper, we introduced the linear programming subject to max-product fuzzy relation inequalities with discrete variables to denote the optimization management model of physical distribution. It is shown that the inequalities constraints can be transformed into the problem of finding the all the potential minimal solution, and expressed by $2m$ linear equations in 0-1 mixed variables and mn inequalities. Then the original problem can converted into a 0-1 mixed-integer linear programming problem and then adopt to the branch-and-bound scheme to find optimal solution.

Keywords: max-product relational inequalities · potential minimal solution · branch-and-bound · mixed-integer linear program

1 Introduction

Since fuzzy relation equations (FRE) with max-min composition were firstly introduced by Sanchez [1–3], it has attracted many fuzzy mathematics researchers' attention. As an extension, fuzzy relation equations or inequalities were studied. In fact, the composition could be replaced by the general max-t-norm composition, although max-product and max-min compositions were the most frequently and commonly used t-norms. As demonstrated in [4], the complete solution set of max-t-norm fuzzy relational equations can be completely determined by a unique maximum solution and a finite number of minimal solutions [5–8]. It's easy to compute the maximum solution, but finding all the minimal solution is NP-hard problem [10]. It is worth to mention that Li and Fang [11] provided a complete survey and detailed discussion on fuzzy relational equations. They studied fuzzy relational equations in a general lattice-theoretic framework and introduced classification of basic fuzzy relational equations.

Meanwhile, optimization problems subject to fuzzy relation equations or inequalities were introduced and studied [9–12, 19–32]. Fang and Li [15] firstly investigated an linear optimization problem with a consistent system of max–min equations. They converted it into a 0-1 integer programming problem and solved this by the branch-and-bound method. Then many research improved this method [25–27]. The linear objective optimization problem with max-min FRI was investigated by Zhang et al. [21]. Guo and Xia [24] proposed a method to accelerate the resolution of this problem. In [16], the other consider a special linear optimization problem in which all the coefficients are fuzzy numbers. In these linear objective optimization problems with FRI constraints, the variables are all continue variables over the interval $[0, 1]$. However the fact is that, the data in real world are always discrete instead of continue. And even the fuzzy information is also come from the discrete data in the real world. So in this paper, we studied modified LP problem of finding a minimization of a linear objective function of n discrete variables in h values, subject to m max-product inequality constraints.

The rest of the paper is organized as follow. Due to the practical application background in the optimization management of physical distribution, we introduce a linear optimization problem subject to max-product fuzzy relation inequalities with discrete variables in Sect. 2. Treatment of max-product inequalities with discrete variables is provided in Sect. 3. In Sect. 4, we illustrate effectiveness of resolution by numerical experiments. Further discussion and simple conclusion are arranged in Sect. 5.

2   Problem Statement

In recent years, with the rapid development of logistics in China, its storage construction and management have become increasingly difficult with the continuous expansion of the scale. Due to the increase of various massive data, how to reasonably build a storage system and manage logistics operations to improve the operational efficiency of logistics has become an important and urgent problem to be solved.

Assume that there are n logistics warehouses, i.e. $A_1, A_2, \ldots, A_n$, located in different places of a specific area, such as a city. The jth logistics warehouse A_j can provide x_j transport capacity. Due to the limitation of the actual transportation means, the transportation capacity value can only take part of the fixed value $\{d_{j_1}, d_{j_2}, \ldots, d_{j_h}\}$. In general, The transportation capacity that the ith logistics warehouse can obtain from the jth warehouse is less than or equal to x_j. Hence it is reasonable to assume that A_i can get transport capacity from A_j with the quality $a_{ij}x_j$, where a_{ij} denote the transportation efficiency. Assume that the minimum capacity required by A_i is b_i, then the highest transport capacity level on which A_i get from $\{A_1, A_2, \ldots, A_n\}$ should be larger than b_i, so we can get constraints:

$$\bigvee_{j=1}^{n} a_{ij}x_j \geq b_i \tag{1}$$

where each x_j may achieve one of n possible nonnegative discrete values $\{d_{j_1}, d_{j_1}, \ldots, d_{j_h}\}$ with $0 \leq d_{j_1} < d_{j_2} < \cdots < d_{j_h}$ for all $j = 1, 2, \ldots, n$.

Since the transport capacity is proportional to the cost, we cannot choose large transport capacity without limitation due to the consideration of management cost. Assume the construction cost of unit transportation capacity as c_i, and the management cost of the transportation between A_i and A_j as c_{ij}.

Assume that the maximum management budget of the A_i is l_i. In order to keep the management cost within the budget, we can get the following conditions:

$$\bigvee_{j=1}^{n} c_{ij}x_j \leq l_i \tag{2}$$

where each x_j may achieve one of n possible nonnegative discrete values $\{d_{j_1}, d_{j_1}, < \cdots >, d_{j_h}\}$ with $0 \leq d_{j_1} < d_{j_2} < \cdots < d_{j_h}$ for all $j = 1, 2, \ldots, n$.

Our goal is to save construction costs on the premise of ensuring transportation capacity and not exceeding the management budget. Consequently, the objective function is:

$$f(x) = \sum_{j=1}^{n} c_j x_j \tag{3}$$

Since variables and the parameters can be normalized into the unit interval $[0, 1]$. After normalizing the variables and the parameters, the problem becomes the following optimization problem:

Minimize the linear objective function :

$$f(x) = \sum_{j=1}^{n} c_j x_j \tag{4}$$

where $c_j \in R$ for all $j = 1, 2, \ldots, n$, subject to the constraints

$$\bigvee_{j=1}^{n} a_{ij}x_j \geq b_i, \qquad i = 1, 2, \ldots, n, \tag{5}$$

$$\bigvee_{j=1}^{n} c_{ij}x_j \leq l_i, \qquad i = n+1, l+2, \ldots, 2n, \tag{6}$$

where $a_{ij}, c_{ij} \in [0, 1]$, and each x_j may achieve one of n possible nonnegative discrete values $\{d_{j_1}, d_{j_1}, \ldots, d_{j_h}\}$ with $0 \leq d_{j_1} < d_{j_2} < \cdots < d_{j_h} \leq 1$ for all $j = 1, 2, \ldots, n$.

For the discrete max-product constraints, we can transform them into max-product relational inequalities with discrete variables: for an upper-bound constraint in (6):

$$\bigvee_{j=1}^{n} a_{ij}x \leq l_i$$

can obviously be written as n constraints:

$$a_{ij}x_j \le l_i, \quad i.e. \quad x_j \le l_i/a_{ij}, \quad j = 1, 2, \ldots, n$$

Combined with the discrete constraints:

$$x_j \in \{d_{j_1}, d_{j_1}, \ldots, d_{j_h}\}$$

we can get new discrete values, which include the information of the upper-bound constraints

$$x_j \in \{d_{j_1}, d_{j_1}, \ldots, d_{j_h}\} \cap [0, \bigwedge_{i=n+1}^{2n} l_i/a_{ij}]$$

So we can transform all the discrete max-product constraints into max-product relational inequalities with discrete variables.

So the problem we considered in this paper, linear programming subject to max-product inequalities with discrete variables can be formulated as follow:

$$\min \quad f(x) = \sum_{j=1}^{n} c_j x_j$$

$$\text{s.t.} \quad \bigvee_{j=1}^{n} a_{ij}x_j \ge b_i \quad i = 1, 2, \ldots, m, \tag{7}$$

where $x_j \in \{d_{j_1}, d_{j_1}, \ldots, d_{j_h}\} \in [0, 1], c_j > 0, a_{ij}, b_i \in [0, 1]$.

3 Representation

3.1 Treatment of Multivalued Discrete Variables

The major difficulty for solving the problem (7) comes from nonconvexity caused by the max-product operation. So we need to convert such a problem into 0-1 mixed-integer linear programming problem and then find the optimal solution by branch-and-bound method.

And the first step is to go from multivalued variables to 0-1 valued by noticing that each multivalued discrete variable $x_j, i = 1, 2, \ldots, n$ in problem (7), can be represented by h binary s in two linear equations as:

$$x_j = \sum_{k=1}^{h} d_{j_k} u_{j_k}, \tag{8}$$

$$\sum_{k=1}^{h} u_{j_k} = 1, \tag{9}$$

where $u_{j_k} \in \{0, 1\}$ for $j = 1, 2, \ldots, n$ and $k = 1, 2, \ldots, h$ It is important to know that the nh binary variables of $\{u_{jk}\}, j = 1, 2, \ldots, n$ and $k = 1, 2, \ldots, h$, induced by (8) and (9) may cause heavy computational burden when n and h become very large. To avoid this burden, Li and Lu [17,18] proposed a logarithmic approach which only need $n[\log_2 x]$ binary variables, nh nonnegative variables, and $n[\log_2 x]$ linear equations to represent the n discrete variables with h values by the following way [13,14]:

$$\lambda_{jr} = \sum_{k=1}^{h} t_{rk} u_{jk}, j = 1, 2, \ldots, n, r = 1, \ldots, [\log_2 h], \tag{10}$$

where $\lambda_{jr} \in \{0, 1\}$ is binary variable; $u_{j_k} \geq 0$ is a nonnegative variable, for $j = 1, 2, \ldots, n, k = 1, 2, \ldots, h$, and $r = 1, \ldots, [\log_2 h]$; and t_{rk} are binary variables by solving the equations of $1 + \sum_{r=1}^{[\log_2 h]} 2^{r-1} t_{rk} = k$ for $k = 1, 2, \ldots, h$.

In this setting, $n[\log_2 h]$ binary variables (λ_{jr}), nh nonnegative variables(u_{j_k}), and $2n+n[\log_2 h]$ linear equations are needed to represent the n discrete variables in h values.

It will reduce the binary variables by use the logarithmic method to replace representation (8) and (9), when a branch-bound method scheme is applied for solving mixed-integer programming problems.

By the 2 methods above, the n discrete variables in h values can be replaced by binary variables and linear equations. The next step is to treat the m fuzzy relation inequalities with max-product.

3.2   Treatment of Max-Product Fuzzy Relation Inequalities

Firstly we introduce some properties of max-product inequalities. As is known to all, the solution set of max-product inequalities is determined by a unique maximum solution and a finite number of minimal solutions. So we denote the the solution set as $X(A, b, \geq)$, the maximum solution as $\hat{x}$ and the minimal solution as $\{\check{x}^1, \check{x}^2, \ldots, \check{x}^t, \}$(where t means the number of minimal solution).

Theorem 1. If $X(A, b, \geq) \neq \emptyset$, then $\hat{x}_j = 1$ for all $j = 1, 2, \ldots, n$. So feasible domain restricted by the max-product fuzzy relation inequalities constraints can be expressed as:

$$x \in X(A, b, \geq) = \bigcup_{l=1}^{t} \{x | \check{x}^l \leq x\}$$

If we know all the minimal solution of max-product fuzzy relation inequalities, then we can get the feasible domain of the problem (7), so the main problem we faced to is how to compute all the minimal solutions. However it is a NP-hard problem, so we introduce a method to find all the potential minimal solution instead of minimal solution.

Definition 1. Matrix $Q = (q_{ij})_m \times n$ called a judgement matrix, where,

$$q_{ij} = \begin{cases} \frac{b_i}{a_{ij}}, & a_{ij} > b_i, \\ 0, & a_{ij} \leq b_i, \end{cases} \tag{11}$$

Theorem 2. *If matrix Q is the judgement matrix of the fuzzy relation inequalities, then inequalities are consistency if and only if for any $1 \leq i \leq m$, there exist at least one $j_i \in \{1, 2, \ldots, n\}$ such that $q_{ij_i} \neq 0$.*

Definition 2. *Assume Q is the judgement matrix of the inequalities, matrix $S = (s_{ij})_m \times n$, where $s_{ij} \in \{0, d_{ij}\}$. We call S a solution matrix if for any $1 \leq i \leq m$, there exist a unique j_i such that $s_{ij} \neq 0$.*

Assume S is a solution matrix of fuzzy relation inequalities (), denote $x^S = (x_1^S, x_2^S, \ldots, x_n^S)$, where

$$x_j^S = \bigvee_{i=1}^{m} s_{ij} \tag{12}$$

Theorem 3. *Assume S is a solution matrix, and x^S defined by (13), then $x^S \in X(A, b, \geq)$ and for any $x^0 \in X(A, b, \leq)$, there exist a solution matrix S and concerned x^S, such that $x^S \leq x^0$.*

Proof. x^S is obvious a solution of inequalities, so we just need to proof the second half. Because $x^0 \in X(A, b, \leq)$, we have:

$$\bigvee_{j=1}^{n} a_{ij} x_j^0 = a_{i1} x_1^0 \vee a_{i2} x_2^0 \vee \cdots \vee a_{in} x_n^0 \geq b_i, i = 1, 2, \ldots$$

So there exist an j_i such that:

$$a_i^{j_i} x_{j_i}^0 \geq b_i$$

Further more;

$$\frac{b_i}{a_{ij_i}} \leq x_{j_i}^0 \leq 1$$

If Q is the judgement matrix, then

$$q_{ij_i} = \frac{b_i}{a_{ij_i}} \neq 0$$

Let $S = (s_i j) m \times n$, where,

$$s_{ij} = \begin{cases} q_{ij}, & j = j_i, \\ 0, & otherwise, \end{cases} \tag{13}$$

Obviously, the ith row of S, there exist a unique nonzero element i.e. $s_{ij_i} = q_{ij_i} \neq 0$, so S is a solution matrix and $x_j^S = \bigvee_{i=1}^{m} s_{ij}$ is the solution concerned.

Next we proof that $x_j^0 \geq x_j^S$ hold for any $j \in J$. Denote $I_j = \{i \in I | j_i = j\}$, then problem can be divided into two cases:

(i) if $I_j = \emptyset$, then there is no $i \in I$ such that $j_i = j$, so $s_{ij} = 0$ and $x_j^S = \bigvee\limits_{i=1}^{m} s_{ij} = 0$. Absolutely $x_j^0 \geq x_j^S$ holds.

(ii) if $I_j \neq \emptyset$, for the $i \in I_j$, there exist an $j_i = j$ so we have:

$$x_j^0 \geq d_{ij} = \frac{b_i}{a_{ij_i}}, \quad s_{ij} = d_{ij} = \frac{b_i}{a_{ij_i}}$$

for the $i \notin I_j$, we have $s_{ij} = 0$, so:

$$
\begin{aligned}
x_j^S = \bigvee_{i=1}^{m} s_{ij} &= \left(\bigvee_{i \in I_j} s_{ij} \right) \vee \left(\bigvee_{i \notin I_j} s_{ij} \right) \\
&= \left(\bigvee_{i \in I_j} s_{ij} \right) \vee \left(\bigvee_{i \notin I_j} s_{ij} \right) \\
&= \left(\bigvee_{i \in I_j} d_{ij} \right) \vee \left(\bigvee_{i \notin I_j} 0 \right)
\end{aligned}
\tag{14}
$$

So $x_j^0 \geq x_j^S$ hold for any $j \in J$.

From theorem 3, it's easy to know that any minimal solution can be denote as x^S. So the set $\{x^S\}$ contain all the potential minimal solution. and we can find all the x^S by solution matrix S. Further more, the max-product fuzzy relation inequalities can be equivalently represented as follow,

$$\sum_{j=1}^{n} p_{ij} v_{ij} = 1 \tag{15}$$

$$\sum_{j=1}^{n} v_{ij} = 1 \tag{16}$$

$$a_{ij} x_j \geq p_{ij} v_{ij} b_i \tag{17}$$

where $v_{ij} \in \{0, 1\}$ and $p_{ij} = \begin{cases} 1, & a_{ij} > b_i, \\ 0, & a_{ij} \leq b_i, \end{cases}$

Proof. From the Theorems 1 and 3, we know that the feasible domain feasible domain restricted by the max-product fuzzy relation inequalities constraints can be expressed equivalently as:

$$x \in X(A, b, \geq) = \bigcup_{l=1}^{t} \{x | \check{x}^l \leq x\} = \bigcup_{S} \{x | \check{x}^S \leq x\}$$

And consider the definition of (16), we know for a certain solution matrix, $x \geq x^S$ equal to the constraints as follow:

$$x_j \geq s_{ij}, i = 1, 2, \ldots, m; j = 1, 2, \ldots, n.$$

at the same time

$$s_{ij} = v_{ij}q_{ij}, \sum_{j=1}^{n} v_{ij} = 1, q_{ij} \in \{0,1\}$$

consider that $q_{ij} = p_{ij}\frac{b_i}{a_{ij}}$, and there exist a unique j_i such that $s_{ij} \neq 0$ each row of S, so we have :

$$\sum_{j=1}^{n} p_{ij}v_{ij} = 1$$

the feasible domain is equal to the constraints: (15), (16), (17).

3.3   Linear Reformulation

Combine the treatment of max-product fuzzy relation inequalities in (17) and the two methods for the expression of multivalued variables, we can convert the original problem (7) into three model respectively:
   Model 1

$$\min \quad f(x, u, v) = \sum_{j=1}^{n} c_j x_j$$

$$subject \quad to \quad (8), (9), (15), (16), (17)$$

$$x_j \geq 0, u_{jk}, v_{ij} \in \{0,1\}$$

$$for \quad i = 1, 2, \ldots, m; j = 1, 2, \ldots, n, k = 1, 2, \ldots, h.$$

In this model, there are nh binary variables of u_{jk}; mn binary variables of v_{ij}; n nonnegative variables of x_j; $2(m+n)$ linear equations (i.e.,((8)+(9)+(15)+(16)); and mn linear inequality constraints (i.e., (17)) involved.
   Model 2

$$\min \quad f(x, u, v) = \sum_{j=1}^{n} c_j x_j$$

$$subject \quad to \quad (8), (9), (10), (15), (16), (17)$$

$$x_j, u_{jk} \geq 0, \lambda_{jr} \in \{0,1\}$$

$$for \quad i = 1, 2, \ldots, m, r = 1, 2, \ldots, [\log 2h]; j = 1, 2, \ldots, n, k = 1,$$

In this model, there are $n[\log 2n]$ binary variables of λ_{jr}; nh nonnegative variables of u_{jk}; mn binary variables of v_{ij}; n nonnegative variables of x_j; $2(m+n) + n[\log 2n]$ linear equations (i.e., ((8)+(9)+(10)+(15)+(16)); and mn linear inequality constraints (i.e., (17)) involved.

4   Computational Experiments

To ensure there exist optimal solution in the problem, we apply the following rules to randomly generate its coefficients:

(1) a_{ij} are generated randomly by the uniform distribution over $[0, 1]$, and c_j are taken from $[1, 10]$ for $i = 1, 2, \ldots, m, j = 1, 2, \ldots, n$.

(2) d_{j_k} are generated randomly by the uniform distribution over $[0, 1]$ such that $d_{j_1} < d_{j_2} < \cdots < d_{j_h}$ for $j = 1, 2, \ldots, n$ and $k = 1, 2, \ldots, h$.

(3) To make sure the test instance is feasible, b_i are generated randomly by the uniform distribution over the interval of $[(\bigvee_{j=1}^{n} a_{ij}d_{j_h}]) \times 0.1, (\bigvee_{j=1}^{n} a_{ij}d_{j_h}]) \times 0.8]$ for $i = 1, 2, \ldots, m$.

This computational experiment consists of 21 cases with different $m \in \{100, 200\}, n \in \{60, 80, 100, 120\}$ and $h \in \{60, 80, 100\}$

All the experiments have been run on a PC equipped with the Intel Coore I5-5257U CPU, 8 GB RAM, and Windows 10 (64 bit) operating system. GUROBI(2016) is the chosen MIP solver for solving all the instances. The computational results are shown in Tables 1 and 2. In the Table, NB denote the number of 0-1 variables, NC denote the number of continuous variables, LE

Table 1. Model 1

m	n	h	NB	NC	LE	LI	T
100	60	60	9600	60	320	6000	6.29 s
100	80	60	12800	80	360	8000	7.39 s
100	100	60	16000	100	400	10000	8.00 s
100	120	60	19200	120	440	12000	11.17 s
100	60	80	10800	60	320	6000	6.80 s
100	80	80	14400	80	360	8000	8.71 s
100	100	80	18000	100	400	10000	10.13 s
100	120	80	21600	120	440	12000	19.80 s
100	60	100	12000	60	320	6000	7.92 s
100	80	100	16000	80	360	8000	9.61 s
100	100	100	20000	100	400	10000	13.17 s
100	120	100	24000	120	440	12000	19.12 s
200	60	60	15600	60	520	12000	22.06 s
200	80	60	20800	80	560	16000	28.03 s
200	100	60	26000	100	6000	20000	72.64 s
200	60	80	16800	60	520	12000	24.62 s
200	80	80	22400	80	560	16000	47.20 s
200	100	80	28000	100	600	20000	74.26 s
200	60	100	18000	60	520	12000	31.17 s
200	80	100	24000	80	560	16000	49.49 s
200	100	100	30000	100	600	20000	73.10 s

Table 2. Model 2

m	n	h	NB	NC	LE	LI	T
100	60	60	6360	3660	680	6000	8.31 s
100	80	60	8480	4880	840	8000	12.74 s
100	100	60	10600	6100	1000	10000	16.44 s
100	120	60	12720	7320	1160	12000	32.15 s
100	60	80	6420	4860	740	6000	9.56 s
100	80	80	8560	6480	920	8000	12.61 s
100	100	80	10700	8100	1100	10000	14.13 s
100	120	80	12940	9720	1280	12000	24.94 s
100	60	100	6420	6060	740	6000	11.06 s
100	80	100	8560	8080	920	8000	18.52 s
100	100	100	10700	10100	1100	10000	16.22 s
100	120	100	12840	12120	1280	12000	21.10 s
200	60	60	1260	3660	880	12000	23.85 s
200	80	60	16480	48800	1040	16000	38.92 s
200	100	60	20600	6100	1200	20000	49.24 s
200	60	80	12420	4860	940	12000	28.81 s
200	80	80	16560	6480	1120	16000	45.66 s
200	100	80	20700	8100	1300	20000	61.15 s
200	60	100	12420	6060	880	12000	62.22 s
200	80	100	16560	8080	1120	16000	36.91 s
200	100	100	20700	10100	1300	20000	54.00 s

denote the number of linear equation constraints, LI denote the number of linear inequality constraints,

Sample Heading (Fourth Level). T denote the average CPU time of solving 10 randomly generated instances of each size.

From the Table 1 and Table 2, we can find that all the cases can be solved in 1.5 min, so the model 1 and model 2 are both efficacious. It is also interesting to see that there is no significant difference between model 1 and 2.

5   Conclusion

Considering the practical application background in the optimization management of physical distribution, we introduce a linear optimization problem subject to max-product fuzzy relation inequalities with discrete variables. For solving the problem, we replaced n discrete variables in h values with binary variables and linear equations by two most common methods. By finding all the potential minimal solutions, we transformed the max-product fuzzy relation inequality constraints into linear equation and inequality constraints with mn 0-1 binary variables. Then we built two models of 0-1 mixed-integer linear programming for the original problem equally. The numerical experiments illustrated the efficiency of the two reformulation models.

It is worth pointing out that fuzzy relation equations or inequalities with other max-t-norm can be treated by the similar way.

Acknowledgements. The authors declare that they have no competing interests.

References

1. Sanchez, E.: Equations de Relations Flous. These Biologie Humaine, France (1972)
2. Sanchez, E.: Resolutions in fuzzy relation equations. Inf. Control **30**, 38–48 (1976)
3. Sanchez, E.: Solution in composite fuzzy relation equations: application to medical diagnosis in Brouwerian logic. In: Gupta, M.M., Saridis, G.N., Gaines, B.R. (eds.) Fuzzy Automata and Decision Processes, pp 221–234. North-Holland, Amsterdam (1993)
4. Nola, A.D., Sessa, S., Pedrycz, W., Sanchez, E.: Fuzzy Relational Equations and Their Applications in Knowledge Engineering. Kluwer Academic Press, Dordrecht (1989)
5. Yang, X.-P., Zheng, G.-Z., Zhou, X.-G., Cao, B.-Y.: Lexicography minimum solution of fuzzy relation inequalities: applied to optimal control in P2P file sharing system. Int. J. Mach. Learn. Cybern. (2016)
6. Yang, X.-P., Zhou, X.-G., Cao, B.-Y.: Multi-level linear programming subject to addition-min fuzzy relation inequalities with application in Peer-to-Peer file sharing system. J. Intell. Fuzzy Syst. **28**, 2679–2689 (2015)
7. Yang, X.-P., Zhou, X.-G., Cao, B.-Y.: Latticized linear programming subject to max-product fuzzy relation inequalities with application in wireless. Inf. Sci. **358–359**, 44–55 (2016)
8. Yang, X.-P., Zhou, X.-G., Cao, B.-Y.: Singal variable term semi-latticized fuzzy relation geomtric programming with max-product operator. Inf. Sci. **325**, 271–287 (2015)
9. Yang, X.-P., Zhou, X.-G., Cao, B.-Y.: Min-max programming problem subject to addition-min fuzzy relation inequalities. IEE Trans. Fuzzy Syst. **24**, 111–119 (2016)
10. Drewniak, J., Matusiewicz, Z.: Properties of max-* relation equations. Soft Comput. **14**, 1037–1041 (2010)
11. Li, P., Fang, S.-C.: A survey on fuzzy relational equations, part I: classification and solvability. Fuzzy Optim. Decis. Making **8**, 179–229 (2009)
12. Li, P., Fang, S.-C.: On the resolution and optimization of a system of fuzzy relation equations with sup-T composition. Fuzzy Optim. Decis. Making **7**, 169–214 (2008)
13. Li, H.-L., Huang, Y.-H., Fang, S.-C.: A logarithmic for reducing binary variables and inequalitiy constraints in solving task assignment problems. INFORMS J. Comput. **25**(4), 643–653 (2013)
14. Li, H.-L., Huang, Y.-H., Fang, S.-C.: Linear reformulation of polynomial discrete programming for fast computation. INFORMS J. Comput. **29**(1), 108–122 (2017)
15. Fang, S.-C., Li, G.-Z.: Solving fuzzy relation equations with a linear objective function. Fuzzy Sets Syst. **103**, 107–113 (1999)

16. Molai, A.A.: Fuzzy linear objective function optimization with fuzzy valued max-product fuzzy relation inequality constraints. Math. Comput. Model. **5**, 1240–1250 (2010)
17. Li, H.-L., Lu, H.-C.: Global optimization for generalized geometric programs with mixed free-sign variables. Oper. Res. **57**(3), 701–713 (2009)
18. Li, H.-L., Hao-chun, L., Huang, C.-H., Nian-Ze, H.: A superior representation method for piecewise linear functions. INFORMS J. Comput. **21**(2), 314–321 (2009)
19. Loetamonphong, J., Fang, S.-S.: Optimization of fuzzy relation equations with max-product composition. Fuzzy Sets Syst. **118**, 509–517 (2001)
20. Wang, P.Z., Wang, D.Z., Sanchez, E., Li, E.S.: Latticized linear programming and fuzzy relation inequalities. J. Math. Anal. Appl. **159**, 72–87 (1991)
21. Zhang, H.T., Dong, H.M., Ren, R.H.: Programming problem with fuzzy relation inequality constraints. J. Liaoning Normal Univ. **3**, 231–233 (2003)
22. Ghodousian, A., Khorram, E.: An algorithm for optimizing the linear function with fuzzy relation equation constraints regarding max-prod composition. Appl. Math. Comput. **178**, 502–509 (2006)
23. Ghodousian, A., Khorram, E.: Fuzzy linear optimization in the presence of the fuzzy relation inequality constraints with max-min composition. Inf. Sci. **178**, 501–519 (2008)
24. Guo, F.-F., Xia, Z.-Q.: An algorithm for solving optimization problems with one linear objective function and finitely many constraints of fuzzy relation inequalities. Fuzzy Optim. Decis. Mak. **5**, 33–47 (2006)
25. Guu, S.-M., Wu, Y.-K.: Minimizing a linear objective function with fuzzy relation equation constraints. Fuzzy Optim. Decis. Mak. **1**(4), 347–360 (2002)
26. Guu, S.-M., Wu, Y.-K.: Minimizing a linear objective function under a max-t-norm fuzzy relational equation constraint. Fuzzy Sets Syst. **161**, 285–297 (2010)
27. Wu, Y.-K., Guu, S.-M., Liu, J.Y.-C.: An accelerated approach for solving fuzzy relation equations with a linear objective function. IEEE Trans. Fuzzy Syst. **10**, 552–558 (2002)
28. Wu, Y.-K., Guu, S.-M.: Minimizing a linear function under a fuzzy max-min relational equation constraint. Fuzzy Sets Syst. **150**, 147–162 (2005)
29. Wu, Y.-K., Guu, S.-M.: A note on fuzzy relation programming problems with max-strict-t-norm composition. Fuzzy Optim. Decis. Mak. **3**, 271–278 (2004)
30. Pandey, D.: On the optimization of fuzzy relation equations with continuous t-norm and with linear objective function. In: Proceedings of the Second Asian Applied Computing Conference, AACC 2004, Kathmandu, Nepal, pp. 41–51 (2004)
31. Shivanian, E., Khorram, E.: Monomial geometric programming with fuzzy relation inequality constraints with max-product composition. Comput. Indust. Eng. **56**, 1386–1392 (2009)
32. Thole, U., Zimmermann, H.-J., Zysno, P.: On the suitability of minimum and product operators for intersection of fuzzy sets. Fuzzy Sets Syst. **2**, 167–180 (1979)

Global Optimization
for the Concave-Concave Multiplicative
Programming with Coefficient

ChangXin Zhu[1] and YongBin OuYang[2]([✉]) [iD]

[1] Department of Basic Course, Guangzhou Vocational and Technical University of
Science and Technology, Guangzhou 510550, China
[2] Guangzhou Huali Science and Technology Vocational College, Guangzhou 511325,
Guangdong, China
15011973027@163.com

Abstract. In this paper, we present a global optimization algorithm for
globally solving the problem (CMPC) of minimizing a concave-concave
multiplicative function with coefficient over a compact convex set. We
firstly convert problem (CMPC) into an equivalent programming problem (CMPC(H)) by introducing $2p$ auxiliary variables. By utilizing the
llinearization technique, initial non-convex nonlinear problem (CMPC)
is reduced to a sequence of convex programming problems through the
successive refinement of a linear relaxation of feasible region and of the
objective function. It has been proved that the algorithm possesses global
convergence. Some numerical examples are given to illustrate validity of
the proposed method.

Keywords: Concave-concave multiplicative programming · Global
optimization · Simplicial partitioning · Duality theory

1 Introduction

The problem of central interest in this article is given by

$$(\text{CMPC}) \quad v = \min h(x) = \sum_{i=1}^{p} c_i f_i(x) g_i(x),$$
$$\text{s.t.} \qquad x \in X = \{x \in \mathrm{R}^n | Ax \le b, x \ge 0, \},$$
(1)

where $p \ge 2$, for any $i = 1, 2, \cdots, p$, f_i, and g_i, are concave functions defined on
R^n, $A \in R^{q \times n}$, $c_i, i = 1, 2, \cdots, p$, are real constant coefficients, X a nonempty
bound and closed set in R^n. We assume also that for each $i = 1, 2, \cdots, p$, $l_i \le
f_i(x) \le u_i$ and $L_i \le g_i(x) \le U_i$ for all $x \in X$, where l_i, u_i, L_i and U_i are
positive scalars that satisfy $l_i \le u_i$ and $L_i \le U_i, i = 1, 2, \cdots, p$. When $c_i =
1(i = 1, 1, \cdots, p)$ and the problem maximize a generalized concave multiplicative
function, the problem (CMPC) is called the generalized concave multiplicative
programming problem studied by Benson [1]. A rectangular, branch-and-bound
algorithm is proposed for solving the resulting problem.

B.-Y. Cao et al. (Eds.): ICFIE 2022, LNDECT 207, pp. 49–63, 2024.
https://doi.org/10.1007/978-981-97-2891-6_4

Multiplicative programming problems have attracted considerable attention in the literature because of their large number of practical applications in various fields of study, including financial optimization [2], plant layout design [3], robust optimization [4], VLISI chip design [5], data mining/ pattern recognition [6].

When f_i and $g_i, i = 1, 2, \cdots, p$, are linear functions defined on R^n, the problem (CMPC) is called linear multiplicative programming (LMP). Since LMP may possess many local minima, it is known to be among the hardest problems [7]. In the past 20 years, many solution algorithms have been proposed for globally solving the problem (LMP). The methods can be classified as branch-and-bound methods [8–12], a primal and dual simplex method [13], an outcome-space cutting plane method [14], heuristic methods [15], decomposition method [16].

When f_i and $g_i, i = 1, 2, \cdots, p$, are convex functions defined on R^n, the problem (CMPC) is called the generalized convex multiplicative programming problem studied by Konno, Kuno, and Yajima [17]. An outer approximation algorithm is proposed for solving the resulting problem.

When $p = 1$, f is quadratic function and g is linear functions with exponents, nonlinear multiplicative problems (CMPC) is studied by R. Cambini and C. Sodini [18]. They show how problem (CMPC) can be solved by means of the optimal level solution approach.

The organization and content of this article can be summarized as follows. In Sect. 2 we present how to derive the equivalent problem (CMPC(H)) and construct the LLBF of the objective function. Then, the convex relaxation programming of (CMPC(H)) is found. In Sect. 3, we propose a branch and bound for globally solving problem (CMPC(H)). And the convergence of the presented algorithm is established. In Sect. 5, Some numerical examples are given to illustrate validity of the proposed method. The some concluding remarks are given in Sect. 6.

2   Equivalent Problem and Convex Relaxation Programming

In this section, we firstly show how to construct initial simplex and convert problem (P) into an equivalent nonconvex programming problem (CMPC(H)) by introducing a $2p$-dimension vector $(t_i, s_i)(i = 1, 2, \cdots, p)$. Then, we present how to construct the LLBF of the objective function and convex relaxation programming of (CMPC(H)).

2.1   Equivalent Problem

In order to construct initial simplex S^0, assume that

$$\gamma = \max_{x \in X} \sum_{j=1}^{n} x_j,$$

and, for all $j = 1, 2, \cdots, n$,

$$\gamma_j = \min_{x \in X} x_j.$$

An initial simplex S^0 in R^p that contains X as follows:

$$S^0 = \left\{ x \in R^n | x_j \geq \gamma_j, j = 1, 2, \cdots, n, \sum_{j=1}^{n} x_j \leq \gamma \right\}.$$

Notice that finding γ and $\gamma_j, j = 1, 2, \cdots, n$, amounts to solving $n + 1$ linear programming problems. Given γ and $\gamma_j, j = 1, 2, \cdots, n$, the simplex S^0 have $n + 1$ vertices. The vertex set of S^0 is $\{v^0, v^1, \cdots, v^n\}$, where

$$v_j^0 = \gamma_j, \quad j = 1, 2, \cdots, n,$$

and, for each $j = 1, 2, \cdots, n, i = 1, 2, \cdots, n$

$$v_j^i = \begin{cases} \gamma_j & \text{if } j \neq i, \\ \gamma - \sum_{j \neq i} \gamma_j & \text{if } j = i. \end{cases}$$

Notice that either $\gamma = \sum_{j=1}^{n} \gamma_j$ or $\gamma > \sum_{j=1}^{n} \gamma_j$, and for any $i = 1, \cdots, n$,

$$(v^i - v^0)^T = (0, 0, \cdots, \gamma - \sum_{j=1}^{n} \gamma_j, \cdots, 0),$$

where $\gamma - \sum_{j=1}^{n} \gamma_j$ is the ith component of $v^i - v^0$. We can easily shows that $S^0 \supseteq X$.

2.2 Equivalent Problem

For any $i = 1, 2, \cdots, p$, since $f_i(x)$ is a concave function on R^n, then $f_i(x)$ is continuous on a compact set $X \subseteq R^n$. Therefore, to globally solve problem (P), for each $i = 1, 2, \cdots, p$, assume that

$$l_i = \min_{x \in X} f_i(x),$$

and

$$L_i = \max_{x \in X} f_i(x).$$

By introducing additional variable vector $y = (y_1, y_2, \cdots, y_p) \in R^p$, we construct a rectangular set D as follows:

$$D =: \{y \in R^p | l_i \leq y_i \leq L_i, i = 1, 2, \cdots, p\}.$$

Note that the set D does not be computed necessarily in the performance of the algorithm (see the proposed algorithm in Sect. 3).

For any simplex $S \subseteq S^0 \subseteq R^n$, define the nonconvex programming problem (P1(S)) as follows:

$$
\text{(P1(S))} \quad v(S) = \min h(x, y) = \sum_{i=1}^{p} g_i(x) y_i,
$$
$$
\text{s.t.} \quad f_i(x) - y_i \leq 0, \quad i = 1, 2, \cdots, p, \tag{2}
$$
$$
Ax - b \leq 0,
$$
$$
x \in S, y \in D.
$$

In order to solve problem (P), the branch and bound algorithm solves problem (P1(S^0)) instead. The validity of solving problem (P1(S^0)) in order to solve problem (P) follows from Theorem 1.

Theorem 1. If (x^*, y^*) is a global optimal solution for problem (P1(S^0)), then x^* is a global optimal solution for problem (P) and $y_i^* = f_i(x^*), i = 1, 2, \cdots, p$. If x^* is a global optimal solution for problem (P), then (x^*, y^*) is a global optimal solution for problem (P1(S^0)) , where $y_i^* = f_i(x^*), \quad i = 1, 2, \cdots, p$. The global optimal values v and $v(S^0)$ of problems (P) and (P1(S^0)) , respectively, are equal.

Proof. The proof of this theorem follows easily from the definitions of problems (P) and (P1(S^0)); therefore, it is omitted.

2.3   Duality Bound

For each n-dimensional simplex S created by the branching process, the algorithm computes a lower bound LB(S) for the optimal value $v(S)$ of problem (P1(S)). The next theorem shows that, by using the Lagrangian weak duality theorem of nonlinear programming, the lower bound LB(S) can be found by solving an ordinary linear program.

Theorem 2. Let $S \subseteq S^0 \subseteq R^n$ be a n-dimensional simplex with vertices $\overline{x}^0, \overline{x}^1, \cdots, \overline{x}^n$, and let $J = \{0, 1, 2, \cdots, n\}$. Then LB(S)$\leq v(S)$, where LB(S) is the optimal value of the linear programming problem (LP(S)) with duality variables $\lambda^T = (\lambda_1, \lambda_2, \cdots, \lambda_q) \in R^q, \theta^T = (\theta_1, \theta_2, \cdots, \theta_p) \in R^p$

$$
\text{LP(S) (LB(S))} = \max \quad -\lambda^T b + t
$$
$$
\text{s.t. } t - \sum_{i=1}^{p} \theta_i f_i(\overline{x}^j) - \lambda^T A \overline{x}^j \leq 0, \quad j = 0, 1, 2, \cdots, n, \tag{3}
$$
$$
0 \leq C, \ \lambda \geq 0, \ \theta \geq 0, t \text{ free.}
$$

where $C \in R^q$ and $C = (\min_{j \in J} g_1(\overline{x}^j), \min_{j \in J} g_2(\overline{x}^j), \cdots, \min_{j \in J} g_p(\overline{x}^j))^T$.

Proof. By the definition of $v(S)$ and the weak duality theorem of Lagrangian duality, $v(S) \geq LB(S)$, where

$$
LB(S) = \max_{\substack{\lambda \geq 0 \\ \theta \geq 0}} \left\{ \min_{\substack{y \in D \\ x \in S}} \left[\sum_{i=1}^{p} g_i(x) y_i + \sum_{i=1}^{p} \theta_i [f_i(x) - y_i] + \lambda^T (Ax - b) \right] \right\}
$$

$$
= \max_{\substack{\lambda \geq 0 \\ \theta \geq 0}} \left\{ -\lambda^T b + \min_{\substack{y \in D \\ x \in S}} \left[\sum_{i=1}^{p} g_i(x) y_i + \sum_{i=1}^{p} \theta_i [f_i(x) - y_i] + \lambda^T Ax \right] \right\}
$$

$$
= \max_{\substack{\lambda \geq 0 \\ \theta \geq 0}} \left\{ -\lambda^T b + \min_{x \in S} \left[\sum_{i=1}^{p} \theta_i f_i(x) + \lambda^T Ax + \min_{y \in D} <G(x), y> \right] \right\}
$$

where $G(x) = (g_1(x) - \theta_1, \cdots, g_p(x) - \theta_p) \in R^p$.

Since

$$
\min_{y \in D} \langle G(x), y \rangle = \begin{cases} 0, & \text{if } G(x) \geq 0, \forall x \in S, \\ -\infty, & \text{otherwise}, \end{cases}
$$

it follows that,

$$
LB(S) = \max \left\{ -\lambda^T b + \min_{x \in S} \left[\sum_{i=1}^{p} \theta_i f_i(x) - \lambda^T Ax \right] \right\}
$$
$$
\text{s.t. } g_i(x) - \theta_i \geq 0, \quad i = 1, 2, \cdots, p, \forall x \in S, \tag{4}
$$
$$
\lambda \geq 0, \quad \theta \geq 0.
$$

Since S is a compact polyhedron with extreme points $\overline{x}^j, j = 0, 1, 2, \cdots, n$, and $g_i(x)$ is concave function, then, for each $\theta \geq 0$ and $\lambda \geq 0$, $g_i(x) - \theta_i \geq 0$ holds for all $x \in S$ if and only if it holds for all $x \in \{\overline{x}^0, \overline{x}^1, \cdots, \overline{x}^n\}$. So, for all $j \in J$, we can get $g_i(\overline{x}^j) - \theta_i \geq 0$, that is,

$$
\theta_i \leq g_i(\overline{x}^j), \qquad \forall j \in J. \tag{5}
$$

Notice that for all $j \in J$, the left-hand-side of (5) is the same linear function of θ, then (5) is equivalent that $\theta \leq C$, where $C \in R^p$ and $C = (c \min_{j \in J} g_1(\overline{x}^j), \cdots, \min_{j \in J} g_p(\overline{x}^j))^T$.

Therefore,

$$
(LB(S)) = \max \left\{ -\lambda^T b + \min_{x \in S} \left[\sum_{i=1}^{p} \theta_i f_i(x) + \lambda^T Ax \right] \right\}
$$
$$
\text{s.t. } \theta \leq C, \lambda \geq 0, \quad \theta \geq 0.
$$

That is,

$$
(LB(S)) = \max -\lambda^T b + t
$$
$$
\text{s.t. } t \leq \sum_{i=1}^{p} \theta_i f_i(x) + \lambda^T Ax, \quad x \in S
$$
$$
\theta \leq C, \lambda \geq 0, \quad \theta \geq 0, t \text{ free}.
$$

For any $\lambda \geq 0$, and $\theta \geq 0$, since $f_i(x)$ are concave function and $A_l x$ a linear function of x, where A_l denotes the lth row of A, one can get that $\sum_{i=1}^{p} \theta_i f_i(x) + \lambda^T A x$ is a concave function of x. Because simplex S is a compact polyhedron with extreme points $\overline{x}^0, \overline{x}^1, \cdots, \overline{x}^n$, this implies for any $\lambda \geq 0$, and $\theta \geq 0$, $t \leq \sum_{i=1}^{p} \theta_i f_i(x) + \lambda^T A x$ holds if and only if

$$\sum_{i=1}^{p} \theta_i f_i(\overline{x}^j) + \lambda^T A \overline{x}^j - t \geq 0, \quad j = 0, 1, 2, \cdots, n.$$

The proof is complete.

Proposition 1. Let $S^1, S^2 \subseteq R^n$ be a n-dimensional subsimplices of S formed by the branching process such that $S^1 \subseteq S^2 \subseteq S^0$. Then

(i) $LB(S^1) \geq LB(S^2)$,
(ii) $LB(S^1) > -\infty$.

Proof. (i) It follows from the proof of Theorem 2 for an arbitrary simplex S.
(ii) It follows from (i) that we need only show $LB(S^0) > -\infty$. From Theorem 2, we can get

$$LB(S^0) = \max_{\substack{\lambda \geq 0 \\ \theta \geq 0}} \left\{ \min_{\substack{y \in D \\ x \in S^0}} \left[\sum_{i=1}^{p} g_i(x)y_i + \sum_{i=1}^{p} \theta_i[f_i(x) - y_i] + \lambda^T(Ax - b) \right] \right\}$$

Let $\lambda = 0, \theta = 0$, then

$$LB(S^0) = \min_{\substack{y \in D \\ x \in S^0}} \sum_{i=1}^{p} g_i(x)y_i.$$

Let $E = S^0 \times D, F(x, y) = \sum_{i=1}^{p} g_i(x)y_i$. We see that $g_i(x)$ is continuous for each $i = 1, 2, \cdots, p$, since $g_i(x)$ is concave function on R^n. Then the function $F(x, y)$ is continuous in (x, y) on M. By the compactness of M, we can get $\min_{\substack{y \in D \\ x \in S^0}} \sum_{i=1}^{p} g_i(x)y_i$ is finite, that is, $LB(S^0) > -\infty$.

Remark 1. The monotonicity property in part (i) of Proposition 1 will be used to help to show the convergence of the algorithm. From part (ii) of Proposition 1, for any p-dimensional simplex S created by the algorithm during the branch and bound search, the duality bounds-based lower bound $LB(S)$ for the optimal value $v(S)$ of problem (P1(S)) is either finite or equal to $+\infty$. When $LB(S) = +\infty$, problem (P1(S)) is infeasible and, as we shall see, S will be eliminated from further consideration by the deletion by bounding process of the algorithm.

Now, we will show how to determine the upper bound of the globally optimal value for (P1(S^0)). For each p-dimensional simplex S generated by the algorithm such that LB(S) is finite, the algorithm generates a feasible solution w to problem (P). As the algorithm finds more and more feasible solutions to problem (P), the upper bound for the optimal value v of problem (P) improves iteratively. These feasible solutions are found from dual optimal solutions to the lower bounding problems (LP(S)) that are solved by the algorithm, as given in the following result.

Proposition 2. Let $S \subseteq S^0 \subseteq R^n$ be a n-dimensional simplex with vertices $\overline{x}^0, \overline{x}^1, \cdots, \overline{x}^n$, and suppose that LB($S$) $\neq +\infty$. Let $w \in R^{n+1}$ be optimal dual variables corresponding to the first $n + 1$ constraints of linear program LP(S). Then $x = \sum_{j=0}^{n} w_j \overline{x}^j$ is a feasible solution for problem (P).

Proof. The dual linear program to problem (LP(S)) is

$$\text{DLP(S)} \quad \text{LB(S)} = \min C^T r$$
$$\text{s.t.} \sum_{j=0}^{n} w_j = 1,$$
$$- \sum_{j=0}^{n} w_j f_i(\overline{x}^j) + r_i \geq 0, \ i = 1, 2, \cdots, p,$$
$$- \sum_{j=0}^{p} w_j A_l \overline{x}^j \geq -b_l, l = 1, 2, \cdots, q,$$
$$w_j \geq 0, \quad j = 0, 1, \cdots, n,$$
$$r_k \geq 0, \quad k = 1, \cdots, p.$$

where A_l denotes the lth row of A, b_l denotes the lth component of $b, l = 1, \cdots, q$. The constraints of problem (DLP(S)) imply that $A \sum_{j=0}^{n} w_j \overline{x}^j \leq b, \sum_{j=0}^{n} w_j \overline{x}^j \geq 0$, since $\overline{x}^j \geq 0, j = 0, 1, \cdots, n$. This implies $x = \sum_{j=0}^{n} w_j \overline{x}^j$ is a feasible solution for problem (P).

3 Global Optimizing Algorithm

To globally solve problem (P1(S^0)), the algorithm to be presented uses a branch and bound approach. There are three fundamental processes in the algorithm, a branching process, a lower bounding process, and an upper bounding process.

3.1 Branching Rule

The branch and bound approach is based on partitioning the n-dimensional simplex S^0 into smaller subsimplicies that are also of dimension p, each concerned with a node of the branch and bound tree, and each node is associated with a linear subproblem on each subsimplicie. These subsimplicies are obtained by the branching process, which helps the branch and bound procedure identify a location in the feasible region of problem (P1(S^0)) that contains a global optimal solution to the problem (P).

During each iteration of the algorithm, the branching process creates a more refined partition of a portion of $S = S^0$ that cannot yet be excluded from consideration in the search for a global optimal solution for problem (P1(S)). The initial partition Q_1 consists simply of S, since at the beginning of the branch and bound procedure, no portion of S can as yet be excluded from consideration.

During iteration k of the algorithm, $k \geq 1$, the branching process is used to help create a new partition Q_{k+1}. First, a screening procedure is used to remove any rectangle from Q_k that can, at this point of the search, be excluded from further consideration, and Q_{k+1} is temporarily set equal to the set of simplices that remain. Later in iteration k, a rectangle S^k in Q_{k+1} is identified for further examination. The branching process is then evoked to subdivide S^k into two subsimplicies S_1^k, S_2^k. This subdivision is accomplished by a process called simplicial bisection.

Definition 1 [19]. Let S be a n-dimensional simplex with vertex set $\{v^0, v^1, \cdots, v^n\}$. Let w be the midpoint of any of the longest edges $[v^r, v^t]$ of S. Then $\{S^1, S^2\}$ is called a *simplicial bisection* of S, where the vertex set of S^1 is $\{v^0, v^1, \cdots, v^{r-1}, w, v^{r+1}, \cdots, v^n\}$ and the vertex set of S^2 is $\{v^0, v^1, \cdots, v^{t-1}, w, v^{t+1}, \cdots, v^n\}$.

3.2   Lower Bound and Upper Bound

The second fundamental process of the algorithm is the lower bounding process. For each simplex $S \subseteq S^0$ created by the branching process, this process gives an lower bound LB(S) for the optimal value $v(S)$ of the following problem (P1(S)),

$$(\text{P1(S)}) \quad v(S) = \min h(x,y) = \sum_{i=1}^{p} g_i(x)y_i,$$
$$\text{s.t.} \quad f_i(x) - y_i \leq 0, \quad i = 1, 2, \cdots, p,$$
$$Ax - b \leq 0,$$
$$x \in S, y \in D.$$

For each simplex S created by the branching process, LB(S) is found by solving a single linear programming LP(S) as follows,

$$(\text{LP(S)}) \ (\text{LB(S)}) = \max -\lambda^T b + t$$
$$\text{s.t.} \ t - \sum_{i=1}^{p} \theta_i f_i(\overline{x}^j) - \lambda^T A \overline{x}^j \leq 0, \quad j = 0, 1, 2, \cdots, n,$$
$$\theta \leq C,$$
$$\lambda \geq 0, \ \theta \geq 0, t \text{ free}.$$

where $\overline{x}^0, \overline{x}^1, \cdots, \overline{x}^n$ denote the vertices of the n-dimensional simplex S.

During each iteration $k \geq 0$, the lower bounding process computes an lower bound LB_k for the optimal value $v(S^0)$ of problem (P1(S^0)). For each $k \geq 0$, this lower process bound LB_k is given by

$$LB_k = \min\{LB(S) | S \in Q^k\}.$$

The upper bounding process is the third fundamental process of the branch and bound algorithm. For each n-dimensional simplex S created by the branching process, this process finds a upper bound for (P1(S)). Let $w \in R^{n+1}$ be optimal dual variables corresponding to the first $n + 1$ constraints of linear program LP(S), and set $x^* = \sum_{j=0}^{n} w_j \overline{x}^{j*}$. Then, from definition of problem (DLP(S)), we have that $Ax^* \leq b, x^* \geq 0$. This implies that x^* is a feasible solution of (P1(S)). Therefore, the upper bound UB(S) of (P1(S)) is $h(x^*)$. In each iteration of the algorithm, this process finds a upper bound for v. For each $k \geq 0$, let $w \in R^n$ be optimal dual variables corresponding to the first n constraints of linear program LP(S), then this upper bound UB_k is given by

$$UB_k = h(x)$$

where x is the incumbent feasible solution for problem (P).

3.3   Deleting Technique

As the branch and bound search proceeds, certain p-dimensional simplices created by the algorithm are eliminated from further consideration. There are two ways that this can occur, either by deletion by bounding or by deletion by infeasibility.

During any iteration $k, k \geq 1$, let UB_k be the smallest objective function value achieved in problem (P) by the feasible solutions to problem (P1(S)) thus far generated by the algorithm. A simplex $S \subseteq S^0$ is deleted by bounding when

$$LB(S) \geq UB_k \tag{6}$$

holds. When (6) holds, searching simplex S further will not improve upon the best feasible solution found thus far for problem (P).

As soon as each n-dimensional simplex S is created by simplicial bisection in the algorithm, it is subjected to the deletion by infeasibility test. Let $\overline{x}^0, \overline{x}^1, \cdots, \overline{x}^n$ denote the vertices of such a simplex S. If, for some $i \in \{1, 2, \cdots, p\}$, we have

$$\min\{A_i \overline{x}^0 - b_i, A_i \overline{x}^1 - b_i, \cdots, A_i \overline{x}^n - b_i\} > 0, \tag{7}$$

then simplex S is said to pass the deletion by infeasibility test and it is eliminated by the algorithm from further consideration. If for each $i \in \{1, 2, \cdots, p\}$, both (7) fail to hold, then simplex S fails the deletion by infeasibility test and it is retained for further scrutiny by the algorithm. The validity of the deletion by infeasibility test follows from the fact that if (7) holds for some i, then for each $x \in S$, there is no $x \in S$ such that

$$A_i x - b_i \leq 0.$$

This implies problem (P1(S)) infeasible.

3.4   Branch and Bound Algorithm

Based on the results and algorithmic process discussed in this section, the branch and bound algorithm for globally solving problem (P) may be stated as follows.

Step 0. (Initialization)

0.1 Initialize the iteration counter $k := 0$; the set of all active node $Q_0 := \{S^0\}$; the upper bound $UB_0 = +\infty$.

0.2 Solve linear program $(LP(S^0))$ for its finite optimal value $LB(S^0)$. Let $w \in R^n$ be optimal dual variables corresponding to the first n constraints of linear program $LP(S^0)$. Set $x^0 = w, UB_0 = h(x^0), LB_0 = LB(S^0)$. Set $k = 1$, and go to iteration k.

Main Step (at iteration k)

Step 1. (Termination) If $UB_{k-1}=LB_{k-1}$, then stop, and x^{k-1} is an global optimal solution for problem (P) and $v=LB_{k-1}$. Otherwise, set $x^k = x^{k-1}$ $LB_k=LB_{k-1}$, $UB_k=UB_{k-1}$. Go to Step 2.

Step 2. (Branching) Let $S^k \in Q^{k-1}$ satisfy

$$S^k \in \arg\min\{LB(S)|S \in Q^{k-1}\}.$$

Use simplicial bisection to divide S^k into S_1^k and S_2^k. Let $\widehat{R} = \{S_1^k, S_2^k\}$.

Step 3. (Infeasiblity test). Delete form $\widehat{R}$ each simplex that passes the deletion by infeasiblity test. Let R represent the subset of $\widehat{R}$ thereby obtained.

Step 4. (Fathoming) For each new sub-simplex $S \in R$, compute the optimal value $LB(S)$ of linear program $(LP(S^0))$. If $LB(S)$ is finite, let $w \in R^n$ be optimal dual variables corresponding to the first $(p+1)n$ constraints of linear program $(LP(S))$.

Step 5. (Updating upper bound) If $h(w) < h(x^k)$, set $x^k = w$, $UB_k=h(x^k)$.

Step 6. (New partition) Let $Q^k = \{Q^{k-1} \setminus \{S^k\}\} \bigcup R$.

Step 7. (Deletion) $Q^k = Q^k \setminus \{S : LB(S) \geq UB_k\}$.

Step 8. (Convergence) If $Q^k = \emptyset$, then stop. UB_k is the optimal value of problem (P), and x^k is an global ε-optimal solution for problem (P). Otherwise, set $k + 1$ and go to Step 1.

4   Convergence of the Algorithm

In this section we give the global convergence of above algorithm. Notice that the algorithm is either finite or infinite. If it is finite, i.e. it terminates at iteration k, then $Q^k = \emptyset$, so that

$$v(S^0) \geq UB_k = h(x^k).$$

Since, by Proposition 1, $v = v(S^0)$ and since $x^k \in X$, this implies that $v = h(x^k)$ and x^k is an global optimal solution for problem (P). Thus, when the algorithm is finite, it globally solves problem (P) as desired. If the algorithm does not terminate after finitely many iterations, then it is easy to show that it generates at least one infinite nested subsequence $\{S^r\}$ of simplices, i.e., where $S^{r+1} \subseteq S^r$ for all r. In this case, the following result is a key to the convergence of the algorithm.

Theorem 3. Suppose that the algorithm is infinite, and that $\{S^r\}$ is an infinite nested subsequence of simplices generated by the algorithm. Let x^* denote any accumulation point of $\{S^r\}$. Then x^* is a global optimal solution for problem (P).

Proof. Suppose that the algorithm is infinite, and let $\{S^r\}$ be chosen as in the theorem. Then, from Horst and Tuy [19], $\bigcap_r S^r = \{x^*\}$ for some point $x^* \in R^n$. Then $x^* \in S^0$.

From Proposition 3, the sequence $LB(S^r)$ is nondecreasing, so the limit $u^* = \lim_{r\to\infty} LB(S^r)$ exists. Then we have $u^* \le v(S^0)$. Next we will show that $u^* \ge v(S^0)$.

Let $G(x,y) : S^0 \times D \to R^{p+q}$ be a vector function formed by the constraints of problem $(P1(S^0))$ as follows:

$$G(x,y) = \begin{cases} f_i(x) - y_i, & i = 1, 2, \cdots, p, \\ Ax - b. \end{cases}$$

Then we can get $\min\{h(x^*, y)|G(x^*, y), y \in D\} \ge v(S^0)$. So, we only need to prove

$$u^* \ge \min\{h(x^*, y)|G(x^*, y) \le 0, y \in D\}. \tag{8}$$

We assume the contrary, that

$$\min\{h(x^*, y)|G(x^*, y) \le 0, y \in D\} > u^*. \tag{9}$$

For any $i = 1, 2, \cdots, p$, since $f_i(x)$ ans $g_i(x)$ are concave functions on R^n, then $f_i(x)$ ans $g_i(x)$ are continuous functions, and hence $h(x,y)$ and $G(x,y)$ are continuous on $M = S^0 \times D$, linear in y for every fixed x. Therefore, for any fixed x, the function $h(x,y) + \lambda^T G(x,y)$ is linear in y and in λ, respectively, where $\lambda \in N = \{R^{p+q}|\lambda \ge 0\}$. Since D is close and bound, then we can get by the classical minimax equality:

$$\min_{y\in D} \max_{\lambda\in N}\{h(x,y) + \lambda^T G(x,y)\} = \max_{\lambda\in N} \min_{y\in D}\{h(x,y) + \lambda^T G(x,y)\}. \tag{10}$$

Since

$$\max_{\lambda\in N}\{h(x,y) + \lambda^T G(x,y)\} = \begin{cases} h(x,y), & \text{if } G(x,y) \le o, \\ +\infty, & \text{otherwise.} \end{cases} \tag{11}$$

we can obtain from (10), for any $x \in S^0$,

$$\min_{y\in D}\{h(x,y)|\lambda^T G(x,y) \le 0, y \in D\} = \max_{\lambda\in N} \min_{y\in D}\{h(x,y) + \lambda^T G(x,y)\}. \tag{12}$$

So, we have from (12)

$$\max_{\lambda\in N} \min_{y\in D}\{h(x^*, y) + \lambda^T G(x^*, y)\} > u^*.$$

Then there exists λ^* satisfying

$$\max_{\lambda\in N} \min_{y\in D}\{h(x^*, y) + (\lambda^*)^T G(x^*, y)\} > u^*.$$

Then, using the continuity of the function $h(x, y) + \lambda^T G(x, y)$ find, for every fixed $y \in D$, there exist an open ball $U_y \subseteq R^n$ around x^* and an open ball $V_y \subseteq R^p$ around y such that

$$h(\hat{x}, \hat{y}) + (\lambda^*)^T G(\hat{x}, \hat{y}) > u^*, \quad \forall \hat{x} \in U_y, \forall \hat{y} \in V_y.$$

Since the sets $V_y (y \in D)$ construct a covering of the compact set D, there exists a finite set $E \subseteq D$ such that the sets $V_y (y \in E)$ still construct a covering of D. Therefor, for every $y \in D$, we have $y \in V_{\hat{y}}$ for some $\hat{y} \in E$. Assume $U = \bigcap_{y \in E} U_y$, hence

$$h(x, y) + (\lambda^*)^T G(x, y) > u^*, \quad \forall x \in U, \forall y \in D.$$

For all sufficiently large r, we have $S^r \subseteq U$, since $\bigcap_r S^r = \{x^*\}$. So, we can get

$$\max_{\lambda \in N} \min_{y \in D} \{h(x, y) + \lambda^T G(x, y) | x \in S^r, y \in D\} > u^*.$$

Therefore, $\text{LB}(S^r) > u^*$. That is a contradiction. Based on the discuss above, we can obtain

$$\begin{aligned} u^* = v(S^0) &= \min\{h(x^*, y) | G(x^*, y) \leq 0, y \in D\} \\ &= \min\{h(x, y) | G(x, y) \leq 0, x \in S^0, y \in D\}. \end{aligned}$$

That implies any optimal solution y^* of the latter linear program then yields an optimal solution (x^*, y^*) for problem (P1(S^0)). Therefore, by Theorem 1, x^* is a global optimal solution for problem (P).

It follows from Theorem 3 that we can easily show two fundamental convergence properties of the algorithm as follows.

Corollary 1. Suppose that the algorithm is infinite. Let w^* denote any accumulation point of $\{\sum_{j=0}^n w_j^r \bar{x}^{r,j}\}_{r=0}^\infty$ where, for each r, $w^r \in R^{n+1}$ denotes any optimal dual variables corresponding to the first $n + 1$ constraints of linear program (LP(S^r)). Then w^* is a global optimal solution for problem (P).

Proof. Suppose that the algorithm is infinite, and let $\{S^r\}$ be chosen as in the theorem. Then, from Horst and Tuy [19], $\bigcap_r S^r = \{x^*\}$ for some point $x^* \in R^n$. Then $x^* \in S^0$. For each simplex S^r, let $\bar{x}^{r,j}, j = 0, 1, \cdots, n$, denote its vertices, then $\lim_{r \in R} \bar{x}^{r,j} = x^*$.

For each r, $w^r \in R^{n+1}$ denotes any optimal dual variables corresponding to the first $n + 1$ constraints of linear program (LP(S^r)). Set $W = \{w \in R^{n+1} | \sum_{j=0}^n w_j = 1, w_j \geq 0, j = 0, 1, \cdots, n\}$. Since W is compact, this implies that there exists an infinite subsequence R' of R such that $\lim_{r \in R'} w^r = w^*$, where $w^* \geq 0, \sum_{j=0}^n w_j^* = 1$. Then we have

$$\lim_{r \in R'} \sum_{j=0}^n w_j^r \bar{x}^{r,j} = \sum_{j=0}^n w_j^* \bar{x}^* = x^*.$$

From Theorem 3, that prove w^* is a global optimal solution for problem (P).

5   Sample Problem

Now we give a example for the proposed global optimization algorithm to illustrate its efficiency.

$$v = \min G(x) = (5 - 0.25x_1^2)(0.125x_2 + 1) + (0.25x_1 + 1)(4 - 0.125x_2^2)$$

$$\text{s.t.} \qquad x \in X \left\{ \begin{array}{l} -5x_1 + 8x_2 \leq 24, \\ 5x_1 + 8x_2 \leq 34, \\ 6x_1 - 3x_2 \leq 15, \\ -4x_1 - 5x_2 \leq -20, \\ x_1 \geq 0, x_2 \geq 0. \end{array} \right\}$$

Prior to initiating the algorithm, from Sect. 2.1, we can determine a simplex S^0 containing X. The vertices S^0 are given by $v^0 = \overline{x}^0 = (0.7018, 1.4286), v^1 = \overline{x}^1 = (4.8697, 1.4286), v^2 = \overline{x}^2 = (0.7018, 4.1429)$.

Initialization.   Solving the following linear programming (LP(S^0)) get LB(S^0)=5.010304 and the dual variables $w^0 = (0.0948597, 0.6770819, 0.2280584)$

max $-24\lambda_1 - 34\lambda_2 - 15\lambda_3 + 20\lambda_4 + t$

s.t. $t - 4.876870_1 - 3.744890_2 - 7.9198\lambda_1 - 14.9378\lambda_2 + 0.075\lambda_3 + 9.9502\lambda_4 <= 0,$

  $t + 0.9284950_1 - 3.744890_2 + 12.9197\lambda_1 - 35.7773\lambda_2 - 24.9324\lambda_3 + 26.6218\lambda_4 <= 0,$

  $t - 4.876870_1 - 1.854550_2 - 29.6342\lambda_1 - 36.6522\lambda_2 + 8.2179\lambda_3 + 23.5217\lambda_4 <= 0,$

  $\theta_1 <= 1.17857, \quad \theta_2 <= 1.17545,$

  $\theta \geq 0, \lambda \geq 0, t$ free.

We set $x^0 = w_1^0 v^0 + w_2^0 v^1 + w_3^0 v^2 = (3.52381, 2.04762)$, UB$_0$ = 8.91891, LB$_0$ = 5.010304, $G^0 = \{S^0\}$, and $k = 1$. Select the convergence tolerance to be equal to $\varepsilon = 10^{-2}$.

Iteration 1. Since UB$_0$−LB$_0$ < ε, S^0 is split by simplicial bisection into S_1^1 and S_2^1 where the vertices of S_1^1 are $(0.7018, 1.4286), (2.78575, 2.78575), (0.7018, 4.1429)$ and the vertices of S_2^1 are $(0.7018, 1.4286), (4.8697, 1.4286), (2.78575, 2.78575)$. Neither S_1^1 nor S_1^1 is deleted by the deletion by infeasibility test. By solving problem (LP(S_1^1)), we obtain the lower bound LB(S_1^1)=7.478248 and the dual variable $w = (0.1040936, 0.8959064, 0)$. Set $x = w_1 v^0 + w_2 v^1 + w_3 v^2 = (2.56882, 2.64448)$, Since $h(x) = 9.59103 > h(x^1)$, where $x^1 = x^0$, we do not update x^1 and UB$_1$=UB$_0$. By solving problem (LP(S_2^1)), we obtain the lower bound LB(S_2^1)=5.717383 and the dual variable $w = (0.0948607, 0.4490240, 0.4561153)$. As with problem (LP($S_1^1$)), the dual to problem (LP(S_2^1)) does not lead to an update of x^1 and UB$_1$. We have $Q^1 = \{S_1^1, S_2^1\}$. and neither S_1^1 nor S_2^1 is deleted by Step 7 from Q^1. At the end of Iteration 1, $x^1 = (3.52381, 2.04762)$, UB$_1$ = 8.91891, LB$_1$ = 5.717383, $G^1 = \{S_1^1, S_2^1\}$.

The algorithm finds a global $\varepsilon-$ optimal value 8.91891 after 8 iterations at the global $\varepsilon-$ optimal solution $x^* = (3.52381, 2.04762)$.

6   Conclusion

We have presented and validated a new simplicial branch and bound algorithm for globally solving the concave-concave multiplicative programming problem (P). The algorithm implements a simplicial, branch and bound search that finds a global optimal solution to the problem that is equivalent to the problem (P). We believe that the new algorithm has several potential practical and computational advantages. Numerical example shows that the proposed algorithm is feasible.

References

1. Benson, H.P.: Global maximization of a generalized concave multiplicative function. J. Optimiz. Theory Appl. **137**(1), 105–120 (2008)
2. Maranas, C.D., Androulakis, I.P., Floudas, C.A., Berger, A.J., Mulvey, J.M.: Solving long-term financial planning problems via global optimization. J. Econ. Dyn. Control **21**, 1405–1425 (1997)
3. Quesada, I., Grossmann, I.E.: Alternative bounding approximations for the global optimization of various engineering design problems. In: Grossmann, I.E. (ed.) Global Optimization in Engineering Design, Nonconvex Optimization and Its Applications, vol. 9. Kluwer Academic Publishers, Norwell (1996)
4. Mulvey, J.M., Vanderbei, R.J., Zenios, S.A.: Robust optimization of large-scale systems. Oper. Res. **43**, 264–281 (1995)
5. Dorneich, M.C., Sahinidis, N.V.: Global optimization algorithms for chip design and compaction. Eng. Optimiz. **25**(2), 131–154 (1995)
6. Bennett, K.P., Mangasarian, O.L.: Bilinear separation of two sets in n-space. Comput. Optimiz. Appl. **2**, 207–227 (1994)
7. Matsui, T.: NP-Hardness of linear multiplicative programming and related problems. J. Glob. Optimiz. **9**, 113–119 (1996)
8. Hong, S.R., Nikolaos, V.S.: Global optimization of multiplicative programs. J. Glob. Optimiz. **26**, 387–418 (2003)
9. Konno, H., Kuno, T.: Generalized linear multiplicative and fractional programming. Annal. Oper. Res. **25**, 147–161 (1990)
10. Shen, P.P., Jiao, H.W.: Linearization method for a class of multiplicative programming with exponent. Appl. Math. Comput. **183**, 328–336 (2006)
11. Zhou, X.G., Wu, K.: A method of acceleration for a class of multiplicative programming problems with exponent. J. Comput. Appl. Math. **223**, 975–982 (2009)
12. Benson, H.P.: A simplicial branch and bound duality-bounds algorithm for the linear sum-of-ratios problem. Eur. J. Oper. Res. **182**, 597–611 (2007)
13. Schaible, S., Sodini, C.: Finite algorithm for generalized linear multiplicative programming. J. Optimiz. Theory Appl. **87**(2), 441–455 (1995)
14. Benson, H.P., Boger, G.M.: Outcome-space cutting-plane algorithm for linear multiplicative programming. J. Optimiz. Theory Appl. **104**(2), 301–322 (2000)
15. Liu, X.J., Umegaki, T., Yamamoto, Y.: Heuristic methods for linear multiplicative programming. J. Glob. Optimiz. **4**(15), 433–447 (1999)
16. Benson, H.P.: Decomposition branch and bound based algorithm for linear programs with additional multiplicative constraints. J. Optimiz. Theory Appl. **126**(1), 41–46 (2005)

17. Konno, H., Kuno, T., Yajima, Y.: Global minimization of a generalized convex multiplicative function. J. Glob. Optimiz. **4**, 47–62 (1994)
18. Cambini, R., Sodini, C.: A finite algorithm for a class of nonlinear multiplicative programs. J. Glob. Optimiz. **26**, 279–296 (2003)
19. Horst, R., Tuy, H.: Global Optimization. Springer, Heidelberg (1993). https://doi.org/10.1007/978-3-662-02947-3

Fish Swarm Algorithm Based Reflecting Surface Adjustment Strategy for Radio Telescopes

Jin-pei Chen[1], Yu-cheng Wu[2], Xi-yue Liu[3], and Guo-dong He[1(✉)]

[1] School of Mathematics and Information Science, Guangzhou University, Guangzhou, China
hguod@gzhu.edu.cn
[2] School of Civil Engineering, Guangzhou University, Guangzhou, China
[3] School of Economics and Statistics, Guangzhou University, Guangzhou, China

Abstract. In this paper, a method for shape adjustment of active reflector panels of radio telescopes under certain constraints is proposed based on the artificial fish swarm algorithm, and a double Gaussian function is introduced to improve the traditional artificial fish swarm algorithm so that the algorithm can quickly escape from local extremes and improve the efficiency of iteration. A method to determine the ideal paraboloid is also proposed by using the coordinate system transformation, and a method to calculate the signal acceptance ratio is given to make the strategy more practical application.

Keywords: improved artificial fish swarming algorithm · coordinate system transformation · reception ratio model

1 Introduction

Radio telescope is an effective tool for mankind to observe the faint signals in the universe. As one of the core structures of radio telescope, the active reflecting surface still has many challenges to be solved in the engineering implementation.

Whether the active reflective surface is close to becoming an ideal paraboloid has an effect on the effectiveness of the reflected signal received by the feeder module. Numerous scholars have studied the optimization of the adjustment algorithm of the active reflecting surface. Shiyun Shen [1] et al. used the MVFSA algorithm for parameter search optimization; Lichun Zhu [2] established an adaptive, self-learning mechanism node displacement control model through the control strategy of the main reflecting surface control system; Jianxing Xue [3] et al. studied the fitting accuracy of the FAST transient paraboloid based on the analysis of the dynamic surface shape of the reflecting surface unit [6].

In this paper, a strategy to determine the ideal paraboloid is proposed under the reflective panel adjustment constraint, and then the reflective surface is adjusted to the working paraboloid by adjusting the radial expansion of the actuator, so that the working paraboloid is as close to the ideal paraboloid as possible to obtain the best reception effect of celestial electromagnetic waves after reflection by the reflective surface. Within the constraint, this paper optimizes the objective function based on the artificial fish swarm algorithm, so that the algorithm is not easy to fall into the local optimum; improves the algorithm to improve the iteration efficiency of the algorithm; and proposes a method to calculate the reception ratio of the reference reflecting spherical surface.

2   Introduction to Active Reflecting Surfaces for Radio Telescopes

2.1   Composition of Active Reflective Surface

The active reflecting surface system is an adjustable spherical surface composed of the main cable network, reflecting panel, lower cable, actuator and support structure. The main cable, the lower cable and the main cable node can be stretched or contracted by the actuator to adjust the position of the reflector panel, thus changing the whole radio telescope from a near spherical surface to a near parabolic surface, which is more conducive to the centralized signal collection. The side-view and top-view structures of the active reflecting surface system are shown in Fig. 1.

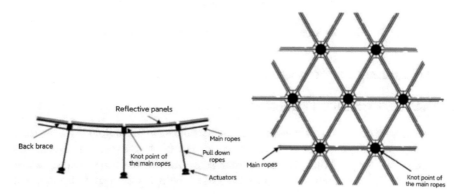

Fig. 1. Schematic diagram of the structure of the active reflective surface system

2.2   Working Principle of Active Reflective Surface System

The active reflecting surface can be divided into two states: the baseline state and the working state. The base state sphere and the focal plane are partially concentric spheres with point C as the circle point, where the radius difference between the two surfaces is F = 0.466R, and the shape of the reflecting surface is adjusted to an approximate rotating paraboloid of 300 m aperture (working paraboloid) in the working state. Figure 2 shows a schematic diagram of the device's profile during the observation. When observing a target S in a certain direction, the center of the receiving plane of the feeder module is moved to the intersection of the straight line SC and the focal plane P. The reflecting panel on the reference sphere is adjusted to form an approximately rotating paraboloid with the straight line SC as the axis of symmetry and P as the focal point, thus reflecting the parallel electromagnetic waves from the target object into the effective area of the feeder module.

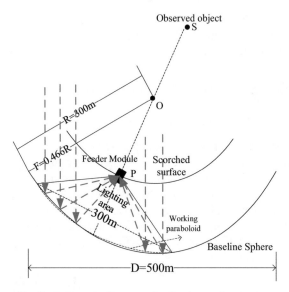

Fig. 2. Schematic diagram of reflective surface system

3   Adjustment Strategy for the Ideal Paraboloid

3.1   Calculating the Ideal Paraboloid

This subsection gives a feasible method to calculate the ideal paraboloid [7]. An approximate ideal paraboloid is calculated in the xoz plane with the z-axis as the axis of symmetry, and then the paraboloid is rotated around the z-axis to obtain the ideal paraboloid. Let the vertex of the parabola be $A(0, -(R + \delta))$, where δ is the vertex displacement, and the equation of the parabola with A as the vertex and F as the focus is

$$z = -R - \delta + \frac{x^2}{4(F + \delta)}$$

Let any point on the parabola Let any point on the parabola (x_p, z_p) that corresponds along the radial direction of the reference circle to the point on the circle (x_b, z_b) on which satisfies the equation.

$$\begin{cases} \frac{z_p}{x_p} = \frac{z_b}{x_b} \\ x_b^2 + z_b^2 = R^2 \end{cases}$$

Vertex displacement δ is mainly caused by the deformation of the reflective surface, and the constraint is to minimize the radial displacement of the main cable node. Since the accuracy of solving this discrete problem depends on the step size, and reducing the step size significantly increases the computational complexity, the problem is equivalently transformed into a continuity problem by calculating the area between the paraboloid and the sphere, thus determining the relative position h of the vertex of the ideal paraboloid to the reference sphere.

$$\delta_{best} = arg_h min \int_{-r}^{r} |-x\sqrt{R^2 - x^2} - \frac{x^2}{4(F+h)} + R + h| dx$$

Then the equation of the ideal paraboloid can be found based on the knowledge of analytic geometry as $x^2 + y^2 = 2p(z+f)$, where $p = 2(F+\delta), f = R+\delta$.

3.2 Coordinate System Conversion

When the position of the celestial body changes, it is necessary to make the rotating paraboloid move on the sphere in real time in order to ensure that the approximate axis of rotation of the rotating paraboloid always passes through the celestial body and the center of the sphere. In this case, we need to define the rotation matrix, let the rotation angle of the coordinate system around the x, y, z coordinate axes are α, β, γ and the rotation matrices are as follows

$$R_z(\theta_1) = \begin{bmatrix} \cos\theta_1 & -\sin\theta_1 & 0 \\ \sin\theta_1 & \cos\theta_1 & 0 \\ 0 & 0 & 1 \end{bmatrix} R_y(\theta_2) = \begin{bmatrix} \cos\theta_2 & 0 & \sin\theta_2 \\ 0 & 1 & 0 \\ -\sin\theta_2 & 0 & \cos\theta_2 \end{bmatrix} R_x(\theta_3) = \begin{bmatrix} 1 & 0 & 0 \\ 0 & \cos\theta_3 & -\sin\theta_3 \\ 0 & \sin\theta_3 & \cos\theta_3 \end{bmatrix}$$

So the new coordinates obtained after rotation are

$$\begin{bmatrix} x_n \\ y_n \\ z_n \end{bmatrix} = W \begin{bmatrix} x \\ y \\ z \end{bmatrix} among\ it\ W = R_X(\theta_3)R_Y(\theta_2)R_Z(\theta_1)$$

Take the object to be observed is located in $\alpha = 36.795°, \beta = 78.169°$. As an example, the coordinates of the ideal parabolic vertices after the transformation are obtained as shown in Table 1, and the comparison results before and after the transformation are shown in Fig. 3.

Table 1. Coordinates of ideal parabolic vertices

x coordinate	y coordinate	z coordinate
−49.259	−36.844	−293.66

3.3 Sole Network Node Regulation Strategy

3.3.1 Determination of the Objective Function

The optimization goal of the cable mesh nodes and active reflective panels is to adjust the reference spherical sheet to be as close as possible to the ideal paraboloid, "close"

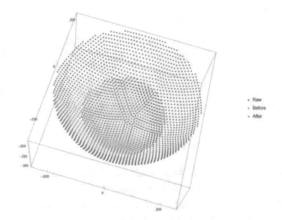

Fig. 3. Comparison of before and after coordinate rotation

can be understood as the distance between the main cable node (x', y', z') and the ideal paraboloid in the z-axis direction is as small as possible, i.e.

$$min \sum_{i=1}^{n} \left(\frac{1}{2p}(x_i'^2 + y_i'^2) - f - z_i' \right)$$

3.3.2 Establishment of Constraint Conditions

There are three main types of constraints in this paper: reflective panel edge length constraint, lower tie length constraint and actuator expansion constraint.

1) Reflective panel edge length constraint

According to related studies, the distance between neighboring nodes may change slightly after the adjustment of the main cable nodes by no more than 0.07%, i.e.

$$(x_i' - x_j')^2 + (y_i' - y_j')^2 + (z_i' - z_j')^2 = e_{ij}'^2 \text{ and } \left| \frac{e_{ij}' - e_{ij}}{e_{ij}} \right| \le 0.07\%$$

2) Lower cable length constraint and actuator expansion constraint

Let the main cable node at the base state (x_i, y_i, z_i), corresponding to the coordinates of the top of the actuator (x_i^c, y_i^c, z_i^c), then the length of the lower cable is

$l^i = \sqrt{(x_i - x_i^c)^2 + (y_i - y_i^c)^2 + (z_i - z_i^c)^2}.$

Let D_i denote the distance between the tip of the actuator and the center of the ball in the reference state, if the actuator i is δ_i then the coordinate constraint of the tip (actuator expansion constraint) is $(x_i^d, y_i^d, z_i^d) = \frac{D_i + \delta_i}{D_i}(x_i^c, y_i^c, z_i^c).$

The lower tie length constraint is $(x_{i'} - x_i^d)^2 + (y_{i'} - y_i^d)^2 + (z_{i'} - z_i^d)^2 = l_i^2, i = 1, 2, \dots, n.$

3.4 Solution of the Optimization Model

3.4.1 Traditional Artificial Fish Swarming Algorithm

The above model is a nonlinear programming model with quadratic objective function and constraints, which can be solved using the artificial fish swarm algorithm [4], which is a kind of only optimization algorithm that simulates the behavior of fish foraging, swarming and tail-chasing, and has the better ability to overcome local extremes and obtain global extremes. The process of fish swarming algorithm is as follows [9].

Step1: Set the individual size of the artificial fish, the maximum step length of the artificial fish movement, the perceived distance, the maximum number of attempts of the fish foraging behavior, the crowding degree of the fish, the maximum number of iterations, and perform the initialization of the fish according to the constraints.

Step2: Determine whether the state position of the artificial fish school is clustering behavior or tail-chasing behavior, and get the food concentration of different artificial fish locations.

Step3: Determine the size of food concentration at the location of the artificial fish of the two behaviors, and get the location of the next artificial fish in different states.

Step4: Repeat Step2 and Step3 until the number of cycles is greater than the individual size of the artificial fish population.

Step5: Repeat the iterations until the number of iterations is greater than the predefined maximum number of iterations, and finally get the optimal solution.

3.4.2 Improved Artificial Fish Swarming Algorithm

In the late iteration of the algorithm, the traditional artificial fish swarming algorithm has a great possibility of distribution around the local extremes, in order to be able to avoid falling into the local extremes, this paper introduces a double Gaussian function [5] foraging update behavior [8].

Assume that the location of the undesirable artificial fish is X_i, the newly generated artificial with the position of X_i^*, then the position of the new artificial fish in the kth dimension X_i^{*k} is $X_i^{*k} = \delta_i^k \times N_1 + |1 - \delta_i^k| \times X_i^k$ where δ_i^k is a random coefficient of 0 or 1. From the central limit theorem, it is known that after several iterative operations, the distance difference of the artificial fish obeys a normal distribution, and the double Gaussian function can be invoked, and the mathematical expression is

$$N_i = (N_{max}^k + \sigma_1 \times e) \times h(r - r_1) + (X_{min}^k + \sigma_2 \times e) \times h(r_1 - r)$$

where $r = \frac{X_i^k - X_{min}^k}{\|X_{max}^k - X_{min}^k\|}$, the X_{max}^k and X_{min}^k represent the upper and lower bounds of the artificial fish population searching in the kth dimensional space, respectively σ_1 and σ_2 represent the mean squared deviation of the double Gaussian function, respectively, r represents the chosen random number with value 0 or 1, e is the random number (conforming to the normal distribution), and h is the step function.

Taking the object to be observed located at $\alpha = 36.795°$, $\beta = 78.169°$ as an example, the transformed principal cable number, coordinates and stretching amount are obtained as shown in Table 2.

Table 2. Number and coordinates of the main cable nodes after adjustment

Node Number	X coordinate (m)	Y coordinate (m)	Z coordinate (m)	Stretching volume
A0	−0.0331	−0.0248	−300.2022	0.1986
B1	6.0914	8.3948	−300.1225	0.9075
......				
A105	18.2436	109.8212	−278.5987	0.4109
B92	30.0327	109.4001	−277.7811	0.3688
......				
B286	−106.9714	96.4505	−264.0001	−0.6000
B287	−102.3404	107.0478	−261.7408	−0.6000
B288	−97.5517	117.4811	−259.0780	−0.6000

The evolutionary curve of the average solution of the objective function of this improved artificial fish swarm algorithm is shown in Fig. 4.

It can be seen that the improved artificial fish swarm algorithm can approach the ideal value faster in the early iterations, but the speed of approach is not obvious in the later iterations.

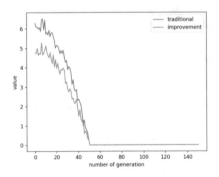

Fig. 4. Comparison of the average iteration curves of the two algorithms

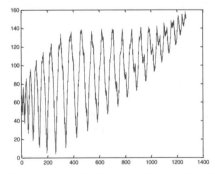

Fig. 5. Schematic diagram of the number of signals received by the feeder compartment

4 Calculation of Feeder Compartment Reception Ratio

4.1 Discrete Spherical Reflection Modeling

The signal from the celestial body to the ground needs to be reinforced by converging the signal to the feeder compartment through the approximate paraboloid. In this section, a solution method for each discrete spherical reflection model in the approximate paraboloid is proposed.

Firstly, considering only the xoy plane, differentiating each reflective panel, approximating each small reflective panel by the center of gravity of each small reflective panel, combining the established coordinate system as well as the angular relationship we can obtain the relationship of all reflected signals after reflection by each center of gravity. $y - y_i = k_i(x - x_i)$, and $\theta_i = \arctan k_i$, where k_i denotes the slope of the reflected signal equation, and θ_i denotes the angle between the reflected panel and the y-axis.

The reception ratio is then calculated based on the tilt angle of each reflective panel and its area. Where, the area of each reflective panel A_i is $A_i = \frac{1}{2} \cdot L_i^2 \cdot \sin\theta_i$, and L_i denotes the side length of each reflective panel, and the area of each small reflective panel is differentiated for each reflective panel $A_i' = \frac{A_i}{n}$, and n denotes the number of differentiated blocks. The number of signals received by each small reflector panel e_i with its area A_i and the tilt angle θ_i. The relationship between $e_i = A_i \cdot \rho \cdot \cos\theta_i$, and ρ indicates the signal strength received per unit area. The total number of signals E is $= \pi \cdot r^2 \cdot \rho$, the number of reflected signal equations that can pass through the circular equation of the feeder compartment is compared to the number of all signal equations as $\alpha\prime = \frac{\sum_{i=1}^{m} e\prime_i}{E}$, and $e\prime_i$ denotes the number of reflected signals in each reflective panel that can pass through the receiving circular surface of the feed module [10].

4.2 Calculation and Comparison of Reception Ratio

Considering the effect of the tilt angle of the reflective panel on the results, the reception ratio of the feeder compartment after the adjustment of the reflective panel is 0.6762 and the reception ratio of the reference sphere is 0.6673. It can be seen that the reception ratio of the adjusted feeder compartment is larger than that of the reference sphere.

In addition, when the number of reflective panels grows spirally upward from the lowest point of the illuminated area, the vertical distance of each reflective panel from the x-axis is shown in Fig. 5 shown, where the red line indicates the ideal parabolic surface and the blue line indicates the actual reflective surface. It can be seen that the receiving ratio obtained by the strategy used in this paper is very close to the ideal value.

5 Concluding Remarks

In this paper, we study the optimization problem of active reflecting surface adjustment for radio telescopes, obtain the calculation method of the ideal paraboloid by using the coordinate system transformation, solve the optimization model of the lower lasso actuator based on the improved artificial fish swarm intelligent algorithm, and give a method to calculate the signal reception ratio. In practical applications, there are few cases where intelligent algorithms are used to calculate the active reflecting surface of radio telescopes. The optimization model proposed in this paper can be extended to the optimization of active reflecting surface adjustment of the rest of radio telescopes, providing a better optimization idea.

Acknowledgements. This work is supported by 2022 Provincial Student Innovation Training Project of Department of Education of Guangdong Province (s202211078195).

References

1. Shen S.: Surface optimization of radio telescope using MVFSA algorithm. J. Chongqing Univ. Technol. (2022)
2. Zhu, L.: The 500-meter spherical radio telescope (FAST) active reflecting surface deformation control. Res. Inf. Technol. Appl. **3** (2012)
3. Xue, J.X., Wang, Q.M., Gu, X.D.: Prediction and improvement of transient parabolic fitting accuracy of 500 m aperture spherical radio telescope. Optic. Precis. Eng. **23**(7), 2051–2059 (2015)
4. Wang, X.H., Zheng, X.M., Xiao, J.M.: An artificial fish swarm algorithm for solving constrained optimization problems. Comput. Eng. Appl. (03), 40–42+63 (2007)
5. Li, J., Liang, X.: Optimization simulation for convergence speed improvement of artificial fish swarm algorithm. Comput. Simulat. (2018) 01-0232-07. 1006–9348
6. Zhang, S., Peng, Z.: Development and application of large devices for Chinese astronomical science (III) -- 500 meter aperture spherical radio telescope (FAST). J. Chin. Acad. Sci. (06), 670–673+688–689 (2009)
7. Zhu, L., Wu, X., Li, J., et al.: Euler rotation transformation and dynamic equation in Cartesian coordinate system. Marine Surv. Mapp. **30**(3), 20–22 (2010)
8. Huang, H., Wang, Q., Zhang, H., et al.: Development of FAST active reflective surface control system based on Ethernet. Comput. Eng. Appl. **42**(5), 97–100 (2006)
9. Tian, G., Hu, J., Pan, J.: Research on the testing method of long focal length and large off axis off axis parabolic reflectors. In: Summary of the 17th National Symposium on Optical Testing (2018)
10. Zhu, L.: Research on measurement and control of 500 m aperture spherical radio telescope (FAST). National Astronomical Observatory, Chinese Academy of Sciences (2006)

Study on Furnace Temperature Curve Based on Heat Conduction Model

Yaping Chai[1], Bingying Song[1], Congrun Zhang[1], and Haichang Luo[2(✉)]

[1] School of Mathematics and Information Science, Guangzhou University, Guangzhou, China
[2] School of Intelligent Manufacturing, Zhanjiang University of Science and Technology, Zhanjiang, China
luohaichang@zjkju.edu.cn

Abstract. Temperature control is very important to product quality in circuit board welding production of reflow furnace. In this paper, the mathematical model is used to analyze the temperature change in the center of the welding area of the circuit board in the reflow furnace, simulate different conditions, determine the optimal furnace passing speed of the conveyor belt, complete the temperature setting of each temperature zone, and gradually study the furnace temperature curve to solve the problem. This paper first discusses the structure and working principle of the reflow furnace. It is clear that the influencing factors of the furnace temperature curve are temperature zone temperature and furnace passing speed. The furnace temperature curve is drawn using the attachment data as the reference standard. The heat conduction model was established. The finite difference method was used to simplify the solution. On this basis, this paper changes the temperature value set in each small temperature zone, applies the heat conduction difference equation, takes the process boundary as the constraint condition, and takes the conveyor belt passing speed as the optimization goal to establish a single objective optimization model. The improved dichotomy search algorithm is used to quickly and accurately optimize the maximum conveyor belt passing speed.

Keywords: heat conduction model · improved difference method · binary search · target optimization model

1 Problem Restatement

1.1 Problem Background

In the production of electronic products such as integrated circuit boards, the printed circuit boards installed with various electronic components need to be placed in the backwelding furnace, and the electronic components are automatically welded to the circuit boards by heating. In this process, it is very important for product quality to keep all parts of the backwelding furnace at the temperature required by the process. At present, much of the work in this area is controlled and adjusted through experimental testing. The purpose of this problem is to analyze and study the mechanism model.

B.-Y. Cao et al. (Eds.): ICFIE 2022, LNDECT 207, pp. 73–95, 2024.
https://doi.org/10.1007/978-981-97-2891-6_6

Several small temperature zones are set inside the backwelding furnace, which can be divided into four large temperature zones functionally: preheating zone, constant temperature zone, reflux zone and cooling zone (as shown in Fig. 1). Both sides of the circuit board on the conveyor belt uniform speed into the furnace for heating and welding.

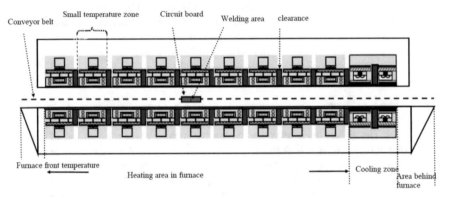

Fig. 1. Section diagram of backwelding furnace

In a backwelding furnace, there are 11 small temperature zones, pre-temperature zones and post-temperature zones (as shown in Fig. 1). The length of each small temperature zone is 30.5 cm, and there is a gap of 5 cm between adjacent small temperature zones. The length of both pre-temperature zone and post-temperature zone is 25 cm.

After starting the backwelding furnace, the air temperature in the furnace will reach stability in a short time, and then the backwelding furnace can be welded. There is no special temperature control for the gap between the front zone, the back zone and the small temperature zone. The temperature is related to the temperature of the adjacent temperature zone, and the temperature near the boundary of each temperature zone may also be affected by the temperature of the adjacent temperature zone. In addition, the temperature of the production workshop is maintained at 25 °C.

After setting the temperature of each temperature zone and the furnace speed of the conveyor belt, the temperature of the center of the welding zone in some positions can be tested by the temperature sensor, which is called the furnace temperature curve (that is, the temperature curve of the center of the welding zone). Attached is the data of furnace temperature curve in an experiment. The temperature set in each temperature zone is 175 °C (small temperature zone 1–5), 195 °C (small temperature zone 6), 235 °C (small temperature zone 7), 255 °C (small temperature zone 8–9) and 25 °C (small temperature zone 10–11). The passing speed of the conveyor belt is 70 cm/min. The thickness of the welding area is 0.15 mm. The temperature sensor starts to work when the temperature in the center of the welding area reaches 30 °C, and the circuit board enters the backwelding furnace to start timing.

The product quality can be controlled by adjusting the set temperature of each temperature zone and the passing speed of the conveyor belt in actual production. On the basis of the above experimental set temperature, the set temperature of each small temperature area can be adjusted within the range of °C. During the adjustment, the temperature in small temperature zones 1–5, 8–9, and 10–11 should be consistent. The furnace speed adjustment range of the conveyor belt is 65–100 cm/min.

In the welding production of circuit board of backwelding furnace, furnace temperature curve should meet certain requirements, which is called process limit (see Table 1).

Table 1. Process limit

Bound name	Minimum value	Maximum value	unit
Temperature rise slope	0	3	°C/s
Temperature drop slope	−3	0	°C/s
The temperature rises from 150 °C to 190 °C	60	120	s
The temperature is greater than 217 °C for the time	40	90	s
Peak temperature	240	250	°C

1.2 Problem Requirement

Question 1: Please establish a mathematical model of temperature variation in the welding area. Assume that the passing speed of the conveyor belt is 78 cm/min, and the temperature values in each temperature zone are 173 °C (small temperature zone 1–5), 198 °C (small temperature zone 6), 230 °C (small temperature zone 7) and 257 °C (small temperature zone 8–9) respectively. Please give the temperature change in the center of the welding zone. List the temperature at the middle point of low-temperature zone 3, 6 and 7 and the center of the welding zone at the end of low-temperature zone 8, draw the corresponding furnace temperature curve, and store the temperature at the center of the welding zone every 0.5 s in the result.csv provided.

Question 2: Assuming that the temperature values of each temperature zone are respectively 182 °C (small temperature zone 1–5), 203 °C (small temperature zone 6), 237 °C (small temperature zone 7) and 254 °C (small temperature zone 8–9). During the welding process, the temperature in the center of the welding area should not exceed 217 °C for too long, and the peak temperature should not be too high. The ideal furnace temperature curve should minimize the area covered by the peak temperature exceeding 217 °C (shaded area in Fig. 2). Please determine the maximum allowable conveyer-belt passing speed.

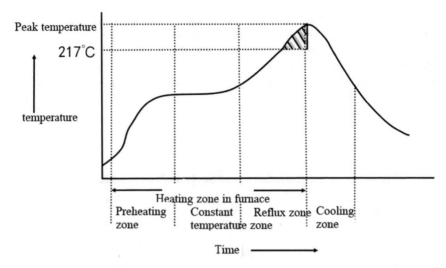

Fig. 2. Schematic diagram of furnace temperature curve

2   Problem Analysis

In the production of electronic products such as integrated circuit board, it is necessary to place the printed circuit board with various electronic components in the backwelding furnace, and weld the electronic components to the circuit board through heating. The temperature control of welding is very important for product quality. Due to the high cost of circuit board production, it is necessary to meet the process limits in the welding production process to ensure product quality. This paper will design the temperature of each temperature zone of the circuit board welding production of the welding furnace and the furnace speed of the conveyor belt.

3   Model Assumption

(1) The data provided in the attachment are true and reliable and can objectively reflect the facts.
(2) It is assumed that the printed circuit board is the same, and the heat conduction to the same layer is the same temperature.
(3) Assume that the heat conductivity is consistent at different temperatures, and the heat conductivity is independent of time, and ignore the influence of temperature and time on the heat conductivity.
(4) It is assumed that only the backwelding furnace provides heat to the welding area, and no external heat source has an effect on the change of heat and temperature.
(5) It is assumed that the heat flux and temperature of the upper and lower surfaces of the circuit board welding area are the same, and there is no internal heat loss.
(6) It is assumed that the influence of heat convection and heat radiation is negligible in this case, and only the effect of heat conduction is considered.

4   Symbol Specification

Symbol	Description
x	The thickness of the printed circuit board
a	Related heat conduction parameters of printed circuit board
v	Speed of conveyer belt passing through furnace
S	Covered area
h	Step size
dQ	Heat flowing out of the cross section
T_{max}	Peak temperature
$k_{(left,i)}$	Peak temperature temperature per dt section of the left-hand curve
$k_{(right,i)}$	Peak temperature temperature per dt section of the right-hand curve

Note: Other unnoted symbols will be specified in the article

5   Model Establishment and Solution

5.1   Question 1

In order to establish the relevant mathematical model, we should first understand the working process and structure area of the backwelding furnace, clear the influencing factors of the furnace temperature curve, draw the furnace temperature curve using the data under the given conditions of the annex, master the basic law of the furnace temperature curve. In order to solve the problem, the heat conduction diagram of backwelding furnace was drawn, and one-dimensional heat conduction equations were established based on thermodynamic knowledge. The difficulty of this problem is that the relevant parameters of heat conduction cannot be determined, so the finite difference method is used to simplify the solving process and reverse derivation to obtain the operational expression of the relevant parameters of heat conduction, and the operational expression is substituted back to the difference model for verification to determine whether its accuracy meets the requirements. The calculation formula is substituted into the finite difference method and the discrete point list is obtained according to the temperature setting of each temperature zone in question 1 and the setting of the furnace passing speed of the conveyor belt. It is the temperature change of the center of the welding zone. The temperature of the center of the welding zone at the middle point of the small temperature zone 3, 6 and 7 and the end of the small temperature zone 8 are listed. Store the temperature at the center of the welding area every 0.5 s in result.csv of the support material.

In view of problem 1, in order to establish the relevant mathematical model to solve the problem, we should first understand the working process and structure area of the backwelding furnace, identify the influencing factors of the furnace temperature curve, draw the furnace temperature curve with the data under the given conditions of the attachment, and master the basic law of the furnace temperature curve. When establishing the model, the heat conduction diagram of the backwelding furnace was drawn to assist understanding, and then one-dimensional heat conduction equations were established according to thermodynamic knowledge. In order to solve the difficulty that the relevant parameters of heat conduction could not be determined, the finite difference method could be used to simplify the solution process and reverse derivation to obtain the operational expression of the relevant parameters of heat conduction. Then the operational expression was substituted back to the difference model for verification to determine whether its accuracy met the requirements. The calculation formula is substituted into the finite difference method, and then the discrete point list is obtained according to the temperature setting of each temperature zone in question 1 and the setting of the furnace passing speed of the conveyor belt. The temperature change of the center of the welding area is given, and the temperature of the center of the welding area at the middle point of small temperature zone 3, 6, 7 and the end of small temperature zone 8 is listed. Then the furnace temperature curve is drawn according to the discrete point list. Finally, the temperature of the center of the welding area every 0.5 s is stored in result.csv of the supporting material.

5.1.1 The Basic Introduction of Back Welding Furnace

5.1.1.1 Work Flow and Structure Area of Back Welding Furnace Are Introduced

In the production of electronic products such as integrated circuit boards, it is necessary to place the printed circuit boards installed with various electronic components in the backwelding furnace. Through heating, the electronic components are automatically welded to the circuit boards. In this process, the temperature control of the backwelding furnace is very important for the product quality.

The welding process in the backwelding furnace refers to the process in which the circuit board is transported at a uniform speed on the conveyor belt of the backwelding furnace and forms a complete welding point after being welded successively through the four large temperature zones of the preheating zone, constant temperature zone, reflux zone and cooling zone of the backwelding furnace.

The schematic diagram of the welding working area of the backwelding furnace is shown in Fig. 3 below. Among the four large temperature zones, the preheating zone, constant temperature zone and reflux zone belong to the heating zone in the furnace.

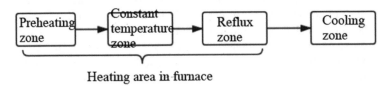

Fig. 3. Schematic diagram of the welding working area of the backwelding furnace

Among them, the working principle of the four major temperature zones of the backwelding furnace [1] is as follows:

1) preheating zone

The purpose of preheating is to reactivate the solder paste and avoid the heating behavior caused by rapid high temperature heating during tin immersion. The main purpose of this area is to heat the PCB components at room temperature as soon as possible, but the heating rate should be controlled within an appropriate range. If it is too fast, it will cause thermal shock, resulting in damage to the circuit board and components; if it is too slow, it will cause insufficient solvent volatilization, affecting the welding quality. Due to the faster heating speed in this area, the temperature difference in the backwelding furnace at the back of the preheating area is large. In order to prevent damage to components by thermal shock, the temperature rise slope is usually set at 1–3 °C/s.

2) constant temperature zone

The main purpose of this area is to make the temperature of the circuit board components in the backwelding furnace tend to be uniform, and to minimize the temperature difference by maintaining a relatively stable temperature so that the flux in the solder paste can play a role and properly distribute. Sufficient time should be given in this area for the temperature of the larger element to catch up with that of the smaller element and for the flux in the solder paste to be fully melted. Until the end of the constant temperature zone, the oxides on the pad, solder ball and component pins are removed under the action of flux, and the temperature of the entire circuit board is balanced. The temperature in this area is generally 140–160 °C, requiring 80–150 s.

3) reflux zone

The temperature in this area is the highest, so that the temperature of the components rises to the peak temperature. The main purpose is to melt the solder paste quickly, and alloy is formed between the pad of the circuit board and the electrode of the component, so that the components are welded to the circuit board. At this stage the reflux time should not be too long, so as not to cause adverse effects on the circuit board. The peak temperature in this section is generally 240–250 °C.

4) cooling zone

The lead tin powder in the solder paste in this area has melted and fully moistened the surface to be connected, and the temperature has cooled to below the solid phase temperature, so that the solder joint solidified. It should be cooled as quickly as possible, which will help to get bright solder joints and good shape. The cooling rate is generally 2–3 °C/s, and the cooling is generally required to be below 100 °C.

As can be seen from the question, after starting the backwelding furnace, the temperature in the furnace will reach stability in a short time, and the temperature in the production workshop will remain at 25 °C. Moreover, in solving the problem below, the permissible setting intervals of temperature in each temperature zone are [165,185] °C (small temperature zone 1–5), [185,205] °C (small temperature zone 6), [225,245] °C (small temperature zone 7), [245,265] °C (small temperature zone 8–9) and 25 °C (small temperature zone 10–11) respectively. The furnace speed adjustment range of the conveyor belt is 65–100 cm/min.

5.1.1.2 Furnace Temperature Curve

Furnace temperature curve (welding area center temperature curve): it is used to record the temperature change of printed circuit board components during the welding process, and is convenient to monitor the temperature of the welding furnace. The temperature condition in the backwelding furnace cannot be seen and perceived, so the furnace temperature curve can be used to measure the actual temperature of the temperature sensor to facilitate the understanding of its condition [2].

The furnace temperature curve is the result of the joint action of several parameters of the backwelding furnace, among which the two parameters that play a decisive role are the speed of the conveyor belt and the temperature setting of the warm zone. The conveyor speed determines the duration of exposure of the circuit board to each temperature zone. Increasing the duration makes the temperature of the components on the circuit board closer to the set temperature of the temperature zone. The sum of the duration of each temperature zone determines the total welding processing time. The temperature setting of each temperature zone affects the temperature of the circuit board when it passes through the temperature zone. The temperature rise slope of the circuit board in the whole welding process is the result of the joint action of two parameters: the passing speed of the conveyor belt and the temperature setting in each temperature zone.

In order to more intuitively understand the temperature change process of the topic, first of all, we use the data of temperature change over time in the center of the welding area in the attachment to draw the furnace temperature curve with MATLAB, and through the calculation of the temperature change slope of the furnace temperature curve and the comparison of the characteristics of each major temperature zone, the furnace temperature curve is divided into small temperature zones, as shown in the Fig. 4 below.

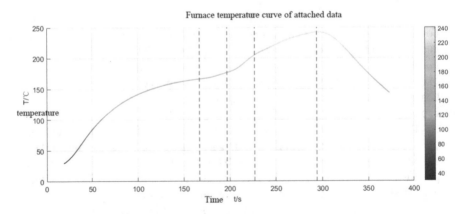

Fig. 4. Attached data furnace temperature curve

Because the furnace temperature graph is drawn based on experimental data, it can provide reference basis for solving the simulated furnace temperature graph.

5.1.2 Establishment of Heat Conduction Model

This paper aims to analyze and study the problem through the mechanism model. Under the given conditions of the topic, according to the interaction mechanism of the main factors, the equilibrium relationship between temperatures is described mathematically. The model built in this question is the mechanism model built by the heat conduction balance, namely, the heat conduction model.

Because the temperature in the furnace has not reached the equilibrium state, the heat transfer from the high temperature to the low temperature place, called heat conduction, which is one of the three heat transfer modes (heat conduction, convection, radiation). Thermal convection mainly transfers heat to gas, but the gas in the backwelding furnace has little influence on the heat transfer of circuit board, and the efficiency of thermal radiation (non-contact radiation) is far lower than that of thermal convection. Therefore, the influence of thermal convection and radiation on temperature changes can be ignored in this paper. In the following, only heat conduction is considered for the heat transfer mode in the welding area of backwelding furnace.

In the backwelding furnace, the upper and lower parts of the welding area conduct heat conduction at the same time. The heat conduction diagram of the backwelding furnace during welding is drawn as Fig. 5 below.

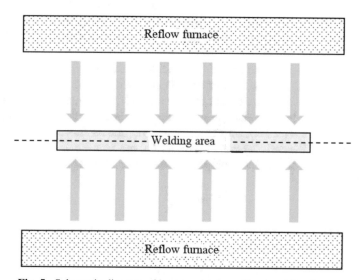

Fig. 5. Schematic diagram of heat conduction of backwelding furnace

Based on the law of conservation of heat and Fourier's law, this paper constructs a heat conduction model [3] of temperature change in the welding area:

$$\begin{cases} \frac{\partial T}{\partial t} = a \cdot \frac{\partial^2 T}{\partial^2 x} \\ a = \frac{k}{c\rho\lambda^2} + w \end{cases} \quad (1)$$

where, x is the thickness of the welding area, k is the heat conduction coefficient, the mass density of the circuit board is ρ, the specific heat is c, λ is the 0.1×10^{-3} mm step length, and a is the related heat conduction parameter of the circuit board. The calculation formula of the heat conduction coefficient and related parameters can be calculated by processing the attached data with the difference equation below. w is the influence factor, which refers to the influence of temperature change on the related parameters of heat conduction. The temperature is higher and the time is longer in the reflux zone after backwelding furnace, and the influence factor is larger here. The heat transfer strength is expressed by the heat passing through the unit cross-sectional area in unit time, which is called heat flux q. The origin of heat conduction is the uneven temperature T, the degree of which is expressed by the temperature gradient ∇T. According to Fourier's Law [4]:

$$q = -k\nabla T = -k\frac{\partial T}{\partial n} \tag{2}$$

In the three-dimensional space, consider the problem of heat conduction in the welding area. The temperature of point (x, y, z) in the welding area at time t is $T(x, y, z)$, assuming that its temperature on the same section is the same, heat transfer occurs on the surface and the surrounding medium, and follows the law:

$$dQ = -k(x, y, z)\frac{\partial T}{\partial n} \tag{3}$$

Among them, the heat conduction coefficient of the welding area is $k(x, y, z)$ in the three-dimensional space. When the heat enters the backwelding furnace, the temperature of the welding area changes.

Since it is assumed that the heat of the welding area is mainly related to the heating area of the backwelding furnace, the external heat source has little influence on it, so the change of the influence of the external heat source on its temperature is not considered. Moreover, since the heat conduction of temperature is carried out in the upper and lower directions of the welding zone at the same time, and its direction is perpendicular to the direction of the welding zone transmission, the model can be simplified from three-dimensional to one-dimensional model. Therefore, it can be simplified as follows:

$$dQ = -k\frac{\partial T}{\partial t}dSdt \tag{4}$$

Take a closed smooth section A in the welding area, and the section area is S, the area enclosed by it is O, and the heat increment from time t_1 to time t_2 is:

$$Q = \int_{t_1}^{t_2}\iint_O k\frac{\partial T}{\partial n}dSdt \tag{5}$$

For the area element dS of each point in the welding area, the heat required from $T(x, t_1)$ to $T(x, t_2)$ is:

$$dQ = c\rho[T(x, t_2) - T(x, t_1)]dS \tag{6}$$

And in the whole region O, with the change of temperature, the energy required is:

$$Q = \iiint_O c\rho[T(x, t_2) - T(x, t_1)]dS \tag{7}$$

From the previous formula (6) and (7), the relation can be derived as follows:

$$\int_{t_1}^{t_2} \iint_A k\frac{\partial T}{\partial n}dSdt = \iiint_O c\rho[T(x, t_2) - T(x, t_1)]dSdt \tag{8}$$

Through Gauss formula [5], it can be concluded that:

$$\int_{t_1}^{t_2} \iiint_O [\frac{\partial}{\partial x}(k\frac{\partial T}{\partial x})]dSdt = \iiint_O c\rho(\int_{t_1}^{t_2}\frac{\partial T}{\partial t}dt)dS \tag{9}$$

From the above formula and the arbitrariness of t_2, t_1, we can know:

$$c\rho\frac{\partial T}{\partial t} = \frac{\partial}{\partial x}(k\frac{\partial T}{\partial x}) \tag{10}$$

When the welding zone is uniform and isotropic, then:

$$\frac{\partial T}{\partial t} = \frac{k}{c\rho}(\frac{\partial^2 T}{\partial^2 x}) \tag{11}$$

According to the law of conservation of heat:

$$\int_{t_1}^{t_2} \iiint_O [\frac{\partial}{\partial x}(k\frac{\partial T}{\partial x})]dSdt = \iiint_O c\rho(\int_{t_1}^{t_2}\frac{\partial T}{\partial t}dt)dS \tag{12}$$

From the above formula and the arbitrariness of t_2, t_1, we can know:

$$c\rho\frac{\partial T}{\partial t} = \frac{\partial}{\partial x}(k\frac{\partial T}{\partial x}) \tag{13}$$

One-dimensional heat conduction equation can be obtained as follows:

$$\frac{\partial T}{\partial t} = a\frac{\partial^2 T}{\partial^2 x} \tag{14}$$

5.1.3 The Solution of Heat Conduction Model

Since the problem requires solving the central temperature of the welding area, and it is difficult to solve the unsteady heat conduction equation to find its analytical solution, we decide to solve the problem by solving the numerical solution of the heat conduction equation [6]. Therefore, the implicit difference method is adopted to solve the model, and then:

It can be seen from the question that the temperature of the production workshop has been maintained at 25 °C, then the initial temperature of the upper and lower outer surfaces of the backwelding furnace is 25 °C, and there are initial conditions:

$$T(x, 0) = 25\,°C \tag{15}$$

When the circuit board enters the backwelding furnace, the timing begins, and the temperature sensor starts to work when the temperature in the center of the welding area reaches. Since the temperature in the production workshop is kept at 25 °C, the air temperature in the backwelding furnace will reach stability in a short time after starting. The time difference is very small, which has little influence on the problem solving and can be ignored.

The temperature set by the first large temperature zone (i.e., small temperature zone 1–5) is 173 °C.

When the circuit board begins to enter the backwelding furnace for welding, the temperature boundary condition of the welding area is:

$$T(x, 0) = 173\,°C \tag{16}$$

The circuit board begins to enter the small temperature zone 6, then the temperature boundary condition of the welding zone is:

$$T(0, t_4) = 198\,°C \tag{17}$$

The circuit board begins to enter the small temperature zone 7, then the temperature boundary condition of the welding zone is:

$$T(0, t_5) = 230\,°C \tag{18}$$

The circuit board begins to enter the small temperature zone 8–9, then the temperature boundary condition of the welding zone is:

$$T(0, t_6) = 257\,°C \tag{19}$$

where, t_3, t_4, t_5, t_6 refers to the time when the circuit board starts to enter the backwelding furnace, small temperature zone 6, small temperature zone 7 and small temperature zone 8–9 respectively.

When the conveyor belt starts to move, the temperatures set in the four large temperature zones in the backwelding furnace are all different. Through multiple simulation of the data, it can be seen that when the circuit board passes through each large temperature zone, the best choice of temperature in the gap between the large temperature zones is:

$$T_{i,j+1} = \frac{T_i + T_{i+1}}{2} \tag{20}$$

$T_{i,i+1}$ refers to the temperature of the gap between i and i+1.

In order to solve the heat conduction equation, the time and space grid of $T(x, t)$ is first defined, and the x coordinate is divided into L_1 and t coordinate into L_2. Let i represent the horizontal axis of position x and j represent the vertical axis of time t, each grid point on the grid corresponds to a temperature value, and use the central difference approximation to replace the partial differentiation of space, namely:

$$\frac{\partial^2 T}{\partial^2 x} = \frac{T_{i-1,j} - 2T_{i,j} + T_{i+1,j}}{\Delta x^2} \tag{21}$$

Substitute forward difference approximation for partial differentiation with respect to time, i.e.:

$$\frac{\partial T}{\partial t} = \frac{T_{i,j+1} - T_{i,j}}{\Delta t} \tag{22}$$

Substituting the first two equations into the heat conduction equation, we get:

$$\frac{T_{i,j+1} - T_{i,j}}{\Delta t} = a^2 \frac{T_{i-1,j} - 2T_{i,j} + T_{i+1,j}}{\Delta x^2} \tag{23}$$

The solution is:

$$T_{i+1,j} = s(T_{i-1,j} + T_{i+1,j}) + (1 - 2s)T_{i,j} \tag{24}$$

Among them:

$$s = \frac{\Delta ta}{\Delta x^2} \tag{25}$$

The diagram [7] of the heat conduction equation is shown as Fig. 6 below:

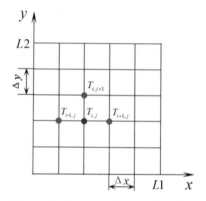

Fig. 6. Schematic diagram of heat conduction equation

By using the reverse difference method of the above formula and MATLAB, we obtained the relevant heat conduction parameters of the circuit board as follows:

$$a = 4.0796 \tag{26}$$

If the j coordinate is known to be different from the temperature value of each lattice point, then the temperature values of $i = 1$ and $i = L_1$ can be known from the boundary conditions, then the temperature value of each lattice point can be calculated. According to the heat conduction equation solved above, starting from the initial condition $j = 1$, the temperature value of each grid point can be calculated step by step using MATLAB.

In order to verify the accuracy of the relevant heat conduction parameter a, the above calculated values of the relevant heat conduction parameter a are substituted into the model 1 heat conduction model, and the simulated furnace temperature curve in the preheating zone is obtained according to the relevant heat conduction parameter a. The preheating section curve in the furnace temperature curve of the attached data in Fig. 6 above is intercepted and compared with it, and the following Fig. 7 is drawn:

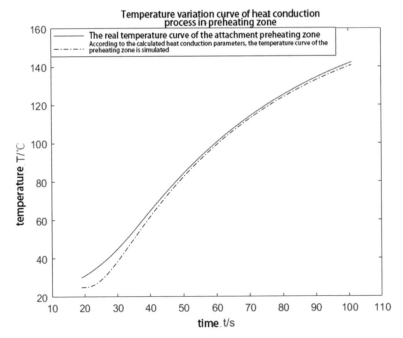

Fig. 7. Variation curve of furnace temperature during heat conduction in preheating zone

As can be seen from the figure above, under the conditions of the conveyor belt passing velocity and temperature zone temperature given in the attached experiment, the real furnace temperature curve of the attached preheating zone is very close to the simulated furnace temperature curve. Therefore, it can be considered that the heat conduction coefficient $a = 4.0796$ calculated by the difference equation above is highly accurate and can be used to solve the difference equation below.

Using the established model heat conduction model, under the assumption that the furnace passing speed of the conveyor belt is 78 cm/min, and the temperature setting values of each temperature zone are 173 °C (small temperature zone 1–5), 198 °C (small temperature zone 6), 230 °C (small temperature zone 7), 257 °C (small temperature zone 8–9) and 25 °C (small temperature zone 10–11) respectively, The model is solved by MATLAB and the corresponding furnace temperature data of the whole process from the beginning of the backwelding furnace to the end of the welding is obtained. The time when the backwelding furnace started heating the circuit board was 19.2308 s. With 0.5 s as the step length and 19 s as the starting point of the circuit board entering the backwelding furnace, the middle point of small temperature zone 3, 6 and 7, the end of small temperature zone 8 and the completion point of the circuit board welding were selected as examples and listed in Table 2 below to show the temperature changes in the center of the welding zone. The temperature of the center of the welding area every 0.5 s is stored in result.csv of the supporting material.

Table 2. Corresponding time point and temperature table of welding position points required by Question 1

Position point	Point in time(s)	Center temperature of welding area (°C)
Starting point	19.0	25.0000
Midpoint of low temperature zone 3	85.5	138.7900
Midpoint of low temperature zone 6	167.5	168.9504
Midpoint of low temperature zone 7	194.0	183.2077
End of low temperature zone 8	233.5	221.1683
Completion point	310.0	140.3751

Calculate the temperature data obtained in the center of the welding area and compare it with the process limit in Table 1 given in the title. The following Table 3 is listed:

Table 3. Process boundary comparison table of solution results of problem 1

Process boundary name	Solution value (maximum value)	Minimum value	Maximum value	unit	Whether the boundary requirements are met
Temperature rise slope	2.8442	0	3	°C/s	yes
Temperature drop slope	−2.7255	−3	0	°C/s	yes
The temperature rises from 150 °C to 190 °C	97.5	60	120	S	yes
The temperature is greater than 217 °C for the time	47	40	90	S	yes
Peak temperature	241.2098	240	250	°C	yes

It can be seen from the above table that all the data are in line with the process limits, so it can be considered that the results obtained by this question meet the requirements of the topic. According to the solution results, the furnace temperature curve of the whole process of the backwelding furnace welding circuit board was drawn, and the range of temperature zones was divided according to the temperature change slope of the furnace temperature curve and the characteristics of each major temperature zone, as shown in Fig. 8 below:

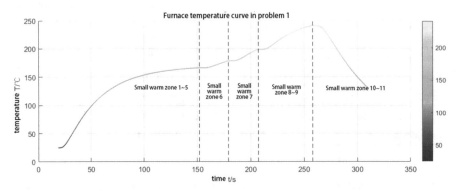

Fig. 8. Furnace temperature curve of the whole welding process of the return welding furnace

Observe the figure above, the color bar on the right side of the figure represents the temperature change of the furnace temperature curve, which is convenient to see the change of the furnace temperature curve in different small temperature areas more directly. The temperature change in the center of the welding zone is analyzed: the temperature in the small temperature zone 1–5 increases with time, which is considered as the preheating zone. The temperature of small temperature zone 6 changes more evenly with time and maintains in a relatively stable state, so this stage can be considered as a constant temperature zone. The temperature of small temperature zone 7 and 8–9 increases with the increase of time, and reaches the peak state of furnace temperature curve at the end of small temperature zone 8–9, so it can be considered that this stage is the reflux zone; The temperature of small temperature zone 10–11 drops rapidly with the increase of time, so it can be considered as a cooling zone. Through the observation and analysis of the furnace temperature curve, we can master the temperature change of circuit board welding process, which is convenient for us to monitor the temperature of the backwelding furnace, and control the completion of the quality of circuit board welding.

To sum up, the temperature in the center of the mid-point welding area of small temperature zone 3 is 138.7900 °C, that of small temperature zone 6 is 168.9504 °C, that of small temperature zone 7 is 183.2077 °C and that of the welding area at the end of small temperature zone 8 is 221.1683 °C. The corresponding furnace temperature curve is shown in the Fig. 8 above.

5.2 Question 2

Obviously, the difference model based on heat conduction in Problem 1 can be applied to problem 2. When solving this question, the setting conditions of the model in Problem 1 are changed, and a single objective optimization model is established. Combined with process limits and speed constraints, the improved binary search algorithm is used to quickly and accurately optimize the maximum passing speed of the conveyor belt, and the maximum allowable passing speed of the conveyor belt is determined.

In connection with the above, it is obvious that the difference model based on heat conduction in Problem 1 can be applied to problem 2. In solving this question, it is only necessary to change the setting conditions of the model in Problem 1 and establish a single objective optimization model. Combined with the process boundary and speed constraint conditions, the improved binary search algorithm is used to quickly and accurately optimize the maximum furnace velocity of conveyor belt. The maximum allowable conveyer belt passing speed can be determined.

5.2.1 Establishment of Single Objective Optimization Model

From the production point of view, the faster the speed of the backwelding furnace is adjusted, the more the number of circuit boards passed in the backwelding furnace per unit time, the higher the production efficiency. But considering that for the circuit board, the speed is too fast or too slow will lead to the circuit board in the furnace experience time is too long or too short, resulting in flux volatilization and solder joint tin-eating changes, and when the temperature rise rate exceeds the circuit board allowed will also cause a certain degree of damage to the circuit board. Therefore, in the actual welding production of circuit board of backwelding furnace, the furnace temperature curve should meet the requirements of certain process limits. Under the premise of meeting the requirements, the maximum passing speed of conveyor belt is solved. In this case, a single objective optimization model is established with the furnace speed as the optimization objective and the process limit as the constraint condition.

Objective function:

$$\max v = \frac{l}{v} \tag{27}$$

Constraints:

$$s.t. \begin{cases} 0 \le k_1 \le 3 \\ -3 \le k_2 \le 0 \\ 60 \le t_7 \le 120 \\ 40 \le t_8 \le 90 \\ 240 \le T_{max} \le 250 \end{cases} \tag{28}$$

where, v is the passing speed; l is the length of the conveyor belt; k_1 is the slope of temperature rise; k_2 is the slope of temperature decline; t_7 is the time between 150 °C–190 °C during the temperature rise; in problem 2, it refers to the time between 1–5 in the low temperature zone, when the temperature t_8 is greater than t_8. T_{max} is the peak temperature.

5.2.2 Solving the Single Objective Optimization Model

In this case, considering the actual production situation, it is necessary to make the quantity of production, the better the benefit, but the speed can not exceed the limit increase, so as to cause damage to the parts. In the case of meeting the process limit, the optimized binary search algorithm is used to solve the maximum conveyor belt passing speed under the required conditions.

Dichotomy [8] is a convenient algorithm for searching optimization problems, which is easy to execute and can be used to search in half to find the optimal solution under the specified precision. Dichotomy search is suitable for quickly finding the exact solution with high precision when the data is large. In this case, it is necessary to search for the optimal furnace speed of 65–100 cm/min in the case of 0.1 precision, so the improved dichotomy method is used to solve the problem.

In solving the single objective optimization problem, the improved binary algorithm [9] is adopted. Based on the binary algorithm, the piecewise function positive and negative transformation process is carried out on the solved one-dimensional heat conduction equation, simplifying the data results and making it easy to use the binary algorithm to solve the results with higher accuracy.

The solution flow chart of dichotomy is shown as in Fig. 9:

MATLAB was used to solve model 2, and the optimal result was that the maximum furnace passing speed of the conveyor belt was 84.8 cm/min when the temperature setting values in each temperature zone were respectively 182 °C (small temperature zone 1–5), 203 °C (small temperature zone 6), 237 °C (small temperature zone 7) and 254 °C (small temperature zone 8–9). The furnace temperature curve under the condition of the maximum passing speed of the conveyor belt in question 2 is drawn by using the obtained data of changes in the center temperature of the welding area along with the welding working time, as shown in Fig. 10 below:

The furnace temperature curve data are statistically and processed, and the solution results of this question are compared with the process limit of Table 1. The list in Table 4 is as follows:

After comparison, it can be seen that under this condition, the solution results meet all constraints of process limits.

In summary, after solving and verifying the established model II, it is shown that under the conditions of 182 °C (small temperature zone 1–5), 203 °C (small temperature zone 6), 237 °C (small temperature zone 7) and 254 °C (small temperature zone 8–9), and under the condition of meeting the process boundary, The maximum allowable conveyer passing speed is 84.8 cm/min.

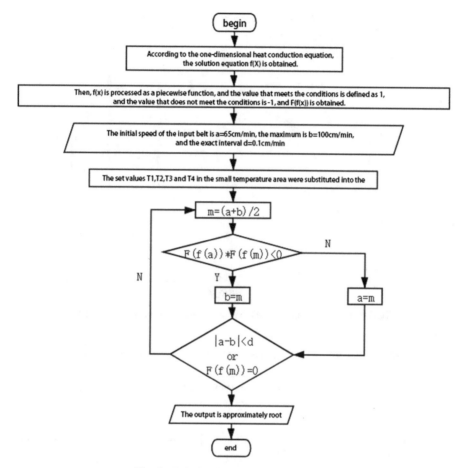

Fig. 9. Solution flow chart of dichotomy

6   Model Test and Error Analysis

6.1   Sensitivity Analysis

In problem 2, the differential model of heat conduction takes many factors into consideration. In order to study whether this model can predict the data sensitively, this paper selects the conveyer belt passing velocity which has a great influence on the differential model of heat conduction for sensitivity analysis, and draws the sensitivity analysis curve diagram for analysis combined with the data.

The furnace passing speed of the conveyor belt was modified as 75 cm/min, 85 cm/min, 70 cm/min and 80 cm/min. The corresponding discrete temperature values were solved according to model I and the sensitivity analysis curve was drawn, as shown in the Fig. 11 below:

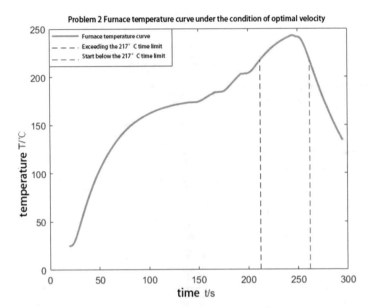

Fig. 10. Furnace temperature curve under optimal transmission speed

Table 4. Process boundary comparison table of solution results of problem 2

Process boundary name	Solution value (maximum value)	Minimum value	Maximum value	unit	Whether the boundary requirements are met
Temperature rise slope	2.8957	0	3	°C/s	yes
Temperature drop slope	-2.7839	-3	0	°C/s	yes
The temperature rises from 150 °C to 190 °C	92	60	120	s	yes
The temperature is greater than 217 °C for the time	49	40	90	s	yes
Peak temperature	240.0562	240	250	°C	yes

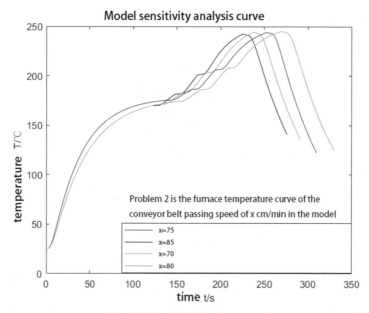

Fig. 11. Sensitivity analysis curve

The figure above shows the temperature in the center of the welding area and the relationship between temperature and time under the conditions of the conveyor belt passing speed of 75 cm/min, 85 cm/min, 70 cm/min and 80 cm/min.

Table 5. The temperature table of the center of the welding area against the passing speed of the conveyor bel

Passing rate	End of low temperature zone 5	End of low temperature zone 6	End of low temperature zone 7	End of low temperature zone 8
75 cm/min	175.4735 °C	186.2369 °C	206.1007 °C	227.2042 °C
85 cm/min	169.8385 °C	181.6378 °C	201.3400 °C	226.4603 °C
70 cm/min	173.9123 °C	186.2701 °C	207.8302 °C	231.6859 °C
80 cm/min	171.2103 °C	183.0924 °C	203.2466 °C	227.8806 °C

According to the data in the Table 5 above, when the passing speed of the conveyor belt is 75 cm/min, 85 cm/min, 70 cm/min and 80 cm/min, the difference of the temperature in the center of the welding zone affected by the small temperature zone changes with the speed, and the results in different areas are more reasonable. There is no temperature change or too drastic change with the change of the speed of the conveyor belt boiler, indicating that the sensitivity of this model is reasonable and can accurately predict the temperature discrete points under different conditions.

6.2 Error Analysis

In this paper, the heat conduction model is established in model 1, and the implicit difference method is used to solve it. Considering the accuracy of the establishment of the model and the solution, the discrete temperature values of the preheating zone stage are solved by using the model and solution method of Problem 1 under the condition of the stability of each small temperature zone known in an experiment of the topic, and the data in the attachment are compared and analyzed.

1. Firstly, the residual error is calculated and then the relative error is obtained by using the residual error:

$$e^{(0)}(k) = x^{(0)}(k) - \hat{x}^{(0)}(k), k = 1, 2, \cdots, n \tag{29}$$

$$\varepsilon(k) = \frac{e^{(0)}(k)}{x^{(0)}(k)}, k = 1, 2, \cdots, n \tag{30}$$

If $\varepsilon(k) < 0.2$, it can be considered to meet the general requirements, if $\varepsilon(k) < 0.1$, it can be considered to meet the higher requirements.

The relative error between the temperature in the center of the welding zone and the temperature under the influence of the preheating zone of the accessories in Model 1 at the temperature set in the experimental preheating zone was calculated by MATLAB. The relative error was calculated by taking 5 s as the step length, as shown in the Table 6:

Table 6. Relative error table of center temperature of welding are

Time (s)	Relative error	Time (s)	Relative error
19.00	0.1674	59.00	0.0135
24.00	0.2305	64.00	0.0117
29.00	0.1599	69.00	0.0107
34.00	0.0889	74.00	0.0100
39.00	0.0506	79.00	0.0097
44.00	0.0317	84.00	0.0097
49.00	0.0220	89.00	0.0099
54.00	0.0167	94.00	0.0101

As can be seen from the analysis of the above table, most of the relative error values of temperature are less than 0.1, and very few of them are greater than 0.1, which indicates that the model I solution is highly accurate, so it is reasonable to establish the fitting condition of model I.

2. Calculate the average relative error

$$y = \frac{\sum \varepsilon(k)}{n-1} \tag{31}$$

In this error result, the average relative error of temperature in the center of the welding area in the preheating zone is calculated as follows:

$$y = 0.0484 \tag{32}$$

3. Find out the accuracy of error analysis of the model

$$P = (1-y) \times 100\% \tag{33}$$

And the index of test accuracy is as follows in Table 7:

The accuracy of the temperature in the center of the welding area in the preheating zone is calculated, and the accuracy is greater than 0.95, indicating that the test index is good. Therefore, the heat conduction model established is reasonable to solve the problem, and the model solving effect is good, which can provide accurate data for the use of the model in the subsequent problems.

Table 7. Error analysis precision index table

Inspection index	good	qualified	Reluctantly	disqualification
Accuracy P	>0.95	>0.80	>0.70	<0.70

Acknowledgements. This work is supported by 2022 School Level Student Innovation Training Project of Guangzhou University (XJ202111078117).

References

1. Feng, J.: Application of reflow welding process and SMT technology in scientific research and production. China Sci. Instrum. **07**, 100–102 (2005)
2. Tang, Z., Xie, B., Liang, G.: Control analysis of reflow furnace temperature curve. Electron. Quality, (08), 15–19+23 (2020)
3. Yang, Y., Jiang, Z., Zhang, X., Chen, W.: On-line dynamic optimization control of heating furnace based on target steel temperature. Metallurgic. Automat. **36**(01), 19–24+52 (2012)
4. Li, A., Wang, Y., Tao, R.: Application of mathematical model of Fourier heat Conduction Equation and Newton Cooling Law in fluid thermal research. Indust. Technol. Innov. **03**(03), 498–502 (2016)
5. Jiang, Q.: Mathematical Modeling. Higher Education Press, Beijing (2011)
6. Xu, J., Tang, B.: Numerical solution of one-dimensional heat conduction equation. J. Huaiyin Norm. Univ. (2004)
7. Zhou, M., Shi, R.: Application of process control system in heat treatment Furnace. Automat. Appl. **12**, 5–6 (2011)
8. Si, S., Sun, Z.: Mathematical Modeling Algorithm and Application, 2nd edn. National Defense Industry Press (2015)
9. Wang, H., Zhu, H.: Improved Dichotomy Search. Comput. Eng. (10), 60–62+118 (2006)

Research on Intelligent Adjustment Technology of Active Reflector of Radio Telescope

Wangwei Zhong[1], Weitong Chen[2], Jialian Li[2], Hongbin Lin[2(✉)], and Yubin Zhong[2]

[1] Guangdong Construction Polytechnic College of Architectural Information, Guangzhou 510440, China

[2] School of Mathematics and Information Science, Guangzhou University, Guangzhou 510006, China

hongbin.lin3589@gmail.com

Abstract. With the development of various technologies, the demand for the accuracy of celestial electromagnetic reception of radio telescops has been significantly improved. In this research, we mainly take "FAST" as an example to examine the intelligent adjustment technology of the radio telescope's active reflector. We propose a model of active reflector adjustment of radio telescope based on mechanism model, and then determine the equation and position of ideal reflector when observing celestial bodies at different positions more accurately, finally, calculate the reception ratio of feed cabin under different situations. To reduce the cost of manual operation, we use the algorithm to intelligently track multiple celestial bodies and realize intelligent automatic reflector adjustment. Our model can provide a certain reference value for the research of radio telescopes, as well as be extended to other cross disciplines.

Keywords: Surface Reflection · Geometrical Optics · Improved Simulated Annealing Genetic Algorithm

1 Introduction

Radio astronomy has made significant contributions to numerous fields of astrophysical research. The entire world is attempting to overcome the technical challenges of the millimeter wave radio telescope in order to construct a larger high-precision antenna and achieve shorter band observation. FAST is a 500-m-diameter spherical radio telescope which is the world's largest and most sensitive single-aperture radio telescope [1]. The active deformable reflector is one of its main innovations, which is critical for China to achieve major original breakthroughs in the scientific frontier and accelerate innovation-driven development. However, the receiving effect of celestial electromagnetic must be further optimized as various technologies advance [2]. Therefore, determining the ideal paraboloid corresponding to the active reflector when the measured object is in different positions is the key to the active reflector technology.

B.-Y. Cao et al. (Eds.): ICFIE 2022, LNDECT 207, pp. 96–107, 2024.
https://doi.org/10.1007/978-981-97-2891-6_7

At the present, research on reflector antennas is limited both at home and abroad. In China, the most common algorithms are the least square method, nonlinear programming, gradient descent optimization algorithm [3], particle swarm optimization algorithm, etc. However, the shape adjustment direction of a cable mesh antenna reflector has been extensively researched in other countries. And the main method is to begin with recursion and iteration, and then modify the sensitivity matrix recursively according to the system's behavior in iterative adjustment, so that the modeling error is automatically corrected in the adjustment process, and the process converges efficiently. Then on this basis, continue to investigate the optimization design method of mesh reflector antenna cable network considering truss deformation and space thermal effect [4], as well as the optimization of intelligent reflectors. However, there have been issues with earlier studies, including poor calculation accuracy, excessive calculation time, and the requirement to establish parameters in advance of observation. Research on reflector antenna surface accuracy analysis and adjustment techniques is still in its early stages. Additionally, the primary cable node and actuator position autonomous adjustment model for FAST currently only supports the point model and is deficient in the dynamic change adjustment model [3]. As a result, at this time, it is mostly dependent on research into the appropriate configuration, reflector adjustment, and reception ratio of the active reflector used in FAST, which is rather straightforward and has not yet considered popularization.

Based on the aforementioned issues, we propose a mechanism-based model for adjusting the active reflector of a radio telescope. We establish a mathematical model to determine the appropriate adjustment strategy of the active reflector when the measured object is in different positions through the study of the operation mechanism model of the "FAST" active reflector panel combined with the actual operation of the reflector panel. We also realize the intellectualization of the adjustment algorithm, that is, to input the celestial coordinates, It is possible to calculate the adjustment strategy automatically. This project is anticipated to advance the research methodologies and concepts to other cross-disciplinary fields, as well as serve as a particular benchmark for the study of radio telescopes.

To maximize the reception of celestial signals, we calculated the reception ratio of the feed cabin under various conditions and derived the equation and location of the optimal reflector when observing celestial bodies at various positions. At the same time, the expense of manual operation is decreased thanks to clever algorithm. Intelligent automatic reflector adjustment for tracking multiple celestial bodies is possible, as is an active reflector shape adjustment approach for multiple restrictions like actuator expansion and main cable shape variability. In addition, the model programming of this project can calibrate the actual adjustment strategy from various dimensions and output directly the coordinates of the deformed main cable and the expansion of each actuator. We investigate the method of adjusting the ideal paraboloid and the calculation of the feed cabin reception ratio, which provides a certain reference value for the study of radio telescopes, in order to extend it to other interdisciplinary subjects, such as satellite communications, wind farm system simulation, paraboloid cable mesh antenna surface design, etc., providing methods for relevant parameter optimization.

The following is a list of the contributions we made:

(1) A method is proposed to identify the optimal active receiving surface when the observation object is in a particular position. The reflecting surface can be as near to the ideal paraboloid as possible by changing the expansion and contraction of pertinent actuators.
(2) The research results on the FAST telescopes will be applied to other radio telescopes with actively movable reflectors.
(3) Reduce the computation time and follow moving objects when implementing the aforementioned functions.

2   Related Work

2.1   FAST Reflector System

The world's largest single aperture and most sensitive radio telescope, FAST, was independently created by China. Active reflector, signal reception system (feed cabin), and associated control, measurement, and support systems make up the majority of FAST [5]. The main cable net, reflective panel, lower cable, actuator, and supporting structure make up the active reflector system, which is an adjustable sphere. The functioning state and reference state of the active reflector are distinct states. In the working state, the reflector is a roughly rotating paraboloid with a diameter of 300 m. In the reference state, the reflecting surface is a sphere with a radius of about 300 m and a diameter of 500 m. It has the capacity to instantly create a paraboloid with a 300-m diameter. The cabin where the feed system is installed is known as the feed cabin. The Centre disc, which has a 1 m diameter, represents the effective area of the signal received. Only a sphere that is concentric with the reference sphere can the center of the receiving plane of the feed cabin move [6].

2.2   Autonomous Control Strategy of Radio Telescope

Autonomous surface adjustment is becoming increasingly significant as large aperture and high frequency radio telescopes are developed. Lian et al. [7] installed the measurement system on the back of the secondary reflector to establish the relationship between the two reflectors, thus measuring the distance and elevation of the target point of the main refractor. This was done for the large double reverse adjustable radio telescope because the secondary reflector must match the main refractor. Two of these are part of autonomous control: Firstly, make sure that the main reflector and sub reflector are both operating in their ideal positions at all times; Secondly, move the main reflector to the best-fitting location and the sub reflector to the best-fitting focus. For the automatic surface adjustment design of large telescopes, this means that research on the calculation procedure for primary and secondary reflector adjustment and the related simulation verification can offer more helpful recommendations.

2.3 Simulated Annealing Genetic Algorithm

A type of random search algorithm known as the simulated annealing genetic algorithm combines genetic algorithms into its operation [8]. By simulating the similarities between the random search optimization problem and the thermal balance problem of solid material annealing process, it is able to search for the global optimum or an approximation of the global optimum. A technique for finding the best solution involves replicating the actual annealing process. This is known as the simulated annealing algorithm. The algorithm can escape the local extreme point in the simulated annealing process because the annealing temperature minimizes the optimization direction of the solution process while at the same time receiving the inferior solution with a fixed probability. Therefore, the likelihood that a suboptimal solution will be accepted decreases over time, increasing the likelihood that the solution will be optimal. A genetic algorithm is a technique for finding the best answer by mimicking the course of natural evolution. By using copying, crossing, mutation, and other operations to create the solution for the following generation, a genetic algorithm gradually eliminates the solution with a low fitness function value and increases the solution with a high fitness function value. Consequently, after N generations of evolution, it is quite likely that people with high fitness function values would arise; in other words, the problem at hand will be modelled as a biological evolution process utilizing the biological evolution theory.

3 Methodology

We first study the ideal paraboloid when the object to be observed is directly above the reference sphere. We reduce the active reflector to a two-dimensional space using the rotating paraboloid's symmetry, and then establish a two-dimensional rectangular coordinate system using the center C of the ground state sphere as the coordinate origin. After that, we can simplify the reference sphere that corresponds to the reference state and the working paraboloid that corresponds to the working state and obtain the corresponding arc and parabola equations:

$$x^2 + y^2 = R^2 \tag{1}$$

$$-x^2 + 4Dfy + 4Df(R + h) = 0 \tag{2}$$

As shown in Fig. 1, the active reflector has a shape that resembles an arc when it is in the reference state. However, when it is in the working state, it takes on the shape of a parabola and satisfies the equation in Formula (2), shifting the focus to an arc that is centered on the reference sphere. When the object to be observed is directly above the reference sphere, the focus is on the y-axis, and the parabola equation is established with the parabola vertex as the origin:

$$x^2 = 4Dfy \tag{3}$$

Set the focus P' as $P'(0, Df)$, and shift the origin to the center C of the reference state. Since the distance between the center of the sphere and the bottom of the reference

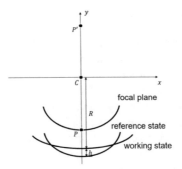

Fig. 1. Focus Diagram

state is radius R, and the distance between the bottom of the reference state and the bottom of the working state is h, then:

$$|CP| = |Df - (R + h)| \tag{4}$$

On the other hand, the center of the receiving plane of the feed cabin can only move on a sphere (focal plane) concentric with the reference sphere, so the radius difference between the sphere and the focal plane corresponding to the reference state is

$$F = 0.466R \tag{5}$$

The object to be observed is directly above the reference sphere and the focus P is directly below the vertical position of the sphere, therefore:

$$|CP| = F(1 - 0.466R) \tag{6}$$

Simultaneous (4) and (6), then

$$Df - (R + h) = -R(1 - 0.466) \tag{7}$$

In this case, the arc equation corresponding to the reference state coincides with the endpoint of the parabolic equation corresponding to the working state. As shown in Fig. 2, taking the spherical center 0 as the coordinate origin, since the radius of the reflecting surface in the reference state is 300 m, the shape of the reflecting surface in the working state is adjusted to an approximate rotating paraboloid with a diameter of 300 m, that is $|DA| = |AC| = |DC| = R$, which means that $\triangle DAC$ is an equilateral triangle and $\angle DAC = 60°$.

Therefore, in order to make the endpoints of the arc equation corresponding to the ground state coincide with those of the parabolic equation corresponding to the working state, both endpoints must satisfy the corresponding equation, which is:

$$\begin{cases} x_1^2 + y_1^2 = R^2 \\ -x_1^2 + 4Dfy_1 + 4Df(R + h) = 0 \end{cases} \tag{8}$$

The actuator's radial expansion distance is $-0.6 \sim +0.6$ m, therefore:

$$\left| \sqrt{x^2 + y^2} - R \right| \leq 0.6 \tag{9}$$

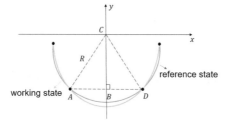

Fig. 2. Schematic Diagram of End Point Coincidence

When the FAST active reflector observes displacement, the control node primarily.has three motion directions [9]:

1) Radius direction (radial) of the reference sphere: the direction from the control node to the center of the reference sphere
2) Longitude direction of the reference sphere: the direction from the vertex to the edge of the surface that is close to the longitude direction of the earth
3) Latitude direction (latitudinal direction) of the reference sphere: the circular direction of equal curvature of the observation paraboloid, which corresponds to the latitude direction of the earth.

We take the center of the receiving plane of the feed cabin to be moved to the focal point to establish an approximate rotating paraboloid. The actuator stroke and the paraboloid's edge condition make up the majority of the main cable point position's considerable radial fluctuation. As a result, when the radial displacement of the paraboloid edge node approaches toward zero, the edge transition becomes more natural; The arc length difference when the active reflector transitions from the reference state to the working state can be interpreted as the radial variation and zonal variation; The change of longitudinal cable length will cause lateral deflection of the node [10]. Based on this, we choose the distance between the bottom of the reference state and the bottom of the working state as well as the paraboloid's focal length to aperture ratio as the goal parameters to optimize. We get the following optimization goals:

1. The radial displacement of the actuator shall be as small as possible

The distance between the parabola and the arc shall be as small as possible. Setting the arc Eq. (1) as f(x), and the parabola Eq. (2) as g(x), then the minimum radial displacement can be expressed as:

$$min\left(\int_{-150}^{150} |f(x) - g(x)| dx \right) \tag{10}$$

2. The longitudinal expansion and contraction of main cable shall be as small as possible

The difference between the reference state and the working state arc length of the corresponding section should be as small as possible. We assume that the arc length corresponding to the reference state is L_1, and the parabola arc length corresponding to the working state is L_2, then:

$$min\left(\int_{-150}^{150}|L_1 - L_2|dx\right) \tag{11}$$

3. The tangent equation of the intersection point of the parabola and the sphere should fit as closely as possible

In order to make the transition between the reference state and the working state more natural, we convert it into the tangent equation corresponding to them as similar as possible. For this reason, we analyze the slopes of the two curves at the intersection of the parabola and the arc. Assuming that the slope of the arc at this intersection is k_1, and the slope of the throwing line at this intersection is k_2. To make the equations of the two fit as closely as possible, the difference between the slopes at the intersection is as small as possible:

$$min(|k_1 - k_2|) \tag{12}$$

Based on the above optimization objectives, we assign different weights to them, convert multiple objectives into single objective constraints, and obtain the following optimization objective functions:

$$minw_1\left(\int_{x_1}^{x_2}|f(x) - g(x)| \mid dx\right) + w_2(|L_1 - L_2|) + w_3(|k_1 - k_2|) \tag{13}$$

In conclusion, we get the nonlinear ideal parabolic programming model based on the reflector adjustment factor:

$$\begin{cases} -x^2 + 4Dfy + 4Df(R + h) = 0 \\ Df - (R + h) = -R(1 - 0.466) \\ x_1^2 + y_1^2 = R^2 \\ -x_1^2 + 4Dfy_1 + 4Df(R + h) = 0 \\ \left|\sqrt{x^2 + y^2} - R\right| \le 0.6 \end{cases} \tag{14}$$

Then, we study the ideal paraboloid when the object to be measured is located on the side of the reference sphere. On the basis of the two-dimensional coordinate system established above, we raise the dimension and transform the original coordinate system so that the apex of the parabola is always on the z axis, as shown in Fig. 3. The following coordinate conversion diagram is obtained:

The corresponding coordinate exchange matrix can be obtained as follows:

$$\begin{bmatrix} x' \\ y' \\ z' \end{bmatrix} = \begin{bmatrix} \cos(\frac{\pi}{2} - \beta) & 0 & -\sin(\frac{\pi}{2} - \beta) \\ 0 & 1 & 0 \\ \sin(\frac{\pi}{2} - \beta) & 0 & \cos(\frac{\pi}{2} - \beta) \end{bmatrix} \begin{bmatrix} \cos\alpha & \sin\alpha & 0 \\ -\sin\alpha & \cos\alpha & 0 \\ 0 & 0 & 1 \end{bmatrix} \begin{bmatrix} x \\ y \\ z \end{bmatrix} \tag{15}$$

$$\begin{bmatrix} x \\ y \\ z \end{bmatrix} = \begin{bmatrix} \cos\alpha & -\sin\alpha & 0 \\ \sin\alpha & \cos\alpha & 0 \\ 0 & 0 & 1 \end{bmatrix} \begin{bmatrix} \cos\alpha & -\sin\alpha & 0 \\ \sin\alpha & \cos\alpha & 0 \\ 0 & 0 & 1 \end{bmatrix}^{-1} \begin{bmatrix} x' \\ y' \\ z' \end{bmatrix} \tag{16}$$

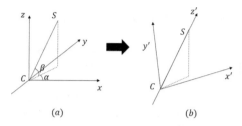

Fig. 3. Schematic diagram of coordinate conversion

wherein, $C - xyz$ is the original coordinate system and $C - x'y'z'$ is the transformed coordinate system,.

When observing targets at different positions, each reflective panel may appear at different positions of the fitting paraboloid. In order to make the fitting degree as high as possible when observing different targets, FAST design requires that the surface curvature of each reflective panel is the same, that is, the radius of curvature is the same as the radius of curvature of the reference billiard surface [11].

4 Experiment

Under the background of FAST active reflector adjustment, we get the following relevant data: $|CA| = R = 300\,\text{m}$, $|AB| = \frac{1}{2}|AD| = \frac{1}{2}|CA| = 150\,\text{m}$ and $|CB| = 150\sqrt{3}\,\text{m}$. Substituting them into Eq. (8) and then we can get:

$$-150^2 + 4Df(-150\sqrt{3}) + 4Df(R+h) = 0 \tag{17}$$

According to the optimization objectives, we combine relevant data to transform the three optimized objectives into

$$min \int_{-150}^{150} \left| \sqrt{x^2 - R^2} - -\frac{x^2}{4Df} + (R+h) \right| dx \tag{18}$$

$$min(|\frac{\alpha}{2\pi} \cdot 2\pi R - \int_{-150}^{150} \sqrt{1 + g'(x)^2} dx|) \tag{19}$$

$$min| -\frac{Df}{150\sqrt{3}} - (-\frac{\sqrt{3}}{3})| \tag{20}$$

Then, the nonlinear ideal paraboloid programming model based on the reflector adjustment factor is obtained as follows:

$$minw_1 \left(\int_{-150}^{150} \left| \sqrt{x^2 - R^2} - \frac{x^2}{4Df} + (R+h) \right| dx \right) +$$

$$w_2(|\frac{\alpha}{2\pi} \cdot 2\pi R - \int_{-150}^{150} \sqrt{1 + g'(x)^2} dx|) \tag{21}$$

$$w_3(| -\frac{Df}{150\sqrt{3}} - (-\frac{\sqrt{3}}{3})|)$$

where, D is the radius of the reflecting surface in the reference state, both of which are 300 m.

We obtain the optimal deformation strategy by continuously optimizing the parameters h and f. Through experiments, we set $w_1 = 0.3$, $w_2 = 0.3$, $w_3 = 0.4$, basing on which, we calculate that $f = 0.46109$, $h = 0.48876$ (f is the ratio of the focal length to the aperture of the paraboloid, h is the distance between the base of the reference state and the base of the working state). Under this condition, the constraint conditions can be met when the objective function is minimum.

In this case, the ground state sphere and parabola can get the best ideal state with four points coincident, and the equation is:

$$\begin{cases} x^2 + y^2 = 300^2 \\ -x^2 + 553.31y + 1.66263 \times 10^5 = 0 \end{cases} \tag{22}$$

After that, the equation of the corresponding ideal paraboloid can be obtained by rotating the parabola equation for one circle:

$$-x^2 - y^2 + 553.31z + 1.662.63 \times 10^5 = 0 \tag{23}$$

Then we study the receiving ratio of the base state feed cabin. Since the observation target is far away from the radio telescope, we regard the electromagnetic wave signal emitted by the observation target as parallel light. It must be noticed that the angle between the incident light and the normal on the reference reflecting sphere is equal to the angle between the reflected light and the normal. We establish a coordinate system with the center of the sphere as the origin. Through simple calculation, it can be seen that the reflected light of the parallel light incident on the sphere cannot be focused. Therefore, it is necessary to calculate the ratio of the signal received in the effective area of the feed cabin to the reflected signal of the reflector within the 300 m aperture.

Considering the boundary conditions of the signal received by the feed cabin, as shown in the figure, the incident point of an incident light is (x_2, y_2), which can just be received by the feed cabin after being reflected by the spherical surface. Set the signal receiving boundary point of the feed cabin as (x_1, y_1), and from the geometric relationship, as shown in Fig. 4, we can get:

$$\frac{x_2 - x_1}{y_2 - y_1} = tan2\theta \tag{24}$$

The light reflected by the circular arc between the symmetrical points of the coordinate points about the y-axis can be received by the feed cabin. Therefore, the ratio of the signal received in the effective area of the feed cabin to the reflected signal of the 300 m aperture internal reflector is converted into the ratio of the area of the curved surface to the area of the 300 m aperture spherical surface. The effective area of the signal received by the feed cabin is the central disc with a diameter of 1 m (red circular arc area in the figure, with an area of S_1). The reflected light is intercepted from the plane where the feed cabin is located, and the corresponding area enclosed by the reflected light is S_2. Then the reception ratio of the reflected sphere after adjustment is:

$$k_1 = \frac{S_1}{S_2} \times 100\% \tag{25}$$

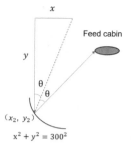

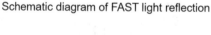

Fig. 4. Tangent diagram of incident light in reference state

Under the actual background, since (x_2, y_2) is on the sphere and (x_1, y_1) is $(-0.5, -160.2)$, the simultaneous Eq. (30) and the spherical equation get:

$$\begin{cases} x^2 + y^2 = 300^2 \\ \frac{x_2 + 0.5}{y_2 + 160.2} = tan2\theta \\ tan\theta = \frac{x_2}{y_2} \end{cases} \quad (26)$$

The solution is $x_2 = -18.325$, and the area ratio formula is simplified into the ratio formula of the circumference angle by using symmetry. Therefore, the receiving ratio formula of the feed cabin is:

$$k_2 = 2arcsin(\frac{x_2}{2R}) \cdot \frac{3}{\pi} \cdot 100\% \quad (27)$$

Through calculation, the receiving ratio of the feed cabin of the reference reflecting sphere is 11.67%.

When the active reflector is in the working state, after the reflector is adjusted, it can be obtained through the deduction of the approximate imaging formula of Fermat's principle on spherical reflection that the reflected light must pass through a certain point.

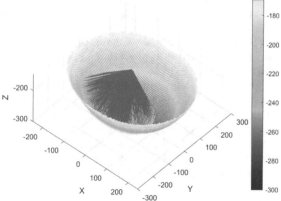

Fig. 5. FAST Ray Reflection Diagram

We can program the diagram of the relationship between the reflected light and FAST in this case, as shown in Fig. 5, where the blue line represents the reflected light of the incident parallel light. It can be observed that the reflected light has a good convergence degree and is concentrated into a type of cone.

This is consistent with our model. According to formula (31), the receiving ratio of the feed cabin in the working state is 67.42%.

5   Conclusion

We mainly study the operation mechanism of the "FAST" active reflector, determine the more accurate equation and position of the ideal reflector when observing celestial bodies at different positions, as well as calculate the receiving ratio of the feed cabin in different situations to achieve the maximum degree of reception of celestial signals. At the same time, the cost of manual operation is reduced. What's more, the automatic adjustment and intelligentization of the reflective surface can be realized for multiple celestial body tracking. The model we built can realize the active reflection surface shape adjustment strategy with multiple constraints such as the expansion and contraction of the actuator and the deformation of the main cable. The expansion and contraction of the device, the actual adjustment strategy is calibrated from multiple dimensions, which increases the convenience and practicality of the actual operation. The method of adjusting the ideal paraboloid and the calculation of the receiving ratio of the feed cabin provided by us provide a certain reference value for the research of radio telescopes, and can be extended to other interdisciplinary subjects and provide methods for related research.

Acknowledgements. This work is supported by 2022 Science Industry Education Integration Practice Base of Guangdong Provincial Department of Education(xj202211078256).

References

1. Tang, T.: From KARST to FAST - the conceptual origin and decision-making process of China's celestial eye. J. Astronomy **64**(01), 36–43 (2023). https://doi.org/10.15940/j.cnki.0001-5245.2023.01.005
2. Zhang, M., Wang, H., Zuo, Y., et al.: Review of the development of radio telescope structure technology. J. Huazhong Univ. Sci. Technol. (Nat. Sci. Edn.) **51**(01), 48–58 (2023). https://doi.org/10.13245/j.hust.239104
3. Guo, X., Pei, Z., Jin, Q., et al.: Research on FAST reflector adjustment and reception ratio based on gradient optimization. J. Taiyuan Univ. Technol. **53**(04), 744–750 (2022). https://doi.org/10.16355/j.cnki.issn1007-9432tyut.2022.04.020
4. Ning, L., Lei, Z., Liu, Y., et al.: Analysis of solar thermal deformation of large aperture radio telescope panel. Technol. Market **29**(12), 78–80+84 (2022)
5. Xu, J., Zhang, W., Hou, X., et al.: China Skyeye - "Fast" celestial signal reflection system design. Electronic Test **480**(03), 62–64 (2022). https://doi.org/10.16520/j.cnki.1000-8519.2022.03.005
6. Zhu, L.: Automatic control system of FAST main reflector. Sci. Technol. Eng. **13**, 1890–1894 (2006)

7. Lian, P., et al.: Surface adjustment strategy for a large radio telescope with adjustable dual reflectors. IET Microwaves, Antennas Propagat. **13**(15), 2669–2677 (2019)
8. Xiao, Q.: Optimization of genetic simulated annealing algorithm. Inform. Recording Mater. **23**(12), 95–98 (2022). https://doi.org/10.16009/j.cnki.cn13-1295/tq.2022.12.052
9. Wang, Z.: Research on FAST Whole Network Control Strategy Based on Iterative Learning Theory. Northeast University (2015)
10. Zhu, L.: Deformation control of active reflector network of 500m aperture spherical radio telescope (FAST). Sci. Res. Inform. Technol. Appli. (04), 69–77 (2012)
11. Gan, H.: FAST reflector error calculation and telescope receiver front-end. University of Chinese Academy of Sciences (2010)

Dynamic Credit Line Decisions for MSMEs

Li Lin[1,2], Zheyu Gong[1], and Yubin Zhong[1(✉)]

[1] School of Mathematics and Information Science, Guangzhou University, Guangzhou, China
zhong_yb@gzhu.edu.cn
[2] School of Mathematical Sciences, South China Normal University, Guangzhou, China

Abstract. This paper mainly studies the credit risk of micro, small and medium-sized enterprises, and establishes the financial credit risk evaluation model and credit limit decision model. 1. Cluster analysis method is used to construct loan quota model; 2. Build a ternary function interest rate model and use the optimal interest rate method; 3. According to the indicators in the wealth ranking method and RQ evaluation method, the reputation rating model is constructed through spearman correlation coefficient detection and multiple regression fitting to evaluate the reputation; 4. Build the optimal loan model, and solve the enterprise loan amount through computer simulation; 5. Establish the expression of stochastic process that includes the influence of corporate profit on systemic risk and make the optimal allocation of loan amount.

Keywords: spearman · coefficient cluster analysis · Pin peak complement model · machine learning · time series prediction

1 Introduction

With the advent of globalization, the development and openness of financial markets continue to improve, and the competition of commercial banks' credit services is more and more intense. Credit loan is the main way of loan in our country for a long time, which is issued through the credit of individuals or enterprises. However, due to the high risk of this type of loan, it generally needs a variety of investigations to reduce the risk. Although the state continues to issue policies to encourage financial institutions to increase the amount of loans, credit evaluation still needs to be further improved due to the small scale of micro, small and medium-sized enterprises, the lack of mortgage assets, and the subjective inspection method. Therefore, this paper selects micro, small and medium-sized enterprises as the research object, adopts the relevant data of 123 enterprises with credit records provided by the 2020 National College Student Mathematical Modeling Question C, and explores scientific and objective methods to study the credit evaluation model of small, medium and micro enterprises from the perspective of banks to determine whether to give loans and the corresponding annual interest rate and loan amount.

2   Credit Risk Model

The technical route is shown in Fig. 1:

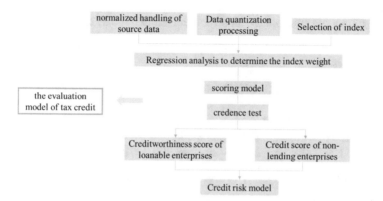

Fig. 1. Technical route of reputation evaluation

2.1   Scoring Model

Iterate to find the optimal coefficient:

$$y = \sum\nolimits_{i=1}^{n} X_i A_i \quad (i = 1, 2, 3) \tag{1}$$

where X_i is the coefficient obtained by iteration [1], A_1 is the determined profit growth rate, A_2 is the amount of output, namely, the enterprise income, and A_3 is the credit rating, namely, the four categories given by the title. A = 87.5, B = 62.5, C = 37.5, D = 12.5. Then the iterative solution is performed as shown in Fig. 2 and Fig. 3, To solve the weight of correlation coefficient:

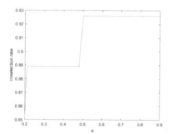
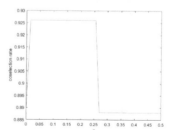

Fig. 2. Conformity of credit rating coefficient   **Fig. 3.** Conformity of profit growth rate coefficient

A1 is 0.115 and A2 is 0.385. Therefore, the credit risk function is obtained as follows:

$$Y = 0.115X_1 + 0.385X_2 + 0.5X_3 \tag{2}$$

The coincidence rate can reach 92.5%. Moreover, through the solution, the credit risk score of lending enterprises can be determined to be -0.3722.

2.2 Quota Model

Cluster analysis was used to classify the loan amount into three categories for processing.

Quota Model Establishment
It is divided into three categories for processing:
① Unified 100,000: enterprise flow less than 200,000 is divided into the first category, a total of group A.
② Unified 1 million: enterprises with flow rate higher than 2,000,000 are divided into the third category, a total of group b.
③ After scoring the reputation risk, the remaining enterprises minus the remaining enterprises of a and b are divided into the second category, which is group c.
Obtain the loan amount of class c and establish the proportion model:

$$W - 10W - 100b \tag{3}$$

$$W_{c_i} = k_i W_c \tag{4}$$

W_c is the total loans of type c enterprises, and W is the fixed annual total credit, W_{c_i} is the loan amount that the type c enterprise can obtain, and k_i is the proportion of the average profit of the type c enterprise in all type c enterprises [2].

Quota Model Solution
Cluster analysis was performed as shown in Fig. 4, Enterprises are divided into three categories according to K-means cluster analysis and the requirements of bank loans for flow are shown in Table 1:

Table 1. Classification of corporate loan amount

Amount available for loan	Enterprise's number
100,000 yuan	21
100,000 yuan ~ 1 million	33
1 million	43

The results of both categories are almost identical to the right table.

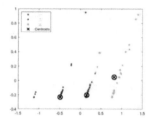

Fig. 4. Clustering diagram

2.3 Interest Rate Model

According to the customer churn rate corresponding to different credit levels [3], an optimization model is built, and the optimal interest rate is determined by the multivariate function extremum method.

Establishment of Interest Rate Model

It is known that when the total amount of annual credit is fixed, if the annual interest rate deviates from the acceptable range of the enterprise itself, it is easy to cause customer loss. Banks also need to encourage more enterprises to have better credit ratings, so the annual interest rate given to enterprises with better ratings is lower [4], and then the linear relationship between the annual total credit and their respective corporate credit is calculated:

$$W_k = (10 + 100n_2 + x_i \sum\nolimits_{i=1}^{n} \frac{k_i}{\sum_{i=1}^{n} k}) \tag{5}$$

where $k = $ a to c, n_1 is the number of enterprises with loans of CNY 100,000 under corresponding k, and n_2 is the number of enterprises with loans of CNY 1 million under corresponding k, x_i is the ith firm.

Solution of Interest Rate Model

Iteratively solve the data, and build a test function as shown in Fig. 5, 6 and 7 to test the annual interest rate corresponding to each A, b and c reputation:

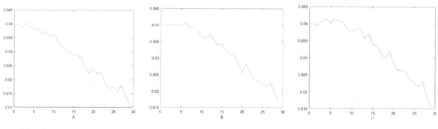

Fig. 5. Turnover rate   **Fig. 6.** Turnover rate   **Fig. 7.** Turnover rate

Table 2 is the final iteration result, indicating the annual loan interest rate and corresponding customer churn rate that a, b and c companies should adopt if they want to obtain the maximum interest rate:

Table 2. Annual loan interest rate and churn rate corresponding to different reputations

	a	b	c
Annual loan interest rate	0.0465	0.0585	0.0585
Churn rate	0.224603354	0.302883401	0.290189098

3 Optimal Loan Limit Model

The technical route is shown in Fig. 8:

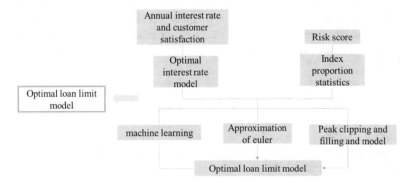

Fig. 8. Technical route for optimizing loans

3.1 Establishment and Solution of Reputation Evaluation Model

Establishment of Reputation Evaluation Model
Based on wealth ranking method and RQ measurement. Based on the invoice data of each enterprise within five years, extract service quality, financial performance and financial performance as evaluation criteria. The percentage of red letter invoices [5] of each enterprise in the total invoices, as well as the average half-year profit of each enterprise and the test index are obtained. Spearman correlation coefficient was used to determine the correlation between the index and reputation level, and the results were shown in Table 3:

Table 3. Correlation discrimination

Index	Correlation
Proportion of red invoices	−0.5893
Amount of output	0.5941
Average half-year profit	0.3373

Find the triple regression function and plug in the data to get the model:

$$Y = 51.3211 - 10.1763x_1 + 4.521x_2 + 0.6433x_3 \qquad (6)$$

Moreover, through the coefficient test of the regression model, the fitting degree is close to 0.6, which is good.

Solving Credit Risk Model

The credit rating score of the enterprise and the established index in the credit evaluation model are substituted into the credit risk model to solve the problem:

$$Y = 0.115X_1 + 0.385X_2 + 0.5X_3 \qquad (7)$$

As a result, the number of companies with access to bank loans is 251, with partial results shown in Table 4:

Table 4. Credit risk scores of 302 companies

Enterprise	Reputation score	Amount of output	Average prof-it growth rate	Credit risk score
1	2.752093695	7.265433137	0.073133885	4.700683038
2	2.752093695	9.362284321	0.073151933	5.749110705
3	2.752093695	4.94086321	0.073264312	3.538413073
...	...	...	...	...

The annual interest rates for loans of different credit lines are shown in Table 5:

Table 5. Annual loan interest rate and customer churn rate of three types of companies

Enterprise	a	b	c
Annual loan interest rate	0.0505	0.0585	0.0585
Churn rate	0.224603354	0.302883401	0.290189098

3.2 Loan Limit Model

Establishment of the Optimal Loan Quota Model

Construct the linear programming model of optimal loan amount according to the constraints:

$$W = \left[1 \times 10^9 - (100y_1 + 10y_2)\right] \times \left(\frac{\sum_{215-y_1}^{215-y_2} X_i}{\sum_{i=1}^{n} X_i}\right) + (100y_1 + 10y_2) \quad (8)$$

s.t

$$X_{y_1} > X_{y_1+1} \quad (9)$$

$$X_{251-y_2} > Y_{251-y_2+1} \quad (10)$$

$$y_1 \in (0, 50) \& y_2 \in (46, 100) \quad (11)$$

Then build the training model according to machine learning:

$$\hat{h}(x) = \sum_{i=1}^{\theta i} \theta_i x^{i-1} \quad (12)$$

$$time = \frac{(ta + tx) - t \times v}{v} \quad (13)$$

$$loss = \frac{1}{2m} \sum_{j=1}^{m} \left(\hat{h}_{\left(x_{j1}\right)} - y_j\right)^2 \quad (14)$$

$$t = 0 \quad (15)$$

Gradient descent

$$\theta_1 = \theta_1 - \alpha \frac{\partial loss(\theta_i)}{\partial \theta_1} \quad (16)$$

$$\theta_2 = \theta_2 - \partial \frac{\partial loss(\theta_1, \theta_1 \ldots)}{\partial \theta_2} \quad (17)$$

$$\theta_n = \theta_{12} - \alpha \frac{\partial (loss\theta_1, \theta_2, \cdots, \theta_n)}{\partial \theta_n} \quad (18)$$

$$n \leq \theta_i \quad (19)$$

$$t = t + 1 \quad (20)$$

$$if \ t < tx \ go \ on \quad (21)$$

$$\theta = (\theta_1, \theta_2, \cdots \theta_n)_{\theta_i} \quad (22)$$

$$X = \left(x^0, x, \ldots, x^n\right)^T \qquad (23)$$

$$loss = \frac{1}{2m} \sum_{j=1}^{m} (\theta X - y_i)^2 \qquad (24)$$

Solution of the Optimal Loan Quota Model

(1) *Machine Learning Optimization Solution*

First, the data is divided into ta group training data and tx test data, and the test data is trained for several times, each time using different training set and test set [6]. Considering that too high learning efficiency may directly lead to the failure of convergence, but too small rate will also reduce the learning speed. Therefore, according to the final error rate, we repeatedly adjust the initial value of θ and the learning rate and select the result with the smallest error rate [7], that is, The final loan allocation results are shown in Table 6.

Table 6. Machine learning allocation results

Enterprise number	Reputation risk score	Profit	Amount available for loan
10	62.5	37874695.51	671107.7972
79	62.5	37630676.98	667457.2648
...	...	...	...
212	62.5	4116529.87	100000
259	62.5	3980259.07	100000

(2) *The Solution of Pin Peak Supplement Model*

As shown in Fig. 9, By constructing the peak clipping and supplement model, the model is tested by the computer cutting and supplement cycle:

①After sorting the profits of enterprises in descending order, it can see that there is a large profit gap between enterprises according to the profit trend.

②Find 84 as the turning point and determine the enterprise with a loan amount of 100,000 yuan. The remaining 127 enterprises are allocated the remaining loan amount through the proportion model.

③Find the singular value. Make up for peak sales and redistribute the remaining amount until there are no more enterprises with more than 1 million.

④The final distribution amount is shown in Table 7 (partial data):

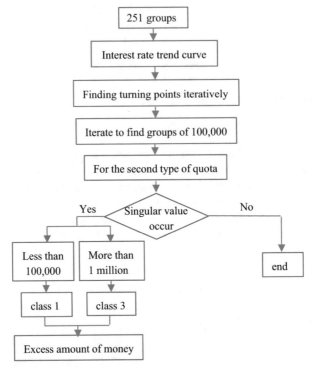

Fig. 9. Pin-peak cmplement model

Table 7. Final loan amount of enterprises

Firm number	risk score	Profit	Amount available for loan
10	62.5	37874695.51	977015.3
...	...	...	...
190	62.5	3988780.47	102894.5
259	62.5	3980259.07	102674.7

4 Credit Line Dynamic Decision Model for MSMEs

The technical route is shown in Fig. 10:

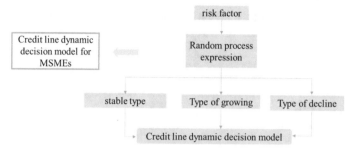

Fig. 10. Final decision route

4.1 Model Establishment and Solve

The enterprises were classified according to (GB/4754–2011), the epidemic variable ρ was added, and a random process expression containing the corporate profits affected by systemic risk was established:

$$\frac{d\pi_t}{\pi_t} = \mu dt + \rho\sigma_i dz_i + \sqrt{1 - \rho^2}\sigma_m dz_m \tag{25}$$

π_t is the enterprise profit at time t, the risk structure is ρ, $\sigma_m dz_m$ represents systemic risk, $\sigma_i dz_i$ represents non − systemic risk, dz is standard Wiener process $dz = \varepsilon\sqrt{dt}$.

At the same time, operational risks in non-systemic risks are mainly considered, and a part of them is proposed as the "epidemic risk coefficient" and set as ρ_i.

$$d\pi_t = \mu\pi_t dt + \sigma_i\pi_t dz_i + \pi_t\rho_i\sqrt{a^2 + b^2} \tag{26}$$

$$\frac{(ln\pi_t - \rho\pi_t) - \sigma_i\varepsilon\sqrt{dt}}{t} = \mu \tag{27}$$

The time series forecast is used to forecast the profits of each enterprise, and a certain amount of skew is given to the declining enterprises.

4.2 Final Decision Model Solution.

The recessionary 2020 expected profits and their profit growth rate are predicted using the time series [8]:

$$y_2 = a * e^{bt} \tag{28}$$

With the exception of companies with an A rating that are both growing or declining, a flat 0.0.4% interest rate is allocated according to expected profit growth. $\mu < r$ Risk-free interest rate, giving 0.04%–0.15% interest rate allocation [9].

Based on the sales peak supplement model, we give special care to declining enterprises, and give the maximum allowable loan to declining enterprises between 100,000 and 1 million yuan. Some of the results are shown in Table 8:

Table 8. Expected profit and loan amount of the enterprise

The name of firm	Expected profit	Loanable limit
*** Building Materials Co., LTD	7802283.3	-26325605.94
......	......	......
*** Trading Co., LTD	7652525.8	837835.9
*** Bidding Agency LTD	1161792.783	943178.4

5 Summary

We use k-means cluster analysis to classify enterprises and avoid the unreasonable situation of borrowing. We have taken into account the epidemic factors in terms of chaos and probability in the prediction analysis, but it does not destroy the balance of the original model for risk assessment, and the solution through machine learning can be further extended to the dynamic solution machine learning and other aspects.

Acknowledgments. This work is supported by 2022 Provincial Student Innovation Training Project of Ministry of Education of P.R.C(S202111078075).

References

1. Wu, L.: Research on credit risk management of Commercial Banks based on Logistic Regression Model. Harbin Institute of Technology (2007)
2. Luo, F., Chen, X.: Credit risk assessment and application of personal small loan based on Logistic regression model
3. Xie, W.: Variable screening method of Spearman correlation coefficient . Beijing University of Technology (2015); Theory Pract. Finance Econ. **38**(01), 30–35 (2017)
4. Guo, L.: Multivariate Distribution Goodness of Fit Test and its application. North China Electric Power University (2014). (in Chinese)
5. Liu, J.: Research on RQ Capital "futures + insurance" risk management model. Jiangxi University of Finance and Economics (2019)
6. Dang, X.L.: Research on Credit Loan risk Assessment of Enterprises based on Analytic Hierarchy Process. Shandong University of Finance and Economics (2016)
7. Zhang, Y., Fan, G.Z.: Credit crunch, firm value and optimal lending rate. Econ. Res. **51**(06), 71–82 (2016)
8. Zhu, W., Zhang, P., Li, P., Wang, Z.: The dilemma of micro, small and medium-sized enterprises and the improvement of policy efficiency under the impact of epidemic: an analysis based on two national questionnaire surveys. Manag. World **36**(04), 13–26 (2020)

Research on the Ordering and Transportation of Raw Materials for Production Enterprises

Jiachong Zheng, Yifeng Zhang, Yuhang Duan, Yang Mao, and Jiexia Yang[✉]

School of Mathematics and Information Science, Guangzhou University, Guangzhou , China
yjxgd@gzhu.edu.cn

Abstract. This paper puts forward solutions to the raw material ordering and transportation problems of a building and decoration plate production enterprise. For the importance of suppliers, Conduct a quantitative analysis of the supply characteristics, Establish a model of the importance evaluation index, The entropy weight method was used to empower each index, Using the TOPSIS method to quantify the importance of suppliers to the enterprise; For the determination of the transport scheme with the least loss, To solve the quantitative relationship between supplier supply materials and production demand using Monte Carlo method based on 0–1 planning, Solve the most economical weekly raw material ordering scheme based on the single-target planning model, Finally, select the index weight ranking high transporter to develop the transport plan with the least loss; For the formulation of ordering plans and transfer plans after reducing production costs, On the basis of the previous question, the multi-objective planning model is used to solve it; Regarding the determination of the ordering and transport programs for the next 24 weeks after the technical modification, By solving the maximum weekly capacity and ordering and transfer scheme of the enterprise, And compared with the fixed capacity of the increase in production capacity.

Keywords: 0–1 planning · entropy weight method · multi-objective planning

1 Introduction

In the whole production and operation process of the enterprise, the raw material ordering activity, as the starting point of the enterprise logistics and capital flow, has a great impact on the subsequent activities of the enterprise. With the acceleration of the trend of economic globalization and the in-depth development of supply chain theory, raw material ordering link itself and its role have gradually changed, for raw material ordering itself, with the rapid development of global procurement and third party logistics, transport transport optimization of the performance of raw material order. It is of great practical significance to consider the selection of [1] in procurement decisions.

© The Author(s), under exclusive license to Springer Nature Singapore Pte Ltd. 2024
B.-Y. Cao et al. (Eds.): ICFIE 2022, LNDECT 207, pp. 119–128, 2024.
https://doi.org/10.1007/978-981-97-2891-6_9

The raw materials used by A building and decorative plate production enterprise are mainly wood fiber and other vegetable fiber materials, which can be divided into three types: A, B and C. The enterprise production by 48 weeks a year, need to advance 24 weeks of raw material ordering and transfer plan, that is, according to the capacity requirements to order the raw material supplier (called "supplier") and the corresponding weekly raw material order quantity (called "quantity"), determine the third party logistics company (called "transshipment") and entrust it to the supplier weekly raw material supply quantity (called "quantity") transfer to the enterprise warehouse [2].

The main purpose of this paper is to solve the corresponding needs of the above enterprises, and the relevant data is taken from the topic C of the 2021 Higher Education Cup National College Students Mathematical Modeling Competition. In order to reduce the influence of uncontrollable factors on the model and make the model more universal, the following assumptions are proposed:

1. Suppose that the unit price of raw material C is 1 and that of raw materials A and B is 1.2 and 1.1, respectively.
2. When evaluating and making suppliers, the differences in product quality, service life, service and pollution degree to the environment between different suppliers are not considered.
3. Suppose the company will produce for 48 weeks a year.
4. Suppose a supplier transports weekly raw materials by a transporter. The enterprise buys all the raw materials actually provided by the supplier.
5. Assuming that the initial raw material storage capacity is 0

2 Quantize the Supply Characteristics of Suppliers, and Establish a Mathematical Model Reflecting the Importance of Ensuring the Production of Enterprises

2.1 Model Building

In order to determine the 50 most important suppliers to the enterprise, this paper quantifies the importance of suppliers according to the characteristic indicators of suppliers, establishes the TOPSIS mathematical model based on entropy weight method, and finally gives the importance of 402 suppliers [3]. First of all, from the order quantity of the enterprise and the characteristics of the supplier. This paper extracts a series of indicators of suppliers, and quantitatively analyzes the supply characteristics of suppliers from three aspects of credit, supply capacity and supply stability.

The specific idea is shown in the following figure (Fig. 1):

Step 1. Define vendor credibility.

The reputation of suppliers is an important indicator of long-term cooperation with them. The reputation reflects the contract spirit of suppliers. The better the reputation, the more likely the supplier is to become an important supplier of the enterprise. Credit rating is measured by the following Formula (1):

$$x = \frac{N_w}{N_{sum}} \tag{1}$$

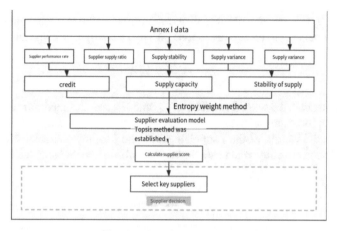

Fig. 1. Modeling flow chart

X is the default rate of the supplier and the number of defaults of the supplier N_w. It defines the order quantity of the enterprise $d_1 > 0$ to the supplier w and the supplier supply quantity $d_2 = 0$ as the supplier default, and the total week N_{sum} is 240 weeks.

Step 2. Define the supplier supply capacity.

The supply capacity of the suppliers is one of the core competitiveness of the suppliers, which reflects the production capacity of the suppliers. Excellent supply capacity is one of the consideration indicators for the long-term cooperation of the enterprises. It is measured by the following Formula (2):

$$g_i = \frac{N_g}{N_d} \tag{2}$$

For the supplier supply capacity g_i, for the supplier N_g, order for the enterprise N_d, defined the ratio of N_g and N_d as the supplier's supply capacity, it is more than 0, because the enterprise for supplier supply all acquisition strategy, considering the loss of goods, and transportation costs, so the i value of g_i is closer to 1 represents the supplier's supply ability more excellent.

Step 3. Define the supplier supply stability.

The supply stability of suppliers is a major indicator that ordering enterprises need to consider, which determines the temporary or stability of the cooperation between supply and demand. It is measured by the following Formula (3):

$$W_i = \frac{\sum_{i=1}^{n} \left(N_{gi} - \overline{N_g} \right)^2}{n} \tag{3}$$

N_{gi} is the weekly supply of each supplier, $\overline{N_g}$ is the average supply of each supplier for 240 weeks, and 402, which is the total number of suppliers. The smaller W_i reflects the higher supply stability of the supplier [4].

Step4. Entropy weight method calculates weights.

First of all, this paper will normalize and normalize the extracted supplier supply characteristic data [5] to ensure the non-negativity of the data, see Formula (4).

$$Z_{ij} = \frac{X_{ij} - X_{min}}{X_{max} - X_{min}} \tag{4}$$

Z_{ij} is the normalized treated variables, X_{max} and X_{min} are the maximum and minimum values of each index, respectively.

Calculate the weight of the i supplier under the j supply characteristic index, and regard it as the probability when calculating the information entropy, see Formula (5).

$$P_{ij} = \frac{Z_{ij}}{\sum_{i=1}^{n} Z_{ij}} \tag{5}$$

The information entropy of the j th supply characteristic index e_j is calculated, and the corresponding information utility value d_j is calculated. The reason for the conversion here is that the larger the information entropy, the less information represents the supply characteristic index, so introducing the information utility value $d_j(d_j = 1 - e_j)$ can measure the information positively, see Formula (6).

$$e_j = -\frac{1}{\ln n} \sum_{n}^{i=1} P_{ij} \ln P_{ij} \tag{6}$$

The entropy weight of each supply characteristic index is finally normalized W_i, see Formula (7).

$$W_i = \frac{d_j}{\sum_{j=1}^{m} dj} \tag{7}$$

Finally, the weight of each index is brought into TOPSIS, and the distance between the i th supplier and the maximum minimum value is calculated, and then the importance score of the i supplier is calculated [5].

Step 5. TOPSIS Methods to the samples.

For this problem, this paper constructs a supplier that achieves the optimal ideal index, and then calculates the actual proximity between each supplier and the ideal supplier. The closer the supplier is, the more important the supplier is to the enterprise.

Find out that each column is the maximum value of each index, written down as the component vector $Z_i^+ (i = 1, 2, \ldots, m)$. This vector represents the ideal supplier, and also finds the minimum value of each column of indicators, denoted as the component vector Z_i^-, see Formula (8).

$$Z^+ = \{Z_1^+, Z_2^+, \cdots, Z_m^+\}, Z^- = \{Z_1^-, Z_2^-, \cdots, Z_m^-\} \tag{8}$$

Define the distance between the i-first sample and the ideal target as D_i^+, the least ideal target as D_i^- and the calculation Formulas (9) are as follow.

$$D_i^+ = \sqrt{\sum_{j=1}^{m}(Z_i^+ - Z_{ij})^2}, D_i^- = \sqrt{\sum_{j=1}^{m}(Z_i^- - Z_{ij})^2} \tag{9}$$

Define the score of the i th enterprise, see the Formula (10).

$$S_i = \frac{D_i^-}{D_i^- + D_i^+}$$ (10)

The closer S_i is to 1, the closer the supplier i is to the ideal target, the more important the supplier is. Otherwise, the closer S_i is to 0, the farther the supplier is from the ideal target, the less important the supplier is. Identify the 50 most important suppliers based on their score [6].

2.2 The Solution of the Model

In the model built in this paper, the supplier default rate, supply capacity and supply stability are very small index, intermediate index and very small index respectively. First of all, the three indicators need to be maximized. The method of transforming very small indicators into very large indicators is the following Formula (11).

$$X' = max - X_i$$ (11)

The method of transforming the intermediate index into a very large index is the following Formula (12).

$$M = max\{|X_i - X_{best}|\}, \tilde{X}_i = 1 - \frac{|X_i - X_{best}|}{M}$$ (12)

The original matrix of each index maximizes the standard matrix. The weights corresponding to the three indexes are calculated by the entropy weight method as shown in the table below (Table 1).

Table 1. The corresponding weight table of each index

index	Vendor default rate	Supply capacity	Supply stability
weight	0.037524	0.192562	0.769914

Define the maximum value and the minimum are the following Formula (13).

$$Z^+ = \{0.031212, 0.115877, 0.399127\} Z^- = \{-0.06859, 0, 0.001722\}$$ (13)

By calculating the distance between each supplier and the most ideal and least ideal suppliers, we get the score table of each supplier, and ranking the scores to get the final 50 most important suppliers is shown in Table 2:

Table 2. The Top40 Supplier Table

140	229	361	108	151	201	340	282	139	275
308	329	330	131	356	268	306	348	194	352
143	307	395	126	247	37	374	284	365	31
338	40	364	55	367	346	86	294	80	244

3   On the Basis of the Above Part, Specify the Best Ordering and Transshipment Programs for the Enterprise for the Next 24 weeks

3.1   Model Building

Step 1. 0–1 programming model.

In this paper, the volume of class A, B and C raw materials is converted into the corresponding capacity volume to determine the subsequent constraints on capacity. In order to select the number of suppliers to meet the production requirements, the 0–1 planning model is established. After the analysis, the objective function was determined as $\min \sum_{i=1}^{50} \omega_i$.

The coefficient corresponding to the first i supplier supply is 0 and 1. The constraints are given by Formula (14).

$$\text{s.t} \begin{cases} 0 \leq min \sum \omega_i \leq 50 (i = 1, 2, \ldots 50) \\ \omega_i \in \{0, 1\} \\ \sum_i \omega_i S_{ij} \geq 2*2.82 (j = 1) \\ \sum_i \omega_i S_{ij} \geq 2.82 (j = 2, 3, \ldots 240) \end{cases} \tag{14}$$

Among them, the supply quantity of the first supplier in week j is the product sum of the weekly supply quantity of the supplier. For the raw material inventory that needs to meet the production demand of two weeks in the first week, the raw material inventory only needs to meet the production demand of one week every week.

Step 2. Determine the most economical raw material ordering scheme by characteristic analysis.

First, 32 suppliers were classified according to their supply characteristics, divided into three categories: periodic, fluctuating and stable. The next 24 weeks of supply are forecast according to the characteristics of each category. Then the cost performance analysis of three raw materials A, B and C, the calculation method is: $\alpha_i = L_i \times M_i$

Among them, the cost performance of raw materials in class I is the volume of class I raw materials per cubic meter, and the unit price of raw materials in class I. For calculation convenience, the unit price of raw material C is set to be 1, so the unit price of raw materials A and B is 1.2 and 1.1 respectively.

In order to achieve the most economical raw material ordering program, the weekly order quantity of various raw materials for the next 24 weeks is determined according to the cost performance of raw materials, which is the most economical raw material ordering program.

Step 3. Transport protocol with the least loss.

Data analysis of 8 transporters obtained the transport quality index of each transporter, see Formula (15).

$$ZS_i = JZ_j^* QZ_i + FC_i^* QZ_j \tag{15}$$

This includes the transport quality index of the i transporter, the mean of the loss rate of the previous 240 weeks, the variance of the previous 240 weeks, and the weights corresponding to the mean and variance.

According to the transport quality index, the 8 transporters are ranked. When selecting the transporters, the top transporters can complete the transport plan with the least loss.

3.2 The Solution of the Model

Step 1. Minimum vendor selection for the solution.

Construct the weekly supply volume expression: $\sum_i \omega_i S_{ij}$

Including the coefficient corresponding to each supplier's supply, whose values are 0 and 1, corresponding to the 0–1 plan. The supply quantity of the first supplier is the weekly supply quantity of the supplier. For the raw material inventory that needs to meet the production demand of two weeks in the first week, the raw material inventory that only needs to meet the production demand of one week each week.

After 10 rounds of million-level simulation search, it is finally concluded that at least 24 suppliers are required to supply raw materials to meet the production demand, as shown in Table 3.

Table 3. Final 24 supplier tables

identifier	S229	S361	S108	S151	S201	S340	S282	S139
identifier	S275	S308	S329	S330	S131	S356	S306	S348
identifier	S194	S352	S143	S307	S395	S126	S374	S284

Step2. The most economical raw material ordering scheme solution.
Cost performance of raw material A/B/C,see Formula (16).

$$\alpha_A = 0.6^*1.2 = 0.72, \alpha_B = 0.66^*1.1 = 0.726, \alpha_C = 0.72^*1 = 0.72 \qquad (16)$$

It can be seen that raw materials A and C are equivalent in terms of cost performance and are better than raw materials B. Therefore, for the sake of economy, raw materials A and C should be limited, and raw material B should be ordered when they cannot meet the production demand. In this paper, the weekly raw material order program [7] is obtained, with the first week as an example is shown in Table 4.

Table 4. The most economical order program schedule for the first week

identifier	S005	S006	S008	S009	S015	S018	S021	S022	S023	S024
order amount	115	282	2	6390	55	169	201	1	945	54

The transport scheme with the least loss is solved by the single-target planning model, with the three suppliers before week 13 as shown in Table5.

Table 5. Week 13 transport protocol table

supplier ID	Week 13							
	T1	T2	T3	T4	T5	T6	T7	T8
S004	0	0	802	0	0	0	0	0
S005	0	0	1049	0	0	0	0	0
S006	0	0	0	0	0	108	0	0

4 Develop New Ordering Plan and Transfer Plan, Reduce the Cost of Transfer and Storage for Enterprises, and Make the Transfer Loss Rate of Transport Ters as Little as Possible

4.1 Model Building

Just change it based on the content of title 3, which is A multi-objective planning problem. The objective function is to make the largest amount of raw material A, the minimum amount of C, the minimum transfer and storage cost and the minimum loss rate[8]. The constraint is to meet the production requirements. Through multi-objective planning solution, the new ordering scheme and transport cost can be solved.

The objective function are as the following formulas (17), (18).

$$min \sum_{i=1}^{32} \sum_{j=1}^{24} C_i X_{ij} \qquad (17)$$

$$max \sum\nolimits_{A} X_{ij}, s.t. L_{ij} \leq X_{ij} \leq U_{ij} \tag{18}$$

Among them, X_{ij} is the quantity of A raw materials, is the lower limit of A raw materials and the upper limit of the quantity of A raw materials.

4.2 The Solution of the Model

To solve the multi-objective planning model, we get a new raw material ordering scheme [9]. The ordering protocol is as follows, taking the week 23 ordering protocol as an example is shown in Table 6.

Table 6. Week 23 Ordering Programme Form

identifier	S005	S008	S018	S024	S229	S275	S329
order amount	5280	1052	5253.559	641.4519	1447.205	572.49	417.0804

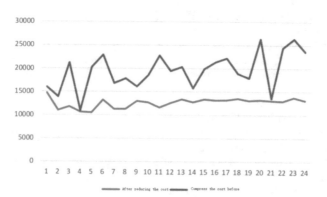

Fig. 2. Cost compression, before and after the comparison chart

As shown in Fig. 2, the strategy of purchasing class A as much as possible and class C raw materials as little as possible has A significant effect on cost control [10].

5 Summary and Outlook

In this paper, the entropy weight method and TOPSIS method are used to score the suppliers, rather than the hierarchical analysis method. The results are more objective and accurate, and various methods are used to test the results. The results show that the method is robust and has high accuracy. For problem 2, the model of 0–1 planning is established, and the model is solved. This paper not only uses the traditional planning method, but also uses the Monte Carlo algorithm, and skillfully classifies the types of suppliers, so as to solve problem 2. For problem three and problem four, continue to use the algorithm of problem two. The difference is that problem three is the progression of problem two, while problem four is the backward solution of problem two. For the whole problem, this paper uses novel methods and tests the model multiple times, so the results of this paper have high accuracy.

Acknowledgements. This work is supported by 2022 School Level Student Innovation Training Project of Guangzhou University (xj202211078265).

References

1. Ding, X.: Research on optimization of purchasing decision considering multiple transportation schemes. Shanghai Jiaotong University (2012)
2. Ge, J.: Research Status of Supplier Selection Decision Based on Quantitative Model J. Nantong Shipping Vocat. Tech. College **20**(02), 42–46 (2021)
3. Liu, B.: Research on Supplier Evaluation and Selection Improvement of A Company based on AHP and TOPSIS. Shanghai International Studies University (2021)
4. Wang, S.: Research on the Selection and evaluation index system of third-party logistics suppliers. China Storage Trans. **04**, 118–119 (2021)
5. Guan, Z., Zhou, B., Ma, Q.: Research on supplier selection and order quantity allocation under multi-product procurement Comput. Integr. Manuf. Syst. **11**, 1626–1632 (2005)
6. Wang, L., Chen, C., Zhang, J., Yi, J.: Supplier optimization and order quantity allocation model based on improved particle swarm optimization. China Manag. Sci. **17**(06), 98–103 (2009)
7. Han, Y.: Research on Multi-cycle Purchasing Quantity Allocation under Uncertain Environment. Northeastern University (2008)
8. Wang, Y.: Supply Chain Procurement Optimization Model under Fuzzy Random Demand Environment. Northeastern University (2009)
9. Chen, M.: Research on Raw Material Ordering Model Considering Raw Material Price Uncertainty. Northeastern University (2014)
10. LNCS Homepage. http://www.springer.com/lncs

Research on Cantonese Cultural and Creative Product Design Strategy Based on Fuzzy Kano Model

Xingyi Zhong[✉]

College of Architectural Design and Art, Guangdong Construction Polytechnic, Guangzhou,
China
xy_zhong2333@163.com

Abstract. From the perspective of consumers, the paper identifies the design requirement attributes of Cantonese cultural and creative products based on the fuzzy Kano model. On the basis of the preliminary in-depth interview method, the fuzzy Kano model is constructed to classify consumer needs according to five demand attributes. According to the results of the fuzzy Kano model, the Better-Worse coefficients for different types of needs are calculated. Combined with the four quadrant model to rank the importance, the paper proposed corresponding design strategies for Cantonese cultural and creative products. The aim is to provide a theoretical basis and design basis for Cantonese cultural and creative product design and other tourism cultural and creative product designs.

Keywords: Cantonese Cultural and creative product · fuzzy Kano model ·
design strategy

1 Introduction

With the increasing attention paid to cultural industries nationwide, Cantonese culture, Cantonese cultural industry, and Cantonese cultural products have entered the market and gained widespread attention gradually. Cantonese cultural and creative products embody cultural inheritance and the spirit of the times. However, in the actual market, the overall quality of the design is not high. There is a lack of in-depth exploration of consumer needs, which fails to meet consumer expectations [1].

In related literature, Guo Chentao and Zhong Lin explored the Cantonese cultural and creative design method by referring to successful cases of cultural and creative products at home and abroad [2]. Ji Wenrui and Li Jing conducted Cantonese cultural and creative product design practice based on the nostalgic emotions [3]. Li Sina summarized and analyzed the characteristic elements of Cantonese architecture and applied them to cultural and creative product design [4]. Wang Haoxu, Liang Shumin, and others combined design cases to propose thinking about integrating Cantonese folk art into the design of Guangzhou tourism cultural and creative products [5]. The above literature mainly discusses Cantonese cultural and creative product design from the perspective of designers. There is a lack of research from the perspective of consumers.

B.-Y. Cao et al. (Eds.): ICFIE 2022, LNDECT 207, pp. 129–139, 2024.
https://doi.org/10.1007/978-981-97-2891-6_10

In the era of experience-oriented design, designers need to start from the perspective of consumers and explore their design needs in depth, so as to allow products to occupy a place in the market. Therefore, the paper conducts research on strategies for Cantonese cultural and creative product design from the perspective of consumers. The research method is based on the fuzzy Kano model. The aim of the paper is to guide the design direction of Cantonese cultural and creative products and enhance consumers' emotional experience.

2   Related Theory

2.1   Cantonese Cultural and Creative Products

Cantonese cultural and creative products are the sum of industrial products and services that combine the unique cultural characteristics of Cantonese with creative design based on the tourism resources of the Cantonese region, which have the market value and meet consumer needs [6, 7]. The paper focuses on the industrial products of Cantonese cultural and creative products for research. Currently, consumers have a low desire to purchase Cantonese cultural and creative products [1]. To develop Cantonese cultural and creative products that meet consumer expectations, it is necessary to start from the root and explore consumer needs for Cantonese cultural and creative products.

2.2   Fuzzy Kano Model

The Fuzzy Kano model integrates the fuzzy comprehensive evaluation method into the traditional Kano model. It can transform qualitative evaluation into quantitative evaluation based on the theory of fuzzy mathematics, which has the characteristics of clear results and strong systematic. It can solve the problem that the uncertainty caused by the complex and variable consumer psychology better than the traditional Kano model [8].

The Kano model divides user needs into five categories: Must-be requirements (M), One-dimensional requirements (O), Attractive requirements (A), Indifferent requirements (I), and Reverse requirements (R) [9]. Must-be requirements are essential product requirements that, if not satisfied, will result in a significant decrease in user satisfaction. If these requirements are met, user satisfaction is not affected greatly. One-dimensional requirements are user expectations. User satisfaction increases as the degree of these requirements satisfied increases. Attractive requirements are needs that users are not aware of in advance. If these needs are satisfied, user satisfaction will increase significantly. If these needs are not met, user satisfaction is not affected greatly. Indifferent requirements are needs that users do not care about, and whether they are satisfied or not will not affect user satisfaction. Reverse requirements refer to needs that users do not need, and if the corresponding functionality is provided, it will reduce user satisfaction.

The traditional Kano model presents users with both positive and negative questions for each given product design demand. Users are required to choose among five qualitative options: 'liked', 'must be', 'neutral', 'live with' and 'dislike'. The fuzzy Kano model requires users to score each of the five qualitative options between 0 and 1, with a total score of 1. By assigning qualitative questions to numerical scores, user needs can be captured more accurately. Then, calculate the fuzzy relationship matrix of positive and negative questions. Finally, the need membership degree vector is calculated referring to the Kano model evaluation table (Table 1). In Table 1, Q is questionable result, usually indicating that user has misunderstood or puzzle with the question. In this paper, users equal to consumers.

Table 1. Kano model evaluation

functional	dysfunctional				
	like	must be	neutral	live with	dislike
like	Q	A	A	A	O
must be	R	I	I	I	M
neutral	R	I	I	I	M
live with	R	I	I	I	M
dislike	R	R	R	R	Q

3 Research Method and Processes

3.1 Requirements Acquisition and Organization

Firstly, in-depth interviews is conducted to investigate the audience of Cantonese cultural and creative products. The core consumer group of Cantonese cultural and creative products is in-service workers with 1–9 years of work experience and university students who have not participated in work but have certain economic strength [10]. Twelve consumers were interviewed, and the data structure is summarized in Table 2.

Table 2. User data structure

item	content	number of people	proportion
gender	male	7	58.33%
	female	5	41.67%
age distribution	25–30	10	83.33%
	31–35	2	16.67%
years of work	4–9	5	41.67%
	1–3	6	50.00%

(*continued*)

Table 2. (*continued*)

item	content	number of people	proportion
	0	1	8.33%
education	undergraduate	4	33.33&
	master	8	66.67%
related experience	native	7	58.33%
	working in Guangzhou	3	25.00%
	traveled to Guangzhou	2	16.67%

The interview mainly revolves around consumers' views on Cantonese cultural and creative products. User journey map is used to sort out the demands for Cantonese cultural and creative products according to the process of consumers' consumption of Cantonese cultural and creative products (Fig. 1).

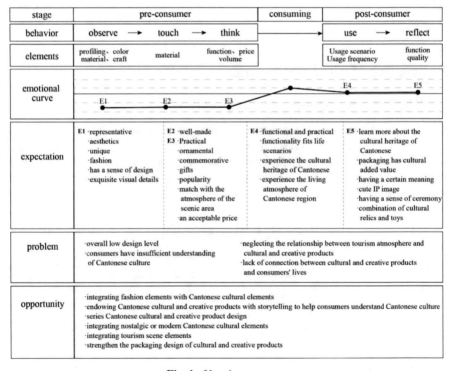

Fig. 1. User journey map

From the Fig. 1, it can be seen that the overall satisfaction of consumers with the design of Cantonese cultural and creative products is relatively low. Based on the process of product design elements and consumer behavior, redundant and similar demand points are eliminated. Finally, 20 product needs are screened and summarized (Table 3).

Table 3. Product needs summarize

culture element	C1	incorporating Cantonese elements
	C2	containing a certain Cantonese cultural story
	C3	containing a certain Cantonese cultural customs
	C4	a certain craft element that embodies Cantonese culture
	C5	reflect a certain historical characteristic of Cantonese culture
	C6	reflect the characteristics of modern life in the Cantonese region
look	L1	beautiful
	L2	fashion
	L3	cute image
	L4	exquisite workmanship
function	F1	practicability
	F2	commemorative
	F3	close to life
	F4	combination of cultural relics and toys
	F5	can be directly used in scenic spots
	F6	collection of check-in attractions to form a series of products
package	P1	integrating the characteristics of Cantonese culture
	P2	suitable for gift giving
	P3	easy to carry
brand	B1	has a certain level of popularity

3.2 Requirement Attributes Based on Fuzzy Kano Model

Based on the Kano model for questionnaire design, each design demand is questioned in both positive and negative directions. Using online questionnaires, a total of 102 questionnaires were distributed, with 79 valid questionnaires.

Taking C4 demand as an example, the corresponding positive and negative questionnaire data are shown in Table 4. If Cantonese cultural products reflect a certain craft element that embodies Cantonese culture, the corresponding matrix is $X = [0.2\ 0.4\ 0.4\ 0\ 0]$. If Cantonese cultural products do not reflect a certain craft element of Cantonese culture, the corresponding matrix is $Y = [0\ 0\ 0.2\ 0\ 0.8]$. The generated fuzzy relationship matrix formula is as follows:

$$S = X^T Y = \begin{bmatrix} 0 & 0 & 0.04 & 0 & 0.16 \\ 0 & 0 & 0.08 & 0 & 0.32 \\ 0 & 0 & 0.08 & 0 & 0.32 \\ 0 & 0 & 0 & 0 & 0 \\ 0 & 0 & 0 & 0 & 0 \end{bmatrix}$$

Table 4. Questionnaire data of C4 based on fuzzy Kano model

demand		like	must be	neutral	live with	dislike
C4	functional	0.2	0.4	0.4	0	0
	dysfunctional	0	0	0.2	0	0.8

Match the values in the matrix one by one according to the Table 3, add the same type of requirements, and obtain the requirement membership vector:

$$T = \left(\frac{0.64}{M}, \frac{0.16}{O}, \frac{0.04}{A}, \frac{0.16}{I}, \frac{0}{R} \right)$$

Introduce a confidence level of 0.4 [8]. When the element value in the demand membership vector T is ≥ 0.4, the value is taken as 1, otherwise it is taken as 0, resulting in $T = (1, 0, 0, 0, 0)$, which indicates that the attribute of C4 from the consumer belongs to Must-be requirement (M).

Calculate the fuzzy relationship matrix for each requirement in each questionnaire, and summarize it finally to obtain the fuzzy Kano model requirement classification of Cantonese cultural and creative products (Table 5). When the cumulative number is equal, the requirement attributes are determined based on the priority M, O, A, I, and R of the requirement attributes [8].

Table 5. The requirement attributes of Cantonese cultural and creative product based on the fuzzy Kano model

need	number of questionnaires						attributes
	M	O	A	I	R	Q	
C1	5	20	29	21	0	4	A
C2	10	18	29	18	0	4	A
C3	10	20	29	19	0	1	A

(*continued*)

Table 5. (*continued*)

need	number of questionnaires						attributes
	M	O	A	I	R	Q	
C4	7	23	26	23	0	0	A
C5	13	21	22	23	0	0	I
C6	4	23	20	31	0	1	I
L1	7	29	21	22	0	0	O
L2	4	16	25	33	0	1	I
L3	2	7	27	43	0	0	I
L4	14	27	22	15	1	0	O
F1	4	16	34	25	0	0	A
F2	9	22	28	19	0	1	A
F3	5	17	33	24	0	0	A
F4	3	10	30	35	0	1	I
F5	7	10	30	30	1	1	A
F6	4	11	27	34	2	1	I
P1	3	21	31	24	0	0	A
P2	6	10	39	24	0	0	A
P3	3	28	27	21	0	0	O
B1	1	11	32	35	0	0	I

3.3 Analysis of Better-Worse Satisfaction Coefficient

Calculate the Better-Worse coefficients for each demand attribute by using the Better-Worse satisfaction coefficient formula. The formula for calculating the Better-Worse satisfaction coefficient is:

$$Better = (A + O)/(A + O + M + I)$$
$$Worse = -1(M + O)/(A + O + M + I)$$

Better coefficient is the satisfaction increase coefficient, while the Worse coefficient is the satisfaction decrease coefficient. If the absolute values of the satisfaction increase coefficient and satisfaction decrease coefficient of a certain design demand approach 0, the impact of the design demand on user satisfaction is small. On the contrary, if they approach 1, there will be a significant impact on user satisfaction [11]. Since I is an Indifferent attribute that does not affect user satisfaction, it is not included in this satisfaction calculation range [12]. Therefore, it is no need to calculate requirements C5, C6, L2, L3, F4, F6, and B1. The Better-Worse coefficients results for each requirement are shown in Table 6.

Table 6. Better-Worse coefficients for each demand attribute

Design demand	attribute	B	|W|
C1	A	0.653	0.333
C2	A	0.627	0.373
C3	A	0.628	0.385
C4	A	0.620	0.380
L1	O	0.633	0.456
L4	O	0.628	0.526
F1	A	0.633	0.253
F2	A	0.641	0.397
F3	A	0.633	0.278
F5	A	0.519	0.221
P1	A	0.658	0.304
P2	A	0.620	0.203
P3	O	0.696	0.392

In order to further investigate the importance of different design requirements shown in Table 6, a four quadrant diagram is constructed by using the mean as the coordinate origin position, |W| as the abscissa, and B as the ordinate (Fig. 2).

According to the four quadrant rule, the design requirements in the first quadrant are important and urgent. They are P3 (easy to carry), F2 (commemorative), and L1 (beautiful).

The design needs in the second quadrant are important but not urgent. They are P1 (packaging incorporates Cantonese cultural characteristics), C1 (incorporating Cantonese language elements), F1 (practicability), and F3 (close to life).

The design needs in the third quadrant are unimportant and non-urgent. They are P2 (suitable for gift giving) and F5 (can be directly used in scenic spots).

The design needs in the fourth quadrant are unimportant but urgent. They are C2 (containing a certain Cantonese cultural story), C3 (containing a certain Cantonese cultural customs), C4 (a certain craft element that embodies Cantonese culture), L4 (exquisite workmanship).

4   Design Strategy

4.1   Design Core

The core of Cantonese cultural and creative product design is: aesthetics, commemorative value, and portability. Aesthetics are the primary foundation for attracting consumers. The design of Cantonese cultural and creative products must conform to the aesthetic characteristics of contemporary mainstream consumer groups, and use the unique image and color to match the characteristics of Cantonese region as the cultural connotation

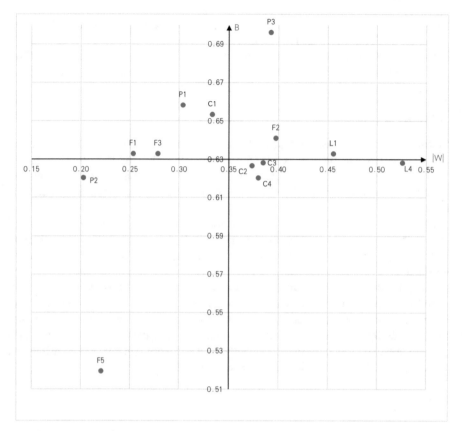

Fig. 2. Quadrant diagram of design requirements for Cantonese cultural and creative products

foundation so that the product can ensure its commemorative value function. Due to the fact that the purchase scenario of Cantonese cultural and creative products is mainly in tourist attractions, portability must also be a design factor that designers need to consider, such as the volume and weight of the product (including packaging), or offering online mailing services, in order to promote consumers' desire to purchase Cantonese cultural and creative products.

4.2 Application of Cantonese Language Elements

Cantonese is the iconic language of Cantonese culture, and the application of Cantonese elements in Cantonese cultural and creative products belongs to the user's expected needs. Although it is not the primary design point, it plays a very important role in attracting users. To incorporate Cantonese language elements, the following design methods can be considered:

1: Product name: Choose a Cantonese name for Cantonese cultural and creative products that reflects a certain story or custom of Cantonese culture, and use Cantonese name to evoke users' cultural perception of Cantonese cultural and creative products.

2: Form expression: Cantonese cultural and creative products can be integrated with Cantonese character images for design.

3: Packaging design: The text introduction of the product can be translated into Cantonese. In addition, the Cantonese can be integrated into packaging design, using packaging as a carrier to convey Cantonese culture.

4.3 Close to User Life

The functions of Cantonese cultural and creative products should be close to the user's life scene, which is actually consistent with practicality. Only by fully considering the usage scenarios of current consumers using Cantonese cultural and creative products can the value of Cantonese cultural and creative products as cultural transmission carriers be enhanced.

Design can be combined with daily necessities to increase the frequency of use of Cantonese cultural products. The use scenarios of gift giving also should be considered. If gift giving is done, the brand's popularity and packaging's sense of ceremony need to be valued.

After meeting the above three points, the indifferent needs of users should be considered to expand the possibility of more usage scenarios for Cantonese cultural and creative products, such as direct use in scenic spots, combination of toys and cultural relics, and collection of series products by checking in scenic spots. Diversified cultural and creative product design types and usage scenarios can enhance user interest, providing consumers with a richer cultural experience.

5 Concluding Remarks

The paper explores the design strategy of Cantonese cultural and creative products based on the fuzzy Kano model. In the early stage, user demand data was obtained through in-depth interviews. In the later stage, questionnaire design was conducted based on the fuzzy Kano model. Interval fuzzy numbers were used to investigate user demand, and the four quadrant model was used for further exploration of the demand results. Finally, a design strategy for Cantonese Culture was proposed. The design strategies obtained in this study have insufficient relevance to Cantonese culture, but provide a certain theoretical basis and strategic reference for the design of other tourism cultural and creative products. Subsequent research can be combined with specific cultural or design cases in Cantonese culture to assist in the design and development of Cantonese cultural and creative products comprehensively.

References

1. Zhong, X., Sun, T.: Research on consumers' cognition of Cantonese cultural products in the new era. Market Weekly **2**(1), 79–85 (2023)
2. Guo, C., Lin, Z.: Based on the basic elements of "emotion" and "meaning" in Cantonese culture. Appreciation **21**, 99–100 (2021)
3. Ji, W., Li, J.: Design of Cantonese cultural creative products under nostalgic representation. Packag. Eng. **42**(06), 306–313 (2021)

4. Li, S.: Application of Cantonese traditional architectural elements in the design of cultural and creative products. Fashion Color **419**(06), 17–19 (2021)
5. Wang, H., Liang, S., Lu, W., et al.: Research on the design of Guangzhou tourism cultural and creative products based on Cantonese folk art. Light Ind. Sci. Technol. **37**(03), 124–126 (2021)
6. Gao, Y., Xu, X.: Research on the cultural and creative tourism products design with full-effect experience: taking Fuchun mountains e-post on the poetry road as an example. Zhuangshi **12**, 101–106 (2022)
7. Zeng, H.: Research on the application of Guangzhou regional cultural elements in tourism cultural product design. Jiangxi Science & Technology Normal University, Nanchang (2022)
8. Meng, Q., He, L.: Fuzzy-Kano-based classification method and its application to quality attributes. Ind. Eng. J. **16**(03), 121–125 (2013)
9. Atlason, R.S., Stefansson, A.S., Wietz, M., et al.: A rapid Kano- based approach to identify optimal user segments. Res. Eng. Des. **29**(3), 459–467 (2018)
10. Hui, Q.: Purchasing intention analysis based on cultural product features from consumer classification. Zhejiang University, Hangzhou (2016)
11. Wang, W., Feng, R., Wei, T.: Emotional design of epidemic prevention products based on fuzzy Kano model. J. Mach. Des. **39**(10), 140–146 (2022)
12. Deng, W., Zhang, Z., Wang, Q.: Modular office storage product design based on fuzzy Kano model. Packag. Eng. **43**(14), 100–106 (2022)

Establishment and Application of Catalyst Optimization Model Based on Genetic Algorithm

Shuyi Wang[1], Weize Zhang[1], Ziqi Zhong[2], Yongtao Li[1], and Zhongyuan Peng[3(✉)]

[1] School of Mathematics and Information Science, Guangzhou University, Guangzhou, China
[2] School of Civil Engineering, Guangzhou University, Guangzhou, China
[3] Department of Social Sciences, Maoming Polytechnic, Maoming 525000, Guangdong, People's Republic of China
tntship@126.com

Abstract. This paper explores the optimal catalyst combination for ethanol preparation of c4 olefin to promote the production of c4 olefin. Firstly, grey correlation analysis was used to analyze the influence of the catalyst composition on ethanol conversion and C4 olefin selectivity. Secondly, with the highest yield of C4 olefin as the goal and temperature and experimental material dosage as the limiting conditions, a single objective optimization model was established. Finally, the traditional genetic algorithm is improved with the idea of multi-population, and the improved algorithm is used to solve the model. This method can solve the immature convergence problem of genetic algorithm. Compared with the traditional genetic algorithm, the result of this algorithm is more accurate.

Keywords: Grey Correlation · Genetic Algorithm · Programming model

1 Introduction

C4 olefin is an important chemical raw material in the production of chemical products and pharmaceutical intermediates. During the preparation process, the catalyst combination and temperature will affect the selectivity and yield of C4 olefin. The traditional production of C4 olefin mainly relies on fossil energy for preparation. In the contemporary era of energy shortage and worsening environmental pollution, it is of great value and significance to study the optimal catalyst combination for the preparation of C4 olefin by ethanol coupling.

The common catalyst combination used in the experiment of ethanol coupling preparation of C4 olefin is the combination of loading capacity, loading ratio and ethanol concentration. The main products are ethylene, olefin, acetaldehyde and fatty alcohol, etc.

At present, there are few researches on catalyst optimization model in China. Lu Shaopei [1] designed a Co/SiO2-HAP catalyst with both acid and base active sites on the surface, and provided the optimal catalyst combination from the perspective of reaction

© The Author(s), under exclusive license to Springer Nature Singapore Pte Ltd. 2024
B.-Y. Cao et al. (Eds.): ICFIE 2022, LNDECT 207, pp. 140–150, 2024.
https://doi.org/10.1007/978-981-97-2891-6_11

mechanism, which has great reference value in the verification of results. Wu Jiayun [2] et al. adopted the improved grey correlation algorithm to consider the difference in the absolute position of various parameters of the catalyst. The new algorithm has a high detection accuracy for olefin selectivity, while the effect on ethanol conversion is not obvious.

The data in this paper come from Question B of 2021 Higher Education Cup National Mathematical Contest in Modeling for College Students. In order to reduce the influence of uncontrollable factors on the model and make the model more general, the following hypotheses are proposed:

(1) The influence of human factors and experimental environment on experimental results should be excluded during the experiment.
(2) After the end of the experiment, the selectivity of each product remains unchanged, that is, the effect of reconversion of product into reactant is not considered.

2 The Effects of Different Catalyst Combinations on Ethanol Conversion and C4 Olefin Selectivity Were Analyzed Based on Grey Correlation

2.1 Model Preparation and Data Processing

Based on the above analysis, the influences of the three independent variables, such as discharge and charging mode, ratio and HAP, on the ethanol conversion and C4 olefin selectivity can be analyzed. Since the types of these three independent variables are few, using them as independent variables will reduce the effectiveness of the calculation results when applying grey correlation analysis. Therefore, we will use the catalyst mass, Co load, ethanol concentration and temperature with valid data as the comparative series of analysis and the reference series composed of ethanol conversion and the selectivity of C4 olefin for calculation.

Through literature review, it is known that under this catalyst and this reaction, when the reaction temperature is less than or equal to 350 °C, the main product is high carbon alcohol. When the reaction temperature is higher than 350 °C, the main product is C4 olefin [7]. At these two temperatures, various factors may have different effects on ethanol conversion and C4 olefin selectivity. Therefore, we consider to divide the attached data into two parts for analysis. One part will analyze the experiment at 250 °C–350 °C, and the other part will analyze the experiment at 400 °C and 450 °C.

2.2 Establish the Grey Relational Analysis Model

STEP1: Determine the analysis sequence.

First, we determine the data sequence that can reflect the behavior characteristics of the system -- the reference sequence, that is, determine the dependent variables. In this case, we have two reference sequences, respectively ethanol conversion and C4 olefin selectivity, whose values can reflect the influence of various factors on them.

Then determine the data sequence of factors that influence the behavior of the system - the comparison sequence. The data processing section has analyzed that the influencing factors are: catalyst mass, Co load, ethanol concentration and temperature.

STEP2: Preprocessing of variables

Before calculating the correlation degree, the original data should be dimensionless and the range of variables should be reduced to simplify the calculation. Set $X_i = (x_i(1), x_i(2), \ldots, x_i(n))$ as the behavior sequence of factors X_i:

$$X_i' = \frac{x_i(k)}{\frac{1}{n}\sum_{k=1}^{n} x_i(k)}, \ (k = 1, 2, \ldots, n) \tag{1}$$

STEP3: Calculation of correlation coefficient [3].

Set the parameter number column after data processing as $X_0' = (x_0'(1), x_0'(2), \ldots, x_0'(n))$, Set the comparison number column as $X_i' = (x_i'(1), x_i'(2), \ldots, x_i'(n))$, $(i = 0, 1, 2, 3, \ldots, m)$:

$$\Delta_i(k) = \left| x_0'(k) - x_i'(k) \right|, \ (k = 1, 2, \ldots, n) \tag{2}$$

The maximum and minimum difference between the two poles:

$$\begin{cases} a = \max_i \max_k \Delta_i(k) \\ b = \min_i \min_k \Delta_i(k) \end{cases} \tag{3}$$

Then, the grey correlation coefficient is set as $\gamma(k)$, which is calculated by the following formula:

$$\gamma_i(k) = \frac{a + \delta \cdot b}{\Delta_i(k) + \delta \cdot b} \tag{4}$$

resolution coefficient δ is used to weaken the influence of excessive b that distorts the correlation coefficient. This coefficient is introduced to improve the difference significance between the correlation coefficients, and 0.75 is adopted.

STEP4: Calculation and comparison of grey correlation degree.

The average of the correlation coefficients between the comparison series and the reference series in each period is used to quantitatively reflect the correlation degree of the two series. The calculation formula is as follows [6]:

$$\gamma_i = \frac{1}{n}\sum_{k=1}^{n} \gamma(k), \ (i = 0, 1, 2, \ldots, m) \tag{5}$$

2.3 Solution of Model

Since there are two reference sequences, we can carry out grey correlation analysis for ethanol conversion and C4 olefin selectivity respectively.

If the ethanol conversion rate is taken as the reference sequence for analysis and the data is preprocessed first, the method of data initialization is adopted for processing, as shown in the Table 1 and Table 2:

Table 1. Before data preprocessing

number	1	2	3	...
Ethanol conversion rate	2.07	5.85	14.97	...
Mass of catalyst	400	400	400	...
Load of CO	1	1	1	...
Concentration of ethanol	1.68	1.68	1.68	...
Temperature	250	275	300	...

Table 2. After data preprocessing

number	1	2	3	...
Ethanol conversion rate	1	2.83	7.24	...
Mass of catalyst	1	0.93	0.80	...
Load of CO	1	0.93	0.80	...
Concentration of ethanol	1	0.93	0.80	...
Temperature	1	0.93	0.80	...

Then, the processed data is substituted into Step 3, the two-stage maximum difference a and two-stage minimum difference b are obtained according to the sequence, and then substituted into the grey correlation coefficient and grey correlation degree model, the Table 3 is obtained:

Table 3. Grey correlation degree of influencing factors at 250 °C–350 °C

Grey correlation degree	Mass of catalyst	Load of CO	Concentration of ethanol	Temperature
Ethanol conversion rate	0.8153	0.8287	0.8231	0.8301
C4 olefin selectivity	0.6595	0.9618	0.9627	0.5142

However, within the temperature range of 350 °C–450 °C, the main product of the reaction is known to be C4 olefin. The influence of catalyst mass, Co loading capacity, ethanol concentration and temperature on ethanol conversion and selectivity of C4 olefin was analyzed by using grey correlation analysis, and reasonable data pretreatment was applied and substituted into the model. Similarly, the Table 4 is obtained:

According to the literature reviewed, the analysis results can be in line with the reality. In this temperature range, C4 olefin is not the main product, so the influence of

Table 4. Grey correlation degree of influencing factors at 350 °C–400 °C

Grey correlation degree	Mass of catalyst	Load of CO	Concentration of ethanol	Temperature
Ethanol conversion rate	0.8475	0.8377	0.8442	0.8836
C4 olefin selectivity	0.9329	0.8478	0.8942	0.8889

temperature on C4 olefin is less than that of the Co load and ethanol concentration of the catalyst.

The results show that in the temperature range from 350 °C to 450 °C also confirm the conclusions of the references, when the main product is C4 olefin at 350 °C–450 °C, the influence of temperature on C4 olefin is significantly higher than that at 250 °C–350 °C, which is similar to other influences of catalyst.

3 Establish Catalyst Combination Optimization Model

3.1 Preparation of the Model

To obtain the catalyst combination and temperature at the highest yield of C4 olefin, the objective function should be determined first. We can take the yield of C4 olefin y_0 as the dependent variable of the objective function. It can be expressed as a function of ethanol conversion y_1 multiplied by a function of C4 olefin selectivity y_2.

As for y_1 and y_2, we will use some factors and temperatures of catalyst combination with effective values to fit ethanol conversion and C4 olefin selectivity respectively. Let the catalyst mass be x_1, Co load be x_2, ethanol concentration be x_3, and temperature be x_4.

First of all, we used the rstool toolbox in MATLAB for fitting, and found that the fitting effect using the complete quadratic method in the toolbox was better than other fitting methods, but the effect could not reach the best. Therefore, we improve the function obtained by the complete quadratic fitting. If the cubic term is included in the function, the function model is as follows [4]:

$$y_k = \beta_1 + \beta_2 x_1 + \beta_3 x_2 + \beta_4 x_3 + \beta_5 x_4 + \beta_6 x_1^2 + \beta_7 x_1 x_2 + \beta_8 x_1 x_3 + \beta_9 x_1 x_4$$
$$+ \beta_{10} x_2^2 + \beta_{11} x_2 x_3 + \beta_{12} x_2 x_4 + \beta_{13} x_3^2 + \beta_{14} x_3 x_4 + \beta_{15} x_3^2 + \beta_{16} x_1^3$$
$$+ \beta_{17} x_2^3 + \beta_{18} x_3^3 + \beta_{19} x_4^3 \tag{6}$$

Then, the data of catalyst mass, Co loading capacity, ethanol concentration, temperature, ethanol conversion rate and selectivity of C4 olefin were substituted to obtain in the Table 5:

Table 5. The polynomial coefficients of the function

y_1	β_1	β_2	β_3	β_4	β_5	β_6	β_7	β_8	β_9	β_{10}
	824.86	0.04	-193	-117	-6.04	0.00	0.07	-0.02	0.00	63.57
	β_{11}	β_{12}	β_{13}	β_{14}	β_{15}	β_{16}	β_{17}	β_{18}	β_{19}	
	36.26	-0.01	51.16	-0.01	0.02	0.00	-8.67	-10.6	0.00	
y_2	β_1	β_2	β_3	β_4	β_5	β_6	β_7	β_8	β_9	β_{10}
	235.24	-0.14	239.83	78.67	-3.76	0.00	-0.05	0.12	0.00	-88.3
	β_{11}	β_{12}	β_{13}	β_{14}	β_{15}	β_{16}	β_{17}	β_{18}	β_{19}	
	-26.1	-0.02	-54.4	-0.02	0.01	0.00	11.42	14.06	0.00	

3.2 Build a Programming Model

According to the function of ethanol conversion y_1 and C4 olefin selectivity y_2 obtained above, the yield of C4 olefin y_0 can be described as [5]:

$$y_0 = y_1 \times y_2 \tag{7}$$

Then the objective function can be determined as:

$$Y = max\ y_0 \tag{8}$$

According to the specific experiment and the data in the attachment, the dependent variables can be determined as follows. Then the optimal model of catalyst combination and temperature at the highest yield of C4 olefin can be established comprehensively:

$$Y = max\ y_0$$

$$s.t. \begin{cases} 20 < x_1 < 400 \\ 0.5 < x_2 < 5 \\ 0.3 < x_3 < 2.1 \\ 250 < x_4 < 400 \end{cases} \tag{9}$$

If the catalyst combination and temperature are explored when the temperature is lower than 350°, the upper bound of the range can be adjusted to 350.

4 Improved Genetic Algorithm

4.1 Introduction to Genetic Algorithms

Genetic algorithm (GA) is a highly parallel, stochastic and adaptive probabilistic search algorithm for global optimization based on natural selection and evolution. For this optimization model, we consider using genetic algorithm to solve it.

Firstly, we set the influencing factors as four different chromosome factors: catalyst mass, Co load, ethanol concentration and temperature. We set the yield of C4 olefin as the fitness of all cases, and the highest yield of C4 olefin is the optimal fitness. Then, the optimal fitness is obtained by initializing the population, calculating the fitness, selecting, crossing and inheriting. The specific process is as Fig. 1:

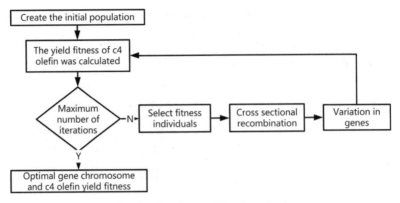

Fig. 1. Genetic algorithm flow chart

4.2 The Solution Result

The genetic algorithm was used to calculate the above optimization model, and the parameters of the genetic algorithm were defined. For example, the maximum number of iterations was set to 500, the crossover probability was set to 0.7, and the mutation probability was set to 0.05, etc. (See the code for specific parameters). The changes of optimization objectives and iteration times are shown as Fig. 2:

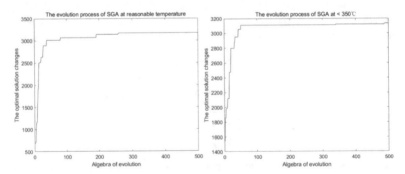

Fig. 2. Graph of SGA optimization objective changing with the number of iterations

It can be seen from the figure that with the increase of the number of iterations, the optimization fitness first increased rapidly and then tended to be flat. In the case of any reasonable temperature, after 110 iterations, the yield of C4 olefin reaches the optimal value. When the temperature is less than 350 °C, the yield of C4 olefin reaches an optimal value after 370 iterations. Better catalyst combination and temperature were obtained in the Table 6.

4.3 The Genetic Algorithm is Improved by Using the Idea of Multi-population

Since genetic algorithm may have the problem of early convergence, immature convergence is a phenomenon that cannot be ignored in genetic algorithm, which is mainly

Table 6. The result of genetic algorithm

Optimum value	C4 olefin selectivity	Mass of catalyst	Load of CO	Concentration of ethanol	Temperature
Any T	59.98%	399.47	1.44	2.09	398.79
T < 350 °C	31.62%	398.71	1.68	2.09	348.24

manifested in the fact that all individuals in the population tend to the same state and stop evolution. We make the following improvements:

(1) Introduce multiple populations to optimize search simultaneously; Different populations are assigned different control parameters to achieve different search purposes.
(2) All populations are connected by migration operators to realize the co-evolution of multiple populations; The optimal solution is the result of co-evolution of multiple populations.
(3) The artificial selection operator is used to save the optimal individuals in each evolutionary generation of various groups and serve as the basis for judging the convergence of the algorithm [8].

The specific method is as follows: first, set the same factor chromosome and fitness as the genetic algorithm, then make the crossover probability and mutation probability change within a certain range, and adjust the termination conditions to achieve the optimized genetic algorithm [8]. The flow chart of its algorithm is shown in the Fig. 3:

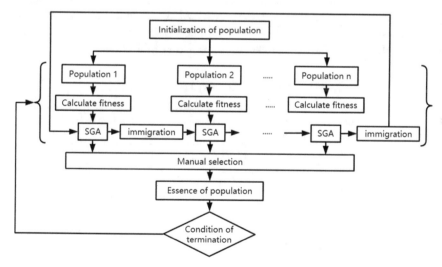

Fig. 3. Improved genetic algorithm flow chart

4.4   The Improved Solution Results

Through the improved genetic algorithm, the necessary algorithm parameters were set, and the range of crossover probability and mutation probability was set. The termination condition was set as "if the optimal value in the population does not change after 10 consecutive operations, it is considered that the optimal value has been found and the optimization is quit". The changes of optimization objectives and iteration times are shown as Fig. 4:

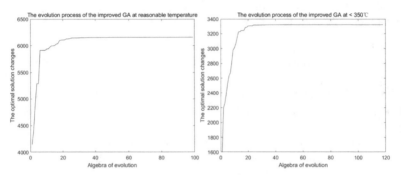

Fig. 4. Graph of the evolution process of the improved GA optimization objective changing with the number of iterations

It can be seen from the figure that with the increase of the number of iterations, the optimization fitness first increased rapidly and then tended to be flat. In the case of optimization at a reasonable temperature, the optimal yield of C4 olefin can be obtained by exiting the optimization at about 130 iterations. Also, the optimization situation when the temperature is lower than 350° is analyzed, and the optimization exits at about 120 iterations. And the results are obtained, as shown in the Table 7:

Table 7.   The result of improved genetic algorithm

Optimum value	C4 olefin selectivity	Mass of catalyst	Load of CO	Concentration of ethanol	Temperature
Any T	63.13%	400	1.44	2.1	400
T < 350°C	33.22%	399.9	1.74	2.1	350

4.5   Final Result Analysis

We will use the results of the improved multi-population genetic algorithm as the optimal result. Combined with the charging mode and ratio ratio scheme of the second question analysis and the result analysis of the third question, it is known that under the same experimental conditions, the charging mode I is used in the catalyst combination, the ratio ratio is 1: 1. When the catalyst weight is 400 mg, the Co load is 1.44wt %, the ethanol concentration is 2.1 ml/min, and the temperature is 400 °C, the yield of C4 olefin can reach the optimal 63.13%.

When the temperature is lower than 350 °C, the loading mode I is used in the catalyst combination, the ratio ratio is 1:1, the catalyst mass is 399.9 mg, the Co load is 1.74wt %, the ethanol concentration is 2.1 ml/min, and the temperature is 350 °C, the C4 olefin yield can reach the optimal 33.22%.

According to literature review, when the mixing ratio of Co/SiO2 and HAP is 1:1, the reaction temperature is 400 °C, and the Co loading is 1wt%, the catalyst performance is optimal [7]. The mixing method has little influence on the performance of catalyst [8]. It is roughly the same as the result obtained by us, and also reflects the accuracy and rationality of the result obtained by us.

5   Summary and Outlook

This paper uses function fitting analysis, control variable method, grey correlation analysis and other methods. The relationship between catalyst combination and temperature on selectivity of C4 olefin and conversion of ethanol was analyzed in detail. When solving the optimization model of C4 olefin yield, the idea of multi-population is used to improve the genetic algorithm, and the problem of premature convergence under the standard genetic algorithm is solved.

This paper also has some limitations. When the fitting function of ethanol conversion and C4 olefin selectivity was calculated in the third part, only the influences of catalyst components and temperature on ethanol conversion and C4 olefin selectivity were considered. In the model solving, it was assumed that the other products outside C4 olefin had no influence on the whole chemical reaction and C4 olefin selectivity, which may cause errors.

Acknowledgments. This work is supported by 2022 National Student Innovation Training Project of Ministry of Education of P.R.C (202211078148).

References

1. Lu, S.P.: Preparation of butanol and C_4 olefin by ethanol coupling. Dalian University of Technology (2018). (in Chinese)
2. Wu, J., Chen, G., Zhou, Z., Zhang, Y., Zou, Q., Su, P.: Study on improved grey relational degree algorithm by adjusting C_4 olefin preparation parameters. Chem. Res. Appl. **34**(10), 2545–2551. 202. (in Chinese)
3. Li, B.L.: Multiple Adaptive least squares curve fitting method and its application. Yangtze University (2014)
4. Wang, J., Qiu, J., Zhao, J., Dai, Q., Ma, Y.: Application of MATLAB CFTool in Graphic sediment grain size parameter calculation. Bull. Oceanol. Limnol. (04), 115–118 (2018)
5. Li, J.: Empirical Analysis and Demand Forecast of railway passenger volume. Hebei University of Economics and Business (2021)

6. MBA Think Tank Encyclopedia: Wiki.mbalib.com/wiki.2015.10 control variable method. (mba lib.com)
7. Tan, X., Deng, J.: Grey correlation analysis: a new method for multivariate statistical analysis. Stat. Res. (03), 46–48(1995). (in Chinese)
8. Cao, M.: Research on grey correlation analysis model and its application. Nanjing University of Aeronautics and Astronautics (2007)

Application and Optimization of Shape Adjustment Design Based on "FAST" Active Reflector Model

Hongwei Tang[1], Xianglong Li[1], and Wanying Wu[2(⊠)]

[1] School of Mathematics and Information Science, Guangzhou University, Guangzhou, China
[2] School of Civil Engineering, Guangzhou University, Guangzhou, China
3011864791@qq.com

Abstract. For the establishment of an ideal parabolic reflector adjustment model and the optimization of the reception effect of celestial electromagnetic waves reflected by the reflector, it is a major problem in the actual operation of the China Sky Eye FAST - 500 m aperture spherical radio telescope. This paper makes reasonable assumptions about the working state of the active reflector, determines the ideal paraboloid, and optimizes the mediation scheme of the reflector in combination with the specific situation, providing theoretical reference for the actual operation and testing.

The ideal paraboloid is determined when the object to be observed is directly above the reference sphere. First of all, the FAST reflection principle is obtained by focusing analysis of concave mirror and working principle of active reflection panel. Combined with the relative displacement of the apex of the paraboloid and the reference sphere, the expansion length of the apex of the paraboloid is obtained, thus enhancing the adjustability of the model. Finally, the parabolic equation in the three-dimensional plane is obtained by rotating the two-dimensional parabola around the z-axis.

When the object S to be observed is located, the ideal paraboloid is determined, and the reflective panel adjustment model is established. First, the virtual time angle coordinate system is established, and then the horizontal right angle coordinate system is established according to the position of FSAT. The paraboloid expression is obtained through the time angle coordinate and the horizontal right angle expression. Using the virtual coordinate system, the paraboloid whose vertex coincides with the center of the reflecting surface is defined as the reference paraboloid in this system, and the polar coordinate equation is obtained.

For determining the movable points covered by the paraboloid, we establish a sphere coverage model and calculate the Euclidean distance to determine the number of specific coverage points. Determine the polar angle corresponding to the point according to the position coordinates of the cable point within the effective aperture. Then calculate the extreme length of the cable point after changing its position to obtain the actual expansion.

Keywords: FAST reflective panel adjustment model · Paraboloid focusing principle · Principle of virtual transformation model

© The Author(s), under exclusive license to Springer Nature Singapore Pte Ltd. 2024
B.-Y. Cao et al. (Eds.): ICFIE 2022, LNDECT 207, pp. 151–162, 2024.
https://doi.org/10.1007/978-981-97-2891-6_12

1 Restatement of the Problem

1.1 Problem Background

Five-hundred-meter Aperture Spherical Telescope, referred to as FAST. As shown in Fig. 1.

Fig. 1. FAST 3D diagram

Among them, the active reflective surface system includes the main cable net, reflective panel, lower cable, actuator and supporting structure. as shown in Fig. 2.

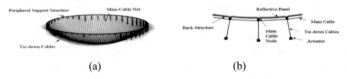

(a)   (b)

Fig. 2. Cable net structure

Active reflector is mainly divided into reference state and working state. Its observation profile is shown in Fig. 3.

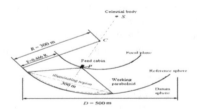

Fig. 3. FAST section diagram

Point C is the spherical center of the reference sphere and the focal diameter ratio is F/R = 0.466.

1.2 Question Raising

1, When the object S to be observed is directly above the reference sphere, that is, $\alpha = 0°$, $\beta = 90°$, the ideal paraboloid is determined by considering the adjustment factors of the reflecting panel.

2, When the object S to be observed is located at $\alpha = 36.795°$, $\beta = 78.169°$, the ideal paraboloid is determined. Establish the adjustment model of the reflecting panel, adjust the expansion and contraction of the relevant actuator, and make the reflecting surface close to the ideal paraboloid as far as possible.

2 Notations

Symbols	Descriptions
F	the radius difference between two concentric spheres (reference sphere and focal plane)
R	the reference spherical radius
α, β	the azimuth and elevation of the celestial body S
s	the distance between the object and the mirror
f	the focal length of curved mirror
d	the distance between two points
R'	the cable displacement vector
h	the relative displacement between the apex of the paraboloid and the reference sphere
D	the paraboloid aperture
r_{sky}	the Earth radius
H	the time angle
t	the time
ω	the polar angle
N_{TR}	the total number of reflection points
N_{rec}	the number of effective reflection points

3 Establishment and Solvation About Mathematical Model

3.1 Establishment and Solution of Problem 1 Model

For problem 1, supposing that the object S to be observed is directly above the reference sphere, $\alpha = 0°$, $\beta = 90°$, the electromagnetic wave from the target celestial body is vertically, parallelly and symmetrically directed to the FAST radio telescope, as shown in Fig. 4:

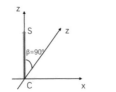 

(a) Orientation diagram   (b) Optical schematic diagram

Fig. 4. Model diagram

In this case, the celestial position s, the origin c, and the focus p are collinear, which is quite ideal for us. We first analyze the principle of concave mirror reflection imaging, then establish parabolic equation models with parameters in two-dimensional space and three-dimensional space respectively, and finally substitute our actual data to solve the parameter values to obtain the equation of ideal paraboloid, as follows:

$$x^2 + y^2 = 559.2z + 1.677844 \times 10^5 \tag{1}$$

The detailed modeling and solution process is as follows.

3.1.1 The Establishment and Solution of Ideal Paraboloid Model

When the object S to be measured is directly above the reference sphere, $\alpha = 0°$, $\beta = 90°$, At 90°, the light of celestial bodies can shine vertically and parallel to the FAST plane, as shown in Fig. 4.

Because paraboloid and sphere are isotropic, when analyzing deformation strategy, we can simplify the problem to two-dimensional space, replace the basic sphere with arc line, and replace paraboloid with paraboloid [1].

The spherical center of the reference sphere is now taken as the coordinate origin, and the arc and parabola equations are as follows:

$$\begin{cases} x^2 + z^2 = R^2 \\ x^2 = 2p[z - (R' + h)] \end{cases} \tag{2}$$

Where, the spherical radius $R = 300\,m$, the cable downward displacement vector $R' = -300\,m$, and h is the relative displacement between the apex of the paraboloid and the reference spherical surface.

Where h can be uniquely determined by the focal ratio f and the two continuous conditions at the junction of the edge of the paraboloid and the sphere [2]. (the apex of the paraboloid is below the neutral sphere):

$$h = \sqrt{R^2 - \left(\frac{D}{2}\right)^2} + \frac{D^2}{16 \times R \times f'} - R \tag{3}$$

Wherein, the parabolic aperture $D = R = 300\,m$, and the focal diameter ratio $f' = \frac{F}{R} = 0.466$.

It can be calculated that h = 0.0437 m. With the Z axis as the center, rotate the parabola on the two-dimensional plane to obtain the three-dimensional paraboloid:

$$\begin{cases} x^2 + z^2 = R^2 \\ x^2 = 2pz + 2p(R + h) \end{cases} \tag{4}$$

Substitute the data $\frac{P}{2} = 0.466R$ to obtain the two-dimensional parabolic equation:

$$x^2 = 559.2z + 1.677844 \times 10^5 \tag{5}$$

The three-dimensional parabolic equation is obtained by rotating around the Z axis, that is, the object S to be measured is located directly above the reference sphere, $\alpha = 0°$, $\beta = 90°$. The three-dimensional ideal paraboloid model at 90° is shown in Fig. 5.

$$x^2 + y^2 = 559.2z + 1.677844 \times 10^5 \tag{6}$$

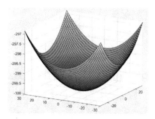

Fig. 5. The schematic diagram of three-dimensional ideal paraboloid model

3.2   Establishment and Solution of Problem 2 Model

Suppose that the object S is slanted towards the base sphere, $\alpha = 36.795°, \beta = 78.169°$, The electromagnetic waves from the target object are directed towards the FAST radio telescope at a specific Angle, as shown in Fig. 6:

(a) Azimuth diagr   (b) Optical schematic diagram

Fig. 6. Model diagram

Considering the adjustment factors of the reflector panel, the expansion of the actuator was adjusted, and a three-dimensional ideal paraboloid model was established to improve its coincidence with the reflector.

3.2.1.1 Real Coordinate System and Virtual Coordinate System

Real coordinate system: the coordinate system with c as the origin and the vertical direction of the ground upward as the z axis.

Virtual coordinate system: For each ideal paraboloid in different positions, c is taken as the origin, and the line between the focus p and c of the ideal paraboloid is taken as the z-axis coordinate system.

Taking the center of the reference sphere as the origin of polar coordinates and setting any direction as the direction of coordinate system, a virtual polar coordinate system is established. Under this condition, there can be infinitely many virtual coordinate systems.

As $y = x^2$ an example, polar coordinate equation under two-dimensional plane virtual coordinate system is constructed, as shown in Fig. 7.

$$\begin{cases} x = \rho\cos\alpha \\ y = \rho\sin\alpha \end{cases} \Rightarrow \begin{cases} x = \rho\cos\left(\alpha - \left(\frac{\pi}{2} - \theta\right)\right) \\ y = \rho\sin\left(\alpha - \left(\frac{\pi}{2} - \theta\right)\right) \end{cases} \tag{7}$$

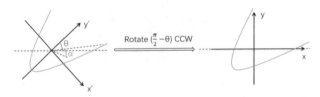

Fig. 7. Parabolic diagram of two-dimensional coordinate system

Map from two-dimensional to three-dimensional space, establish three-dimensional virtual coordinates, and then transform from three-dimensional virtual coordinate system to real three-dimensional coordinate system, as shown in Fig. 8 and 9.

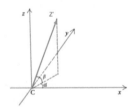

Fig. 8. Virtual coordinate diagram

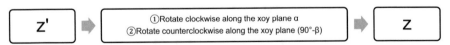

Fig. 9. Transformation flow chart

Rotating square matrix

$$
\begin{bmatrix}
1 & \frac{-\sin\beta}{\sin\varphi}\cos(\varphi-\alpha) & \cos\beta\sin(\varphi-\alpha) \\
\frac{-\sin\beta}{\sin\varphi}\sin(\varphi-\alpha) & 1 & \cos\beta\sin(\varphi-\alpha) \\
\cos\beta\cos\varphi & \cos\beta\sin\varphi & \sin\beta
\end{bmatrix}
$$

The transformation process is:

$$
\begin{cases}
z = \rho\sin\theta \\
x = \rho\cos\theta\cos\varphi \\
y = p\cos\theta\sin\varphi
\end{cases}
\Rightarrow
\begin{cases}
x = \rho[\sin(\theta-\beta)\cos(\varphi-\alpha)+\cos\theta\cos\varphi)] \\
y = \rho[\sin(\theta-\beta)\sin(\varphi-\alpha)+\cos\theta\sin\varphi)] \\
z = \rho\cos(\theta-\beta)
\end{cases}
\tag{8}
$$

3.2.1.2 The Polar Coordinate Equation Representation of Surface Equation of a Reference Paraboloid in Real Coordinate System

For the convenience of discussion, the paraboloid in different states is defined first. When the position of the celestial body S changes, the position and opening direction of the benchmark paraboloid also change, but the shape remains unchanged. We call this kind of paraboloid virtual paraboloid.

Based on the reference paraboloid obtained in problem 1, the equation of the paraboloid is transformed into the polar coordinate equation, and the polar coordinate equation is taken as the polar coordinate equation of the reference paraboloid:

$$
\begin{cases}
z = \rho\sin\theta \\
x = \rho\cos\theta\cos\varphi \\
y = p\cos\theta\sin\varphi
\end{cases}
\tag{9}
$$

$$
\rho^2\cos^2\theta = 559.2\rho\sin\theta + 1.677844 \times 10^5 \tag{10}
$$

3.2.2.1 Introduce the Time Angle Coordinates and Fix the Time Angle

Based on the above equations, the celestial coordinate system, the instant angular coordinate system, can be established to express the position of celestial bodies by quantitative method $\{x_t, y_t, z_t\}$. The basic circle of the time-angle coordinate system is set as the celestial equator, and the basic point is the celestial north and the celestial South Pole.as shown in Fig. 10 [3].

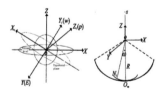

Fig. 10. Control coordinate system diagram

Where, the pointed direction is the north celestial pole, the pointed direction is the west point and satisfies the left-handed rule. Horizontal rectangular coordinate system {x, y, z}, Where z points to the zenith, x points to the north point, y points to the east point, O is the center of the Earth, also is the origin of the rectangular coordinate system.

Establish virtual time Angle coordinate expression:

$$\begin{cases} x_t = r_{sky}\cos\beta\cos H \\ y_t = r_{sky}\cos\beta\sin H \\ z_t = r_{sky}\sin\beta \end{cases} \tag{11}$$

Among them, r_{sky} is the radius of the earth. Is the Angle between the observation direction and the equator, namely the elevation Angle. H is an Angle, which is a function of t.

$$H = f(t) \tag{12}$$

In the terrestrial cartesian coordinate system {x, y, z}, when the vertex of the paraboloid is the positive center, the equation of the paraboloid expressed in polar coordinates is:

$$\rho^2\cos^2\theta = 2p\sin\theta + 2p(R+h) \tag{13}$$

$$\frac{p}{2} = 0.466R \tag{14}$$

Where, h is the relative displacement between the vertex of the paraboloid and the reference sphere [4].

the spherical coordinates of any point on a paraboloid with the center of the reflecting surface as its vertex are:

$$\rho^2\cos^2\theta\cos^2\varphi + \rho^2\cos^2\theta\sin^2\varphi = 559.2\rho\sin\theta + 1.677844 \times 10^5 \tag{15}$$

Convert to rectangular coordinates to obtain:

$$x^2 + y^2 = 559.2z + 1.677844 \times 10^5 \tag{16}$$

3.2.2.2 Solve the Parabolic Polar Coordinate Expression

When the vertex of the paraboloid deviates from the center of the reflection panel, virtual coordinates are used for transformation to obtain:

$$\begin{cases} x = \rho[\sin(\theta - \beta)\cos(\varphi - \alpha) + \cos\theta\cos\varphi)] \\ y = \rho[\sin(\theta - \beta)\sin(\varphi - \alpha) + \cos\theta\sin\varphi)] \\ z = \rho\cos(\theta - \beta) \end{cases} \tag{17}$$

the polar coordinate expression of the paraboloid can be obtained.

$$\rho^2\left(\sin^2(\theta - \beta) + \cos^2\theta + 2\sin(\theta - \beta)\cos\theta\cos\alpha\right)$$
$$= 559.2\rho\cos(\theta - \beta) + 1.677844 \times 10^5 \tag{18}$$

Among them, α, β is parameter, θ is Polar Angle, $\alpha = 36.795°$, $\beta = 78.169°$. Put the plug in type α, β can be calculated when the observation object S located in $\alpha = 36.795°$, $\beta = 78.169°$ Is an ideal paraboloid.

$$\rho^2 \left(sin^2(\theta - 78.169°) + cos^2\theta + 2\,sin(\theta - 78.169°)cos\,\theta\,cos\,36.795° \right)$$
$$= 559.2\rho\,cos(\theta - 78.169°) + 1.677844 \times 10^5 \tag{19}$$

3.2.2.3 Solve the Parabolic Polar Angle Expression

We define ω as the Angle between any node N_i on the paraboloid and paraboloid vertex O_m, then ω is called the polar Angle, as shown in Fig. 11.

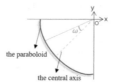

Fig. 11. Schematic diagram of pole angle

When the vertex of the paraboloid deviates from the center of the spherical crown, the radial radius corresponding to any node on the ideal paraboloid is [5]:

$$\rho = \frac{-2Fcos\omega + 2\sqrt{F^2cos^2\omega + FRsin^2\omega}}{sin^2\omega} \tag{20}$$

Where, F can be seen as the focal length, $F = 0.466R$.

The   vertex coordinate of the ideal paraboloid is $(-49.26152, -36.84562, -293.66973)$, and the unit is meter.

3.2.3.1 Determine the Position of the Panel Adjustment Deformation Node

According to the analysis of the above requirements, when the caliber is 300 m, each node through a certain displacement to form such an ideal paraboloid. As for the nodes needed to determine the deformation of the paraboloid, the spherical coating model can be established by taking advantage of the symmetry of the three-dimensional plane, as shown in Fig. 12 [6].

When $|d| \leq r$, N_i The point set contains all the moving nodes of the sphere.

The coordinates of point C can be solved by the polar coordinate equation of the sphere:

$$\begin{cases} x_c = Rcos30°cos\beta cos\alpha \\ y_c = Rcos30°cos\beta sin\alpha \\ z_c = Rcos30°sin\beta \\ \alpha = 78.169° \\ \beta = 36.795° \end{cases} \tag{21}$$

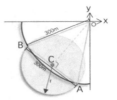

Fig. 12. Schematic diagram of sphere cladding model

3.2.3.2 Determine the Radial Displacement

According to the working principle of the actuator, the leading motion direction of the main cable node should be radial. In this case, the radial displacement of a node is used to represent the expansion of a node [7].

As shown in Figure x, point C is the vertex of the paraboloid. Set the point C (x_c, y_c, z_c) and use the polar coordinates of the sphere to solve:

$$\begin{cases} x_c = R\cos30°\cos\beta\cos\alpha \\ y_c = R\cos30°\cos\beta\sin\alpha \\ z_c = R\cos30°\sin\beta \\ \alpha = 78.169° \\ \beta = 36.795° \end{cases} \tag{22}$$

For a node $N_j\left(x_j, y_j, z_j\right)$ in its paraboloid, it can be obtained by the cosine theorem of a triangle

$$d_{cN_j} = R^2 + R^2 - 2R^2\cos\omega \tag{23}$$

$$\cos\omega = \frac{-d_{cN_j}^2 + 2R^2}{2R^2} \tag{24}$$

The nonlinear constraint function can be obtained by combining the above equations:

$$\begin{cases} \sin^2\omega = 1 - \cos^2\omega \\ \cos\omega = \frac{-d_{cN_j}^2 + 2R^2}{2R^2} \\ \rho = \frac{-2F\cos\omega + 2\sqrt{F^2\cos^2\omega + FR\sin^2\omega}}{\sin^2\omega} \end{cases} \tag{25}$$

Consider the theoretical expansion quantity Δr and set the objective function:

$$\Delta r = R - \rho \tag{26}$$

Through conditional transformation, set point $N_j\left(x_j', y_j', z_j'\right)$, and obtain from spherical polar coordinates:

$$\begin{cases} x_j' = (R - \Delta r)\cos\beta\cos\alpha \\ y_j' = (R - \Delta r)\cos\beta\sin\alpha \\ z_j' = (R - \Delta r)\sin\beta \end{cases} \tag{27}$$

Set the actual node expansion quantity as dir, and give the equations of expansion quantity change:

$$\text{dir} = \begin{cases} -0.6, \ \Delta r < -0.6 \\ \Delta r, \ -0.6 \leq \Delta r \leq 0.6 \\ 0.6, \ \Delta r > 0.6 \end{cases} \tag{28}$$

4 Conclusions

The inspiration of this article comes from the National College students Mathematical modeling contest, the title is to model the "FAST" eye, and then according to the celestial bodies in different positions, so as to adjust the different paraboloid, to achieve the effect of observing celestial bodies. For the establishment of an ideal parabolic reflector adjustment model and the optimization of the reception effect of celestial electromagnetic waves reflected by the reflector, it is a major problem in the actual operation of the China Sky Eye FAST - 500 m aperture spherical radio telescope. Based on the specific situation, this paper makes reasonable assumptions about the working state of the active reflector, determines the ideal paraboloid, optimizes the reflector mediation scheme.

When the object S to be observed is directly above the reference sphere, the ideal paraboloid is determined. Finally, the parabolic equation under the three-dimensional plane is obtained by rotating the two-dimensional parabola around the z-axis:

$$x^2 + y^2 = 559.2z + 1.677844 \times 10^5$$

When the object S to be observed is located at $\alpha = 36.795°$, $\beta = 78.169°$, the ideal paraboloid is determined, the reflection panel adjustment model is established, and the adjusted node data are calculated. Which satisfies

$$\rho = \frac{-2F\cos\omega + 2\sqrt{F^2\cos^2\omega + FR\sin^2\omega}}{\sin^2\omega}$$

For determining the movable points covered by the paraboloid, we establish a sphere coverage model. The space rectangular coordinates after the position change of the cable node can be obtained from the known reference state rectangular coordinates.

Acknowledgments. This work is supported by 2022 National Student Innovation Training Project of Ministry of Education of P.R.C (202211078163).

References

1. Li, M., Zhu, L.: Optimization analysis of FAST instantaneous paraboloid deformation strategy. J. Guizhou Univ. (Nat. Sci. Edn.) **2012**, 30–34 (2012)
2. Sun, S.: Research on modeling and simulation of FAST node control system based on MRAC theory. Northeast University (2015)

3. Sun, C., Zhu, L., Yu, D.: FAST main reflector node motion control algorithm. Sci. Technol. Eng. (2012)
4. Guo, Y.: T-S fuzzy modeling of node displacement of FAST cable network. Northeast University (2014)
5. Ji, X., Wang, H., Zheng, F., Li, P.: Study on the convergence performance of triangular element thin film assembled solar concentrator. J. Solar Energy (2021)
6. Liu, J., Liu, G., Chen, D.: Fresnel formula discusses the amplitude change of light wave reflection and refraction. Phys. Bull. (2021)
7. Nan, R.: 500m spherical reflector radio telescope FAST. Chin. Sci. Ser. G Phys. Mech. Astronomy **35**, 449–466 (2005)

Research and Practice of Internet+ Innovation and Entrepreneurship Ability Cultivation Model Based on Cooperative Learning

Yubin Zhong, Weitong Chen, Jialian Li, and Hongbin Lin[✉]

School of Mathematics and Information Science, Guangzhou University, Guangzhou 510006, China
zhong_yb@gzhu.edu.cn, hongbin.lin3589@gmail.com

Abstract. With the development of society, it has become increasingly evident that the knowledge possessed by university graduates cannot meet the needs of society. Exploring what abilities students need to possess in order to adapt to the demands of social development after graduation is a very realistic research topic. This paper proposes a cooperative learning model based on Internet+ and combines it with project-based learning, demonstrating through empirical research that the model is effective in enhancing the innovation and entrepreneurship abilities of college students. The paper suggests that in order to produce high-level innovation and entrepreneurship works, it is necessary for talented individuals with professional, innovative, and technical skills to undergo effective cooperative learning based on Internet+. This is an optimized solution that can be used for cultivating the innovation and entrepreneurship abilities of students.

Keywords: Cooperative learning · Innovation and entrepreneurship abilities · Ability cultivation model · Practice · optimized cultivation plan

1 Introduction

"Internet+" refers to the combination of traditional industries with internet-based information technology, completing the transformation and upgrading of the economy through optimizing production factors, updating business systems, and restructuring business models. As of June 2022, the number of internet users in China reached 1.051 billion, with an internet penetration rate of 74.4% [1]. Its characteristics of low cost, market expansion, and industrial intelligence are driving the continuous evolution of economic ideology. With the continuous research and promotion of "Internet+" by the industry and academia, on July 4, 2015, the State Council of China officially released the "Guiding Opinions on Actively Promoting the 'Internet+' Action," proposing 11 key actions, including "Internet+" entrepreneurship and innovation, "Internet+" collaborative manufacturing, and "Internet+" inclusive finance [2]. To this day, "Internet Plus" has deeply rooted in various fields of society and economy and driven the growth of a new economic model. The "Internet+" College Students' Innovation and Entrepreneurship

B.-Y. Cao et al. (Eds.): ICFIE 2022, LNDECT 207, pp. 163–176, 2024.
https://doi.org/10.1007/978-981-97-2891-6_13

Competition in China was personally proposed by Premier Li Keqiang and co-organized by twelve national departments, including the Ministry of Education, the National Development and Reform Com-mission, and the Ministry of Industry and Information Technology. The competition adopts a three-level system of campus-level preliminaries, provincial-level semi-finals, and national finals, and the participating teams are generally composed of full-time students from various stages of universities, ranging from three to fifteen members or full-time students who have graduated no more than four years. The participating projects usually combine the new generation of information technology with various fields such as the new engineering, medical, and agricultural sciences closely, and leverage the Internet's role in social services while also combining the characteristics of university majors. The "Internet+" College Students' Innovation and Entrepreneurship Competition has currently become the top-ranked competition on the list of university competitions published by the Ministry of Education. It has not only received high attention from various universities and teachers and students but also provided a good publicity channel and incubation environment for start-up teams.

2 Principles and Models of Collaborative Learning Based on Internet+

2.1 Principles of Collaborative Learning Based on Internet+

Collaborative learning is an instructional activity centered around group work, where students work together under the guidance of a teacher and engage in cooperative and competitive interactions among peers. It has a positive impact on teaching across various disciplines, and helps students to engage in a process of "re-creation" under the guidance of a teacher [2]. The idea of collaborative learning has a long history in China, with the basic concept of "lonely self-study leads to ignorance" mentioned in the Xueji. However, formalized collaborative learning originated in the early 1970s in the United States, and has been recognized as one of the most significant educational reforms of modern times due to its outstanding effects on student academic performance and learning quality. [3] In general, the focus of collaborative learning is on the interaction and cooperation between peers, with peer collaboration being the foundation of this approach.

Collaborative learning is an instructional activity that is primarily conducted through group work, often involving mutual assistance between peers. When combined with "Internet+", collaborative learning expands the functions of learning groups to a certain extent by combining students' individual actions and learning processes with the trending network environment. The advantages of hypermedia information representation and processing, virtualization, and informal interpersonal communication and interaction methods compensate for the shortcomings of traditional collaborative learning, and enhance the innovation, entrepreneurial, practical, and teamwork abilities of learning groups.

2.2 Model of Collaborative Learning Based on Internet+

In this era of rapid internet development, leveraging internet resources can greatly promote collaborative learning among group members. We can easily access rich electronic materials, high-quality open courses, and learning websites through online means, and also communicate and collaborate extensively with other learners with different levels of knowledge, experiences, and perspectives. This not only bridges the information and knowledge gap among group members, but also provides the team with more varied research directions. It even changes the situation where group collaborative learning is limited to collaboration among members within the group, making the form and content of group collaboration more flexible and diverse.

Let the cooperative learning system includes individual n learners, namely $L_1, L_2, ..., L_N$, each learner L_i is equipped with some individual learning abilities s_i and some ability to collaborate with others c_i. The learning outcomes of the learner L_i can be described by a learning variable x_i that can be a measure of some ability or knowledge level. Assuming information exchange and knowledge sharing among learners, this can be described by a social network graph. The social network graph $G = (V, E)$ represents the connection between learners, where $V = \{L_1, L_2, ..., L_N\}$ represents the set of learners, E represents the set of connections between learners, i.e., a group of ordered pairs (L_i, L_j) and represents the connection between the learners L_i and L_j.

In this model, the learning outcomes of each learner x_i can be characterized by the following equation:

$$x_i = f(s_i, c_i, x_j), j \in N_i \tag{1}$$

Where f is a function of the learning outcome $x_j (j \in N_i)$ of the individual learning ability s_i, cooperative learning ability c_i, and the learner L_j related to the learner L_i, and N_i represents the collection of learners related to the learner L_i.

This equation describes how a learner's learning outcomes are influenced by their individual learning ability, collaborative learning ability, and interactions with others. With this equation, the process and results of collaborative learning can be mathematically modeled and analyzed, guiding the practice and improvement of collaborative learning. The schematic diagram of the model is shown in Fig. 1.

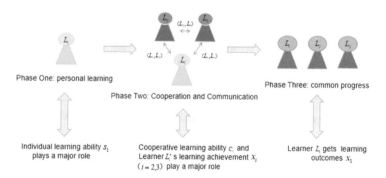

Fig. 1. Illustration of collaborative learning model

Building a platform for collaborative learning among group members and between groups through the internet. In traditional collaborative learning, all members are required to communicate face-to-face at the same time and place. However, with the advent of the internet, we can now discuss and communicate online, free from the constraints of time and space, and collaborate more flexibly. This not only allows for the freedom to choose the time for collaborative learning, but also saves commuting time for attending meetings. In the event that a member is unable to attend a meeting, there is no need to reschedule the meeting time, as the team's progress can be synchronized through the sharing of meeting records, shared resources, and even meeting recordings of other attending members on the online platform [5].

3 Application of the Cooperative Learning Model Based on Internet+ in Improving College Students' Innovation and Entrepreneurship Abilities

3.1 Application in the "Internet+" Competition

In the process of participating in the "Internet+" competition, it is necessary to analyze practical problems, explore customer needs, and constantly innovate products or business models. This not only requires students to have good innovation abilities to upgrade and transform product processes, but also requires students to have sufficient "business acumen". As shown in Fig. 2, Only good innovation abilities combined with a reasonable business model can make the innovation and entrepreneurship projects of participating teams stand out and win the favor of judges and investors. The former can be summarized as the innovative application of professional knowledge learned by students during their college studies, which is the stage of transforming knowledge into products; the latter is the stage of using products to generate greater commercial value and social benefits by exploring social needs and product commercial attributes [8].

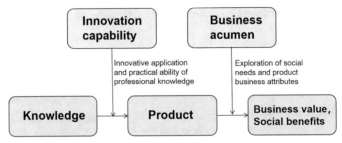

Fig. 2. Students innovate business models through the "Internet Plus" competition

For example, as energy resources become increasingly scarce and the demand for energy conservation and environmental protection in human society rises, the development of new energy vehicles has become an important issue facing the world's automotive industry. A certain company has provided market research data on electric vehicles. Through the "Internet+" project, many teams of three college students have established mathematical models to propose sales strategy recommendations based on the data. One

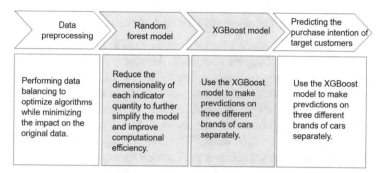

Fig. 3. Mind map of predicting target customers' purchase intention

team member analyzed the market research data of electric vehicles and judged the target customer's purchase intention prediction model based on the results, and gave sales strategy recommendations based on the model. The specific solution is shown in Fig. 3:

Due to the fact that machine learning algorithms do not work well on imbalanced datasets, in order to optimize the model, the team used Python software to balance the data while minimizing the impact on the original data. Then, considering the large number of selected indicators, in order to evaluate the importance of each indicator and further simplify the model by reducing its dimensions, the team used the random forest algorithm to calculate the importance of each indicator and sorted them in descending order based on their importance. This allowed the indicators to more accurately reflect customer purchasing intentions and better fit the training of the model. Finally, the top ten features in terms of importance ranking for each brand were selected as the final indicators, and XGBoost models and neural networks were built using these final indicators as inputs to predict customer purchasing intentions.

According to research findings, the sales team provided the following sales suggestions to the company's sales department: Firstly, identify the sales market. Based on the results of descriptive statistical analysis, the population with the intention to buy a car is mainly concentrated in urban areas and has a certain level of economic strength.

Therefore, in the initial stage of sales, it is advisable to start from large cities' official markets (such as Nanjing, Beijing, and Guangzhou) with high purchasing intentions, as well as private markets in medium and small cities. Secondly, explore potential customers. The predictive model built can be used to predict the buying intentions of target customers, and comprehensive and multi-faceted promotion of the company's cars can deepen the involvement of customers with weak buying intentions to enhance their willingness to purchase.

It can be seen that through the "Internet+" competition training, students can apply their learned knowledge to practice, bring commercial value to enterprises, and bring certain benefits to society. At the same time, in this process, college students' comprehensive abilities, such as innovation, teamwork spirit, business thinking, and practical skills, can be effectively improved.

3.2   Application of Team Collaboration Spirit Cultivation

The "Internet+" Competition sets a minimum team size of three members, which reflects the requirement for a reasonable team structure in the actual innovation and entrepreneurship process. Even if one person's innovation ability is strong, it is still limited. Only through team collaboration and continuous brainstorming during the product development process, can the constant polishing of an innovation and entrepreneurship project be achieved. Therefore, collaborative learning in the team is indispensable. From the preliminary rounds to the final rounds, the "Internet+" Competition has a long schedule. The works selected from the school-level preliminary round teams require three to four months of continuous polishing before they can truly compete in the national finals. It is almost impossible to complete the project solely through the efforts of one person. Therefore, it also requires different members of the same team to divide the work reasonably, give full play to their strengths, and take the essence and discard the dross. For example, students majoring in science and engineering can be responsible for improving and developing the product, while business students can conduct more market research and case analysis of the product's current market. Computer science students can use their professional knowledge to better integrate the product with the Internet, and use the Internet as a communication channel to help the team's product gain more market and potential resources.

In addition, during the innovation process, team members should actively discuss and analyze problems from different perspectives, so that the product can be modeled and executed from a public perspective. This way, the team will not have one person's flights of fancy, and will not face challenges and difficulties in the implementation stage. Therefore, under the encouragement of the "Internet Plus" Competition, participating teams will continuously discuss and collaborate, and in this process, cultivate students' team awareness and collaboration spirit. As shown in the Fig. 4, in the modeling national competition of 2018, Problem B introduced an intelligent processing system consisting of eight CNC machines, an RGV, an upper material conveyor belt, and a lower material conveyor belt:

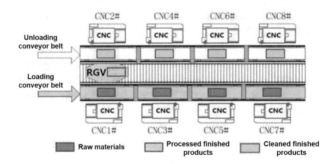

Fig. 4. Schematic diagram of intelligent machining system

The RGV's function is to place raw materials from the loading conveyor onto the corresponding CNC, place the processed materials from the CNC into the cleaning tank, and put the cleaned finished products onto the unloading conveyor.

The problem sets three scenarios and requires participating teams to solve the following two tasks based on the three scenarios:

Task 1: Establish a dynamic scheduling model for RGV and provide corresponding algorithms.

Task 2: Based on the system's operating parameters, verify the practicality of the model proposed in Task 1 and the effectiveness of the algorithm. Then provide reasonable scheduling strategies for RGV and calculate the work efficiency of the intelligent processing system under this strategy.

This problem involves multiple disciplines, including mathematics, physics, and computer science, and greatly tests the cross-disciplinary skills of students. Therefore, to better solve this problem, students from different majors need to coordinate and cooperate to solve the problem together. In this process, students can not only apply their own professional knowledge to practical research but also communicate and complement each other with students from different disciplines. It is evident that this model is of great significance in cultivating students' spirit of unity and cooperation.

3.3 Application in Improving Innovation Capability

Each individual may have a different understanding and solution for the same problem. The same applies to product innovation, where different people may propose different solutions to the same pain point. A good product or solution cannot exist without creating value for others and society. This requires new products to have differentiated advantages compared to existing ones, as well as the ability to sustainably operate and support the product from its inception to maturity and widespread use in the commercial model.

Therefore, collaborative learning is a very suitable solution. Through collaborative learning, everyone can propose and share their ideas and find more different solutions. Innovative individuals in the group can explore social pain points with a more unique perspective, and through collaborative learning, group members can have a more thorough understanding of each idea and ultimately choose the most suitable one together.

Encouraging university students to use collaborative learning to expand their thinking and innovate boldly to solve practical problems is an inevitable trend in participating in the "Internet+" competition. For example, in the "Small and Medium-sized Enterprise Supply Chain Automatic Allocation Platform" project of our school, the innovation is mainly reflected in the use of the GRA-TOPSIS method in modeling to comprehensively evaluate the overall strength of each supplier, formulate allocation strategies, and obtain evaluation results with a wider adaptation range.

The GRA-TOPSIS method is an improved comprehensive integration evaluation method that fully integrates the advantages of two evaluation methods and achieves an effective combination of qualitative and quantitative methods. Using this method can obtain more reasonable evaluation results compared to the original method. The specific steps are as follows:

Let the evaluation index set $P = (P_1, P_2, \cdots, P_N), N = \{1, 2, \cdots, n\}$, optional supplier scheme $M = (1, 2, \cdots, m)$, evaluation matrix as $X = (x_{ij})_{m*n}, i \in M, j \in N$. Where xij is the evaluation value of the ith supplier under the jth indicator.

Step 1: Standardization.

Vector normalization is used to standardize the value of feature supply index and obtain $Y = (y_{ij})_{m*n}$:

$$y_{ij} = \frac{x_{ij}}{\sqrt{\sum_{k=1}^{n} x_{kj}^2}} (i = 1, 2, \cdots, 402; j = 1, 2, \cdots 5) \tag{2}$$

Step 2: calculate the weighted normalized judgment matrix $Z = (z_{ij})_{m*n}$, where

$$z_{ij} = w_j * y_{ij}, w = (w_1, w_2, \cdots w_n) \tag{3}$$

Step 3: the positive ideal solution $Z^+ = (z_1^+, z_2^+, \cdots, z_n^+)$ and the negative ideal solution $Z^- = (z_1^-, z_2^-, \cdots, z_n^-)$, where when Zj is a very large index, When the Zj is a very small indicator,

$$Z_j^+ = max_i Z_{ij}, Z_j^- = min_i Z_{ij} \tag{4}$$

When the Zj is a very small indicator,

$$Z_j^+ = min_i Z_{ij}, Z_j^- = max_i Z_{ij} \tag{5}$$

Step 4: Calculate the Euclidean distance between the alternative solution and the negative ideal solution d_i^+, d_t^-, where

$$d_i^+ = \sqrt{\sum_{j=1}^{n} \left(z_{ij} - z_j^+\right)^2}, d_i^- = \sqrt{\sum_{j=1}^{n} \left(z_{ij} - z_j^-\right)^2} \tag{6}$$

Step 5: Grey correlation coefficient matrix of alternative to positive and negative ideal solutions $R^+ = (r_{ij}^+)_{m*n}$ and $R^- = (r_{ij}^-)_{m*n}$, where

$$r_{ij}^+ = \frac{min_i min_j \left|z_j^+ - z_{ij}\right| + \rho max_i max_j \left|z_j^+ - z_{ij}\right|}{\left|z_j^+ - z_{ij}\right| + max_i max_j \left|z_j^+ - z_{ij}\right|} = \frac{\rho \omega_j}{\omega_j - z_{ij} + \rho \omega_j} \tag{7}$$

$$r_{ij}^- = \frac{min_i min_j \left|z_j^- - z_{ij}\right| + \rho max_i max_j \left|z_j^- - z_{ij}\right|}{\left|z_j^- - z_{ij}\right| + max_i max_j \left|z_j^- - z_{ij}\right|} = \frac{\rho \omega_j}{z_{ij} + \rho \omega_j} \tag{8}$$

$\rho \in (0, \infty)$ is called the resolution coefficient, the greater the ρ, the smaller the resolution, generally speaking, the value in the (0, 1) interval, when $\rho \leq 0.5463$, the resolution is the best, generally take $\rho = 0.5$.

Step 6: calculate the grey correlation with positive and negative ideal solutions r_i^+, r_i^-.

$$r_i^+ = \frac{1}{n} \sum_{j=1}^{n} r_{ij}^+, r_i^- = \frac{1}{n} \sum_{j=1}^{n} r_{ij}^- \tag{9}$$

Step 7: Using dimensionless method on $d_i^+, d_i^-, r_i^+, r_i^-$, the results are as follows:

$$D_i^+ = \frac{d_i^+}{maxd_i^+}, D_i^- = \frac{d_i^-}{maxd_i^-}, R_i^+ = \frac{r_i^+}{maxr_i^+}, R_i^- = \frac{r_i^-}{maxr_i^-} \tag{10}$$

Step 8: Combine D_i^+, D_i^- with R_i^+, R_i^-. Construct formulas based on their numerical values and the degree of closeness to the ideal solution. The larger the value of S_I^+, the more optimal the supplier's solution; the smaller the value of S_I^-, the more optimal the supplier's solution.

$$S_I^+ = \alpha D_i^- + \beta R_i^+, S_I^- = \alpha D_i^+ + \beta R_i^- \tag{11}$$

The parameters α and β reflect the degree of preference of enterprise management personnel for Euclidean distance and grey correlation degree. When assigning values, $\alpha + \beta = 1$ and $\alpha, \beta \in [0, 1]$.

Step 9: Calculate the matching degree of optional suppliers' proposals.

$$C_i^* = \frac{S_i^+}{S_i^+ + S_i^-} \tag{12}$$

The available suppliers can be ranked based on the degree of match with the plan. The greater the degree of match, the better the corresponding supplier, and the lower the degree of match, the worse the corresponding supplier.

This algorithm offers more globally-oriented and dynamically flexible deployment plans that can be updated with new data, as compared to traditional linear programming algorithms. It provides better solutions for enterprises to customize their real-time ordering and transportation plans for raw materials.

3.4 Application of Entrepreneurial Skills Enhancement

The goal of "Internet+" is to combine the internet with traditional industries to find new business opportunities and development. In the process of developing a startup product or team, a series of complex problems need to be faced. Students are no longer dealing with the simple entrepreneurial conditions assumed in textbooks, but rather with a complex and constantly changing real operating environment. Therefore, it is particularly important to use internet resources to facilitate collaboration and learning among group members. In this process, team members will gain knowledge and experience of the process of enterprise operation and management activities, main business processes and their interrelationships through various channels. This also promotes knowledge integration, enabling students to truly experience success and failure, and deeply understand the importance of competitiveness, teamwork, and professional skills [9].

In recent years, the projects that have won national gold and silver awards in the "Internet+" Innovation and Entrepreneurship Competition are not only highly competitive in product innovation, but more importantly, their business model design is reasonable, allowing their projects to sustain long-term development and ultimately create value for society. Therefore, the "Internet+" Competition is different from other discipline-specific competitions in that it places more emphasis on students' business capabilities, which cannot be learned solely through in-school knowledge acquisition but rather must be achieved through online channels and collaborative learning. This also helps university students to have earlier exposure to society and promotes their independent development of innovative entrepreneurial skills through practice.

Therefore, the "Internet+" project can help students better utilize their professional knowledge to understand market demands, identify customer pain points, and thus enhance their entrepreneurial abilities. The radar chart is shown in Fig. 5.

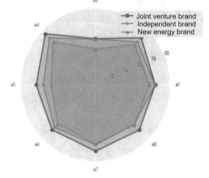

Fig. 5. Radar chart of customer satisfaction for three brands

3.5 The Application of Comprehensive Abilities (Including Developing Skill Special-Ties, in the Improvement Process is Crucial)

Innovation and entrepreneurship are collaborative processes that require each team member to utilize their strengths and work together [10]. If there are differences in opinions or missing members with specific skills, cooperation and learning will be difficult, and the project may not be completed.

In general, based on the development process of innovation and entrepreneurship, the roles of startup team members can be roughly divided into product design and development, business strategy formulation, and product operation. Different roles have different requirements for students' abilities and future development directions. For example, product design and development are generally responsible for students with a strong theoretical foundation in the relevant field. They are responsible for creating, updating, and iterating products, which can only be completed by students who possess professional knowledge. On the other hand, students who are responsible for formulating business strategies and product operations are generally business majors. They possess professional knowledge of product, market, and target customer analysis and can better explore potential development opportunities in the market, providing specific strategies and directions for project implementation. Therefore, each member of the team needs to choose a suitable direction based on their characteristics and knowledge to achieve collaborative learning, accumulate experience, and enhance their abilities.

In addition, participating in innovation and entrepreneurship competitions is an excellent way for students to exercise their thinking skills. For example, in the "Step-by-Step Fingerprint Matching Model Based on the Central Area Triangle Model" project, students need to first analyze the factors that affect fingerprint retrieval and determine the main objects for screening before establishing the model. Secondly, in the process of building the model, students need to conduct extensive literature review and research the cutting-edge knowledge in the relevant field, establish a mathematical model to solve the problem through analysis. Finally, for some models, students also need to consider some special cases and add robust mechanisms to protect or verify the feasibility of the model.

During the initial screening phase of model construction, the known point data is linked to a central point by introducing the mean point, and the distance and angle standard deviation of the point data are calculated as fingerprint evaluation factors, denoted as $\sigma_{d(m)}$ and $\sigma_{\theta(m)}$. These two factors are weighted and combined to obtain the characteristic score of $S(m)$ as the overall fingerprint feature score. The process is shown in Fig. 6:

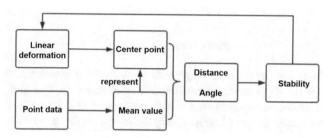

Fig. 6. Scatter Plot with Linear Transformation Relationship

For the same minutiae group of a fingerprint, there is always a linear transformation as follows:

$$P_i = R \times P_j + T = \begin{pmatrix} cos\theta & -sin\theta \\ sin\theta & cos\theta \end{pmatrix} \times P_j + \begin{pmatrix} x_i \\ y_i \end{pmatrix} \tag{13}$$

The above formula represents the mapping relationship between the same minutia group $P_i(x_i, y_i)$ and $P_j(x_j, y_j)$, where R is a rotation linear transformation and T is a translation linear transformation along the x and y axes. However, due to the interference of non-linear deformation in fingerprint acquisition, the mapping is not one-to-one. Therefore, distance standard deviation $\sigma_{d(m)}$ and direction standard deviation $\sigma_{\theta(m)}$ are proposed as indices.

Regarding the standard deviation of distance, the x-axis pixel coordinate $\overline{x_n} = \frac{\sum_{i=0}^{m} x_i}{n}$ and y-axis pixel coordinate $\overline{y_n} = \frac{\sum_{i=0}^{m} y}{n}$ of the average point can be calculated from the given data, and thus the coordinates of the average point can be obtained. The distance formula can be used to calculate the distance from each fine point $(\overline{x_n}, \overline{y_n})$ to the coordinate point, and then m effective fine points can be selected to represent the degree of centralization between them and the average point, that is, the distance of fine points in the fingerprint from the average point. The formula is as follows:

$$\sigma_{d(m)} = \sqrt{\frac{\sum_{i=0}^{m}[(x_i - \overline{x_n})^2 + (y_i - \overline{y_n})^2]}{m}} \tag{14}$$

For the selected m points, it is necessary to further observe their relative positions. Introducing the directional standard deviation, since the direction information of each fine node has already been provided in the data, it is only necessary to calculate the directional standard deviation within the m points to represent the relative rotation angles of the fingerprint fine nodes around the average point, as follows:

$$\sigma_{\theta(m)} = \sqrt{\frac{\sum_{i=0}^{m}[(\theta_i - \theta_n)^2]}{m}} \tag{15}$$

Next, the overall feature score $S(m)$ of the model is introduced to represent the overall feature information of the fingerprint. During screening, it is only necessary to check the matching degree of $S(m)$, which is matching the fingerprint with the $S(m)$ value of each fingerprint. If there is a fingerprint whose difference from its $S(m)$ value is within a certain range, a preliminary judgment of identity can be made, and the initial screening is completed. The specific formula is as follows:

$$|S_T(m) - S_i(m)| \leq \phi \tag{16}$$

$$S(m) = a\sigma_{\theta(m)} + b\sigma_{d(m)} \tag{17}$$

Among them, T represents the fingerprint to be tested, i represents the fingerprint in the template library, a is the parameter that determines the standard deviation of the orientation of the finer points, and b is the parameter that determines the standard deviation of the distance. a and b must satisfy the condition that $a + b = 1$, and the size of the effects of the two factors in the component on the finer point matching needs to be considered.

In the parameter tuning stage of the preliminary screening, a stability evaluation score $\omega(m)$ is introduced to represent the stability of the finer points of the same fingerprint under different conditions, where $\omega(m)$ values lower indicates greater stability of the model. The parameter m represents the number of selected finer points. The formula for the stability evaluation score $\omega(m)$ is as follows:

$$\omega(m) = \frac{\sum_{i=0}^{m}(a\sigma_{\theta(i)} + b\sigma_{d(i)})}{m} \tag{18}$$

During the fine screening stage, the difference between the minimum angle, the second minimum angle, and the maximum side length of a fine node is calculated based on its information. Further, a similarity function Si is constructed to represent the similarity of the triangle structure.

To extract triangle α_i from the retrieval vector of the fingerprint template library and triangle β_i from the input fingerprint retrieval vector, the differences δ_1, δ_2, and δ_3 between the minimum angle α_{i-min}, β_{i-min}, the second minimum angle α_{i-med}, β_{i-med}, and the maximum side length $\alpha_{i-\tau}$, $\beta_{i-\tau}$ of the two triangles are calculated using the following formula:

$$\delta_1 = |\alpha_{i-min} - \beta_{i-min}| \leq \delta_{min} \tag{19}$$

$$\delta_2 = |\alpha_{i-med} - \beta_{i-med}| \leq \delta_{med} \tag{20}$$

$$\delta_3 = |\alpha_{i-\tau} - \beta_{i-\tau}| \leq \delta_{\tau} \tag{21}$$

Further determination of the matching degree of triangles is done by calculating the structural similarity function of the triangles and introducing the Si function.

$$Si = \frac{\frac{\delta_{min} - \delta_1}{\delta_{min}} + \frac{\delta_{med} - \delta_2}{\delta_{med}} + \frac{\delta_{\tau} - \delta_3}{\delta_{\tau}}}{3} \tag{22}$$

Usually, the larger the value of the *Si* function, the more similar the two triangles are in their structural composition.

To protect the system from special cases that may make matching difficult during runtime, a robust mechanism has been added. In the coarse screening stage, the lower limit of the number of effective fine nodes in the central domain is set to m. In the fine screening stage, the lower limit of the number of effective details in the sub-central domain is set to q. Additionally, a loop mechanism is included, such as expanding the search area again and continuing the fine screening loop to search again. After verification, the accuracy of this model is above 95%, reaching the practical stage.

Throughout the competition, students need to continuously analyze problems, study them, and finally solve them. Their thinking ability will be greatly exercised.

4   Conclusion

Our school, the School of Mathematics and Information Science at Guangzhou University, has implemented a phased training program for students participating in the "Internet+" entrepreneurship competition. The program aims to cultivate composite talents with innovative entrepreneurial consciousness, unique business vision, innovative spirit, and teamwork collaboration ability through three aspects: innovation and entrepreneurship basic skills, team collaboration based on cooperative learning, and innovation consciousness and teamwork spirit. This effort has achieved remarkable results.

Based on the principles of cooperative learning and the model of cooperative learning based on the Internet Plus, the practical application of the cooperative learning-based Internet Plus innovation and entrepreneurship training model has promoted students' development in various aspects. Under this model, students' entrepreneurship abilities have been exercised, their teamwork spirit has been developed, their innovation abilities have been strengthened, and their comprehensive abilities have also been improved. In this process, students can truly connect with society and constantly improve their practical innovation and entrepreneurship skills through practice and communication.

References

1. The 50th "statistical report on the development of china's internet" was released by the China Internet Network Information Center. J. Natl. Libr. Sci. **31**(05), 12 (2022)
2. Guiding opinions of the state council on actively promoting the "Internet Plus" action. Chin. Ministry Educ. Bull. (Z2), 6–19 (2015)
3. Wang, T.: The Concept and Implementation of Cooperative Learning. China Personnel Press, Beijing, China (2002)
4. Sheng, Q., Zheng, S.: Collaborative Learning Design. Zhejiang Education Press, Hangzhou (2006)
5. Sheng, Q.: What kind of teaching tasks are suitable for cooperative learning. People's Educ. (2004)
6. Slavin, R.E., Tan, W.: A research on cooperative learning. International Perspectives (1994)
7. Sharen, S.: Collaborative learning theory. Shandong Education Research, May 1996
8. Cao, J., Shao, L., Yu, A., Feng, L., Wen, J.: Exploration of students' innovative thinking ability based on "Internet+" college students' innovation and entrepreneurship competition. Sci. Consult. (Sci. Technol. Manage.) (08), 47–49 (2022)

9. Xie, Y.: Analysis of the effect of campus business simulation challenge on high school students' understanding of society. Inspir. Mag. (02), 265 (2017)
10. Tan, D., Liu, X., Jiang, W., Zhan, L.: Research on the mechanism and path of cultivating college students' innovation and entrepreneurship ability under the background of internet plus. Commer. Exhibit. Econ. (20), 106–108 (2022). https://doi.org/10.19995/j.cnki.CN10-1617/F7.2022.20.106

Research on the Optimization Model of Ethanol Coupling to C4 Olefins Based on Regression Analysis

Ying Xie, Qi Li, Wenya Zhu, Qiwen Wu, Jun Wan, and Hong Mai[✉]

School of Mathematics and Information Science, Guangzhou University, Guangzhou, China
yjxgd@163.com

Abstract. China is mainly prepared with fossil raw materials, C4 olefin, ethanol is a good raw material for the preparation of low-carbon olefins, but due to the preparation of ethanol coupling, the reaction principle of C4 olefins is more complicated, and a lot of research has not been invested in China, on this basis, the comprehensive use of BP neural network, genetic algorithm, orthogonal test and other methods, combined with SPSSpro, MATLAB and other software to process the data, explore the influence of catalyst combination and temperature on C4. Olefin yield. The results showed that when the Co load was 2 wt%, the Co/SiO2 content was 120 mg, the HAP content was 200 mg, the ethanol concentration was 2.1ml/min and the temperature was 400, the yield of C4. Olefin was about 62%. The obtained ethanol coupling preparation, the optimal control scheme of C4. Olefins is more efficient, can effectively promote the production of C4 olefins, in order to make up for the traditional preparation, C4 olefins shortcomings. And it can provide additional researchers with certain ideas.

Keywords: Fitting · multiple linear regression · genetic algorithms · BP neural networks · orthogonal experimental design

1 Background of the Issue

1.1 Question Restatement

In recent years, with the rise crude oil prices in international and the rising production costs of olefins, the central government pro-posed to accelerate the development of methanol, ethanol, dimethyl ether, coal-to-liquid, etc. as oil substitutes, and listed coal chemical technology as one of the key points of national scientific and technological research. According to the research prospect of ethanol coupling to prepare C4olefins [1] and the feedback information of technology development, this technology has great research value and application level.

1.2 Ask Questions

1. The established and well-established catalyst combinations and temperature control scheme were explored to explore their effects on ethanol conversion and selective size of C4 olefins.
2. The effects of these variables and temperatures on ethanol conversion and C4 olefin selectivity were explored for catalyers with different ethanol concentration, Co loading, Co/SiO2 content and HAP content.
3. How to select the catalyst n combinations and temperature so that ethanol coupling to prepare C4 olefins yields most and obtain the optimal control scheme.
4. The research on the optimal control scheme of ethanol-coupled C4 olefins, combined with the actual production, carried out many experiments to provide more complete production schemes for industrial C4 olefins.

2 Problem Analysis

Aiming at the problem1: In order to study the relationship between temperature and different variable in each group of catalyst combinations, using the temperature of distinct groups as the independent variable, and performing a linear regression [2] on ethanol change rate and C4 olefin selectivity.

In response to Question 2: Obtaining the linear regression equations between ethanol conversion and C4 olefin selectivity and all variables. In addition, using the BP neural network model to train the collated data to obtain a neural network for ethanol change rate and C4 olefin selectivity prediction, and observe its predictive effect.

For problem 3: Using fitting equations to model the information about the yield of C4 olefin. For the trained neural network. Data generated according to certain rules, and then the trained neural network is used to make predictions.

To the fourth question: An analytical model based on orthogonal experimental design [3] was established, with Co loading, CoSiO2 dose, HAP dose, and ethanol concentration as independent variable factors, orthogonalizing it, and generating no interaction with the orthogonal test table of overlapping influences.

3 Model Assumptions

1. It is assumed that the temperature and the configuration of the catalyst in this experiment are accurate, and there is no error in temperature and catalyst.
2. The charge ratio of Co/SiO2 to HAP in the catalyst is precise and the catalyst contains no other impurities.

4 Symbol Description

symbol	meaning
Z_1	Ethanol conversion rate
Z_2	C4 olefin selectivity
x_1	Co load
x_2	Co/SiO2 amount
x_3	HAP volume
x_4	Ethanol concentration
x_5	temperature
ε	Random error term
$\beta_0, \beta_1, \beta_2, \beta_3, \beta_4, \beta_5$	Regression coefficient

5 Data Preprocessing

First, we simplified the established data into a relatively complete catalyst combination and temperature control scheme, then deleted the abnormal data, and analyzed whether the two different charging methods in the experiment would lead to the error of the experimental results. By drawing the relationship between ethanol conversion and temperature under different charging methods of the two groups of same charging materials, as well as the relationship between C4 olefin selectivity and temperature, we found that the change trend and degree of the two groups were almost the same. Therefore, we can conclude that the two charging modes have little influence on the experimental results, and the data of group A and group B can be combined for analysis.

6 Model Establishment and Solution

6.1 Problem 1: Model Establishment and Solution

Establishment of Model
Based on the analysis of relatively complete catalyst combination and temperature control scheme, the relevant data are sorted out, and part of the table is shown in Table 1:

Table 1. Data required for the analysis of Question 1

Catalyst combination number	temperature	Ethanol conversion rate	C4 olefin selectivity
A1	250	0.0207	0.34
	...		

(continued)

Table 1. (*continued*)

Catalyst combination number	temperature	Ethanol conversion rate	C4 olefin selectivity
	350	0.3680	0.47
...	...	...	...
B7	250	0.0440	0.04
	...		
	400	0.6940	0.38

Unvariable linear regression equation was used to analyze the relationship between ethanol conversion and temperature, and the selectivity of C4 olefin and temperature under different catalyst combinations, and the following models were obtained:

$$A_i = \hat{\beta}_0 + \hat{\beta}_1 T$$
$$\tilde{A}_i = \hat{\beta}'_0 + \hat{\beta}'_0 T \tag{1}$$

The experiment proves that the multivariate variables are not always linearly correlated. This paper further selects some unitary linear regression equations of ethanol change rate with respect to temperature and the results of goodness of fit and probability p for analysis, as shown in Table 2:

$$\begin{cases} y_1 = 0.15436T - 3.242 \\ y_2 = 0.2216T - 41.546 \\ y_3 = 0.2612T - 59.1949 \\ y_4 = 0.2266T - 52.4105 \\ y_5 = 0.2297T - 57.813 \\ y_6 = 0.2864T - 72.820 \\ y_7 = 0.2683T - 65.888 \\ \quad\quad \ldots\ldots \end{cases} \tag{2}$$

Table 2. Linear regression for each group: goodness-of-fit and probability p (partial).

combination	Goodness-of-fit R2	Probability p
A1	0.78691	0.04476
A2	0.83592	0.02973
A3	0.91281	0.00078
...	...	...

The above table shows that the fitting effect is not good. In this paper, the unvariable linear regression is not used, but the cftool fitting toolbox [4] in MATLAB [5] is used to carry out multiple fitting for the temperature, ethanol conversion and selectivity of C4 olefin in each catalyst combination, and the optimal fitting equation is obtained.

Solution of Model

Cftool fitting toolbox was used to establish two sets of data of temperature and ethanol conversion, temperature and C4 olefin selectivity for each catalyst combination. On the basis of observing the scatter diagram of each group of data, make proper fitting analysis for each group of data, and determine the adjusted R-squared size to evaluate the fitting effect, among which 8 groups are exponential fitting, 40 groups are Gaussian fitting [6]. After comparing 40 groups, it is found that the adjusted R-squared value of the 16th group is closest to 1, indicating the best fitting effect.

The Gaussian fitting function model [7] is as follows:

$$y = ae^{-\left(\frac{x-b}{c}\right)^2} \tag{3}$$

The fitting results obtained by using the matlab fitting toolbox are shown in Table 3 and Table 4.

Table 3. Group parameters for the conversion rate and temperature of ethanol satisfying Gaussian fitting under different catalyst combinations

Constituencies	R^2 (goodness-of-fit)	a	b	c
A1	0.9633	315.6	569.3	149.2
A3	0.9848	88.13	441.1	129.1
B1	0.9998	97.87	520.4	133.9
B3	0.9989	344.7	647.6	148.2

Table 4. Table1parameters for the conversion rate and temperature of ethanol satisfying Gaussian fitting under different catalyst combinations

Constituencies	R^2 (goodness-of-fit)	a	b	c
A1	0.8716	48.84	331.9	128.8
A3	0.9947	54.91	416.1	107.6
A4	0.9528	88.23	575.9	202.2
B1	0.9962	59.01	495.2	158.6
B2	0.9991	46.21	450	119
B3	0.9614	47.84	582	202.4

By observing the above table, it is found that the effects of each group are very significant under Gaussian fitting.

Exponential fitting equation model:

$$y = ae^{bx} \tag{4}$$

The fitting results obtained by using the matlab fitting toolbox are shown in Table 5 and Table 6.

Table 5. Group parameters for fitting ethanol conversion and temperature satisfaction index under different catalyst combinations

Constituencies	R	a	b
A5	0.9869	0.394	0.01313
A9	0.9994	0.006922	0.0217
B2	0.999	0.01647	0.01978
B5	0.9998	0.01256	0.02046

Table 6. Olefin selectivity and temperature satisfaction index fitting under different catalyst combinations

Constituencies	R	a	b
A2	0.8882	1.235	0.009808
A6	0.9464	0.02233	0.01849
A10	0.9166	0.01622	0.01605
B4	0.9114	0.3054	0.0106

According to the above table, the goodness of fit of each group is very significant under the index fitting, that is, the fitting effect is good.

6.2 Problem 2: Model Establishment and Solution

Establishment of Model
To Analyze the data. Then, The effect of the multiple linear regression equation [8] is roughly shown in the following equation:

$$\begin{cases} Z_1 = \beta_0 + \beta_1 x_1 + \beta_2 x_2 + \beta_3 x_3 + \beta_4 x_4 + \beta_5 x_5 + \varepsilon \\ Z_2 = \beta_0' + \beta_0' x_1 + \beta_2' x_2 + \beta_3' x_3 + \beta_4' x_4 + \beta_5' x_5 + \varepsilon \end{cases} \tag{5}$$

Use BP neural networks to fit and predict the related ethanol conversion and C4olefin selectivity. The Fig. 1 below shows a neural network with two hidden layers:

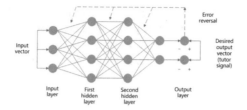

Fig. 1. Simplified representation of a neural network

The related process is shown in the following Fig. 2.

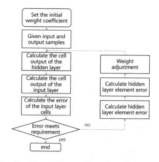

Fig. 2. Flowchart of a neural network

After training the model, substituting the relevant data to be predicted to view its prediction results, and comparing the predicted results with the values given in the original attachment.

Solution of Model

After multiple linear regression analysis of the data in Table 1, we can obtain the equations:

$$\begin{cases} Z_1 = -0.8257 + 0.0014x_1 - 0.0004x_2 + 0.00145x_3 - 0.0874x_4 + 0.0034x_5 \\ Z_2 = -0.5007 - 0.0321x_1 + 0.0015x_2 - 0.0006x_3 + 0.0278x_4 + 0.0019x_5 \end{cases} \quad (6)$$

we use the goodness-of-fit R^2 to evaluate these two equations.

Table 7. Goodness-of-fit for the two equations

equation	Z_1	Z_2
Goodness-of-fit R^2	0.7967	0.7331

As can be seen from the Table 7, it fit of the multiple linear regression model to the data in Table 1 is not ideal, and there is a large error.

We use Bayesian regularization methods to train neural networks. Y is the ethanol conversion rate corresponding to each group of data, Y_1 is the C4 olefin selectivity corresponding to each group of data, and $NewX$ is the data needed to predict the ethanol conversion rate and C4 olefin selectivity. Their goodness-of-fit are shown in Fig. 3 and Fig. 4.

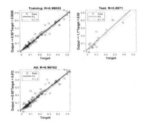

Fig. 3. Fitting curve of neural networks to genetic conversion rate

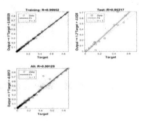

Fig. 4. Fitting 1 C4olefins

We use the trained neural network and predict data, to get their predictions of ethylene conversion and C4 olefin selectivity in the data.

Table 8. For neural networks predicting ethanol conversion and C4olefin selectivity

Catalyst combination code	Ethanol conversion rates	Predicted ethanol conversion rate	C4 olefin selectivity	Predicted C4olefin selectivity
A1	0.0207	0.0113	0.34	0.3441
	...			
	0.3680	0.3455	0.47	0.5064
...				
B7	0.0440	0.0513	0.04	0.0477
	...			
	0.6940	0.6959	0.38	0.3798

From the above Table 8, its prediction is better. But the neural network model can only predict the value.

6.3   Problem 3: Establishment and Solution of the Model

Establishment of Model

From the information given in the title, we can know that the yield of C4olefin is related to ethanol conversion and C4olefin selectivity, and their relationship is shown in the following formula:

$$Yield\ of\ C4\ olefin = Ethanol\ conversion\ rate \times C4\ olefin\ selectivity \tag{7}$$

Therefore, we can established in problem 2 to build an optimization model [9] as follows:

$$S = \max Z_1 \times Z_2 \tag{8}$$

$$\begin{cases} 0.5 \leq x_1 \leq 2 \\ 10 \leq x_2 \leq 200 \\ 10 \leq x_3 \leq 200 \\ 0.3 \leq x_4 \leq 2.1 \\ 250 \leq x_5 \leq 450 \end{cases} \tag{9}$$

We then use the code to generate 5 factor values at appropriate step sizes for various possible catalyst combinations and temperatures. The specific values are shown in the following Table 9:

Table 9. Randomly generate the range and step size of the experimental data

	minimum	Step	maximum	The number of groups with the value
Cload capacity	0.5	0.5	2	4
Co/SiO2 content	10	10	200	19
HAP content	10	10	200	19
Ethanol concentration	0.9	0.2	2.1	6
temperature	250	50	450	3

Solution of Model

In the genetic toolbox, the fitness function value and the optimal individual are selected. Then run the genetic algorithm to get the following data in Table 10 and graph results in Fig. 5.

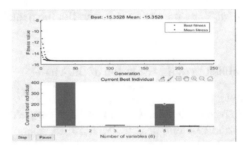

Fig. 5. Optimal C4olefin yield without temperature limitation

From the operating results, we found its corresponding catalyst combination and temperature, as shown in the following Table 10:

Table 10. Combination and temperature at optimal C4 yield

factor	temperature	Co load	Co/Sio content	HAP content	Ethanol concentration
No temperature limit	400	0.3	10	2.1	200

BP neural network prediction solution: According to the data processing method mentioned above, we can get the top five catalyst combinations and temperatures for C4 olefin yields, as shown in Table 11.

Table 11. Top 5 optimal catalyst combinations and temperatures

serial number	Co load (wt%)	Co/SiO2 content (mg)	HAP content (mg)	Ethanol concentration (ml/min)	temperature	C4olefin yield
1	2	120	200	2.1	400	0.620
2	2	130	200	2.1	400	0.619
3	2	110	200	2.1	400	0.611
4	2	140	200	2.1	400	0.609
5	2	170	200	1.3	400	0.596

6.4   Problem 4: Establishment and Solution of Model Model

Establishment of Model

Four factors were selected for orthogonal experimental design, and the results were as shown in Table 12:

Table 12. Six factors and the number of levels in their factors

numbering	Co load (wt%)	Ethanol concentration (ml/min)	Co/SiO2 content (mg)	HAP content (mg)
1	0.5	0.3	10	10
2	1	0.9	25	25
3	2	1.68	33	33
4	5	2.1	50	50
5			67	67
6			75	75
7			100	100
8			200	200

Solution of Model

Step1: A table of orthogonal experimental design is generated by using SPSS at different levels for 4 factors.

Step2: Some data in the orthogonal design table obtained by MATLAB programming query, screen out the repeated data, and get the final orthogonal experimental design table.

Step3: Bring the final orthogonal experimental design table into the bp neural network trained in problem 3, and obtain the top five experimental groups, as shown in the following Table 13.

Table 13. Experimental groups of the first five groups

Co load (wt%)	Ethanol concentration (ml/min)	Co/SiO2 content (mg)	HAP content (mg)	Yield of C4olefin
0.5	0.9	100	90	0.528621417
0.5	0.9	67	67	0.503507337
0.5	0.3	75	25	0.485287094
0.5	1.68	100	75	0.484539639
0.5	0.9	100	25	0.460703203

7 Model Evaluation

7.1 Advantages of the Model

1. Analyze the data with visual graphics to get the model
2. A variety of models were used to fit the relationship between variables and optimize the model

7.2 Disadvantages of the Model

1. Since the amount of experimental data is not too much, the trained neural network *net* and *net*1 cannot predict the randomly generated data we give well, and there are some errors, and the model is not universal.

8 Conclusions

In order to solve the problem of the preparation of C4 olefin, we established a linear regression and stablished model about different temperature. We found BP neural network had a better fitting degree. In order to create more C4 alkenes, we establish the optimal model equation, and use algorithms to solve the variables. Analytical model of an orthogonal experimental was disigned to provide a perfect scheme.

Acknowledgments. This work is supported by 2022 Provincial Student Innovation Training Project of Ministry of Education of P.R.C (S202211078214).

References

1. Shaopei, L.Y.U.: Ethanol Coupling to Prepare Butanol and C4 Olefin. Dalian University of Technology, Liaoning (2018)
2. He, X.: Applied Statistical Regression Analysis: R Language Edition, Publishing House of Electronics Industry, Beijing (2017)
3. Zou, Y.: Research on the application of orthogonal experimental design preferred teaching method. J. Sci. Teach. (16), 108–111 (2018)
4. Cheng, J., Zheng, M., Lou, J.: Comparison of common experimental optimization design methods laboratory. Res. Explor. **31**(7), 7–11(1) (2012)
5. Shoukui, S., Zhaoliang, S.: Mathematical Modeling Algorithms and Applications. National Defense Industry Press, Beijing (2015)
6. Qinglong. a Gaussian function (Gaussian function), a detailed analysis of [DB/OL]. https://blog.csdn.net/qinglongzhan/article/details/82348153, 03 Sep 2018
7. Simple but extraordinary. Gauss curve fitting principle and implementation [DB/OL]. https://blog.csdn.net/c914620529/article/details/50393238, 24 Dec 2015
8. Li, B.L.: Multiple adaptive least squares curve fitting method and its application. Yangtze University (2014)
9. Probability Theory and Mathematical Statistics. Higher Education Press, Beijing, pp. 314–373 (2019)

An Empirical Study on Eco-Tourism Network Integral Problem

Jiasheng Wu[1], Yongru Cen[1], Ziqing Li[1], Weiyu Liu[1], Lanxi Bai[1], and Shuitian Wu[1,2(✉)]

[1] School of Mathematics and Information Science, Guangzhou University, Guangzhou, China
Suidenn@gzhu.edu.cn
[2] School of Management, Guangzhou University, Guangzhou, China

Abstract. With the improvement of people's living standard, the demand for ecotourism is becoming more and more vigorous, which provides the impetus for the implementation of ecotourism network points system. Taking South China Botanical Garden in Guangzhou as the research object, through questionnaire survey and interview methods, this paper analyzes the tourists' cognition of eco-tourism network point system. The results show that the proportion of tourists who love nature and pay attention to ecological protection is relatively high, but the proportion of tourists who participate in ecological protection through actual behavior is low, and the awareness of ecological and environmental protection is not synchronized with the behavior. It is of practical significance to implement the eco-tourism points system.

Keywords: ecotourism · Network points system · South China Botanical Garden

1 Related Research Progress

The 19th National Congress of the Communist Party of China put forward that the major contradiction in the new era has been transformed into "the contradiction between unbalanced and inadequate development and people's ever-growing needs for a better life". People have a deeper understanding of health, and tourists are more inclined to choose green and healthy tourism methods. Therefore, the implementation of eco-tourism network points system is possible and will promote the development of eco-tourism. Ecotourism is a form of tourism in line with the pursuit of a better life, based on the premise of not destroying the ecological environment and aiming at strengthening the body of tourists and being friendly to the ecological environment. However, compared with the traditional way of tourism, people have different understandings of ecotourism. Tourism agencies lack ecotourism products, ecotourism destinations still lack a modern management system, and tourism service facilities and incentive mechanism need to be improved. In this regard, the relevant personnel have carried on the exploration to the ecotourism industry, and put forward the sustainable development of the method, the development of ecotourism to provide substantive suggestions [1]. In Gwangju, South Korea, "carbon banking system" has been implemented. Tourists can

B.-Y. Cao et al. (Eds.): ICFIE 2022, LNDECT 207, pp. 189–200, 2024.
https://doi.org/10.1007/978-981-97-2891-6_15

obtain points through relevant means, and the points have the functions of consumption and discount, forming a relatively complete point system and operation system. Combined with network technology, the advantages of the point system can be more perfectly played, and at the same time, the long-term development of ecotourism and the sound development of marketization can be promoted [2]. On January 18, 2022, the National Development and Reform Commission and other seven departments jointly issued the Implementation Plan to Promote Green Consumption, which put forward a green consumption points system to encourage green consumption by exchanging points for related products and some discounts. The South China Botanical Garden of the Chinese Academy of Sciences (in Guangzhou, hereinafter referred to as the South China Botanical Garden), as the second botanical garden in China, is of high ecological and environmental quality, making it suitable for eco-tourism. According to the reality of tourism in South China Botanical Garden, this study explores the effect of network points system on eco-tourism.

Ecotourism takes sustainable development as the concept and ecological environment protection as the premise. It is a modern tourism mode that conforms to the harmonious coexistence between man and nature and returns to ecology and nature. The term "ecotourism" is often used as a synonym for green and sustainable tourism. Ecotourism is a subsector of tourism, as a specific type of tourism, designed within the framework of an industrial age worldview. Originally defined as: "Travel to relatively undisturbed or unpolluted natural areas for the specific purpose of studying, appreciating and enjoying the landscapes of these areas and their wildlife, as well as any existing (past and present) cultural expressions." The ecotourism industry became a key promoter of innovative products - nature reserves. National parks and nature reserves have emerged as innovative ecotourism products, and one task for tourism managers is to bring them to the tourism market through promotional marketing strategies. The demand for ecotourism products is based on an individual's desire to experience nature and the diversity and richness that wild nature has to offer; Travel to places such as various parks and reserves has also become a legitimate response to people's needs [3].

Ecotourism has become a dominant industry in promoting ecological civilization construction in the new era [4]. The development of ecotourism resources has become an important way to promote regional economic development, rural revitalization and improve the sustainability of farmers' livelihoods, and is one of the important subjects to promote the development of ecotourism industry in ethnic minority areas [5]. Region-wide eco-tourism construction can significantly improve the level of economic development and environmental protection [6]. In recent years, the development of Chinese traditional tourism was challenged greatly, and the transformation of traditional tourism became inevitable [7]. China is shifting from the phase of pursuing quantity development to the phase of pursuing quality development. The 20th National Congress of the Communist Party of China proposed that "Chinese-style modernization is the modernization of harmonious coexistence between man and nature". Eco-tourism has also become an important part of the 2030 vision goal, and eco-tourism is in line with the policy of high-quality development and the new concept of "green" development [8]. In the past three years, the long distance tourism market has experienced great volatility, and the outbound tourism market has just recovered. However, short distance travel is a major way of travel, and looking for short distance and uncrowded tourist destinations has become a hot spot in some markets, which directly stimulates the interest of eco-tourism lovers. Therefore, it is of practical significance to make full use of the Internet for information exchange and develop the eco-tourism points system.

2 Research Object and Method

Ecotourism network point system is a network point system which is established on the basis of ecotourism and with the help of Internet and computer technology. South China Botanical Garden is an ideal case study of eco-tourism network points system. Located in Tianhe District, Guangzhou, South China Botanical Garden is the second national botanical garden in China and the largest South Asian tropical botanical garden in the world. Covering a total area of 333 hectares, the park is divided into 38 special parks. It is a comprehensive theme park integrating science popularization, culture, education, sightseeing, leisure and entertainment, with beautiful landscape and rich scientific and cultural connotations. Each plant in the park is hung with science popularization cards printed with the name and introduction of the plant, and each plant is hung with two-dimensional code. Visitors can scan the code to see the introduction of the plant and listen to the voice explanation. It is a new way of ecological knowledge popularization supported by digital technology. At present, the South China Botanical Garden has successfully realized the return of 36 species of rare and endangered plants in South China, such as rhododendron and red mountain tea, so that 95% of the rare and endangered plants in South China have been effectively protected. At present, the garden has 17,168 ex situ protected plants, 643 rare and endangered plants, and 337 state key protected plants. The abundant species and unique landscape design attract a large number of tourists to the park for ecological tourism. Relying on its rich ecological and environmental resources, South China Botanical Garden has initiated scientific and technological innovation activities, such as "Internet of Things + Horticulture" popular science experience course, promoting research cooperation with primary and secondary schools, increasing camping and other sites in the park, etc., to promote economic development of the park while protecting natural resources, which is in line with the purpose of eco-tourism.

This study adopts the methods of interview survey and observation. An online questionnaire survey was carried out in May and June 2022. In September 2022, the research team went to South China Botanical Garden for field research, observing and taking photos to record the environment of South China Botanical Garden and tourists' eco-tourism. Recorded the vegetation environment and landscape design of the South China Botanical Garden, the local tourist flow, the age group of tourists, the consumption of tourists in the South China Botanical Garden, and the interview dialogue of tourists. At the later stage of the survey, the interview content and questionnaire of tourists were statistically analyzed, and finally the views of tourists of South China Botanical Garden on the network point system were obtained.

Social and economic background of the respondents: the social and economic background and play characteristics of the tourists can be obtained through relevant research questionnaires. The age group of the respondents is 18–45 years old, accounting for

70.9%. Bachelor's degree and junior college degree were mainly included (69.4%). Students (35.4%), teachers, scientific researchers (14.6%), and enterprise workers are a relatively large occupation. In the family structure, the majority lived with their parents (39.6%) and with primary school students (24.3%). Only 51.4% of the tourists visited for the first time, which shows that the revisit rate of the botanical garden is relatively high. Units travel less (5.6%), and several other travel methods accounted for a similar proportion. Tourism, acquisition of ecological knowledge and close to nature are the top three travel purposes, and the travel time was mainly 0.5 d (52.1%).

Therefore, it can be found that the survey objects of South China Botanical Garden are diversified, the main audience age group is young and middle-aged people, the audience group structure is mainly family travel, and the revisit rate of the botanical garden is high, indicating that the audience group is stable and long-term. Young and middle-aged people are more willing to accept new things and are more concerned about the ecological environment in China, so they are willing to understand and try the eco-tourism points system, which has a positive impact on the implementation of the eco-tourism network points system. In addition, the audience structure is given priority to with family travel, most families focus on the family education of children, in the process of family education, parents are the supporters of children's education resources and the creator of the education environment, therefore, parents for education children learning environment protection, ecological protection and other related knowledge support attitude, for the implementation of ecological tourism system provide good conditions. Therefore, the investigation of different social attributes.

In addition, south China botanical garden has always attaches great importance to science communication and popular science education, as early as in 1999, south China botanical garden became the first "national popular science education base", held popular science education activities, make full use of the base of south China botanical garden resources, popular science education activities main audience for family, followed by age in 6–15 years old middle school students and adult groups. Thus it can be seen that the science education system of South China Botanical Garden has been relatively perfect, with a solid foundation of science education, and can fully support the development of ecological points system in South China Botanical Garden.

In order to more intuitively understand the comprehensive perception characteristics of tourists on the eco-tourism network integration system, and explore the connection between high-frequency words, the micro word cloud is used to generate the network semantic relationship map (see Fig. 1). Through observation, it can be seen that the center of the whole map is "ecotourism", "Internet", "system", "and" tourists ", and these four words are very closely related, and these four words are closely related to:" plant "," tourism "," botanical garden "," ecology " and so on. At the same time, many words related to ecotourism, such as "park", "environment", "green" and "resources", are also present, which shows that people are particularly impressed by the green and sustainable development of ecotourism, and the green park has become a symbol of ecotourism. It can be seen that ecotourism has gradually become one of the common ways for people to travel, spreading the excellent traditional culture of various regions and driving the consumption growth of local products, not only injecting new vitality into the tourism industry, but also providing new impetus for economic development.

To sum up, the semantic network map mainly contains positive or neutral words about green development and economic development, indicating that tourists have a good overall image perception of eco-tourism and eco-tourism network integration system.

Fig. 1. Text semantic relationship network related to the ecotourism network integration system (From 微词云)

3 Research Results

A total of 143 valid questionnaires were collected in the online questionnaire survey. The interviewees included young people, middle- aged people, elderly people, traveling families and local workers, which ensured the diversity of survey samples and the availability of survey data.

The vast majority of tourists are willing to do some small tasks related to ecology in the process of ecotourism to get points, and use the points to exchange for tourist souvenirs, scenic spot tickets coupons, hotel coupons and so on to improve ecotourism the desire of Table 1.

Table 1. Tourists' willingness to engage in ecotourism

Question	Option	Proportion
If you can get points for ecotourism, which can be exchanged for tourist souvenirs, ticket coupons, coupons for hotel consumption and feedback ecology (planting flowers and plants in your name), are you willing to increase the number of ecotourism?	Be Willing	95.1%
	Unwillingness	4.9%

Most tourists are willing to get points by watching ecological protection videos and completing relatively simple ecological problems, while some tourists are willing to get points by punching in and taking photos of scenic spots, taking video publicity and completing small tasks offline. A very few tourists hope to get points through other ways. Most tourists tend to obtain ecological points while traveling in simple and convenient ways in Table 2.

Table 2. What tourists prefer to do

Question	Option	proportion
If you earn points for small tasks, what do you want it to do?	Watch ecological conservation videos	71.33%
	Answer the little video about ecological protection	65.73%
	Take photos and punch in on scenic spots or shoot promotional videos	53.85%
	Complete small tasks offline	32.17%
	Other	0.7%

Most tourists say that the ecotourism system is generally attractive to them, and a few say that they are not interested in ecotourism. The survey results are also closely related to the majority of young and middle-aged and middle-aged people in Table 3.

Table 3. Tourists' ratings of the point system

Question	Option	Proportion
How attractive is the ecotourism network points system to you?	1 point	2.8%
	2 point	4.2%
	3 point	1.4%
	4 point	4.2%
	5 point	15.38%
	6 point	19.58%
	7 point	16.08%
	8 point	16.78%
	9 point	5.59%
	10 point	13.99%

Tourists tourism purpose for the following: ① just like ecotourism destination and enjoy ② like to ecotourism destination and strict with their behavior is not damage to the environment ③ like to ecological tourism destination and do some public things, few is like to go and keen to protect the ecological environment in Table 4.

Table 4. Ecological tourism methods preferred by tourists

Question	Option	Proportion
Which way do you tend to travel to?	Just like the ecotourism destination and enjoy it	30.07%
	Like to go to an eco-tourism destination and be strict with their own behavior is not harmful to the environment	32.87%
	Like to go to eco-tourism destinations and do some public welfare things	32.87%
	Like to go and keen to protect the ecological environment	4.7%

As for the form of points cash, ① exchange tourist souvenirs ② tourism consumption vouchers (scenic spot tickets, hotel consumption coupons, etc.) ③ feedback ecology (planting flowers and plants in the name of individuals); people like the form of points cash to the same degree, and all show great enthusiasm. Tourists all like to enjoy the convenience or access to tourism-related resources to improve the quality of tourism in Table 5.

Table 5. The preferred form of point redemption

Question	Option	Proportion
What kind of points exchange form do you prefer?	Exchange for tourist souvenirs	65.03%
	Travel consumption deduction coupons (scenic spot tickets, hotel consumption coupons)	62.94%
	Give back to ecology (plant flowers plants)	55.94%
	Other	0%

The vast majority of tourists think that the ecotourism network points platform is helpful to improve their awareness of ecological and environmental protection in Table 6.

Table 6. Tourists' attitudes towards ecotourism

Question	Option	Proportion
Do you think the ecotourism network points platform is helpful to enhance your awareness of ecological and environmental protection?	Yes	95.8%
	No	4.2%

Most of the tourists support determining their eco-tourism destinations and travel routes through online media such as TikTok, Xiaohongshu and "b", and a small number of tourists believe that Kuaishou is also a very good platform. A very small number of tourists hope to determine the ecotourism destinations and travel routes through other platforms. This also reflects the high acceptance of most people to the eco-tourism points system in Table 7.

Table 7. What social media are tourists more inclined towards

Question	Option	Proportion
What platforms do you think is more appropriate to identify eco-tourism behavior?	Tiktok	74.83%
	The Little Red Book	56.64%
	Bilibili	52.45%
	Kwai	15.38%
	Other	4.9%

A total of 143 valid questionnaires were collected from the online questionnaire survey. The interviews included young people, middle-aged people, elderly people, traveling families and local workers, which ensured the diversity of the survey sample and the availability of the survey data.

The survey found that 95.1% of the respondents expressed their support for the ecotourism network points system, and believed that the system is of great help to ecological protection, and were willing to increase the number of ecotourism through the ecotourism network points system. To some extent, this shows that tourists generally hold a positive attitude towards the ecotourism network points system.

In terms of attraction, most tourists say that the eco-tourism system is generally attractive to them, while a few tourists say that they are not interested in the eco-tourism themselves or do not understand the tourism forms of eco-tourism. The survey results show that most tourists tend to immerse themselves in the beautiful experience of eco-tourism, and are not willing to spend time to understand or participate in ecological protection. The purpose of the investigators' participation in ecotourism is to enjoy the process of ecotourism and not cause damage to the ecology. To some extent, this shows that it is difficult to implement the ecotourism network points system, and it is necessary to overcome the stereotype of the masses on the concept of ecotourism.

In terms of access to ecotourism network points, 71.3% of respondents wanted to get points by watching ecological protection videos; 65.7% wanted to get points by answering ecology-related questions. Both of these investigators are willing to learn ecological protection and are willing to understand and participate in ecological protection. The results show that there are still most people who love nature and pay attention to ecological protection, and most of the tourists have a high awareness of ecological protection. In addition, 53.8% of the respondents hope through photo clock in ecological tourism attractions or promotional video access to integral, with the help of Internet media expression and transfer the love of ecology, the results reflect the Internet has high interactivity, immediacy, efficiency and convenience characteristics for ecological tourism integral system created more possibilities and increase the stronger attraction. The survey found that the number of people willing to complete small tasks of ecological protection in the process of ecological tourism is relatively small, 32.1%, reflecting that the proportion of tourists who participate in ecological protection through actual behavior is low, the awareness and behavior of ecological and environmental protection are not synchronized, and the motivation to participate in ecological and environmental protection is insufficient.

In the form of points redemption, 65% of respondents tend to exchange points for travel souvenirs; 62.9% of respondents tend to exchange points for travel consumption deduction vouchers, and 55.9% of respondents tend to use points to plant flowers and plants in their own name to give back to the ecology. These data show that tourists all like to enjoy it while traveling, and can obtain some material sustenance with certain ecological tourism value.

In terms of the network point system to improve tourists' awareness of ecological and environmental protection, 95.8% of tourists believe that the point system can improve their awareness of ecological and environmental protection to different degrees, which has good practical significance. The above data show that the ecotourism network points system is good for cultivating people's good habits of ecological protection and improving people's interest in ecotourism.

In order to further analyze the research results of the eco-tourism network integration system, investigate the focus of the project research survey and visualize the data obtained by the questionnaire survey, use WordArt online to generate word clouds with visual impact, and intuitively and efficiently display high-frequency words (see Fig. 2). Through observation, it is found that the key high-frequency words of the word cloud are "ecology", "tourism", "ecological tourism", "network", "points system", "tourists" and "ecological environment". In addition, "consciousness", "consumption", "economy", "experience", "group", "enjoy" enjoy ",", reward ",", Internet " and other words closely related to the eco-tourism network points system are also presented. Above key high frequency vocabulary focus illustrates the ecological tourism network integral system is established on the basis of ecological tourism, fully using the Internet and computer technology and establish a kind of integral and exchange integral double tube form of network integral system, has promote the sustainable development of ecological tourism, provide new economic development for ecological tourism, meet the demand of contemporary tourists ecological tourism and enhance the consciousness of tourists to protect the ecological environment of realistic significance and research value.

Through further analysis of the questionnaire results, we found that the promotion of ecotourism network points system is of high significance. The ecotourism network points system can increase the awareness of tourists to protect the ecological environment, and promote more tourists to choose environmental protection tourism routes and ecological tourism modes, so as to reduce the damage to the ecological environment. Through the establishment of the network points system, tourists can effectively encourage them to participate in more public welfare activities to protect the ecological environment, and improve the tourism experience and satisfaction of tourists. At the same time, the

Fig. 2. Research results of ecotourism network integral system (From WordArt)

ecotourism network points system can attract more tourists to participate in the eco-tourism consumption, promote the development of the ecotourism market and drive the economic growth of related industries. In contrast, the ecotourism network points system also faces some hidden problems. The ecotourism network points system needs appropriate technical support and management system, and needs to deal with tourist information, points statistics and reward distribution, which is difficult to implement. The ecotourism network points system also needs a very reasonable reward mechanism to better promote tourists to protect the ecological environment and participate in social public welfare activities, otherwise it may reduce the enthusiasm of tourists. In addition, the ecotourism network points system involves a large number of user information collection and processing work, which puts forward higher requirements for information security and privacy protection.

As Geoffrey Wall argued in the sustainability of the development of ecotourism in his Is Ecotourism Sustainable paper: If ecotourism is to be sustainable, if it is to be sustainable, it must be economically feasible, environmentally appropriate, and socio-culturally acceptable [9]. Without a positive experience, then tourists will not come to —— and there will be no tourism. If ecotourism is not economically feasible, then the facilities and services needed by most ecological tourists will not be provided, and the potential economic benefits of ecotourism to industry providers and local residents will not be realized. If the environment and its treasures are not maintained, then the resource base of ecotourism will be destroyed. —— If tourism continues to develop, it is impossible to become ecotourism unless someone can persuade tourists to restore the seriously degraded environment. If ecotourism is not culturally acceptable and the locals cannot benefit from the existence of ecotourism, they will become hostile to ecotourism and may destroy ecotourism. Economy, the environment, and the culture are all involved. Both of them are crucial to the successful introduction, operation and sustainable development of ecotourism. At the same time, many places are developing the virtual tourism of the existing tourism landscape, which can also be used as a module in the ecotourism network points system. The construction of the ecotourism network points system can not only publicize the scenic spots, increase their influence and attract tourists, but also meet some requirements of tourists for tourist attractions. Second, it can develop virtual tours for tourist landscapes that no longer exist or will soon disappear, which can be replicated. These digital landscapes have collectible value and are a natural and effective way to protect, replicate and disseminate cultural heritage [10]. Therefore, we conclude that the ecotourism system can only be organically combined with the local culture, economy and ecology.

Acknowledgments. The project is supported by the 2022 innovative training project of Guangzhou University "Exploration and Research on the Point System Platform of Eco-Tourism in the Post-Epidemic Period"(Number xj202211078253);the research results of the phase of the higher Education Bureau of Guangzhou University scientific research project "Guangzhou Red Cultural Resources Digital Protection and Tourism Development" (Number 202235162).

References

1. Zhang, Y.F.: Eco-tourism planning and sustainable development of tourism. Environ. Sci. Manag. **47**(05), 147–150 (2022)
2. Hongying, Z., Yuanfei, K., Lan, T.: Integration of green credits and carbon inclusive development to promote green consumption. World Environ. **3**, 48–51 (2022)
3. Dalia, V., Zivile, G., Rita, V., et al.: Barriers to start and develop transformative ecotourism business. Eur. Countryside **13**(04), 39–44 (2021)
4. Mingyue, Y., Jiaxin, C.: Eco-tourism practice of ecological civilization construction: theoretical logic and policy suggestions. Price Theor. Pract. **10**, 87–91 (2022)
5. Hangli, Z., Xinnan, A.: The impact of ecotourism on farmers' sustainable livelihood in minority areas: a case study of Lichuan City, Hubei Province. Forestry Econ. **44**(11), 22–39 (2022)
6. Zhang, Y.L., Xu, T.: Can the whole-region eco-tourism construction help the region to achieve both "lucid waters and lush mountains" and "gold and silver mountains"? Based on the experience of Fujian Province. J. Arid Land Resourc. Environ. **37**(01), 185–193 (2023)

7. Yinan, Z.: Exploring the transformation of traditional tourism in the era of internet e-commerce. Western Tourism. **08**, 1–3 (2022)
8. Ping, H., et al.: Study on high-quality development of ecotourism in the yellow river basin (Henan section). Coop. Econ. Technol. **03**, 33–35 (2023)
9. Guanmei, H., Zhen, W.: The application of virtual reality technology in the coordination and interaction of regional economy and culture in the sustainable development of ecotourism. Math. Probl. Eng. **2022** 2022
10. Yuan, Z., Shi, J.: Research on ecotourism resource attraction -- a case study of Hebei province. Prod. Res. **20**, 61–62+80 (2008)

Study on Iterative Desert Crossing Based on Game Theory

Jiekai Cao[1], Wenzhu Wang[1], Shengqi Deng[1], and Qingping He[2(✉)]

[1] School of Mathematics and Information Science, Guangzhou University, Guangzhou, China
[2] School of Physics and Materials Science, Guangzhou University, Guangzhou, China
hqping@gzhu.edu.cn

Abstract. Consider the following mini-game: With a map, players use initial funds to buy a certain amount of water and food (including food and other daily necessities), and walk through the desert from the starting point. Depending on the weather you will encounter on the way, you can also replenish funds or resources in mines and villages, with the goal of reaching the finish line within the allotted time and keeping as much money as possible. During the game, sometimes the player does not know all the situations, and sometimes does not know all the situations, at this time how to determine the walking route; Sometimes there are multiple players playing the game, and you need to determine the walking route.

Keywords: Crossing the desert · Optimal strategy · Linear programming model · Dynamic programming model · Cyclic iteration model

1 Introduction

Consider the following mini-game: With a map, the player uses the initial funds to buy a certain amount of water and food (including food and other daily necessities), and walks through the desert from the starting point. Depending on the weather you will encounter on the way, you can also replenish funds or resources in mines and villages, with the goal of reaching the finish line within the allotted time and keeping as much money as possible.

The basic rules of the game are as follows:

(1) With days as the basic unit of time, the start time of the game is day 0, and the player is located at the starting point. A player must reach the end point on or before the deadline, after which the player's game ends.
(2) Water and food are needed to cross the desert, and their smallest unit of measurement is a box. The sum of the water and food quality that the player has per day cannot exceed the maximum weight limit. If the end is not reached and the water or food is exhausted, the game is considered a failure.
(3) The weather every day is one of the three conditions of "sunny", "high temperature" and "sandstorm", and the weather is the same in all areas of the desert.
(4) Every day, players can reach another area adjacent to one area in the map, or they can stay in place. Sandstorm days must stay where they are.

(5) The amount of resources consumed by the player for one day in place is called the base consumption, and the number of resources consumed by walking for one day is multiple of the basic consumption.

(6) On Day 0, players can use their initial funds to purchase water and food at the base price. Players can stay at or return to the starting point, but cannot purchase resources at the starting point more than once. Players can return the remaining water and food when they reach the finish line, and each box is returned at half the base price.

(7) When players stay in the mine, they can obtain funds through mining, and the amount of funds obtained by mining for one day is called the basic income. If mining, the amount of resources consumed is multiple of the base consumption; If not mining, the amount of resources consumed is based on consumption. You cannot mine on the day you arrive at the mine. Sandstorm days can also be mined.

(8) Players can purchase water and food at any time when passing through or staying in the village with the remaining initial funds or funds obtained from mining, and the price per box is 2 times the base price.

1. Assuming that there is only one player, and the weather conditions are known in advance every day during the entire game period, try to give the optimal strategy of the player in general. Solve the "first level" and "second level" in the attachment, and fill in the corresponding results in the Result.xlsx respectively.

2. Suppose there is only one player, the player only knows the weather conditions of the day, you can decide the action plan of the day, try to give the best strategy of the player in general, and discuss the "third level" and "fourth level" in the attachment.

3. There are n existing players who have the same initial bankroll and start from the starting point at the same time. If any $k(2 \leq k \leq n)$ of them walks from zone A to zone B on a given day, any one of them will consume $2k$ times the amount of resources consumed by the base consumption; If any $k(2 \leq k \leq n)$ players mine in the same mine on a given day, the amount of resources consumed by any of them is 3 times the base consumption, and the funds obtained by each player through mining in a day are $\frac{1}{k}$ of the basic income; If any $k(2 \leq k \leq n)$ players purchase resources in the same village on a given day, the price per box is n6 times the base price. Otherwise, the number of resources consumed and the price of resources are the same as in single-player.

(1) Assuming that the weather conditions are known in advance for each day of the entire game period, the action plan of each player shall be determined on the day and cannot be changed thereafter. Try to give the strategy that players should adopt in general, and discuss the "fifth level" in the attachment.

(2) Assuming that all players only know the weather conditions of the day, from day onwards, each player will know the action plan of the remaining players for the day and the amount of resources remaining after the end of the day's action, and then determine their respective action plan for the next day. Try to give the strategy that players should adopt in general, and discuss the "sixth level" in the attachment.

For the first question, the question requires to find the player's optimal strategy, that is, to reach the end within the specified time, and the most funds are left at this time. According to the topic, income can only be obtained through mining, so the optimal strategy has three situations: first, through the shortest route to reach the end, at this time the least energy is consumed, so that you can also get a certain amount of remaining funds; Second, after mining in the mine, you can earn a certain amount of money, but after simple calculations, you can only go to the mine to dig once or twice. Considering these two scenarios, several questions arise: how much water and food the player needs to buy at the starting point; How much water and food the player needs to buy in the village; The player passes through several mines; How long each visit to the mine lasts. With these issues in mind, this paper uses a linear programming model to build a model to arrive at the best strategy on how to get the most money left.

For problem two, the question requires to find the player's optimal strategy, that is, to reach the end within the specified time, and the most funds left at this time, but compared to problem one, the problem deepens because the player does not know the weather conditions every day, so the situation to be considered in question one still needs to be considered in this problem, the solution of the problem focuses on how to solve the problem of weather, for the third level, because the map is relatively simple, you can use the exhaustive method to solve this problem; For the fourth level, because the map is more complex, it cannot be solved by exhaustive methods, and because the behavior of the day depends on the weather conditions of the day, and the behavior of the day affects the subsequent behavior decision, this paper uses dynamic programming to establish a model, so as to derive the best strategy of how to go to make the most funds left.

For the first question of question three, because the map of the question is the same as the map of the third level, the conclusion of the third level can be applied to the conclusion of the problem, but the problem that this problem deepens relative to the third level is that there are already two players at this time, in order to reach the end of the most funds left, so the player must consider the walking route of another player, and the other player will consider the walking route of the player, so that the cycle will continue, The cycle does not end until the player knows which route to take to reach the end and has the most money left. With this in mind, this article uses an iterative loop model to build a model to arrive at the best strategy for how the player can go to maximize the amount of money left. In response to the second question of question 3, because the map and conditions of this question are consistent with the fourth level, some of these conclusions are adopted, and multiple players play the game, so the loop iteration model is still used to solve the problem and the optimal strategy is obtained.

2 Model Building

(1) Suppose that in the first and second levels, the food and water left after arriving at the village are almost exhausted;

(2) In the third and fifth levels, it is assumed that the probability of high temperature weather is the same as the probability of sunny weather, that is, each of them, sandstorms will only appear for 1–2 days and will not appear sandstorms for two consecutive days;

(3) In the fourth and sixth levels, because the probability of sandstorm weather is small, the probability of sandstorm weather is ignored in non-extreme weather conditions;

(4) In the fifth and sixth levels, it is assumed that each player is sane [1], that is, they will not choose a route that will lose money if they walk alone;

(5) In the fifth and sixth levels, it is assumed that all players will consider the existence of other players and change their walking path accordingly [2, 3];

(6) In levels 3, 4, and 6, it is assumed that all players' strategies can at least guarantee that they will reach the end [1, 2].

3 Algorithm Implementation

Model Preparation

By analyzing the map of the first level, the shortest route from the starting point to the village is $1 \to 25 \to 24(26) \to 23 \to 22 \to 9 \to 15$(village) From the shortest route of the mine. 12(mine) $\to 14 \to 16 \to 17 \to 21 \to 27$ Or, furthermore, it can be found that the shortest distance from the starting point to the mine is unrelated to whether or not it passes through the village. Therefore, the final cost depends on the number of days of mining and the number of round trips between the village and the mine. 12(mine) $\to 14 \to 15$(village) $\to 9 \to 21 \to 27$

By analyzing the map of the second level, it can be determined that the distance is the same whether or not the mine is passed through. Furthermore, it is found that the remaining funds are less when the route that does not pass through the mine reaches the destination. Therefore, the question to consider is how many mines need to be passed through, resulting in two scenarios: one is to pass through only one mine, and the other is to pass through two mines. Among them, there are only two simple cases for passing through one mine: one is to purchase food and water at the starting point and complete the entire journey, and the other is to replenish food and water at a village along the way after purchasing supplies at the starting point, and then complete the entire journey. Through calculation and comparison, it is known that when reaching the destination by passing through two mines, there is more remaining funds. Therefore, only the detailed discussion of the scenario of passing through two mines is necessary.

Model Establishment and Solution
Level One

Consider the case of only going to the mine once to mine.
According to the question, the initial capital is:

$$g_0 = 10000 \tag{1}$$

Starting capital. g_0
Remaining funds after purchasing food and water at the starting point.

$$g_1 = g_0 - (10f_0 + 5w_0) \geq 0 \tag{2}$$

Number of boxes of food and water purchased at the starting point, and remaining funds after purchasing food and water at the starting point. f_0 w_0 $.g_1$

Due to the weight limit, there are restrictions.

$$2f_0 + 3w_0 \leq 1200 \tag{3}$$

During the process of going from the starting point to the village, there are the following restrictions:

$$W = 3 \times 8 \times 2 + 3 \times 5 \times 2 + 2 \times 10 \times 1 = 98 \leq w_0 \tag{4}$$

$$F = 3 \times 7 \times 2 + 3 \times 6 \times 2 + 2 \times 10 \times 1 = 98 \leq f_0 \tag{5}$$

During the rest of the journey, specifically the process towards the end of the village mine, it can be determined through analysis that the range of mining stoppage time is:

$$2 \leq d_0 \leq 10, d_0 \in N_+ \tag{6}$$

Number of days for mining. d_0

It can be concluded that the limits for calculating the total number of boxes of food and water consumed in the process of the village → the end of the mine → are:

$$f(d_0) = f_0 - F + f_1 \tag{7}$$

$$w(d_0) = w_0 - W + w_1 \tag{8}$$

Due to the weight limit, the restriction on the number of boxes to purchase food and water in the village is:

$$2(f_0 + f_1 - F) + 3(w_0 + w_1 - W) \leq 1200 \tag{9}$$

So money for the purchase of food and water in the village

$$g_2 = g_1 - 2(10f_1 + 5w_1) \tag{10}$$

The income from mining is:

$$g_3 = g_2 + 1000d_0 \tag{11}$$

Profits from mining, funds for purchasing food and water in the village, number of days for mining. g_3 g_2 .d_0

Due to the number of food and water boxes purchased can only be a positive integer, i.e.

$$f_0, f_1, w_0, w_1 \in N_+$$

$$max \ g = g_3(f_0 + f_1 - F - f(d_0))(w_0 + w_1 - W - w(d_0)) \tag{12}$$

In summary, the model for the entire linear programming is:

$$max \ g = g_3(f_0 + f_1 - F - f(d_0))(w_0 + w_1 - W - w(d_0))$$

$$s.t. \begin{cases} 98 \le f_0 \\ g_0 - (10f_0 + 5w_0) \ge 0 \\ 2f_0 + 3w_0 \le 1200 \\ 98 \le w_0 \\ 2 \le d_0 \le 10, d_0 \in N_+ \\ 2(f_0 + f_1 - F) + 3(w_0 + w_1 - W) \le 1200 \\ f(d_0) = f_0 - F + f_1 \\ w(d_0) = w_0 - W + w_1 \\ g_1 - 2(10f_1 + 5w_1) \ge 0 \\ f_0, f_1, w_0, w_1 \in N_+ \end{cases} \tag{13}$$

Second scenario: Make a round trip between the village and the mine, that is, perform two mining operations at the mine.

Comparing the first and second scenarios, it can be seen that the consumption of food and water from the starting point to the mine is the same, that is, formulas (1) to.

During the subsequent journey, specifically in the process of village and mining village, the number of boxes of food and water consumed in this process is completely determined, thus the limiting condition for calculating the total consumption of food and water in the process of village and mining village can be derived.

$$f_2(d_1) \le f_0 - F + f_2 \tag{14}$$

$$w_2(d_1) \le w_0 - W + w_2 \tag{15}$$

Due to the player's weight limit, the restriction on the number of food and water crates that can be purchased in the village is:

$$2(f_0 + f_2 - F) + 3(w_0 + w_2 - W) \le 1200 \tag{16}$$

First time purchasing food and water in the village funds:

$$g_2 = g_1 - 2(10f_2 + 5w_2), g_5 \ge 0 \tag{17}$$

Remaining funds after the first mining:

$$g_3 = g_2 + 1000d_1 \tag{18}$$

$$f_3(d_1, d_2) = f_0 - F + f_2 - f_2(d_1) + f_3 \tag{19}$$

$$w_3(d_1, d_2) = w_0 - W + w_2 - w_2(d_1) + w_3 \tag{20}$$

Due to the player's weight limit, the restriction on the number of food and water crates that can be purchased in the village is:

$$2(f_0 + f_2 + f_3 - F - f_2(d_1)) + 3(w_0 + w_2 + w_3 - W - w_2(d_1)) \le 120 \tag{21}$$

So the funds after the second purchase of food and water in the village:

$$g_4 = g_3 - 2(10f_3 + 5w_3), g_4 \geq 0 \qquad (22)$$

Remaining funds after the second mining:

$$g_5 = g_4 + 1000d_2 \qquad (23)$$

Due to the number of food and water boxes purchased can only be a positive integer, i.e.

$$f_0, f_2, f_3, w_0, w_2, w_3 \in N_+ \qquad (24)$$

By calculation, the time constraints for the first and second mining are as follows:

$$0 \leq d_1 + d_2 \leq 8, d_1, d_2 \in N_+ \qquad (25)$$

$$max\ g = g_5(f_0 + f_1 + f_3 - F - f(d_0) - f_3(d_1, d_2))$$

$$+2.5(w_0 + w_2 + w_3 - W - w_2(d_1) - w_3(d_1, d_2)) \qquad (26)$$

The model for the entire linear programming is:

$$max\ g = g_5(f_0 + f_1 + f_3 - F - f(d_0) - f_3(d_1, d_2))$$

$$+2.5(w_0 + w_2 + w_3 - W - w_2(d_1) - w_3(d_1, d_2))$$

$$s.t. \begin{cases} g_0 - (10f_0 + 5w_0) \geq 0 \\ 2f_0 + 3w_0 \leq 1200 \\ 98 \leq w_0 \\ 98 \leq f_0 \\ f_2(d_1) \leq f_0 - F + f_2 \\ w_2(d_1) \leq w_0 - W + w_2 \\ 2(f_0 + f_2 - F) + 3(w_0 + w_2 - W) \leq 1200 \\ g_1 - 2(10f_2 + 5w_2) \geq 0 \\ f_3(d_1, d_2) = f_0 - F + f_2 - f_2(d_1) + f_3 \\ w_3(d_1, d_2) = w_0 - W + w_2 - w_2(d_1) + w_3 \\ 2(f_0 + f_2 + f_3 - F - f_2(d_1)) + 3(w_0 + w_2 + w_3 - W - w_2(d_1)) \leq 120 \\ g_3 - 2(10f_3 + 5w_3) \geq 0 \\ f_0, f_2, f_3, w_0, w_2, w_3 \in N_+ \\ 0 \leq d_1 + d_2 \leq 8, d_1, d_2 \in N_+ \end{cases} \qquad (27)$$

Third scenario: the case of going directly to the endpoint

$$g = 10000 - [(8 + 8 + 5) \times 2] \times 5 + [(6 + 6 + 7) \times 2] \times 10 = 9410 \qquad (28)$$

Level Two

Remaining funds after purchasing food and water at the starting point.

$$g_1 = g_0 - \left(10f_0' + 5w_0'\right) \geq 0 \tag{29}$$

Due to the weight limit, there are the following restrictions.

$$2f_0' + 3w_0' \leq 1200 \tag{30}$$

During the process of going to the mine from the starting point, there are the following constraints:

$$W' = 114 \leq w_0' \tag{31}$$

$$F' = 110 \leq f_0' \tag{32}$$

Limiting condition for the total number of food and water boxes consumed during the mining process.

$$f_1'\left(d_1'\right) \leq f_0' - F' \tag{33}$$

$$w_1'\left(d_1'\right) \leq w_0' - W' \tag{34}$$

Purchase of food and water in the village funds.

$$g_1 = g_0 + 1000d_1' - 5\left(w_0' + 2w_1'\right) - 10\left(f_0' + 2f_1'\right) \tag{35}$$

Due to the player's weight limit, the restriction on the number of boxes of food and water that can be purchased in the village is:

$$2\left(f_{0'} + f_{1'} - F' - f_{1'}(d_{1'})\right) + 3\left(w_{0'} + w_{1'} - W' - w_{1'}(d_{1'})\right) \leq 1200 \tag{36}$$

The limits on the number of boxes of food and water consumed during the final journey, i.e. the village $\rightarrow$ the end of the mine $\rightarrow$ the terminus are:

$$f_2'\left(d_1', d_2'\right) = f_0' - F' + f_1' - f_1'(d_1') \tag{37}$$

$$w_2'\left(d_1', d_2'\right) = w_0' - W' + w_1' - w_1'(d_1') \tag{38}$$

Number of purchased food and water boxes can only be a positive integer.

$$f_0', f_1', w_0', w_1' \in N_+ \tag{39}$$

By calculation, the time constraints for the first and second mining are as follows:

$$0 \leq d_{1'} + d_{2'} \leq 15, d_{1'}, d_{2'} \in N_+ \tag{40}$$

$$max\, g = g_{012}^{''}\left(f_0^{'} + f_1^{'} - F^{'} - f_1^{'}\left(d_1^{'}\right) - f_2^{'}(d_1^{'}, d_2^{'})\right)$$
$$+ 2.5(w_0^{'} + w_1^{'} - W^{'} - w_1^{'}(d_1^{'}) - w_3(d_1^{'}, d_2^{'})) - 5\left(w_0^{'} + 2w_1^{'}\right) - 10\left(f_0^{'} + 2f_1^{'}\right) \quad (41)$$

The model for the entire linear programming is:

$$max\, g = g_{012}^{''}\left(f_0^{'} + f_1^{'} - F^{'} - f_1^{'}\left(d_1^{'}\right) - f_2^{'}(d_1^{'}, d_2^{'})\right)$$
$$+ 2.5(w_0^{'} + w_1^{'} - W^{'} - w_1^{'}(d_1^{'}) - w_3(d_1^{'}, d_2^{'})) - 5\left(w_0^{'} + 2w_1^{'}\right) - 10\left(f_0^{'} + 2f_1^{'}\right)$$

$$s.t. \begin{cases} g_0 - \left(10f_0^{'} + 5w_0^{'}\right) \geq 0 \\ 2f_0^{'} + 3w_0^{'} \leq 1200 \\ W^{'} = 114 \leq w_0^{'} \\ F^{'} = 110 \leq f_0^{'} \\ f_1^{'}\left(d_1^{'}\right) \leq f_0^{'} - F^{'} \\ w_1^{'}\left(d_1^{'}\right) \leq w_0^{'} - W^{'} \\ g_0 + 1000d_1^{'} - 5\left(w_0^{'} + 2w_1^{'}\right) - 10\left(f_0^{'} + 2f_1^{'}\right) \geq 0 \\ 2\left(f_0^{'} + f_1^{'} - F^{'} - f_1^{'}\left(d_1^{'}\right)\right) + 3\left(w_0^{'} + w_1^{'} - W^{'} - w_1^{'}\left(d_1^{'}\right)\right) \leq 1200 \\ f_2^{'}\left(d_1^{'}, d_2^{'}\right) = f_0^{'} - F^{'} + f_1^{'} - f_1^{'}\left(d_1^{'}\right) \\ w_2^{'}\left(d_1^{'}, d_2^{'}\right) = w_0^{'} - W^{'} + w_1^{'} - w_1^{'}\left(d_1^{'}\right) \\ f_0^{'}, f_1^{'}, w_0^{'}, w_1^{'}, d_1, d_2 \in N_+ \\ 0 \leq d_1 + d_2 \leq 15 \end{cases} \quad (42)$$

Level Three

Due to the daily profit of mining being 200 yuan, it can be concluded that mining for a day in high temperature weather actually results in a loss, while mining for a day in sunny weather can yield a net profit of 35 yuan.

If the weather is clear these two days, the number of food and water boxes that players need to consume will take at least 7 days of mining to recover the funds.

Thus, at this point, the player is unable to return to the finish line within the specified time. Therefore, if the weather for the extra two days is different, it is even less likely for the player to reach the finish line or earn enough funds to cover the cost of the food and water boxes needed by the player. This indicates that in the third level map, the player can only choose the second route [4].

Choosing the second route, due to the higher cost of food and water supplies required for walking in hot weather, it is necessary to consider whether to walk in hot weather. There are two options for the player.

First scenario: When encountering sunny weather, walking is required. However, when encountering hot weather, stay in place for one day, but ensure that the journey is completed within 10 days. If encountering hot weather during the following three days, continue walking. Second scenario: Disregard weather conditions and walk every day.

Level Four

Through analysis, the basis for the appearance of the strategy is as follows:

During hot weather, the daily basic consumption is 9 boxes of water and 9 boxes of food; during sunny weather, the daily basic consumption is 3 boxes of water and 4 boxes of food. Therefore, the average daily basic consumption is 6 boxes of water and 6.5 boxes of food.

$$\frac{f}{w} = \frac{6.5}{6} \tag{43}$$

$$3w + 2f = 1200 \tag{44}$$

Due to the ample amount of time, players can go to the mine to mine. Through simple calculations, it can be determined that if, starting from the starting point, the remaining funds after mining at the mine and then going to the endpoint are less than the situation where resources can be replenished by traveling back and forth between the village and the mine when resources are insufficient, the following will discuss in detail the situation where resources can be replenished by traveling back and forth between the village and the mine when resources are insufficient.

Considering extreme weather conditions, namely 4 days of high temperatures and one day of sandstorm, a round trip between the mine and the village requires 82 boxes of water and 82 boxes of food; if it is not extreme weather, the round trip between the mine and the village requires 48 boxes of water and 52 boxes of food, calculated using the mean value.

The number of boxes of food and water consumed per day during mining in the mine is converted into funds according to the price of the village: in the case of sandstorm weather, 900 yuan per day is required; In hot weather conditions, 810 yuan per day is required; In clear weather, $330 per day is required.

Due to the low probability of sandstorms, the difference in expenditure of funds between high temperatures and sandstorm weather is small, and sandstorm weather does not affect mining, so sandstorm weather can be approximately regarded as high temperature weather.

Net daily earnings from mining in the mine can reach 430 yuan.

Thus, at least two days of mining are needed to make up for the cost of a round trip. Due to the limited time of only 30 days, it is possible to make a maximum of two round trips. However, after calculation, the cost of two round trips is too high, and there is even a possibility of not being able to reach the destination. Therefore, the case of making one round trip is considered [4, 5].

The number of food and water crates that need to be purchased when arriving at the village should satisfy the demand for returning from the village to the mine, the demand for mining at the mine, and the demand for going from the mine to the destination. The sum of the time it takes to return from the village to the mine and the time it takes to go from the mine to the destination is 5 days. The number of food and water crates consumed in these five days is 65 and 60. t

So the time to mine for the second time is obtained:

$$30 - t - 5 = 25 - t \tag{45}$$

Thus, the number of boxes of food and water consumed in the second mining operation are:

$$(25 - t) \times 3 \times 6.5 = 487.5 - 19.5t \tag{46}$$

$$(25 - t) \times 3 \times 6 = 450 - 18t \tag{47}$$

So the number of boxes of food and water to buy in the village is:

$$65 + 487.5 - 19.5t = 552.5 - 19.5t \tag{48}$$

$$60 + 450 - 18t = 510 - 18t \tag{49}$$

Level Three

Since only the second route, which does not pass through the mine, can be taken directly to the destination, and this route only requires three steps to reach the destination, there are not many weather conditions. This article uses the exhaustive method to draw the following conclusions: the expected remaining funds for walking using the first scenario is 9318.75 yuan, and the expected remaining funds for walking using the second scenario is 9430 yuan. Therefore, it is better to walk using the second method, that is, to ignore the weather conditions and walk every day. At this time, it is necessary to purchase 54 boxes of water and 54 boxes of food at the starting point.

Level Four

Based on the decision basis above, make the following decision.

Buy 232 boxes of water and 252 boxes of food at the starting point, and go to the mine according to the route. $1 \to 2 \to 3 \to 8 \to 13 \to 18$

Second scenario; if the day is not the fourth day counting backwards from a 30-day period, then it is necessary to determine whether the remaining amount of substance satisfies the following conditions:

$$w \geq 82 + 3w_0 f \geq 82 + 3f_0 \tag{50}$$

By analyzing the map of level five, and based on the known conclusion of level three, and finally according to the requirements of the question, the following conclusions can be drawn: the player will only search for routes with a step count less than or equal to 4 [4], therefore there are 10 routes available, as shown in Fig. 1.

By analyzing the map of level six, find all the routes from the starting point that can reach the mine within five steps, as shown in Fig. 2.

Level Five

Translated content: The iterative change Diagram is shown in Fig. 3.

From the table and the graph, it can be seen that the expected remaining funds for the first route are the highest, and the probability is also the highest, so the player should choose that route. $1 \to 2 \to 4 \to 6 \to 13$

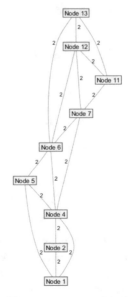

Fig. 1. 10 routes represented in the figure.

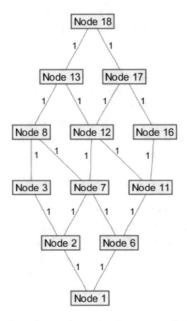

Fig. 2. All roadmaps from the starting point that can reach the mine within 5 steps

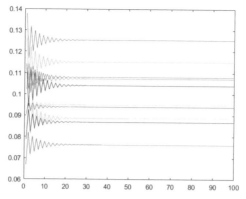

Fig. 3. Iterative Change Diagram

Level Six

Based on the decision basis above, make the following decisions:

translated content: The player's walking route starts from the mining village and ends at the mine. After calculating with software, the optimal route for the player to go to the mine is:. $1 \rightarrow 2 \rightarrow 3 \rightarrow 8 \rightarrow 13 \rightarrow 18$

Second scenario; if the day is not the fourth day counting backwards from a period of 30 days, it is necessary to determine whether the remaining amount of substance satisfies the following conditions $18 \rightarrow 13 \rightarrow 14$.

Otherwise, evaluate the following conditions (57). If satisfied, consider the quantity of other players in the mine who met the conditions.

4 Model Verification and Error Analysis

Model verification

From a model perspective, because the question requires the maximum amount of funds and there are many constraints, the use of linear programming as a model is appropriate; from the model results, when the player starts, the purchased supplies have already reached the weight limit, and when they arrive at the village, the water resources are almost completely depleted, and when they reach the end point, they either consume almost all or completely consume the supplies purchased at the village. Therefore, from the results, this is already the optimal solution. Otherwise, there would be overweight at the starting point or more supplies left at the end point, so the establishment of this model is correct.

Probability of changing the occurrence of high temperature weather, comparing the expected loss of high temperature staying and not staying, using software to obtain different probabilities of high temperature weather occurrence, and the expected loss of high temperature staying and not staying are shown in Fig. 4.

From the graph, it can be seen that regardless of the probability of high temperature occurrence, the expected loss of high temperature staying will be higher than the expected

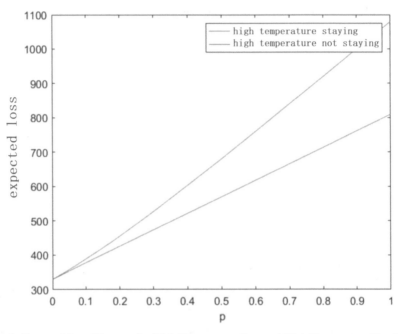

Fig. 4. Expected Loss Diagram for High Temperature Stay and High Temperature Non-Stay

loss of high temperature not staying. Therefore, the results of this model are correct, which indirectly demonstrates the correctness and applicability of this model.

Error Analysis

When calculating these constraints, sometimes the results are not integers. In this article, we handle them by adding or subtracting one from the obtained results, so this is one source of error. When buying supplies in the village, the consideration is actually that it must be less than or equal to the weight limit. However, in this article, we only consider the case of being equal to the weight limit, so this is also a source of error.

In the third level, due to considering every situation, there is no error; in the fourth level, under non-extreme weather conditions, this article uses the average value to calculate the daily consumption, which is one source of error; during the mining process, the occurrence of sandstorms is ignored and treated as high-temperature weather conditions, which is the second source of error; due to the small probability of sandstorm weather, this article assumes a probability of 0.5 for the occurrence of high-temperature weather conditions, which is the third source of error [6].

The source of error in Level 5: Due to the map of this question being the same as Level 3, the conclusion of Level 3 was adopted from the beginning without considering other situations, which resulted in certain errors. The source of error in Level 6: Due to the use of the conclusion from Level 4 in the process of solving the problem, the source of error in Level 4 remains the source of error in Level 6. Additionally, this article assumes that every player is rational and does not consider some special situations, which leads to certain errors.

5   Model Evaluation and Promotion

5.1   Advantages

Linear programming models are relatively easy to understand, and each formula can be found in the problem statement.

Dynamic programming models can adapt to the current situation.

The iterative loop model can clearly represent each player considering the walking route of other players, thereby determining the walking route of this player.

5.2   Model Promotion

Linear programming models can also be applied to transportation problems. Sometimes each carriage or cabin may have a maximum weight limit and require a balance, while also aiming to achieve the maximum quality of transported goods. The use of linear programming models can solve these types of problems.

The iterative model can also be applied in scenarios of game theory, such as the problem of taxi drivers picking up passengers [7].

Acknowledgments. The project is supported by Provincial college student Innovation Training Program of Guangzhou University.

References

1. Xiao, J.: Exploration on the typological design of battle royale type online game. Knowl. Base (23), 38+36 (2018)
2. Gao, R., Zhang, H.: Research on multiplayer game design based on game theory. Popular Lit. Art (07), 104–105 (2020)
3. Hu, K.: Study on the Cooperative Behavior of Conforming Individuals in the Network Prisoner's Dilemma Game. Yunnan University of Finance and Economics (2019)
4. Fu, S.: Strategy Regulation and Optimization of Network Evolution Game with Risk and Memory. Shandong University, China (2018)
5. Li, J., Zheng, R., Tang, Y., Yuan, L., Chen, Y.: Incomplete information game solving based on SMT. Comput. Syst. Appl. **29**(01), 261–265 (2020)
6. Xu, Y.: Dynamics Analysis of Website Competition Model with Multiple Delays. Wuhan University of Technology, Wuhan (2019)
7. Van Benthem, J., Liu, F.: The design of game theory and the development of modal logic. J. Tsinghua Univ. (Philos. Soc. Sci. Edn) **34**(02), 131–139+201 (2019)

**Joint International Conference
for "Four Meetings and a Celebration"**

Research on Hamacher Operations for q-rung Orthopair Fuzzy Information

Wen Sheng Du$^{(\boxtimes)}$

School of Business, Zhengzhou University,
Zhengzhou 450001, People's Republic of China
wsdu@zzu.edu.cn

Abstract. Research on q-rung orthopair fuzzy information aggregation serves as a significant part of the q-rung orthopair fuzzy set theory. Note that the aggregation operators are built on operations over q-rung orthopair fuzzy values. Recently, the Hamacher operations, generalizations of algebraic and Einstein operations, on q-rung orthopair fuzzy information were proposed. In this paper, we give an exhaustively theoretical analysis of these developed operations for q-rung orthopair fuzzy values. Particularly, we investigate the operational rules of q-rung orthopair fuzzy Hamacher operations, and the results are consistent with our expectation. Thus the meaningfulness of these operations is confirmed, which provides a solid fundamental basis for further studies. Moreover, the monotonicity and limiting cases of these operations with respect to parameters contained therein are examined.

Keywords: Hamacher operation · operational rule · q-rung orthopair fuzzy value

1 Introduction

Professor Yager [1] proposed the concept of q-rung orthopair fuzzy sets (q-ROFSs) in 2017 as extensions of fuzzy sets [2], intuitionistic fuzzy sets [3], Pythagorean fuzzy sets [4] and other types of orthopair fuzzy sets, Fermatean fuzzy sets [5], for example. The basic components of q-ROFSs are q-rung orthopair fuzzy values (q-ROFVs), which are characterized by degrees of membership and nonmembership. For any q-ROFV, we require the sum of qth powers of its membership degree and nonmembership degree being bounded by one.

Research on operations for q-ROFVs attracted attentions of academic scholars all over the world. Liu and Wang [6] developed the algebraic addition and multiplication operations for q-ROFVs. Based on the Archimedean t-norms and t-conorms, Liu and Wang [7] further introduced a family of operations over q-ROFVs, which greatly expanded the research scopes. Especially, Akram et al. [8] presented the Einstein operations, a good alternative to algebraic operations, on q-ROFVs. Jana et al. [9] put forward the Dombi operations for q-ROFVs which involve a general parameter reflecting the flexibility, and Du [10] gave some

B.-Y. Cao et al. (Eds.): ICFIE 2022, LNDECT 207, pp. 219–231, 2024.
https://doi.org/10.1007/978-981-97-2891-6_17

indepth characteristics of these operations. Seikh and Mandal [11] proposed the Frank sum and Frank product of q-ROFVs by the use of Frank t-norms and t-conorms. With the aid of Hamacher operational laws, Darko and Liang [12] investigated the Hamacher operations over q-ROFVs.

Hamacher [13] introduced the Hamacher operations which are generalizations of the algebraic and Einstein operations. The parameter within Hamacher operations makes them more general and flexible than algebraic and Einstein operations. Huang [14] developed the Hamacher operations for intuitionistic fuzzy values. Tan et al. [15] proposed some Hamacher operations for fusing hesitant fuzzy elements. Wu and Wei [16] utilized the Hamacher operations to generate aggregation operators for Pythagorean fuzzy values. Hadi et al. [17] derived the Hamacher operations on Fermatean fuzzy values. In 2020, to aggregate q-ROFVs, Darko and Liang [12] put forward the corresponding Hamacher operations and applied them to the problem of multi-attribute group decision making. Liu et al. [18] combined these operations with the entropy-based GLDS method to handle issues with q-rung orthopair fuzzy information. Zhu [19] presented the improved MULTIMOORA method with the help of Hamacher operations on q-ROFVs. Kakati and Rahman [20] proposed some Hamacher aggregation operators based on the generalized Shapley index and Choquet integral operator to reflect correlations among the attributes.

Although the research on q-rung orthopair fuzzy Hamacher aggregation operators goes far over the past years, some fundamental problems related to Hamacher operations have not been considered yet. The present study checks whether the Hamacher operations are suitable for introduction into q-rung orthopair fuzzy environment. We discuss the basic operational rules of q-rung orthopair fuzzy Hamacher operations so as to testify the reasonability of these operations. What is more, taken into account the parameter in Hamacher operations, the monotonicity of q-rung orthopair fuzzy Hamacher operations with respect to the parameter deserves to be analyzed.

2   Preliminaries

For later use, in this section, the basic concepts and operations defined on q-rung orthopair fuzzy values are recalled.

The q-ROFSs are relatively new generalizations of intuitionistic fuzzy sets, whose basic components are q-ROFVs.

Definition 1. *[1] An orthopair $\langle a, b \rangle$ is said to be a q-rung orthopair fuzzy value if*

(1) $0 \leq a, b \leq 1$,
(2) $0 \leq a^q + b^q \leq 1$, where $q \geq 1$.

From Definition 1, we can see that

- if $q = 1$, then $\langle a, b \rangle$ is an intuitionistic fuzzy value due to the requirement of $a + b \leq 1$ [3];

- if $q = 2$, then $\langle a, b \rangle$ is a Pythagorean fuzzy value because of $a^2 + b^2 \leq 1$ [4];
- if $q = 3$, then $\langle a, b \rangle$ is a Fermatean fuzzy value by the restriction of $a^3 + b^3 \leq 1$ [5].

Therefore, the previous three types of fuzzy values are special cases of q-ROFVs.

The order relation $\preceq$ between two q-ROFVs $x = \langle a_1, b_1 \rangle$ and $y = \langle a_2, b_2 \rangle$ is given as [1]

$$x \preceq y \text{ iff } a_1 \leq a_2 \text{ and } b_1 \geq b_2. \tag{1}$$

The basic (algebraic) operations on q-ROFVs x and y are presented as follows [6]:

$$x \oplus y = \left\langle \left(a_1^q + a_2^q - a_1^q a_2^q\right)^{1/q}, b_1 b_2 \right\rangle, \tag{2}$$

$$x \otimes y = \left\langle a_1 a_2, \left(b_1^q + b_2^q - b_1^q b_2^q\right)^{1/q} \right\rangle. \tag{3}$$

The Einstein sum and Einstein product of x and y are, respectively [8, 21]

$$x \oplus_\varepsilon y = \left| \left(\frac{a_1^q + a_2^q}{1 + a_1^q a_2^q} \right)^{1/q}, \left(\frac{b_1^q b_2^q}{1 + \left(1 - b_1^q\right)\left(1 - b_2^q\right)} \right)^{1/q} \right|, \tag{4}$$

$$x \otimes_\varepsilon y = \left| \left(\frac{a_1^q a_2^q}{1 + \left(1 - a_1^q\right)\left(1 - a_2^q\right)} \right)^{1/q}, \left(\frac{b_1^q + b_2^q}{1 + b_1^q b_2^q} \right)^{1/q} \right|. \tag{5}$$

Moreover, the λ times and λth power of $x = \langle a, b \rangle$ are respectively formulated by [6]: $\lambda > 0$,

$$\lambda x = \left\langle \left(1 - \left(1 - a^q\right)^\lambda\right)^{1/q}, b^\lambda \right\rangle, \tag{6}$$

$$x^\lambda = \left\langle a^\lambda, \left(1 - \left(1 - b^q\right)^\lambda\right)^{1/q} \right\rangle. \tag{7}$$

While, based on Einstein operations, the above operations are given by [8, 21]:

$$\lambda \cdot_\varepsilon x = \left| \left(\frac{(1 + a^q)^\lambda - (1 - a^q)^\lambda}{(1 + a^q)^\lambda + (1 - a^q)^\lambda} \right)^{1/q}, \left(\frac{2b^{\lambda q}}{(2 - b^q)^\lambda + b^{\lambda q}} \right)^{1/q} \right|, \tag{8}$$

$$x^{\varepsilon \lambda} = \left| \left(\frac{2a^{\lambda q}}{(2 - a^q)^\lambda + a^{\lambda q}} \right)^{1/q}, \left(\frac{(1 + b^q)^\lambda - (1 - b^q)^\lambda}{(1 + b^q)^\lambda + (1 - b^q)^\lambda} \right)^{1/q} \right|. \tag{9}$$

3   Hamacher Operations on q-rung Orthopair Fuzzy Values

In this section, some fundamental properties of Hamacher operations for q-ROFVs are investigated to manifest the rationality of the developed operations.

Hamacher [13] introduced a parameterized family of t-conorms and t-norms named the Hamacher sum and Hamacher product. The Hamacher t-(co)norms are the only strict t-(co)norms that can be expressed as rational functions.

Definition 2. *[13] The addition and multiplication of x and y in the closed unit interval $[0, 1]$ based on Hamacher operations are, for $\gamma > 0$,*

$$x +_H^\gamma y = \frac{x + y + (\gamma - 2)xy}{1 + (\gamma - 1)xy}, \tag{10}$$

$$x \times_H^\gamma y = \frac{xy}{1 + (\gamma - 1)(1 - x)(1 - y)}. \tag{11}$$

Darko and Liang [12] and Liu et al. [18] generalized the Hamacher operations to the q-rung orthopair fuzzy circumstance, separately.

Definition 3. *[12, 18] Let $x = \langle a_1, b_1 \rangle$ and $y = \langle a_2, b_2 \rangle$ be two q-ROFVs. Then we can calculate the Hamacher sum and Hamacher product of x and y, which are expressed as follows: $\gamma > 0$,*

$$x \oplus_H^\gamma y = \left| \left(\frac{a_1^q + a_2^q + (\gamma - 2)a_1^q a_2^q}{1 + (\gamma - 1)a_1^q a_2^q} \right)^{1/q}, \left(\frac{b_1^q b_2^q}{1 + (\gamma - 1)(1 - b_1^q)(1 - b_2^q)} \right)^{1/q} \right|, \tag{12}$$

$$x \otimes_H^\gamma y = \left| \left(\frac{a_1^q a_2^q}{1 + (\gamma - 1)(1 - a_1^q)(1 - a_2^q)} \right)^{1/q}, \left(\frac{b_1^q + b_2^q + (\gamma - 2)b_1^q b_2^q}{1 + (\gamma - 1)b_1^q b_2^q} \right)^{1/q} \right|. \tag{13}$$

It can be seen that the addition and multiplication operations on q-ROFVs via Hamacher operations are generalizations of those via algebraic and Einstein operations [6,8]. In terms of parameter q, these operations are generalizations of Hamacher operations on intuitionistic fuzzy values, Pythagorean fuzzy values and Fermatean fuzzy values [14,16,17]. Note also that these operations were further extended to the complex q-rung orthopair fuzzy, interval-valued q-rung orthopair fuzzy and q-rung orthopair trapezoidal fuzzy settings [22–24].

It is obvious that Hamacher operations on q-ROFVs are monotonic with respect to parameter γ, that is, if $\gamma_1 \geq \gamma_2$, then $x \oplus_H^{\gamma_1} y \succeq x \oplus_H^{\gamma_2} y$ and $x \otimes_H^{\gamma_1} y \preceq x \otimes_H^{\gamma_2} y$. Specially, we have $x \oplus_\varepsilon y \succeq x \oplus y$ and $x \otimes_\varepsilon y \preceq x \otimes y$.

Theorem 1. *Let $\oplus_H^\gamma$ and $\otimes_H^\gamma$ be the Hamacher addition and multiplication operations. Then, for q-ROFVs x, y, z,*

(1) $x \oplus_H^\gamma y = y \oplus_H^\gamma x$,　　　　　　(2) $x \otimes_H^\gamma y = y \otimes_H^\gamma x$,

(3) $x \oplus_H^\gamma (y \oplus_H^\gamma z) = (x \oplus_H^\gamma y) \oplus_H^\gamma z$,　　(4) $x \otimes_H^\gamma (y \otimes_H^\gamma z) = (x \otimes_H^\gamma y) \otimes_H^\gamma z$.

Proof. We only prove (3). Assume that $x = \langle a_1, b_1 \rangle$, $y = \langle a_2, b_2 \rangle$, $z = \langle a_3, b_3 \rangle$, it follows from Definition 3 that

$$x \oplus_H^\gamma (y \oplus_H^\gamma z)$$

$$= \langle a_1, b_1 \rangle \oplus_H^\gamma \left| \left(\frac{a_2^q + a_3^q + (\gamma - 2)a_2^q a_3^q}{1 + (\gamma - 1)a_2^q a_3^q} \right)^{1/q}, \left(\frac{b_2^q b_3^q}{1 + (\gamma - 1)(1 - b_2^q)(1 - b_3^q)} \right)^{1/q} \right|$$

$$= \left| \left(\frac{a_1^q + a_2^q + a_3^q + (\gamma - 1)a_1^q a_2^q a_3^q + (\gamma - 2)(a_1^q a_2^q + a_1^q a_3^q + a_2^q a_3^q) + (\gamma - 2)^2 a_1^q a_2^q a_3^q}{1 + (\gamma - 1)(a_1^q a_2^q + a_1^q a_3^q + a_2^q a_3^q) + (\gamma - 1)(\gamma - 2)a_1^q a_2^q a_3^q} \right)^{1/q}, \right.$$

$$\left. \left(\frac{b_1^q b_2^q b_3^q}{1 + (\gamma - 1)(2 - b_1^q - b_2^q - b_3^q + b_1^q b_2^q b_3^q) + (\gamma - 1)^2 (1 - b_1^q)(1 - b_2^q)(1 - b_3^q)} \right)^{1/q} \right|$$

$$= \left| \left(\frac{a_1^q + a_2^q + (\gamma - 2)a_1^q a_2^q}{1 + (\gamma - 1)a_1^q a_2^q} \right)^{1/q}, \left(\frac{b_1^q b_2^q}{1 + (\gamma - 1)(1 - b_1^q)(1 - b_2^q)} \right)^{1/q} \right| \oplus_H^\gamma \langle a_3, b_3 \rangle$$

$$= (x \oplus_H^\gamma y) \oplus_H^\gamma z.$$

$\square$

In light of Theorem 1, we obtain that operations $\oplus_H^\gamma$ and $\otimes_H^\gamma$ are commutative and associative. For q-ROFVs x_i, $i = 1, 2, \ldots, n$, the notations $x_1 \oplus_H^\gamma x_2 \oplus_H^\gamma \cdots \oplus_H^\gamma x_n$ and $x_1 \otimes_H^\gamma x_2 \otimes_H^\gamma \cdots \otimes_H^\gamma x_n$ can thus be shortly represented by, without doubt, $\bigoplus_{i=1}^n {}_H^\gamma x_i$ and $\bigotimes_{i=1}^n {}_H^\gamma x_i$, respectively.

Theorem 2. *Let* $\{x_i = \langle a_i, b_i \rangle\}_{i=1}^n$ *be a collection of q-ROFVs. Then we have*

$$\bigoplus_{i=1}^n {}_H^\gamma x_i = \left| \left(\frac{\prod_{i=1}^n (1 + (\gamma - 1)a_i^q) - \prod_{i=1}^n (1 - a_i^q)}{\prod_{i=1}^n (1 + (\gamma - 1)a_i^q) + (\gamma - 1) \prod_{i=1}^n (1 - a_i^q)} \right)^{1/q}, \right.$$

$$\left. \left(\frac{\gamma \prod_{i=1}^n b_i^q}{\prod_{i=1}^n (1 + (\gamma - 1)(1 - b_i^q)) + (\gamma - 1) \prod_{i=1}^n b_i^q} \right)^{1/q} \right|, \quad (14)$$

$$\bigotimes_{i=1}^n {}_H^\gamma x_i = \left| \left(\frac{\gamma \prod_{i=1}^n a_i^q}{\prod_{i=1}^n (1 + (\gamma - 1)(1 - a_i^q)) + (\gamma - 1) \prod_{i=1}^n a_i^q} \right)^{1/q}, \right.$$

$$\left. \left(\frac{\prod_{i=1}^n (1 + (\gamma - 1)b_i^q) - \prod_{i=1}^n (1 - b_i^q)}{\prod_{i=1}^n (1 + (\gamma - 1)b_i^q) + (\gamma - 1) \prod_{i=1}^n (1 - b_i^q)} \right)^{1/q} \right|. \quad (15)$$

Proof. The validity of Eq. (14) for all positive integers n follows by mathematical induction. Specifically, for $n = 2$,

$$x_1 \oplus_{\mathrm{H}}^{\gamma} x_2 = \left| \left(\left(\frac{a_1^q + a_2^q + (\gamma - 2)a_1^q a_2^q}{1 + (\gamma - 1)a_1^q a_2^q} \right)^{1/q}, \left(\frac{b_1^q b_2^q}{1 + (\gamma - 1)\left(1 - b_1^q\right)\left(1 - b_2^q\right)} \right)^{1/q} \right) \right|$$

$$= \left\langle \left(\left(\frac{\left(1 + (\gamma - 1)a_1^q\right)\left(1 + (\gamma - 1)a_2^q\right) - \left(1 - a_1^q\right)\left(1 - a_2^q\right)}{\left(1 + (\gamma - 1)a_1^q\right)\left(1 + (\gamma - 1)a_2^q\right) + (\gamma - 1)\left(1 - a_1^q\right)\left(1 - a_2^q\right)} \right)^{1/q}, \right. \right.$$

$$\left. \left. \left(\frac{\gamma b_1^q b_2^q}{\left(1 + (\gamma - 1)(1 - b_1^q)\right)\left(1 + (\gamma - 1)(1 - b_2^q)\right) + (\gamma - 1)b_1^q b_2^q} \right)^{1/q} \right) \right\rangle.$$

Hence the conclusion is true for $n = 2$. Suppose that Eq. (14) holds for $n = k$, namely,

$$\bigoplus_{i=1}^{k}{}^{\gamma}_{\mathrm{H}} x_i = \left| \left(\left(\frac{\prod_{i=1}^{k}\left(1 + (\gamma - 1)a_i^q\right) - \prod_{i=1}^{k}\left(1 - a_i^q\right)}{\prod_{i=1}^{k}\left(1 + (\gamma - 1)a_i^q\right) + (\gamma - 1)\prod_{i=1}^{k}\left(1 - a_i^q\right)} \right)^{1/q}, \right. \right.$$

$$\left. \left. \left(\frac{\gamma \prod_{i=1}^{k} b_i^q}{\prod_{i=1}^{k}\left(1 + (\gamma - 1)(1 - b_i^q)\right) + (\gamma - 1)\prod_{i=1}^{k} b_i^q} \right)^{1/q} \right) \right|.$$

Then, when $n = k + 1$, we have

$$\bigoplus_{i=1}^{k+1}{}^{\gamma}_{\mathrm{H}} x_i = \bigoplus_{i=1}^{k}{}^{\gamma}_{\mathrm{H}} x_i \oplus_{\mathrm{H}}^{\gamma} x_{k+1}$$

$$= \left| \left(\left(\frac{\prod_{i=1}^{k}\left(1 + (\gamma - 1)a_i^q\right) - \prod_{i=1}^{k}\left(1 - a_i^q\right)}{\prod_{i=1}^{k}\left(1 + (\gamma - 1)a_i^q\right) + (\gamma - 1)\prod_{i=1}^{k}\left(1 - a_i^q\right)} \right)^{1/q}, \right. \right.$$

$$\left. \left(\frac{\gamma \prod_{i=1}^{k} b_i^q}{\prod_{i=1}^{k}\left(1 + (\gamma - 1)(1 - b_i^q)\right) + (\gamma - 1)\prod_{i=1}^{k} b_i^q} \right)^{1/q} \right) \oplus_{\mathrm{H}}^{\gamma} \langle a_{k+1}, b_{k+1} \rangle$$

$$= \left| \left(\left(\frac{\left(1 + (\gamma - 1)a_{k+1}^q\right)\prod_{i=1}^{k}\left(1 + (\gamma - 1)a_i^q\right) - \left(1 - a_{k+1}^q\right)\prod_{i=1}^{k}\left(1 - a_i^q\right)}{\left(1 + (\gamma - 1)a_{k+1}^q\right)\prod_{i=1}^{k}\left(1 + (\gamma - 1)a_i^q\right) + (\gamma - 1)\left(1 - a_{k+1}^q\right)\prod_{i=1}^{k}\left(1 - a_i^q\right)} \right)^{1/q}, \right. \right.$$

$$\left. \left(\frac{\gamma b_{k+1}^q \prod_{i=1}^{k} b_i^q}{\left(1 + (\gamma - 1)(1 - b_{k+1}^q)\right)\prod_{i=1}^{k}\left(1 + (\gamma - 1)(1 - b_i^q)\right) + (\gamma - 1)b_{k+1}^q \prod_{i=1}^{k} b_i^q} \right)^{1/q} \right)$$

$$= \left| \left(\left(\frac{\prod_{i=1}^{k+1}\left(1 + (\gamma - 1)a_i^q\right) - \prod_{i=1}^{k+1}\left(1 - a_i^q\right)}{\prod_{i=1}^{k+1}\left(1 + (\gamma - 1)a_i^q\right) + (\gamma - 1)\prod_{i=1}^{k+1}\left(1 - a_i^q\right)} \right)^{1/q}, \right. \right.$$

$$\left. \left(\frac{\gamma \prod_{i=1}^{k+1} b_i^q}{\prod_{i=1}^{k+1}\left(1 + (\gamma - 1)(1 - b_i^q)\right) + (\gamma - 1)\prod_{i=1}^{k+1} b_i^q} \right)^{1/q} \right) \right|,$$

i.e., Eq. (14) holds for $n = k+1$. Therefore, Eq. (14) holds for all positive integers. Equation (15) holds similarly. $\qquad \square$

For a positive integer n, let us introduce the following two denotations $n \cdot_H^\gamma x = \overbrace{x \oplus_H^\gamma x \oplus_H^\gamma \cdots \oplus_H^\gamma x}^{n \text{ times}}$ and $x^{\gamma_H n} = \overbrace{x \otimes_H^\gamma x \otimes_H^\gamma \cdots \otimes_H^\gamma x}^{n \text{ times}}$. Then from Theorem 2, we can easily deduce the following corollary by taking every $x_i = x$.

Corollary 1. *Let* $x = \langle a, b \rangle$ *be a* q-ROFV. *The Hamacher* n-*times and Hamacher* n*th-power of* x *are formulated by, for a positive integer* n:

$$n \cdot_H^\gamma x = \left| \left(\left(\frac{\left(1 + (\gamma - 1)a^q\right)^n - \left(1 - a^q\right)^n}{\left(1 + (\gamma - 1)a^q\right)^n + (\gamma - 1)\left(1 - a^q\right)^n} \right)^{1/q} \right., \right.$$
$$\left. \left(\frac{\gamma b^{nq}}{\left(1 + (\gamma - 1)\left(1 - b^q\right)\right)^n + (\gamma - 1)b^{nq}} \right)^{1/q} \right|,$$

$$x^{\gamma_H n} = \left| \left(\left(\frac{\gamma a^{nq}}{\left(1 + (\gamma - 1)\left(1 - a^q\right)\right)^n + (\gamma - 1)a^{nq}} \right)^{1/q} \right., \right.$$
$$\left. \left(\frac{\left(1 + (\gamma - 1)b^q\right)^n - \left(1 - b^q\right)^n}{\left(1 + (\gamma - 1)b^q\right)^n + (\gamma - 1)\left(1 - b^q\right)^n} \right)^{1/q} \right|.$$

The following corollary reveals the operational laws of Hamacher cumulative operations with multiple index sets.

Corollary 2. *Let* $\left\{ x_{ij} = \langle a_{ij}, b_{ij} \rangle \right\}_{i,j=1}^{n}$ *be a collection of* q-ROFVs. *Then we have*

$$\bigoplus_{i=1}^{n}{}^\gamma_H \bigoplus_{j=1}^{n}{}^\gamma_H x_{ij} = \left| \left(\frac{\prod_{i=1}^{n}\prod_{j=1}^{n}\left(1 + (\gamma-1)a_{ij}^q\right) - \prod_{i=1}^{n}\prod_{j=1}^{n}\left(1 - a_{ij}^q\right)}{\prod_{i=1}^{n}\prod_{j=1}^{n}\left(1 + (\gamma-1)a_{ij}^q\right) + (\gamma-1)\prod_{i=1}^{n}\prod_{j=1}^{n}\left(1 - a_{ij}^q\right)} \right)^{1/q} \right.,$$
$$\left. \left(\frac{\gamma \prod_{i=1}^{n}\prod_{j=1}^{n} b_{ij}^q}{\prod_{i=1}^{n}\prod_{j=1}^{n}\left(1 + (\gamma-1)(1 - b_{ij}^q)\right) + (\gamma-1)\prod_{i=1}^{n}\prod_{j=1}^{n} b_{ij}^q} \right)^{1/q} \right|, \quad (16)$$

$$\bigotimes_{i=1}^{n}{}^\gamma_H \bigotimes_{j=1}^{n}{}^\gamma_H x_{ij} = \left| \left(\frac{\gamma \prod_{i=1}^{n}\prod_{j=1}^{n} a_{ij}^q}{\prod_{i=1}^{n}\prod_{j=1}^{n}\left(1 + (\gamma-1)(1 - a_{ij}^q)\right) + (\gamma-1)\prod_{i=1}^{n}\prod_{j=1}^{n} a_{ij}^q} \right)^{1/q} \right.,$$
$$\left. \left(\frac{\prod_{i=1}^{n}\prod_{j=1}^{n}\left(1 + (\gamma-1)b_{ij}^q\right) - \prod_{i=1}^{n}\prod_{j=1}^{n}\left(1 - b_{ij}^q\right)}{\prod_{i=1}^{n}\prod_{j=1}^{n}\left(1 + (\gamma-1)b_{ij}^q\right) + (\gamma-1)\prod_{i=1}^{n}\prod_{j=1}^{n}\left(1 - b_{ij}^q\right)} \right)^{1/q} \right|. \quad (17)$$

Proof. Denote $\alpha_{i1} = \prod_{j=1}^{n}\left(1 + (\gamma-1)a_{ij}^q\right)$, $\alpha_{i2} = \prod_{j=1}^{n}\left(1 - a_{ij}^q\right)$, $\beta_{i1} = \prod_{j=1}^{n}\left(1 + (\gamma-1)(1 - b_{ij}^q)\right)$ and $\beta_{i2} = \prod_{j=1}^{n} b_{ij}^q$. Then by Eq. (14), we have

$$\bigoplus_{i=1}^{n}{}_{\mathrm{H}}^{\gamma}\bigoplus_{j=1}^{n}{}_{\mathrm{H}}^{\gamma} x_{ij} = \bigoplus_{i=1}^{n}{}_{\mathrm{H}}^{\gamma}\left\langle \left(\frac{\alpha_{i1}-\alpha_{i2}}{\alpha_{i1}+(\gamma-1)\alpha_{i2}}\right)^{1/q}, \left(\frac{\gamma\beta_{i2}}{\beta_{i1}+(\gamma-1)\beta_{i2}}\right)^{1/q}\right\rangle$$

$$= \left|\left(\frac{\prod_{i=1}^{n}\left(1+\frac{(\gamma-1)(\alpha_{i1}-\alpha_{i2})}{\alpha_{i1}+(\gamma-1)\alpha_{i2}}\right) - \prod_{i=1}^{n}\left(1-\frac{\alpha_{i1}-\alpha_{i2}}{\alpha_{i1}+(\gamma-1)\alpha_{i2}}\right)}{\prod_{i=1}^{n}\left(1+\frac{(\gamma-1)(\alpha_{i1}-\alpha_{i2})}{\alpha_{i1}+(\gamma-1)\alpha_{i2}}\right) + (\gamma-1)\prod_{i=1}^{n}\left(1-\frac{\alpha_{i1}-\alpha_{i2}}{\alpha_{i1}+(\gamma-1)\alpha_{i2}}\right)}\right)^{1/q}, \right.$$

$$\left. \left(\frac{\gamma\prod_{i=1}^{n}\frac{\gamma\beta_{i2}}{\beta_{i1}+(\gamma-1)\beta_{i2}}}{\prod_{i=1}^{n}\left(1+(\gamma-1)(1-\frac{\gamma\beta_{i2}}{\beta_{i1}+(\gamma-1)\beta_{i2}})\right) + (\gamma-1)\prod_{i=1}^{n}\frac{\gamma\beta_{i2}}{\beta_{i1}+(\gamma-1)\beta_{i2}}}\right)^{1/q}\right|$$

$$= \left\langle \left(\frac{\prod_{i=1}^{n}\alpha_{i1}-\prod_{i=1}^{n}\alpha_{i2}}{\prod_{i=1}^{n}\alpha_{i1}+(\gamma-1)\prod_{i=1}^{n}\alpha_{i2}}\right)^{1/q}, \left(\frac{\gamma\prod_{i=1}^{n}\beta_{i2}}{\prod_{i=1}^{n}\beta_{i1}+(\gamma-1)\prod_{i=1}^{n}\beta_{i2}}\right)^{1/q}\right\rangle.$$

The other equation follows similarly.   □

Extending the parameter n in Corollary 1 to any positive real number λ comes the following definition.

Definition 4. *[12, 18] Let $x = \langle a, b\rangle$ be a q-ROFV. The Hamacher scalar multiplication and Hamacher exponentiation operations on x are given by: for $\lambda > 0$,*

$$\lambda \cdot_{\mathrm{H}}^{\gamma} x = \left|\left(\frac{\left(1+(\gamma-1)a^q\right)^{\lambda} - \left(1-a^q\right)^{\lambda}}{\left(1+(\gamma-1)a^q\right)^{\lambda} + (\gamma-1)\left(1-a^q\right)^{\lambda}}\right)^{1/q}, \right.$$
$$\left. \left(\frac{\gamma b^{\lambda q}}{\left(1+(\gamma-1)(1-b^q)\right)^{\lambda} + (\gamma-1)b^{\lambda q}}\right)^{1/q}\right|, \tag{18}$$

$$x^{\gamma_{\mathrm{H}}\lambda} = \left|\left(\frac{\gamma a^{\lambda q}}{\left(1+(\gamma-1)(1-a^q)\right)^{\lambda} + (\gamma-1)a^{\lambda q}}\right)^{1/q}, \right.$$
$$\left. \left(\frac{\left(1+(\gamma-1)b^q\right)^{\lambda} - \left(1-b^q\right)^{\lambda}}{\left(1+(\gamma-1)b^q\right)^{\lambda} + (\gamma-1)\left(1-b^q\right)^{\lambda}}\right)^{1/q}\right|. \tag{19}$$

Obviously, the Hamacher scalar multiplication and exponentiation operations would degenerate respectively to their counterparts generated by algebraic and Einstein operations when taking $\gamma = 1$ or $\gamma = 2$. These operations would reduce respectively to the Hamacher counterparts of intuitionistic fuzzy values, Pythagorean fuzzy values and Fermatean fuzzy values when taking $q = 1$ or $q = 2$ or $q = 3$.

The monotonicity of $\lambda \cdot_{\mathrm{H}}^{\gamma} x$ and $x^{\gamma_{\mathrm{H}}\lambda}$ with respect to γ depends on the value of λ. In fact, Eq. (18) can be rewritten as

$$\lambda \cdot_{\mathrm{H}}^{\gamma} x = \left|\left(1 - \frac{\left(1-a^q\right)^{\lambda}}{\frac{1}{\gamma}\left(1+(\gamma-1)a^q\right)^{\lambda} + \left(1-\frac{1}{\gamma}\right)\left(1-a^q\right)^{\lambda}}\right)^{1/q}, \right.$$
$$\left. \left(\frac{b^{\lambda q}}{\frac{1}{\gamma}\left(1+(\gamma-1)(1-b^q)\right)^{\lambda} + \left(1-\frac{1}{\gamma}\right)b^{\lambda q}}\right)^{1/q}\right|. \tag{20}$$

Take $g(\gamma) = \frac{1}{\gamma}\left(1 + (\gamma - 1)a^q\right)^\lambda + \left(1 - \frac{1}{\gamma}\right)\left(1 - a^q\right)^\lambda$, then its derivative $g'(\gamma) = \frac{(1+(\gamma-1)a^q)^{\lambda-1}}{\gamma^2}\left(\lambda\gamma a^q - \left(1 + (\gamma - 1)a^q\right) + \left(1 - a^q\right)^\lambda(1 + (\gamma - 1)a^q)^{1-\lambda}\right)$. Thus the monotonicity of $g(\gamma)$ is determined by the sign of $h(\gamma) = \lambda\gamma a^q - \left(1 + (\gamma - 1)a^q\right) + \left(1 - a^q\right)^\lambda(1 + (\gamma - 1)a^q)^{1-\lambda}$. Subsequently, we have $h(0) = 0$ and $h'(\gamma) = (\lambda - 1)a^q\left(1 - \left(\frac{1-a^q}{1+(\gamma-1)a^q}\right)^\lambda\right)$. Then the followings hold.

- If $0 < \lambda \leq 1$, then $h'(\gamma) \leq 0$, which gives $h(\gamma) \leq h(0) = 0$. Hence, $g(\gamma)$ is monotonic decreasing with respect to γ.
- If $\lambda > 1$, then $h'(\gamma) \geq 0$, which gives $h(\gamma) \geq h(0) = 0$. Hence, $g(\gamma)$ is monotonic increasing with respect to γ.

Thus, the membership degree of $\lambda \cdot_{\mathrm{H}}^\gamma x$ is monotonic decreasing when $0 < \lambda \leq 1$ and monotonic increasing when $\lambda > 1$. Similarly, the nonmembership degree of $\lambda \cdot_{\mathrm{H}}^\gamma x$ is monotonic increasing when $0 < \lambda \leq 1$ and monotonic decreasing when $\lambda > 1$.

More precisely,

- For $0 < \lambda \leq 1$, $\lambda \cdot_{\mathrm{H}}^\gamma x$ is monotonic decreasing with respect to γ, while $x^{\gamma_{\mathrm{H}}\lambda}$ is monotonic increasing with respect to γ. Namely, for any q-ROFV x, if $\gamma_1 \geq \gamma_2$, then $\lambda \cdot_{\mathrm{H}}^{\gamma_1} x \preceq \lambda \cdot_{\mathrm{H}}^{\gamma_2} x$ and $x^{\gamma_1}{}_{\mathrm{H}}{}^\lambda \succeq x^{\gamma_2}{}_{\mathrm{H}}{}^\lambda$. Specially, we have $\lambda \cdot_\varepsilon x \preceq \lambda x$ and $x^{\varepsilon\lambda} \succeq x^\lambda$.
- For $\lambda > 1$, $\lambda \cdot_{\mathrm{H}}^\gamma x$ and $x^{\gamma_{\mathrm{H}}\lambda}$ are monotonic increasing and monotonic decreasing with respect to γ, respectively. That is, for any q-ROFV x, if $\gamma_1 \geq \gamma_2$, then $\lambda \cdot_{\mathrm{H}}^{\gamma_1} x \succeq \lambda \cdot_{\mathrm{H}}^{\gamma_2} x$ and $x^{\gamma_1}{}_{\mathrm{H}}{}^\lambda \preceq x^{\gamma_2}{}_{\mathrm{H}}{}^\lambda$. Specially, we have $\lambda \cdot_\varepsilon x \succeq \lambda x$ and $x^{\varepsilon\lambda} \preceq x^\lambda$.

Moreover, if $\lambda \to 0^+$, then $\lambda \cdot_{\mathrm{H}}^\gamma x \to \langle 0, 1\rangle$ ($x \neq \langle 1, 0\rangle$), and $x^{\gamma_{\mathrm{H}}\lambda} \to \langle 1, 0\rangle$ ($x \neq \langle 0, 1\rangle$). In addition, if $\lambda \to +\infty$, then $\lambda \cdot_{\mathrm{H}}^\gamma x \to \langle 1, 0\rangle$ ($x \neq \langle 0, 1\rangle$) and $x^{\gamma_{\mathrm{H}}\lambda} \to \langle 0, 1\rangle$ ($x \neq \langle 1, 0\rangle$). In fact,

$$
\lim_{\lambda \to +\infty} \lambda \cdot_{\mathrm{H}}^\gamma x = \left|\lim_{\lambda \to +\infty}\left(\frac{\left(1 + (\gamma - 1)a^q\right)^\lambda - \left(1 - a^q\right)^\lambda}{\left(1 + (\gamma - 1)a^q\right)^\lambda + (\gamma - 1)\left(1 - a^q\right)^\lambda}\right)^{1/q},\right.
$$

$$
\left.\lim_{\lambda \to +\infty}\left(\frac{\gamma b^{\lambda q}}{\left(1 + (\gamma - 1)\left(1 - b^q\right)\right)^\lambda + (\gamma - 1)b^{\lambda q}}\right)^{1/q}\right|
$$

$$
= \left|\lim_{\lambda \to +\infty}\left(1 - \frac{\gamma}{\left(1 + \frac{\gamma a^q}{1-a^q}\right)^\lambda + (\gamma - 1)}\right)^{1/q},\right.
$$

$$
\left.\lim_{\lambda \to +\infty}\left(\frac{\gamma}{\left(1 + \frac{\gamma(1-b^q)}{b^q}\right)^\lambda + (\gamma - 1)}\right)^{1/q}\right|
$$

$$
= \langle 1, 0\rangle.
$$

Theorem 3. *Let x, y be two q-ROFVs and $\lambda, \lambda_1, \lambda_2 > 0$. Then we have the followings:*

(1) $\lambda \cdot_{H}^{\gamma} (x \oplus_{H}^{\gamma} y) = \lambda \cdot_{H}^{\gamma} x \oplus_{H}^{\gamma} \lambda \cdot_{H}^{\gamma} y,$　　(2) $(x \otimes_{H}^{\gamma} y)^{\gamma\lambda}_{H} = x^{\gamma\lambda}_{H} \otimes_{H}^{\gamma} y^{\gamma\lambda}_{H},$

(3) $(\lambda_1 + \lambda_2) \cdot_{H}^{\gamma} x = \lambda_1 \cdot_{H}^{\gamma} x \oplus_{H}^{\gamma} \lambda_2 \cdot_{H}^{\gamma} x,$　　(4) $x^{\gamma(\lambda_1+\lambda_2)}_{H} = x^{\gamma\lambda_1}_{H} \otimes_{H}^{\gamma} x^{\gamma\lambda_2}_{H},$

(5) $\lambda_1 \cdot_{H}^{\gamma} (\lambda_2 \cdot_{H}^{\gamma} x) = (\lambda_1\lambda_2) \cdot_{H}^{\gamma} x,$　　(6) $x^{\gamma(\lambda_1\lambda_2)}_{H} = (x^{\gamma\lambda_1}_{H})^{\gamma\lambda_2}_{H}.$

Proof. (1) Suppose that $x = \langle a_1, b_1 \rangle$ and $y = \langle a_2, b_2 \rangle$, we have

$$
\lambda \cdot_{H}^{\gamma} (x \oplus_{H}^{\gamma} y) = \lambda \cdot_{H}^{\gamma} \left| \left(\frac{a_1^q + a_2^q + (\gamma - 2)a_1^q a_2^q}{1 + (\gamma - 1)a_1^q a_2^q} \right)^{1/q}, \left(\frac{b_1^q b_2^q}{1 + (\gamma - 1)(1 - b_1^q)(1 - b_2^q)} \right)^{1/q} \right|
$$

$$
= \left\langle \left(\frac{(1 + (\gamma - 1)a_1^q)^{\lambda}(1 + (\gamma - 1)a_2^q)^{\lambda} - (1 - a_1^q)^{\lambda}(1 - a_2^q)^{\lambda}}{(1 + (\gamma - 1)a_1^q)^{\lambda}(1 + (\gamma - 2)a_1^q)^{\lambda} + (\gamma - 1)(1 - a_1^q)^{\lambda}(1 - a_2^q)^{\lambda}} \right)^{1/q}, \right.
$$

$$
\left. \left(\frac{\gamma b_1^{\lambda q} b_2^{\lambda q}}{(1 + (\gamma - 1)(1 - b_1^q))^{\lambda}(1 + (\gamma - 1)(1 - b_2^q))^{\lambda} + (\gamma - 1)b_1^{\lambda q} b_2^{\lambda q}} \right)^{1/q} \right\rangle
$$

$$
= \left| \left(\frac{(1 + (\gamma - 1)a_1^q)^{\lambda} - (1 - a_1^q)^{\lambda}}{(1 + (\gamma - 1)a_1^q)^{\lambda} + (\gamma - 1)(1 - a_1^q)^{\lambda}} \right)^{1/q}, \left(\frac{\gamma b_1^{\lambda q}}{(1 + (\gamma - 1)(1 - b_1^q))^{\lambda} + (\gamma - 1)b_1^{\lambda q}} \right)^{1/q} \right|
$$

$$
\oplus_{H}^{\gamma} \left| \left(\frac{(1 + (\gamma - 1)a_2^q)^{\lambda} - (1 - a_2^q)^{\lambda}}{(1 + (\gamma - 1)a_2^q)^{\lambda} + (\gamma - 1)(1 - a_2^q)^{\lambda}} \right)^{1/q}, \left(\frac{\gamma b_2^{\lambda q}}{(1 + (\gamma - 1)(1 - b_2^q))^{\lambda} + (\gamma - 1)b_2^{\lambda q}} \right)^{1/q} \right|
$$

$$
= \lambda \cdot_{H}^{\gamma} x \oplus_{H}^{\gamma} \lambda \cdot_{H}^{\gamma} y.
$$

(3) Suppose that $x = \langle a, b \rangle$, we have

$$
\lambda_1 \cdot_{H}^{\gamma} x \oplus_{H}^{\gamma} \lambda_2 \cdot_{H}^{\gamma} x
$$

$$
= \left| \left(\frac{(1 + (\gamma - 1)a^q)^{\lambda_1} - (1 - a^q)^{\lambda_1}}{(1 + (\gamma - 1)a^q)^{\lambda_1} + (\gamma - 1)(1 - a^q)^{\lambda_1}} \right)^{1/q}, \left(\frac{\gamma b^{q\lambda_1}}{(1 + (\gamma - 1)(1 - b^q))^{\lambda_1} + (\gamma - 1)b^{q\lambda_1}} \right)^{1/q} \right|
$$

$$
\oplus_{H}^{\gamma} \left| \left(\frac{(1 + (\gamma - 1)a^q)^{\lambda_2} - (1 - a^q)^{\lambda_2}}{(1 + (\gamma - 1)a^q)^{\lambda_2} + (\gamma - 1)(1 - a^q)^{\lambda_2}} \right)^{1/q}, \left(\frac{\gamma b^{q\lambda_2}}{(1 + (\gamma - 1)(1 - b^q))^{\lambda_2} + (\gamma - 1)b^{q\lambda_2}} \right)^{1/q} \right|
$$

$$
= \left| \left(\frac{(1 + (\gamma - 1)a^q)^{\lambda_1+\lambda_2} - (1 - a^q)^{\lambda_1+\lambda_2}}{(1 + (\gamma - 1)a^q)^{\lambda_1+\lambda_2} + (\gamma - 1)(1 - a^q)^{\lambda_1+\lambda_2}} \right)^{1/q}, \right.
$$

$$
\left. \left(\frac{\gamma b^{q(\lambda_1+\lambda_2)}}{(1 + (\gamma - 1)(1 - b^q))^{\lambda_1+\lambda_2} + (\gamma - 1)b^{q(\lambda_1+\lambda_2)}} \right)^{1/q} \right|
$$

$$
= (\lambda_1 + \lambda_2) \cdot_{H}^{\gamma} x.
$$

(5) Suppose that $x = \langle a, b \rangle$, we have

$$
\lambda_1 \cdot_{H}^{\gamma} (\lambda_2 \cdot_{H}^{\gamma} x)
$$

$$
= \lambda_1 \cdot_{H}^{\gamma} \left| \left(\frac{(1 + (\gamma - 1)a^q)^{\lambda_2} - (1 - a^q)^{\lambda_2}}{(1 + (\gamma - 1)a^q)^{\lambda_2} + (\gamma - 1)(1 - a^q)^{\lambda_2}} \right)^{1/q}, \left(\frac{\gamma b^{q\lambda_2}}{(1 + (\gamma - 1)(1 - b^q))^{\lambda_2} + (\gamma - 1)b^{q\lambda_2}} \right)^{1/q} \right|
$$

$$
= \left| \left(\frac{(1 + (\gamma - 1)a^q)^{\lambda_1\lambda_2} - (1 - a^q)^{\lambda_1\lambda_2}}{(1 + (\gamma - 1)a^q)^{\lambda_1\lambda_2} + (\gamma - 1)(1 - a^q)^{\lambda_1\lambda_2}} \right)^{1/q}, \left(\frac{\gamma b^{q(\lambda_1\lambda_2)}}{(1 + (\gamma - 1)(1 - b^q))^{\lambda_1\lambda_2} + (\gamma - 1)b^{q(\lambda_1\lambda_2)}} \right)^{1/q} \right|
$$

$$
= (\lambda_1\lambda_2) \cdot_{H}^{\gamma} x.
$$

The proofs of the others are analogues and so are omitted.　　　□

From Theorem 3, one can see that the Hamacher scalar multiplication and Hamacher exponentiation operations on q-ROFVs are associative (with respect to coefficients) and distributive.

4   Conclusion

Hamacher operations with the parameter γ are extensions of algebraic (if $\gamma = 1$) and Einstein (if $\gamma = 2$) operations that are usually applied to the theory of fuzzy sets. Darko and Liang and Liu et al. proposed four Hamacher operations of q-ROFVs. In this paper, we give a comprehensive study of q-rung orthopair fuzzy Hamacher operations and the main conclusions are summarized as follows.

- We have listed some special cases of Hamacher operations for q-ROFVs. More precisely, with different values of γ, q-rung orthopair fuzzy (algebraic) operations and q-rung orthopair fuzzy Einstein operations are special cases of the developed operations; with different values of q, intuitionistic fuzzy Hamacher operations, Pythagorean fuzzy Hamacher operations and Fermatean fuzzy Hamacher operations are special cases of these operations.
- We have discussed the operational rules of Hamacher operations over q-ROFVs. It is shown that these operations perform much like those established for real numbers, which demonstrates that they are indeed well-defined.
- We have investigated the monotonicity of q-rung orthopair fuzzy Hamacher operations with respect to parameter γ, and the limiting cases of Hamacher scalar multiplication and Hamacher exponentiation operations in terms of the multiplier and exponent, respectively.

Since there are many generalizations of q-ROFSs, for example, complex interval-valued q-rung orthopair fuzzy sets, q-rung orthopair hesitant fuzzy sets and q-rung orthopair fuzzy linguistic sets, Hamacher operations on these types of fuzzy sets are to be considered. On the other hand, one can combine Hamacher operations with mean operators, Bonferroni mean, Maclaurin symmetric mean, Muirhead mean [7, 12, 25, 26], etc., to develop aggregation operators on q-ROFVs.

Acknowledgements. This research was supported by the National Natural Science Foundation of China (Nos. 12271493, 61806182), the Innovation Team Support Program for Excellent Young Talents of Zhengzhou University (Grant No. 32320292) and the Natural Science Foundation of Henan (Grant No. 242300421154).

References

1. Yager, R.R.: Generalized orthopair fuzzy sets. IEEE Trans. Fuzzy Syst. **25**(5), 1222–1230 (2017)
2. Zadeh, L.A.: Fuzzy sets. Inf. Control **8**(3), 338–353 (1965)
3. Atanassov, K.T.: Intuitionistic fuzzy sets. Fuzzy Sets Syst. **20**(1), 87–96 (1986)
4. Yager, R.R., Abbasov, A.M.: Pythagorean membership grades, complex numbers, and decision making. Int. J. Intell. Syst. **28**(5), 436–452 (2013)
5. Senapati, T., Yager, R.R.: Fermatean fuzzy sets. J. Ambient. Intell. Humaniz. Comput. **11**(2), 663–674 (2020)
6. Liu, P., Wang, P.: Some q-rung orthopair fuzzy aggregation operators and their applications to multiple-attribute decision making. Int. J. Intell. Syst. **33**(2), 259–280 (2018)
7. Liu, P., Wang, P.: Multiple-attribute decision making based on Archimedean Bonferroni operators of q-rung orthopair fuzzy numbers. IEEE Trans. Fuzzy Syst. **27**(5), 834–848 (2019)
8. Akram, M., Shahzadi, G., Shahzadi, S.: Protraction of Einstein operators for decision-making under q-rung orthopair fuzzy model. J. Intell. Fuzzy Syst. **40**(3), 4779–4798 (2021)
9. Jana, C., Muhiuddin, G., Pal, M.: Some Dombi aggregation of Q-rung orthopair fuzzy numbers in multiple-attribute decision making. Int. J. Intell. Syst. **34**(12), 3220–3240 (2019)
10. Du, W.S.: More on Dombi operations and Dombi aggregation operators for q-rung orthopair fuzzy values. J. Intell. Fuzzy Syst. **39**(3), 3715–3735 (2020)
11. Seikh, M.R., Mandal, U.: Q-rung orthopair fuzzy Frank aggregation operators and its application in multiple attribute decision-making with unknown attribute weights. Granular Comput. **7**(3), 709–730 (2022)
12. Darko, A.P., Liang, D.: Some q-rung orthopair fuzzy Hamacher aggregation operators and their application to multiple attribute group decision making with modified EDAS method. Eng. Appl. Artif. Intell. **87**, 103259 (2020)
13. Hamacher, H.: Über logische Verknüpfungen unscharfer Aussagen und deren zugehörige Bewertungsfunktionen. In: Trappl, R., Klir, G.J., Ricciardi, L. (eds.) Progress Cybern. Syst. Res., vol. 3, pp. 276–288. Hemisphere, Washington (1978)
14. Huang, J.Y.: Intuitionistic fuzzy Hamacher aggregation operators and their application to multiple attribute decision making. J. Intell. Fuzzy Syst. **27**(1), 505–513 (2014)
15. Tan, C., Yi, W., Chen, X.: Hesitant fuzzy Hamacher aggregation operators for multicriteria decision making. Appl. Soft Comput. **26**, 325–349 (2015)
16. Wu, S.J., Wei, G.W.: Pythagorean fuzzy Hamacher aggregation operators and their application to multiple attribute decision making. Int. J. Knowl.-Based Intell. Eng. Syst. **21**(3), 189–201 (2017)
17. Hadi, A., Khan, W., Khan, A.: A novel approach to MADM problems using Fermatean fuzzy Hamacher aggregation operators. Int. J. Intell. Syst. **36**(7), 3464–3499 (2021)
18. Liu, L., Wu, J., Wei, G., Wei, C., Wang, J., Wei, Y.: Entropy-based GLDS method for social capital selection of a PPP project with q-rung orthopair fuzzy information. Entropy **22**(4), 414 (2020)
19. Zhu, Y.: Multiple-attribute decision-making of q-rung orthopair fuzzy sets based on Hamacher norm and improved MULTIMOORA. In: Proceedings of the 4th International Conference on Computer Science and Application Engineering (CSAE 2020), vol. 160, pp. 1–6. ACM, New York (2020)

20. Kakati, P., Rahman, S.: The q-Rung orthopair fuzzy Hamacher generalized Shapley Choquet integral operator and its application to multiattribute decision making. EURO J. Decis. Processes **10**, 100012 (2022)

21. Du, W.S.: A further investigation on q-rung orthopair fuzzy Einstein aggregation operators. J. Intell. Fuzzy Syst. **41**(6), 6655–6673 (2021)

22. Mahmood, T., Ali, Z.: A novel approach of complex q-rung orthopair fuzzy hamacher aggregation operators and their application for cleaner production assessment in gold mines. J. Ambient. Intell. Humaniz. Comput. **12**(9), 8933–8959 (2021)

23. Donyatalab, Y., Farrokhizadeh, E., Shishavan, S.A.S., Seifi, S.H.: Hamacher aggregation operators based on interval-valued q-rung orthopair fuzzy sets and their applications to decision making problems. In: Kahraman, C., Cevik Onar, S., Oztaysi, B., Sari, I.U., Cebi, S., Tolga, A.C. (eds.) INFUS 2020. AISC, vol. 1197, pp. 466–474. Springer, Cham (2021). https://doi.org/10.1007/978-3-030-51156-2_54

24. Gayen, S., Sarkar, A., Biswas, A.: Development of q-rung orthopair trapezoidal fuzzy Hamacher aggregation operators and its application in MCGDM problems. Comput. Appl. Math. **41**, 263 (2022)

25. Qin, Y., Cui, X., Huang, M., Zhong, Y., Tang, Z., Shi, P.: Archimedean Muirhead aggregation operators of q-rung orthopair fuzzy numbers for multicriteria group decision making. Complexity **2019**, 3103741 (2019)

26. Rawat, S.S., Komal: Multiple attribute decision making based on q-rung orthopair fuzzy Hamacher Muirhead mean operators. Soft Comput. **26**(5), 2465–2487 (2022)

Healthcare Referral and Coordination in a Two-Tier Service System via Government Subsidy Scheme

Caimin Wei[1], Zhiyuan Tong[1], Zongbao Zou[2], and Zhongping Li[3(✉)]

[1] Department of Mathematics, Shantou University,
Shantou 515063, People's Republic of China
{cmwei,20zytong}@stu.edu.cn
[2] School of Business, Shantou University, Shantou 515063, China
zbzou1@stu.edu.cn
[3] School of Business, Anhui University, Hefei 230601, China
20223@ahu.edu.cn

Abstract. Healthcare referral in a two-tier service system, transferring patients from an overcrowded comprehensive hospital to an idle community hospital has become an important means of Chinese healthcare reform for balancing the supply of and demand for care service at different level hospitals. To effectively implement such a healthcare referral program, the government may offer a subsidy to patients or a comprehensive hospital. The paper examines the operating efficiency of such a subsidy scheme associated with the healthcare referral program. By establishing a game-theoretic queuing model involving patients, hospitals, and government, we first present the optimal strategy in terms of referral rate and capacity level under the centralized and decentralized systems, respectively. We find that both the optimal referral rate and the optimal capacity level for the decentralized system are never more than those for the centralized system. Then when the objective of the government is to maximize the referral rate of patients subject to the constraint that the optimal referral rate for the centralized system is as a maximum (upper bound), we explore a feasible subsidy mechanism in coordinating the overall healthcare system. Our results indicate that for an inadequate budget, the equilibrium subsidy strategy is suboptimal for the two-tier healthcare system; for a relatively medium budget, the equilibrium subsidy strategy is optimal only when subsidizing the comprehensive hospital; for a sufficient budget, there always exists an optimal subsidy strategy for maximizing the patient welfare as well as coordinating the overall healthcare system.

Keywords: Healthcare operations · Coordination · Subsidy scheme · Queueing-game · Referral rate capacity

1 Introduction

The healthcare system in some countries such as China is composed of comprehensive and community hospitals. High-level hospitals provide some specialty

B.-Y. Cao et al. (Eds.): ICFIE 2022, LNDECT 207, pp. 232–250, 2024.
https://doi.org/10.1007/978-981-97-2891-6_18

services to seriously ill and emergency patients, whereas low-level hospitals provide basic medical care to common and chronic illness patients. Such a system is called a two-tier healthcare system, whose purpose is to balance the supply of service capacity level and demand for healthcare service at different level hospitals. To redress these problems, the low-level hospitals need to redesign their capacities for serving referred patients; the Chinese government requires to establish the efficient subsidy mechanisms for implementing this healthcare referral program. So, the policy maker and hospital managers need to address the following some questions:

(1) how to determine the optimal strategy in terms of healthcare referral rate and low-level hospital's capacity level under the centralized and decentralized systems, and what are the relationships between optimal strategies under the two different systems?

(2) Which subsidy scheme is preferable, subsidizing patients, high-level hospitals, or both patients and high-level hospitals for the government?

To answer these questions, we consider a two-tier Chinese healthcare system that comprises a high-level hospital provider (HHP), a low-level hospital provider (LHP), and the patients. By establishing a game-theoretic model within a queuing framework, we capture the interactions among the patients, two hospitals, and government and then obtain the optimal strategies in terms of healthcare referral rate and capacity level under the centralized and decentralized systems. Therefore, it is related to several research streams including healthcare referral, capacity level, and subsidy scheme. We concisely review each research stream subsequently.

With the healthcare referral aspect, some researchers have focused on the issue of referral rate in a two-level healthcare service system over the past decade [1–4]. In their studies, upstream referral process of patients is popular, i.e. LHPs serve as the first-level gatekeepers and either treat patients' problem or refer them to HHPs. In contrast to the upstream referral, in some countries such as China, the patient's downstream referral flow which is greatly advocated from the HHPs to the LHPs in order to balance the supply of public service capacity and demand for medical services at high- and low-level hospitals [5,6]. However, most of these studies focus on the minimization objective of the service providers' operating costs using queuing models. In contrast, our work is to capture the dynamic interactions among the HHP, the LHP, and the patients by establishing the sequential game for the decentralized and centralized systems.

As regard, service capacity level, most relevant studies consider the situation in which one service provider or two (or multiple asymmetric providers) determines (determine) capacity level to maximize their own profits or revenues, in which customers are heterogeneous or homogeneous in terms of the service valuation that is subject to an additive or a generalized delay cost [7,8]. All of these studies mostly focus on the capacity level in the case of the service provider's profit- or revenue-maximization.

There are some researches on the issues of government subsidy in a two-tier healthcare system. Some empirical literature study on subsidizing private or public hospital provider. Their results show that subsidizing the private hospital would reduce social welfare. Recently, some scholars develop a game model based on a queuing framework in a two-tier healthcare system [9,12]. Chen et al. [13] discuss that the government subsidizes both the free system and the toll system' customers. Zhou et al. [14] use a sequential game approach to discuss patients' choice among hospitals in subsidizing patients. Adida et al. [15] evaluate the impact of a variety of payment schemes on the performance for healthcare services in the presence of patient selection.

Based on above analysis, we consider a two-tier Chinese healthcare system that comprises a HHP, a LHP, and the patients. Our results show that: (1) both the optimal referral rate and the optimal capacity level for the decentralized system are never more than those for the centralized system. (2) The government could induce the implementation of the referrals program either via subsidizing comprehensive hospitals or patients. The optimal combination mainly depends on the financial budget, the HHP's capacity level, and the ratio of treatment price to cost.

The remainder of this paper is organized as follows. Sections 2 reviews the description of model. In Sect. 3, the optimal strategies in terms of healthcare referral rate and capacity level are analyzed in the decentralized and centralized systems, and we compare the optimal strategies between decentralized and centralized systems, and Sect. 4 focuses on the government subsidy mechanisms associated with the healthcare referral program in a two-tier healthcare system. Section 5 provides conclusions and future work.

2   Description of Models

We now consider a two-tier healthcare system which is comprised of heterogeneous patients, an HHP, and an LHP. The total potential patients randomly arrive at the HHP (denoted as hospital 1) and LHP (denoted as hospital 2) according to a Poisson process at the rate λ_1 and λ_2 per unit time, respectively. Each hospital serves (diagnoses or treats) patients based on the first-come-first-served (FCFS) queuing principle. After the diagnosis, the patients would be categorized into two groups according to their illness severity: M-type (mild illness) patients and S-type (severe illness) patients, as depicted in Fig. 1. More specifically, the severity of a patient's problem is described by a real number $x(x \in [0,1])$, which is denoted as the severity's fractile and is uniformly distributed (Lee et al. [3]). The greater the value of x indicates the more severity of the patient's illness. Generally, if a patient's severity is less than a threshold α, she would be categorized into M-type; otherwise, i.e., the patient's severity is greater than the threshold, she would be categorized into S-type. Thus, the arrival rate of M-type patients entering the system for a diagnosis satisfies α. Following the work of Lee et al. [3] and Li et al. [5], we neglect the diagnosing time of every patient at a hospital because the service time is sufficiently short in reality. HHP can offer comprehensive services to two types of patients, but LHP can only offer basic services to M-type patients. For tractability, we assume that in the first diagnosis process, only the M-type patients enter the LHP directly for seeking medical care. Although in some countries such as in China, the upstream referral process of S-type patients is popular, the assumption still holds for some LHPs because of the relatively low arrival rate, which was also studied by Li et al. [5]. Some notations in our model are listed in Table 1.

Service times of the HHP and the LHP are independent and identically distributed exponential random variables with the service rate given by the service capacity μ_1 and μ_2, respectively. For analytical tractability, we assume that the HHP provides an identical service time for S- and M-type patients. We model the HHP and the LHP as $M/M/1$ queues. For each S-type patient, the corresponding service must be offered by the HHP because her illness is serious. By contrast, each M-type patient at the HHP can choose to stay at the HHP or downstream refer to the LHP for follow-up care. We denote the referral rate θ as the number of M-type patients at the HHP who are referred to the LHP for receiving treatment service. For coordinating the two-tier healthcare system to achieve the maximization of the overall patient welfare goal, the government provides referral subsidy to patients or HHP. We assume r_1 and r_2 denotes subsidy rate to HHP and patient per unit patient of referral, respectively.

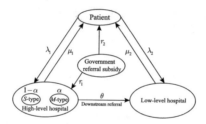

Fig. 1. Patients referral process

Patient Utility and Choice: we consider that the patients' utility depends on the service value, the waiting times, and the service price charged by HHP or LHP. Following Wang et al. [16], we assume that the HHP and LHP charge the same treatment price p for every M-type patient. This is because in China, the government generally does not allow a disparate charge on the same type patients. Generally, HHP provides a heterogeneous service values for M- and S-type patients, which are denoted as v_1 and v_S, respectively, where $v_S > v_1 > 0$. Naturally, each S-type patient is charged a higher treatment price (p_S) compared with M-type patient (p) by the HHP, i.e., $p_S > p > 0$. Furthermore, the service value of the HHP is always higher than that of the LHP (v_2). We follow the common assumption that the patient's waiting cost is denoted as d per unit of time. For the simplification, we assume that the treatment cost is zero. Thus, the net utility of each M-type patient who is waiting time-sensitive at the hospital i ($i = 1, 2$) is given as follows:

$$U_i = v_i - p - dW(\lambda_i, \mu_i),\tag{1}$$

where $W(\lambda_i, \nu_i) = \frac{1}{\mu_i - \lambda_i}$.

Table 1. Notation summary.

Notations	Definition
λ_1, λ_2	Patient arrival rate at the HHP and LHP per unit time, respectively
α	The proportion of M-type patients in the total patients arriving at the HH
μ_1, μ_2	The service capacity of the HHP and LHP per unit time, respectively
r_1, r_2	The subsidy amount of the HHP and patient receiving from the government per unit patient of referral, respectively
θ_1	The referral rate of the patients permitted by the HHP
θ_2	The referral rate of the patients' willingness
θ	The realized referral rate of M-type patients at the HH
v_1, v_2	The value of the HHP's service and LHP's service, respectively
v	The gap between the values of the HHP and LHP's service
p	The hospital's unit price of the service
c	The hospital's unit cost of an increase in the service rate
d	The waiting cost per patient
U_1, U_2	The utility of an M-type patient who seeks care at the HHP and LHP, respectively
SU_1, SU_2	The total utility of the HHP and LHP patients, respectively
SW	The total utility of the overall healthcare system patients

The net utility of each S-type patient at the HHP is given by

$$U_S = [(v_S - v_1) - (p_S - p)] + v_1 - p - \frac{1}{\mu_i - \lambda_i}.$$

Similar to Andritsos and Aflaki [4], we consider that all v_i and v_S are fixed and high enough that all patients obtain non-negative utility from treatment services at either HHP or LHP. Therefore, each M-type patient who acts as a utility maximizer chooses the HHP for receiving treatment when if $U_1 = U_2$; otherwise, the patient refers to the LHP. Following Li et al. [5], we consider the case in which a patient can only choose to whether shift to the LHP or not based on her expected utility maximization when the HHP permits her referral. More specifically, when the HHP permits θ_1 portion patients to transfer, the rate of the patients who are willing to refer θ_2 ultimately will not exceed θ_1, i.e. $\theta_2 \leq \theta_1$. We explore the patients' choice behavior as follows:

Case (i) : $v_1 - W(\lambda_1 - \theta_1, \mu_1) \leq v_2 - W(\lambda_2 + \theta_1, \mu_2)$.
Case (ii) : $v_1 - W(\lambda_1 - \theta_1, \mu_1) > v_2 - W(\lambda_2 + \theta_1, \mu_2)$ and
              $v_1 - W(\lambda_1, \mu_1) < v_2 - W(\lambda_2, \mu_2)$.
Case (iii) : $v_1 - W((1 - \alpha)\lambda_1, \mu_1) < v_2 - W(\alpha\lambda_1 + \lambda_2, \mu_2)$.
Case (iv) : $v_1 - W(\lambda_1, \mu_1) \geq v_2 - W(\lambda_2, \mu_2)$.

Cases (i) and (ii) correspond to the situations in which all and only a portion of the M-type patients permitted by the HHP choose to shift to the LHP, respectively. Case (i) represents that each permitted patient at the HHP who chooses to downstream refer for receiving treatment services would always obtain a higher net utility. Case (ii) represents that only a portion of the permitted patients can obtain a higher net utility compared to staying in the HHP while choosing to downstream refer for receiving treatment service. Particularly, if the waiting time satisfies Case (iii) or Case (iv), all or none of the M-type patients always will expect to shift to the LHP for follow-up treatment services. In practice, some M-type patients at the HHP should refer to the LHP because of excessively long waiting times of the HHP, however, the patients are generally unable to fully transfer. A portion of the M-type patients is often referred downstream, thus the equilibrium decision should satisfy Case (i) or Case (ii). To avoid the trivial case, we choose the Case (ii) to study in the following analysis.

3 Healthcare Referral and Capacity Strategy

In this section, we explore the optimal strategy in terms of healthcare referral rate and capacity level for the optimization objective of each hospital provider's and the overall healthcare system's welfare-maximization, respectively.

3.1 Optimal Strategy Under the Decentralized System

Sequential Model. Under the decentralized system, we shall consider a three-stage Stackelberg game model that involves the HHP, the LHP, and the patients as follows.

Stage 1: The HHP as the leader determines the optimal referral rate for M-type patients who need to be referred to LHP θ_1.

Stage 2: The LHP as a follower determines the optimal capacity level μ_2 according to the referral rate of M-type patients.

Stage 3: Each M-type patient permitted by the HHP decides to shift to the LHP or stay at the HPP, and the corresponding equilibrium referral rate is denoted as θ_2 ($\leq \theta_1$).

We use backward induction to analyze the referral strategy. First, each M-type patient decides whether or not transfer to the LHP, then we anticipate her equilibrium referral rate θ_2^* (μ_2) for a given μ_2 based on her expected utility. Second, we shall derive the capacity level decision of the HHP.

Given θ_1, μ_2, we get $\theta = \min\{\theta_1, \arg\max_{\theta_2}\{U_2(\theta_2) - U_1(\theta_2) \geq 0\}\}$, where

$$U_1(\theta_2) = v_1 - p - dW(\lambda_1 - \theta_2, \mu_1), \quad U_2(\theta_2) = v_2 - p - dW(\lambda_2 + \theta_2, \mu_2).$$

If a patient choice satisfies Case (ii), then θ_2 ($< \theta_1$) portion of the M-type patients at the HHP would choose to divert. In this case, we obtain the following Lemma 1.

Lemma 1. *For Case (ii), the M-type patients' optimal referral rate $\theta_2^*(\mu_2)$ is given as*

$$\theta_2^*(\mu_2) \triangleq \frac{2 + \Delta[(\mu_2 - \lambda_2) - (\mu_1 - \lambda_1)] - \sqrt{4 + \{\Delta[(\mu_2 - \lambda_2) + (\mu_1 - \lambda_1)]\}^2}}{2\Delta},$$

$$(2)$$

where $\Delta = \frac{v}{d}$, $v = v_1 - v_2$, v explains the reduction of her acceptable service value when the patient chooses to shift to the LHP.

Lemma 1 implies when obtaining the referral permission of the HHP, the M-type patients choose only to receive the treatment service from the LHP or the HHP, based on her expected utility maximization.

LHP's Optimal Capacity Level. Given any referral rate θ, the HHP needs to determine its service capacity level to maximize the total expected patient utility $SU_2(\mu_2)$ including M-type referral patients subject to a non-negative profit constraint. In treating the capacity level as a decision variable, we take a long-run approach to the analysis. Hence, under the decentralized system, the LHP's optimization problem can be written as

$$\max_{\mu_2} SU_2(\mu_2) = (\lambda_2 + \theta)U_2(\theta, \mu_2)$$
$$s.\,t.\ \Pi_2(\mu_2) = (\lambda_2 + \theta)p - c\mu_2 \geq 0, \tag{3}$$
$$\lambda_2 + \theta < \mu_2,$$

where

$$SU_2(\mu_2) = \begin{cases} (\lambda_2 + \theta_1)[v_2 - p - \frac{d}{\mu_2 - (\lambda_2 + \theta_1)}], & \text{if } \theta = \theta_1, \\ (\lambda_2 + \theta_2)[v_2 - p - \frac{d}{\mu_2 - (\lambda_2 + \theta_2)}], & \text{if } \theta = \theta_2. \end{cases}$$

$$\Pi_2(\mu_2) = \begin{cases} (\lambda_2 + \theta_1)p - c\mu_2, & \text{if } \theta = \theta_1, \\ (\lambda_2 + \theta_2)p - c\mu_2, & \text{if } \theta = \theta_2. \end{cases}$$

Proposition 1. *Under the decentralized system, given any θ, $SU_2(\mu_2)$ is increasing in μ_2.*

Proposition 1 suggests that the LHP's optimization problem (3) is equivalent to maximizing its own capacity level with its constraints.

Based on (2) and (3), we get the following Lemma 2.

Lemma 2. *Under the decentralized system, given any θ_1, θ is equal to θ_1 or θ_2, which depends on μ_2, namely:*

(i) if $V < \frac{1}{\mu_1 - \lambda_1}$ and $\mu_2 \leq \bar{\mu}_2^0$, or $V \geq \frac{1}{\mu_1 - \lambda_1}$, we have $\theta = \theta_2 = 0 \leq \theta_1$;

(ii) if $V < \frac{1}{\mu_1 - (\lambda_1 - \theta_1)}$ and $\mu_2 \geq \bar{\mu}_2^1$, we have $\theta = \theta_1 \leq \theta_2$;

(iii) if $V < \frac{1}{\mu_1 - (\lambda_1 - \theta_1)}$ and $\bar{\mu}_2^0 < \mu_2 < \bar{\mu}_2^1$, or $\frac{1}{\mu_1 - \lambda_1} > V \geq \frac{1}{\mu_1 - (\lambda_1 - \theta_1)}$ and $\mu_2 > \bar{\mu}_2^0$, we have $0 < \theta = \theta_2 < \theta_1$, where $\bar{\mu}_2^0 = \lambda_2 + \frac{(\mu_1 - \lambda_1)}{1 - \Delta(\mu_1 - \lambda_1)}$, $\bar{\mu}_2^1 = \lambda_2 + \frac{\mu_1 - (\lambda_1 - \theta_1)}{1 - \Delta[\mu_1 - (\lambda_1 - \theta_1)]}$, and $V = v/d$ represents the longest time for an M-type patient at the HHP who is unwilling to refer for receiving follow-up treatment.

Lemma 2 identifies conditions whether none, all, or a portion of the permitted patients choose to refer to the LHP, as shown in Fig. 2. The following proposition provides the optimal capacity level of the LHP.

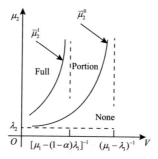

Fig. 2. Optimal referral structure

Proposition 2. *Under the decentralized system, given any θ_1, for a self-financing LHP, the optimal referral rate $\theta^*(\theta_1)$ and optimal capacity level $\mu_2^*(\theta_1)$ are given by:*

(i) for the $V \geq \frac{1}{\mu_1 - \lambda_1}$ case, then $\theta^*(\theta_1) = 0$ and $\mu_2^*(\theta_1) = \frac{p}{c}\lambda_2$,

(ii) for the $\frac{1}{\mu_1 - (\lambda_1 - \theta_1)} \leq V < \frac{1}{\mu_1 - \lambda_1}$ case,

  (a) if $\frac{1}{\Delta}(1 - \frac{1}{\sqrt{\frac{p}{c}-1}})^2 + (\lambda_1 + \lambda_2) \geq \mu_1 \geq (\lambda_1 + \lambda_2)$, then

$$\begin{cases} \theta^*(\theta_1) = \tilde{\theta}_2 \text{ and } \mu_2^*(\theta_1) = \mu_2^{1*}, & \text{for } \max\{\varphi_1^1, 1\} < \frac{p}{c} < \max\{\varphi_1^2, 1\} \\ \theta^*(\theta_1) = 0 \text{ and } \mu_2^*(\theta_1) = \frac{p}{c}\lambda_2, & \text{for } 1 < \frac{p}{c} < \max\{\varphi_1^1, 1\} \text{ or } \frac{p}{c} > \max\{\varphi_1^2, 1\}; \end{cases}$$

  (b) if $\mu_1 < (\lambda_1 + \lambda_2)$, then

$$\begin{cases} \theta^*(\theta_1) = \tilde{\theta}_2 \text{ and } \mu_2^*(\theta_1) = \mu_2^{1*}, & \text{for } \frac{p}{c} > \max\{\varphi_1^0, 1\} \\ \theta^*(\theta_1) = 0 \text{ and } \mu_2^*(\theta_1) = \frac{p}{c}\lambda_2, & \text{for } 1 < \frac{p}{c} < \max\{\varphi_1^0, 1\}, \end{cases}$$

(iii) for the $V < \frac{1}{\mu_1 - (\lambda_1 - \theta_1)}$ case,

  (a) if $\frac{1}{\Delta}(1 - \frac{1}{\sqrt{\frac{p}{c}-1}})^2 + (\lambda_1 + \lambda_2) \geq \mu_1 \geq (\lambda_1 + \lambda_2)$, then

$$\begin{cases} \theta^*(\theta_1) = \tilde{\theta}_2 \text{ and } \mu_2^*(\theta_1) = \mu_2^{1*}, & \begin{array}{l}\text{for } \max\{\varphi_1^1, 1\} < p/c < \max\{\varphi_2^2, 1\} \text{ or} \\ \max\{\varphi_2^2, 1\} < p/c < \max\{\varphi_1^1, 1\}\end{array} \\ \theta^*(\theta_1) = \theta_1 \text{ and } \mu_2^*(\theta_1) = \frac{p}{c}(\lambda_2 + \theta_1), & \text{for } \max\{\varphi_2^2, 1\} < \frac{p}{c} < \max\{\varphi_2^2, 1\} \\ \theta^*(\theta_1) = 0 \text{ and } \mu_2^*(\theta_1) = \frac{p}{c}\lambda_2, & \text{for } \frac{p}{c} > \max\{\varphi_1^2, 1\} \text{ or } 1 < \frac{p}{c} < \max\{\varphi_1^1, 1\}; \end{cases}$$

  (b) if $\mu_1 < (\lambda_1 + \lambda_2)$, then

$$\begin{cases} \theta^*(\theta_1) = \theta_1 \text{ and } \mu_2^*(\theta_1) = \frac{p}{c}(\lambda_2 + \theta_1), & \text{for } \frac{p}{c} > \max\{\varphi_2^0, 1\} \\ \theta^*(\theta_1) = \tilde{\theta}_2 \text{ and } \mu_2^*(\theta_1) = \mu_2^{1*}, & \text{for } \max\{\varphi_1^0, 1\} < \frac{p}{c} < \max\{\varphi_2^0, 1\} \\ \theta^*(\theta_1) = 0 \text{ and } \mu_2^*(\theta_1) = \frac{p}{c}\lambda_2, & \text{for } \frac{p}{c} < \max\{\varphi_1^0, 1\}, \end{cases}$$

  where

$$\tilde{\theta}_2 = \frac{2 + \Delta[(\mu_2^{1*} - \lambda_2) - (\mu_1 - \lambda_1)] - \sqrt{4 + \{\Delta[(\mu_2^{1*} - \lambda_2) + (\mu_1 - \lambda_1)]\}^2}}{2\Delta},$$

$$\mu_2^{1*} = \left\{\begin{array}{l} \frac{1}{2\Delta}\frac{p}{c}\{(1 - \frac{1}{\frac{p}{c}-1}) - \Delta[(\mu_1 - \lambda_1) - \lambda_2]\} \\ + \sqrt{\{[1 - (\frac{p}{c} - 1)^{-1}] + \Delta[(\mu_1 - \lambda_1) - \lambda_2]\}^2 - 4\Delta[(\mu_1 - \lambda_1) - \lambda_2]\}} \end{array}\right\},$$

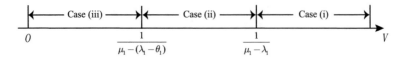

Fig. 3. The impact of V on the equilibrium referral

$$\varphi_1^0 = \underset{\frac{p}{c}}{\arg}\{\mu_2^{1*}(\tfrac{p}{c}) = \lambda_2 + \tfrac{(\mu_1-\lambda_1)}{1-\Delta(\mu_1-\lambda_1)}, \Delta[(\mu_1 - \lambda_1) - \lambda_2] < 0\},$$

$$\varphi_2^0 = \underset{\frac{p}{c}}{\arg}\{\mu_2^{1*}(\tfrac{p}{c}) = \lambda_2 + \tfrac{\mu_1-(\lambda_1-\theta_1)}{1-\Delta[\mu_1-(\lambda_1-\theta_1)]}, \Delta[(\mu_1 - \lambda_1) - \lambda_2] < 0\},$$

$$\varphi_1^1 =$$

$$\underset{\frac{p}{c}}{\min\arg}\left\{\mu_2^{1*}(\tfrac{p}{c}) = \lambda_2 + \tfrac{(\mu_1-\lambda_1)}{1-\Delta(\mu_1-\lambda_1)}, (1 - \tfrac{1}{\sqrt{\frac{p}{c}-1}})^2 \geq \Delta[(\mu_1 - \lambda_1) - \lambda_2] \geq 0\right\},$$

$$\varphi_1^2 =$$

$$\underset{\frac{p}{c}}{\max\arg}\left\{\mu_2^{1*}(\tfrac{p}{c}) = \lambda_2 + \tfrac{(\mu_1-\lambda_1)}{1-\Delta(\mu_1-\lambda_1)}, (1 - \tfrac{1}{\sqrt{p/c-1}})^2 \geq \Delta[(\mu_1 - \lambda_1) - \lambda_2] \geq 0\right\}.$$

Remark 1. Proposition 2 presents three cases in terms of referral structures, which depends on the value of V. As shown in Fig. 3, Case (i) corresponds to a sufficiently high value of V such that no patient chooses to transfer out of the HHP. As the value of V increases and is relatively medium, i.e., $\frac{1}{\mu_1-(\lambda_1-\theta_1)} \leq V < \frac{1}{\mu_1-\lambda_1}$, we show that in equilibrium, none or a portion of the permitted patients would shift from the HHP to the LHP. In general, for the situation of $\frac{1}{\Delta}(1 - \frac{1}{\sqrt{p/c-1}})^2 + (\lambda_1 + \lambda_2) \geq \mu_1 \geq (\lambda_1 + \lambda_2)$, if $\max\{\varphi_1^1, 1\} < \frac{p}{c} < \max\{\varphi_1^2, 1\}$ (namely, the ratio of treatment price to cost is located in medium range), there exists the optimal capacity level of the LHP such that a portion of the permitted patients choose to shift. Otherwise, i.e., $1 < \frac{p}{c} < \max\{\varphi_1^1, 1\}$ or $\frac{p}{c} > \max\{\varphi_1^2, 1\}$, then no patient would opt to refer; see Case (ii). In Case (iii), we observe that for a relatively low value of V, none, a portion or all of the permitted patients may transfer downstream. More specifically, there exist four thresholds with $\varphi_1^1 \leq \varphi_2^1 \leq \varphi_2^2 \leq \varphi_1^2$ (in the case that $\frac{1}{\Delta}(1 - \frac{1}{\sqrt{p/c-1}})^2 + (\lambda_1 + \lambda_2) \geq \mu_1 \geq (\lambda_1 + \lambda_2)$). When the ratio of treatment price to cost is in the range of medium, i.e., $\max\{\varphi_2^1, 1\} < \frac{p}{c} < \max\{\varphi_2^2, 1\}$, all of the permitted patients would refer downstream, and consequently, the LHP's optimal capacity level would satisfy $\mu_2^*(\theta_1) = \frac{p}{c}(\lambda_2 + \theta_1)$. When the ratio of treatment price to cost is in the range of low-medium or medium-high, i.e., $\max\{\varphi_2^2, 1\} < \frac{p}{c} < \max\{\varphi_1^2, 1\}$, only a portion of the patients would refer downstream and $\mu_2^* = \mu_2^{1*}$. Otherwise, i.e., the ratio of treatment price to cost is in the range of low or high (that is, $\frac{p}{c} > \max\{\varphi_1^2, 1\}$ or $1 < \frac{p}{c} < \max\{\varphi_1^1, 1\}$), all of the patients choose to stay in the HHP for receiving follow-up treatment.

HHP's Optimal Referral Rate. In this section, we study HHP's decision regarding M-type patients' referral rate. That is, the HHP selects θ_1 to maximize its total expected utility $SU_1(\theta_1)$, i.e., the optimization problem follows as

$$\max_{\theta_1} SU_1(\theta_1) = (\lambda_1 - \alpha\lambda_1)U_S(\theta_1) + (\alpha\lambda_1 - \theta_1)U_1(\theta_1)$$

$$s.\,t.\ 0\ \leq\ \theta_1 \leq \alpha\lambda_1,$$
$$\Pi_1(\theta_1) = (\lambda_1 - \alpha\lambda_1)p_S + (\alpha\lambda_1 - \theta_1)p - c\mu_1 \geq 0, \qquad (4)$$
$$\lambda_1 < \mu_1 \leq \tfrac{p_S}{c}(1 - \alpha)\lambda_1.$$

The first formula in optimization problem (4) represents the total expected patient utility of the HHP when the M-type patients shift to the LHP. The second formula in (4) ensures all the S-type patients to stay in the HHP for receiving follow-up treatment service. The third formula in (4) requires that the HHP's profit is always nonnegative in the presence of patient referral. In view of a limited capacity of the HHP, we consider that μ_1 is not higher than a threshold, which corresponds to the fourth formula in (4).

Proposition 3. *Under the decentralized system, the HHP's optimal referral rate is given by:*

(i) if $V_1 \geq \frac{1}{(1-\rho_1)(\mu_1-\lambda_1)}$, then $\theta_1^* = 0$;

(ii) if $V_1 \leq \frac{1}{[1-(1-\alpha)\rho_1][\mu_1-(1-\alpha)\lambda_1]}$, then $\theta_1^* = \alpha\lambda_1$;

(iii) if $\frac{1}{[1-(1-\alpha)\rho_1][\mu_1-(1-\alpha)\lambda_1]} < V_1 < \frac{1}{(1-\rho_1)(\mu_1-\lambda_1)}$, then $\theta_1^* = \tilde{\theta}_1^*$,

where $\tilde{\theta}_1^* = \sqrt{\frac{\mu_1}{V_1}} - (\mu_1 - \lambda_1),\, 0 < \rho_1 = \frac{\lambda_1}{\mu_1} < 1,\, V_1 = \frac{v_1-p}{d}$ represents the longest time that the HHP is unwilling to downstream refer an M-type patient to the LHP for treatment services.

The results presented in Proposition 3 is intuitive, that is, the HHP shifts one patient to the LHP if and only if a decrease in marginal expected waiting cost is at least as great as the net reward (minus treatment price) even for the patient. Otherwise, the HHP has no incentive to admit the referral of any M-type patients. Particularly, if $v_1 - p \leq \frac{d\mu_1}{[\mu_1-(1-\alpha)\lambda_1]^2}$ (i.e., $V_1 \leq \frac{1}{[1-(1-\alpha)\rho_1][\mu_1-(1-\alpha)\lambda_1]}$), it is optimal decision for the HHP to send all of the M-type patients to the LH.

From Propositions 2 and 3, we can obtain the HHP's optimal referral rate θ^* and the LHP's optimal capacity level μ_2^* under the decentralized system.

3.2   Optimal Strategy Under the Centralized System

In this section, we consider an idealized centralized system, that is, the HHP and the LHP cooperate completely to maximize the patient welfare of the overall two-tier service system. For the centralized two-tier healthcare system, the optimization problem is to determine θ, μ_2 to maximize patient welfare, with the conditions that LHP's profit to be nonnegative. Thus, we can write the optimization problem as

$$\max_{\theta,\mu_2} SW(\theta, \mu_2) = (\lambda_1 - \alpha\lambda_1)U_S(\theta_1) + (\alpha\lambda_1 - \theta)U_1(\theta) + (\lambda_2 + \theta)U_2(\theta, \mu_2)$$

$$s.\,t.\ \Pi_2 = (\lambda_1 + \theta)p - c\mu_2 \geq 0, \qquad (5)$$
$$0 \leq \theta\ \leq\ \alpha\lambda_1.$$

The first formula in (5) represents total patient welfare of the centralized system in the patient referral service setting. The second formula in (5) is to ensure the nonnegative value of the LHP's profit. The third formula in (5) ensures that any S-type patient will not be downstream referred for seeking treatment.

For any given θ, the patient welfare is increasing in μ_2. Hence, the optimal capacity level satisfies the condition that $(\lambda_1 + \theta)p - c\mu_2 = 0$. For simplification, we obtain the solution of the optimization problem by solving θ for given μ_2. Substituting the optimal referral rate θ^o into the first formula in (5), we can obtain the optimal capacity level μ_2^o. The following proposition describes the optimal solution under the centralized system.

Proposition 4. *Under the centralized system, the HHP's optimal referral rate* θ^o *and the LHP's optimal capacity level* μ_2^o *are given by:*

(i) *if* $V \geq \frac{1}{(1-\rho_1)(\mu_1-\lambda_1)}$, *then* $\theta^o = 0$, $\mu_2^o = \frac{p}{c}\lambda_2$;

(ii) *if* $V \leq \frac{1}{[1-(1-\alpha)\rho_1][\mu_1-(1-\alpha)\lambda_1]}$, *then* $\theta^o = \alpha\lambda_1$, $\mu_2^o = \frac{p}{c}(\lambda_2 + \alpha\lambda_1)$;

(iii) *if* $\frac{1}{[1-(1-\alpha)\rho_1][\mu_1-(1-\alpha)\lambda_1]} < V < \frac{1}{(1-\rho_1)(\mu_1-\lambda_1)}$, *then* $\theta^o = \tilde{\theta}^o$, $\mu_2^o = \tilde{\mu}_2^o$,

where $\tilde{\theta}^o = \sqrt{\frac{\mu_1}{V}} - (\mu_1 - \lambda_1)$, $\tilde{\mu}_2^o = \frac{p}{c}[\lambda_2 + \sqrt{\frac{d\mu_1}{v}} - (\mu_1 - \lambda_1)]$.

Proposition 4 indicates when a reduction in service value exceeds a reduction in marginal expected waiting cost pertaining to patient referral service, no M-type patient should be transferred out of the HHP for the centralized system's optimum outcome; see Case (i). The reason is that in this case, referring every patient always reduces the utility of the overall healthcare system. Otherwise, all or only a portion of the M-type patients should be referred to the LHP.

3.3   Comparison Between Decentralized and Centralized Systems

According to Propositions 3 and 4, we see that the centralized optimal referral rate θ^o and optimal capacity level μ_2^o do not coincide with the decentralized optimal referral rate θ^* and optimal capacity level μ_2^*. However, we can show that $\theta^o \geq \theta^*$, $\mu_2^o \geq \mu_2^*$. In other words, the decentralized optimal strategy is less than or equal to the centralized optimal strategy.

These results can be summarized as the following theorem.

Theorem 1. *The decentralized optimal referral rate and optimal capacity level are not more than the centralized optimal referral rate and optimal capacity level:* $\theta^o \geq \theta^*$, $\mu_2^o \geq \mu_2^*$. *Moreover,*

(i) *if* $V \geq \frac{1}{(1-\rho_1)(\mu_1-\lambda_1)}$, *then* $\theta^o = \theta^* = 0$, $\mu_2^o = \mu_2^* = \frac{p}{c}\lambda_2$;

(ii) *if* $V_1 \leq \frac{1}{[1-(1-\alpha)\rho_1][\mu_1-(1-\alpha)\lambda_1]}$, *then* $\theta^o = \alpha\lambda_1 = \theta_1^* \geq \theta^*$, $\mu_2^o = \frac{p}{c}(\lambda_2 + \alpha\lambda_1) \geq \mu_2^*$;

(iii) *otherwise, then* $\alpha\lambda_1 > \theta^o > \theta^* > 0$, $\mu_2^o = \frac{p}{c}(\lambda_2 + \theta^o) > \mu_2^* = \frac{p}{c}(\lambda_2 + \theta^*)$.

As shown in Fig. 4, we find a sequence of equilibrium referral rate with the change of two hospitals' service value, which can be divided into six regions. Region (1): the service value of the HHP is so high that a decrease in service value is higher than that in marginal expected waiting cost for every referred patient, so referring patients would not benefit both the HHP and the overall healthcare system. Region (2): when the value of v_1 is a medium-high range, we show that the optimal referral rate of the HHP (under the decentralized system) is always lower than that of the overall healthcare system (under the centralized system). More specifically, referring patient is always not beneficial for the HHP, but only a portion referral of the M-type patients is beneficial for the overall healthcare system. Region (3): the service value of the HHP is medium-low, from the perspective of both the HHP and the overall healthcare system, only a portion of M-type patients should be referred from the HH to the LH. Region (4): the service values of both the two hospitals are low enough that for the overall healthcare system, all the M-type patients should be transferred downstream to the LHP (for the HHP, only a portion of M-type patients should be referred). Region (5): the service value of the high- and low-level hospitals is respectively medium-high and high such that for the overall healthcare system, a reduction in service value is lower than that in marginal expected waiting cost for a referral patient (including all the referral M-type patients), i.e., transferring all the M-type patients would be beneficial; but referral of any patient always harms the benefit of the HHP. Region (6): the service values of both the HHP and LHP are so high that the benefit of both the HHP and the overall healthcare system would always be improved in relation to referring all the M-type patients.

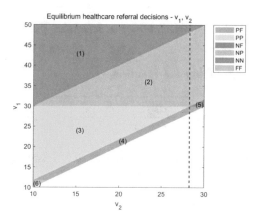

Fig. 4. Effect of v_1, v_2 on equilibrium referral rate for the decentralized and centralized systems ($\lambda_1 = 10$, $\mu_1 = 12.5$, $\lambda_2 = 5$, $\alpha = 0.6$, $c = 5$, $p = 10$, $d = 10$.)

4   Coordination of the Two-Tier Service System via Government Subsidy Scheme

The government's aim is to determine the HHP's optimal subsidy rate r_1 and patients' optimal subsidy rate r_2 to maximize the number of patients who are downstream referred to the LHP. We suppose that the government public fund is limited and denoted by R. Thus, the government's optimization decision model can be written as

$$
\begin{aligned}
\max_{r_1,r_2}\{\theta(r_1,r_2) &= \min\{\theta'_1(r_1), \theta'_2(r_2)\}\} \\
s.t. \ \ \theta'_1(r_1) &\le \theta^o, \\
\theta'_2(r_2) &\le \theta'_1(r_1), \\
r_1\theta'_1(r_1) &+ r_1\theta'_2(r_2) \le R,
\end{aligned}
\tag{6}
$$

where

$$
\begin{aligned}
\theta'_1(r_1) &= \arg\max_{\theta_1}\{SU'_1(\theta_1) = SU_1(\theta_1) + r_1\theta_1\}, \\
\theta'_2(r_2) &= \arg\max_{\theta_2}\{U_1(\theta_2) \ge U_2(\theta_2, \mu_2^o) + r_2\}.
\end{aligned}
$$

The first formula in (6) corresponds to maximize the number of patients who are downstream referred to the LHP but at most centralized optimal referral rate. The second formula in (6) is to guarantee that the optimal referral rate of the HHP does not exceed that of the centralized system. The third formula in (6) ensures that the optimal referral rate of the patients is not higher than that of the HHP. To represent the government's optimal subsidy strategy, the following lemmas are needed.

Lemma 3. *There are four thresholds of government public budget in terms of subsidy $R^{1,2,3,4}$ such that:*

(i) for $R \in (0, R^2)$, the following equation

$$
r_1\sqrt{\frac{d\mu_1}{v_1 - p - r_1}} - (\mu_1 - \lambda_1) = R
\tag{7}
$$

has a unique positive solution with respect to r_1, denoted by r_1^0;
(ii) for $R \in (0, R^4)$, the following equations

$$
\begin{cases}
(r_1 + r_2)[\sqrt{\frac{d\mu_1}{v_1 - p - r_1}} - (\mu_1 - \lambda_1)] = R > v_1 - p - \frac{d\mu_1}{(\mu_1 - \lambda_1)^2} \\
r_2 = v - \sqrt{\frac{d}{v_1 - p - r_1}} + \frac{dc}{(p - c)(\lambda_2 - \mu_1 + \lambda_1 + \sqrt{\frac{d\mu_1}{v_1 - p - r_1}})}
\end{cases}
\tag{8}
$$

has a unique positive solution with respect to r_1, r_2, denoted by (r_1^1, r_2^1), where $0 < R^1 < R^2$ and $0 < R^3 < R^4$,

$$R^1 = (v_2 - p)\left[\sqrt{\frac{d\mu_1}{v}} - (\mu_1 - \lambda_1)\right], R^3 = \alpha\lambda_1\left\{v_1 - p - \frac{d\mu_1}{[\mu_1 - (1-\alpha)\lambda_1]^2}\right\},$$

$$R^2 = \left\{(v_1 - p) - \sqrt{\frac{dv}{\mu_1}} + \frac{dc}{(p-c)[\lambda_2 + \sqrt{\frac{d\mu_1}{v}} - (\mu_1 - \lambda_1)]}\right\}\left[\sqrt{\frac{d\mu_1}{v}} - (\mu_1 - \lambda_1)\right],$$

$$R^4 = \alpha\lambda_1\left\{v_1 - p + v - \frac{d\mu_1}{[\mu_1 - (1-\alpha)\lambda_1]^2} - \frac{d}{\mu_1 - (1-\alpha)\lambda_1} + \frac{dc}{(p-c)(\lambda_2 + \alpha\lambda_1)}\right\}.$$

In fact, r_1^0, r_1^1, r_2^1 are optimal subsidy rate when government's budget is relatively low such that the suboptimum of the centralized system can be achieved via subsidy scheme. Based on Lemma 3, the government's optimal subsidy strategy is represented as the following theorem.

Theorem 2. *The government's optimal subsidy strategy r_1^*, r_2^* can be represented subject to the following conditions:*

(i) for the $V \geq \frac{1}{(1-\rho_1)(\mu_1 - \lambda_1)}$ case, then $r_1^* = r_2^* = 0$.

(ii) For the $\frac{1}{[1-(1-\alpha)\rho_1][\mu_1 - (1-\alpha)\lambda_1]} < V < \frac{1}{(1-\rho_1)(\mu_1 - \lambda_1)}$ case,

    (a) when $R \in (0, R^1)$,

$$\begin{cases} r_1^* = r_1^0, r_2^* = 0, & \text{if } (\mu_1, \frac{p}{c}) \in \Theta_1 \cup \Theta_2 \text{ with } v \leq \sqrt{\frac{d(v_1 - p - r_1^0)}{\mu_1}} \\ r_1^* = r_1^1, r_2^* = r_2^1, & \text{otherwise;} \end{cases}$$

    (b) when $R \in [R^1, R^2]$,

$$\begin{cases} r_1^* = v_2 - p, r_2^* = 0, & \text{if } (\mu_1, \frac{p}{c}) \in \Theta_1 \cup \Theta_2 \text{ with } v \leq \frac{d}{\mu_1} + \sqrt{\frac{d\mu_1}{v}} - \mu_1 \\ r_1^* = r_1^1, r_2^* = r_2^1, & \text{otherwise;} \end{cases}$$

    (c) when $R \in (R^2, \infty)$,

$$\begin{cases} r_1^* = v_2 - p, r_2^* = 0, & \text{if } (\mu_1, \frac{p}{c}) \in \Theta_1 \cup \Theta_2 \text{ with } v \leq \sqrt{\frac{dv}{\mu_1}} \\ r_1^* = v_2 - p, \ r_2^* = v - \sqrt{\frac{dv}{\mu_1}} + \frac{dc}{p-c}\left[\lambda_2 + \sqrt{\frac{d\mu_1}{v}} - (\mu_1 - \lambda_1)\right]^{-1}, & \text{otherwise.} \end{cases}$$

(iii) For the $V \leq \frac{1}{[1-(1-\alpha)\rho_1][\mu_1 - (1-\alpha)\lambda_1]} < V_1$ case,

    (a) when $R \in (0, R^3)$,

$$\begin{cases} r_1^* = r_1^0, r_2^* = 0, & \text{if } (\mu_1, \frac{p}{c}) \in \Theta_1 \cup \Theta_2 \text{ with } v \leq \sqrt{\frac{d(v_1 - p - r_1^0)}{\mu_1}} \\ r_1^* = r_1^1, r_2^* = r_2^1, & \text{otherwise;} \end{cases}$$

    (b) when $R \in [R^3, R^4]$,

$$\begin{cases} r_1^* = v_1 - p - \frac{d\mu_1}{[\mu_1 - (1-\alpha)\lambda_1]^2}, r_2^* = 0, & \text{if } (\mu_1, \frac{p}{c}) \in \Theta_1 \cup \Theta_2 \text{ with } v \leq \frac{d}{\mu_1 - (1-\alpha)\lambda_1} \\ r_1^* = r_1^1, r_2^* = r_2^1, & \text{otherwise;} \end{cases}$$

(c) when $R \in (R^4, \infty)$,

$$\begin{cases} r_1^* = v_1 - p - \frac{d\mu_1}{[\mu_1 - (1-\alpha)\lambda_1]^2}, r_2^* = 0, & \text{if } (\mu_1, \frac{p}{c}) \in \Theta_1 \cup \Theta_2 \text{ with } v \leq \frac{d}{\mu_1 - (1-\alpha)\lambda_1} \\ r_1^* = v_1 - p - \frac{d\mu_1}{[\mu_1 - (1-\alpha)\lambda_1]^2}, \quad r_2^* = v - \frac{d}{\mu_1 - (1-\alpha)\lambda_1} + \frac{dc}{(p-c)(\lambda_2 + \alpha\lambda_1)}, & \text{otherwise.} \end{cases}$$

(iv) For the $V_1 \leq \frac{1}{[1-(1-\alpha)\rho_1][\mu_1 - (1-\alpha)\lambda_1]}$ case,

$$\begin{cases} r_1^* = 0, r_2^* = 0, & \text{if } (\mu_1, \frac{p}{c}) \in \Theta_1 \cup \Theta_2 \text{ with } v \leq \frac{d}{\mu_1 - (1-\alpha)\lambda_1} \\ r_1^* = 0, r_2^* = \min\left\{v - \frac{d}{\mu_1 - \lambda_1(1-\alpha)} + \frac{dc}{(p-c)(\lambda_2 + \alpha\lambda_1)}, \frac{R}{\alpha\lambda_1}\right\}, & \text{otherwise,} \end{cases}$$

where

$$\Theta_1 = \left\{(\mu_1, \frac{p}{c}) \Big| \frac{1}{\Delta}\left(1 - \frac{1}{\sqrt{p/c - 1}}\right)^2 \geq \mu_1 - (\lambda_1 + \lambda_2) \geq 0 \text{ and } \max\{\varphi_2^1, 1\} < \frac{p}{c} < \max\{\varphi_2^2, 1\}\right\},$$

$$\Theta_2 = \left\{(\mu_1, \frac{p}{c}) \Big| \mu_1 < (\lambda_1 + \lambda_2) \text{ and } \frac{p}{c} > \max\{\varphi_2^0, 1\}\right\}.$$

Remark 2. After solving r_1^0, r_1^1, r_2^1 from Eqs. (7)-(8), the optimal subsidy strategy can be made based on Theorem 2. The four conditions in Theorem 2 are complete and mutually exclusive. Hence, the optimal subsidy strategy is unique.

As we can see, the optimal strategy is not complicated. In fact, with the condition that $V \geq \frac{1}{(1-\rho_1)(\mu_1 - \lambda_1)}$, implying that the net benefit driven by patient referral is non-positive, thus the centralized optimal decision is not to downstream transfer any patient from the HHP to the LHP. Consequently, the government should neither subsidize patients nor HHP. By contrast, when the condition that $V_1 \leq \frac{1}{[1-(1-\alpha)\rho_1][\mu_1 - (1-\alpha)\lambda_1]}$ is satisfied, the referral of any M-type patient never reduce the utility of the HHP, and therefore the hospital provider (but some patients) need not (may require) to be subsidized by the government. However, in other situations, subsidizing the HHP is always indispensable for maximizing patient welfare of the centralized healthcare system.

The results presented in Part (ii) (or Part (iii)) of Theorem 2 can be interpreted as follows. When the public budget is insufficient, i.e., $0 < R < R^1$ (or $0 < R < R^3$), the centralized system optimum could not be achieved with subsidizing the HHP or subsidizing both the patients and the HHP, and thus the equilibrium subsidy strategy is only a suboptimal scheme. When the public budget is relatively medium, i.e., $R^1 \leq R \leq R^2$ (or $R^3 \leq R \leq R^4$), the equilibrium subsidy strategy is an optimal outcome for the centralized system only when subsidizing the HHP, otherwise, is still a suboptimal outcome. The reason can be attributed to the fact that a lower decrease of every referral patient's service value $(v_1 - v_2)$ as well as a higher capacity level of the LHP induce the patients to more willingly transfer to the LHP. Consequently, only the HHP should be subsidized by the government with the aim of coordinating the healthcare referral program. For other cases, we show that both the HHP and the patients should be subsidized, and therefore only a medium subsidy is not to guarantee the system optimum to be achieved. In contrast, when the government budget is

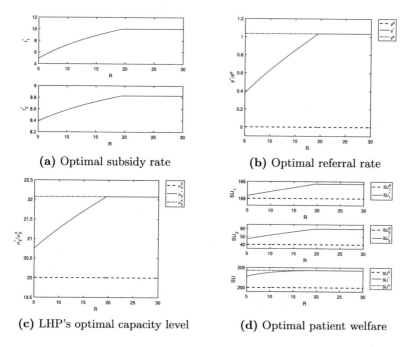

Fig. 5. Effect of R on equilibrium outcomes for the decentralized and centralized systems ($\lambda_1 = 10, \mu_1 = 12.5, \lambda_2 = 5, \alpha = 0.6, c = 5, p = 10, d = 10, v_1 = 30, v_2 = 20.$)

sufficient, i.e., $R > R^2$ (or $R > R^4$), there always exists optimal subsidy strategy such that the patient welfare of the centralized system can be maximized via subsidizing the HHP or both patients and HHP. In addition, we find when the HHP's capacity level and the ratio of treatment price to cost are medium, or the HHP's capacity level is low and the ratio of treatment price to cost is high, the patients should not be subsidized by the government. This is because in this case, the LHP can receive the referral patient by its own self-financing to invest the service capacity.

5   Numerical Analysis

In this section, we focus on the effects on the optimal outcomes (subsidy rate, referral rate, capacity level, and patient welfare) for the decentralized and centralized systems, as shown in Figs. 5-6 (where superscript 0, *, and o denotes the case of no patient referral, and the optimal outcomes for the decentralized and centralized systems, respectively).

As the government subsidy fund increases, the optimal outcomes for both the decentralized and centralized systems first increase and then reach a constant level in Fig. 5 or Fig. 6. More specifically, there exists a threshold on government subsidy. If the government subsidy fund R is limited and lower than the threshold, we find that the optimal outcomes under the decentralized system are closer

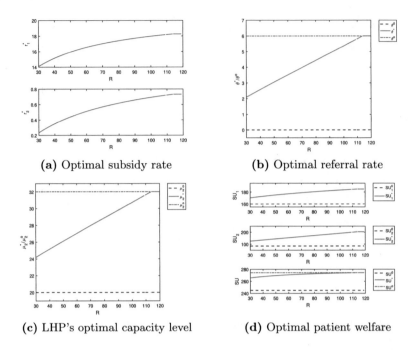

(a) Optimal subsidy rate

(b) Optimal referral rate

(c) LHP's optimal capacity level

(d) Optimal patient welfare

Fig. 6. Effect of R on equilibrium outcomes for the decentralized and centralized systems ($\lambda_1 = 10, \mu_1 = 12.5, \lambda_2 = 5, \alpha = 0.6, c = 5, p = 10, d = 10, v_1 = 30,$

to those under the centralized system with an increase of R. If R is sufficient and higher than the threshold, we observe that the coordination of the two-tier service system via government subsidy scheme would be achieved. As presented in Fig. 5-(d) or Fig. 6-(d), by comparing the patient welfare in the case of patient referral (under the decentralized or centralized system) with that in the case of no patient referral, we see that the patient referral would improve the patient welfare among the HHP, the LHP, and the overall healthcare system. Furthermore, we see that for a sufficient budget, the *win-win* situation among the HHP, the LHP, and the patients can be achieved via the government subsidy scheme.

Figure 5 and Fig. 6 compare the optimal outcomes when $v_2 = 20$ and $v_2 = 29$, and they show that for a higher value of v_2, the optimal referral rate and capacity level for both the decentralized and centralized systems are higher. The increase of v_2 raises the referral rate of patients who willingly choose to be downstream transferred from the HHP to the LHP, and also increases the optimal referral rate for the centralized system. Therefore, the optimal subsidy rate of the HHP would be increasing but the optimal subsidy rate of the patients would be decreasing as v_2 increases. As a result, the patient welfare of the HHP, the LHP, and the overall healthcare system is also increasing in v_2. This indicates that a higher value of v_2 is always beneficial for the HHP, the LHP as well as the overall healthcare system.

6   Conclusions

This paper considers a two-tier healthcare system which is comprised of heterogeneous patients, a comprehensive hospital provider, and a community hospital provider. By constructing a queuing-game model, we capture the dynamic interactions among the three participants and analyze the optimal healthcare referral rate and the optimal capacity level strategy of the centralized and decentralized systems. Then we establish a decision model subsidizing participants under the constraints of the centralized optimal referral rate being as the upper bound, wherein the objective of the government is to maximize the referral rate of patients. Our model and method can be applied to the current work of COVID-19. Severe diseases can be treated in HHP, and ordinary asymptomatic patients can be rehabilitated in LHPs. Therefore, the study model has certain guiding significance for the prevention and treatment of the current work of COVID-19. However, due to space, the specific model will be studied in the next step.

Acknowledgements. This research was supported by the Humanities and Social Science Fund of Ministry of Education of China (No. 21YJC630186), the Natural Science Foundation of Guangdong Province (No. 2022A1515012034), the Philosophy and Social Science Planning Project of Guangdong Province (No. GD20CGL19), and the Startup Fund from Shantou University (STF19024).

References

1. Shumsky, R.-A., Pinker, E.-J.: Gatekeepers and referrals in services. Manag. Sci. **49**, 839–856 (2003)
2. Hasija, S., Pinker, E.-J., Shumsky, R.-A.: Staffing and routing in a two-tier call center. Int. J. Oper. Res. **1**, 8–29 (2005)
3. Lee, H.-H., Pinker, E.-J., Shumsky, R.-A.: Outsourcing a two-level service process. Manag. Sci. **58**, 1569–1584 (2012)
4. Adida, E., Bravo, F.: Contracts for healthcare referral services: coordination via outcome-based penalty contracts. Manag. Sci. **65**, 1322–1341 (2019)
5. Li, N., Kong, N., Li, Q., Jiang, Z.: Evaluation of reverse referral partnership in a tiered hospital system - a queuing-based approach. Int. J. Prod. Res. **55**, 5647–5663 (2017)
6. Liu, X., Cai, X., Zhao, R., Lan, Y.: Mutual referral policy for coordinating health care systems of different scales. Int. J. Prod. Res. **53**, 7411–7433 (2015)
7. Jiang, H., Pang, Z., Savin, S.: Performance-based contracts for outpatient medical services. Manuf. Serv. Oper. Manag. **14**, 654–669 (2012)
8. Andritsos, D.-A., Aflaki, S.: Competition and the operational performance of hospitals: the role of hospital objectives. Prod. Oper. Manag. **24**, 1812–1832 (2015)
9. Vaithianathan, R.: Will subsidizing private health insurance help the public health system? Econ. Rec. **78**, 277–283 (2002)
10. Qian, Q., Guo, P., Lindsey, R.: Comparison of subsidy schemes for reducing waiting times in healthcare systems. Prod. Oper. Manag. **26**, 2033–2049 (2017)
11. Hua, Z., Chen, W., Zhang, Z.G.: Competition and coordination in two-tier public service systems under government fiscal policy. Prod. Oper. Manag. **25**, 1430–1448 (2016)

12. Qian, Q., Zhuang, W.: Tax/subsidy and capacity decisions in a two-tier health system with welfare redistributive objective. Eur. J. Oper. Res. **260**, 140–151 (2017)
13. Chen, W., Zhang, Z.-G., Hua, Z.: Analysis of two-tier public service systems under a government subsidy policy. Comput. Ind. Eng. **90**, 146–157 (2015)
14. Zhou, W., Wan, Q., Zhang, R.-Q.: Choosing among hospitals in the subsidized health insurance system of China: a sequential game approach. Eur. J. Oper. Res. **257**, 568–585 (2017)
15. Adida, E., Mamani, H., Nassiri, S., Impact of payment scheme on performance: Bundled payment vs. fee-for-service. Manag. Sci. **63**, 1606–1624 (2017)
16. Wang, J.-J., Li, Z.-P., Shi, J., , A.-C.: Hospital referral and capacity strategies in the two-tier healthcare systems. Omega **100**, 102229 (2021)

An Indirect Solution to WPLTS Group Decision Making Problem

Guo-Cheng Zhu[✉] and Jian Xu

School of General Education, Guangdong Innovation Technical College, Dongguan 523960, Guangdong, China
569141518@qq.com

Abstract. To study the problem of weighted probabilistic linguistic term set (WPLTS) multi-attribute group decision making. Firstly, WPLTS and WPHFS are defined, and a method to convert WPLTS to WPHFS is given. Secondly, the membership, probability and expert weight in WPHFS are described in the form of three-dimensional point coordinates, on whose basis, a three-dimensional score value model, a three-dimensional deviation model, a distance model for measuring two WPHFEs, and a rule for comparing the size of two WPHFEs are established. Thirdly, the calculation methods of objective weight, comprehensive weight and attribute external weight, internal weight and comprehensive weight of decision experts are set respectively. Finally, under the three weight categories of the three weights and attributes of the decision experts, the maclaurin symmetric average operator is used to aggregate WPHFE to obtain the comprehensive decision value of each scheme and then rank the alternatives. It is not only feasible to transform the WPLTS decision-making problem into the WPHFS decision-making problem to solve it, but also the decision-making model established according to the weight and attribute weight of different types of decision-making experts and the "r" parameter can more objectively optimize the scheme from multiple dimensions.

Keywords: Weighted Probabilistic Linguistic Term Set · Weighted Probabilistic Hesitant Fuzzy Set · 3D Point Coordinates · Maclaurin Symmetric Average Operator

1 Introduction

Since membership in probabilistic linguistic term sets (PLTS) is characterized by linguistic terms, it has good application effect for some qualitative problems. Therefore, since Pang Qi et al. [1] defined it, this theory has been well developed and practiced. However, when PLTS is applied to decision-making problems, we find that although the linguistic term is given the possibility of occurrence, that is, probability, it does not reflect the importance of the linguistic term, so it is easy to cause distortion of decision-making information when aggregating the comprehensive attribute values of the scheme. In order to solve such problems, this paper injects the weight of decision experts into PLTS, and then defines weighted probabilistic linguistic term sets (WPLTS). The probability and weight of decision experts in WPLTS are used as annotations of linguistic

© The Author(s), under exclusive license to Springer Nature Singapore Pte Ltd. 2024
B.-Y. Cao et al. (Eds.): ICFIE 2022, LNDECT 207, pp. 251–267, 2024.
https://doi.org/10.1007/978-981-97-2891-6_19

terms to show the possibility and authority of their occurrence. Compared with PLTS, WPLTS obviously contains more information. To solve the WPLTS decision-making problem, we need to comprehensively consider the blending of decision-making information in the context of qualitative probabilistic linguistic terms [2], which puts forward higher requirements for establishing information integration methods, it is worth exploring whether it is feasible to solve qualitative problems with quantitative methods by transforming WPLTS decision problems into WPHFS decision problems.

Rao Yi et al. [3] determined the expert weight according to the overall consistency principle of the decision-making expert group in terms of the calculation method of decision-making expert weight and attribute weight in the probabilistic hesitant fuzzy set (PHFS) decision-making problem, and then calculated the attribute weight using the fuzziness and hesitancy of the evaluation information; Gao Jianwei et al. [4] calculated attribute weights based on minimized fuzzy entropy and maximized utility value; Wang Xinfan et al. [5] combined the PHFE new Langmuir distance with the maximum deviation value to determine the attribute weight; Luo Hua et al. [6] also calculated the attribute weight based on the maximum deviation method. At present, in the PHFS decision-making problem, the objective weights of decision experts are calculated after the subjective weights of decision experts are known, and the research literature on the final calculation of the comprehensive weights of decision experts is relatively few; In determining attribute weights, entropy method [7] or maximum deviation method [8] calculate attribute weights based on the information difference degree of decision data on attributes of the scheme. However, research literature on calculating attribute weights based on the internal evaluation information difference degree of the scheme on attributes is still scarce.

On the algorithm to solve the PHFS decision-making problem, Su Bingjie et al. [9] established a multi-attribute decision-making model by combining the PHFS correlation coefficient with the probabilistic dual hesitant fuzzy set; Wang Zhiping et al. [10] introduced prospect theory on the basis of TOPSIS method, and the decision results show that the method has good application effect; Zhu Feng et al. [11] set up a set of group decision algorithm according to the similarity of PHFS and the improved radar chart, and verified and analyzed the scientificity of the algorithm through specific cases; Cao Qian et al. [12] used entropy method to determine attribute weights and adopted compromise ratio method to achieve ranking of schemes; Wu Jian et al. [13] proposed a probabilistic hesitant fuzzy multi-attribute decision-making method based on Bonferroni average operator. Through literature review, it can be seen that: ① There is no literature in PHFS that adds the corresponding decision-making expert weight to the membership. In fact, the probability of membership reflects the possibility of its occurrence, and the added weights of decision experts reflect the importance of the membership. Compared with the classical PHFS, the PHFS experts can obviously describe the decision information more objectively after adding weights of decision. ② When describing PHFS, the existing literature does not reflect information data with different dimensions of membership and probability.

In view of the above analysis, the main work of this paper is as follows: First, WPLTS and WPHFS are defined, and a method to convert WPLTS to WPHFS is given. Secondly, the membership degree, probability and decision-making expert weight in weighted probabilistic hesitant elements (WPHFE) are described in the form of three-dimensional point coordinates, and on this basis, the relevant operation of WPHFE is defined. According to the relevant operation, the attribute weight is calculated by entropy method (① the attribute weight (external weight) is determined according to the difference in the attribute scores of each scheme). ② The attribute weight (internal weight) is calculated based on the internal score difference of each scheme in attributes. ③ The comprehensive weight of an attribute is determined according to the degree of separation between its external weight and its internal weight). Thirdly, combining the comprehensive weight of decision experts and the comprehensive weight of attributes, an algorithm for candidate ranking is established, and the algorithm is analyzed according to different weight types of decision expert and attributes. Finally, the theoretical knowledge of this paper is combed in detail through a numerical example.

2 Basic Knowledge

Definition 1. [14] Nonempty set X, and the binary group $H = \{< x, h_x(p_x) > | x \in X\}$ is called PHFS on set X, in which $h(p) = h_x(p_x) = \{\gamma_l(p_l), l = 1, 2, \cdots, |h(p)|, \sum_{l=1}^{h(p)} p_l = 1\}$ will be called PHFE. γ_l in PHFE represents the degree of membership about element x to set H, p_l is the probability of γ_l occurrence, and $|h(p)|$ represents the number of elements in PHFE.

    In order to reflect the importance of membership in PHFS, the WPHFS concept (Definition 2) is defined by adding corresponding decision-making expert weights to each membership.

Definition 2. Nonempty set X, and the binary group. $H = \{< x, h_x(p_x) > | x \in X\}$ is called WPHFS on set X, in which $h(p) = h_x(p_x, \omega_x) = \{\gamma_l(p_l, \omega_l), l = 1, 2, \cdots, |h(p, \omega)|, \sum_{l=1}^{h(p,\omega)} p_l = 1, \sum_{l=1}^{h(p,\omega)} \omega_l = 1\}$ will be called WPHFE. γ_l represents the degree of membership about element x to set H in WPHFE, p_l is the probability of γ_l occurrence, ω_l is the weight of decision experts supporting the degree of membership γ_l occurrence, and $|h(p, \omega)|$ represents the number of elements in WPHFE.

Definition 3. [1] In the MAGDM problem, A is the decision expert set, X is the evaluation scheme set, C is the attribute set, a linguistic term set $S = \{s_\alpha | \alpha = 0, 1, 2, \cdots, \tau\}$, and a PLTS of the scheme $x \in X$ on attribute $c \in C$ is defined as: $L(P) = \{L^{(k)}(P^{(k)}) | L^{(k)} \in S, P^{(k)} \geq 0, k = 1,2,\cdots,\#L(P), \sum_{k=1}^{\#L(P)} P^{(k)} \leq 1\}$ among which, $L^{(k)}(p^{(k)})$ is the linguistic term and the $P^{(k)}$ is corresponding probability of $L^{(k)}$, $\#L(P)$ is the number of probabilistic linguistic terms in $L(P)$.

Definition 4. On the basis of Definition 3, the PLTS with decision expert weights added is represented by WPLTS, which is defined as:

$$L(P,W) = \{L^{(k)}(P^{(k)}, \omega^{(k)}) | L^{(k)} \in S, P^{(k)} \geq 0, k = 1,2,\cdots,\#L(P,W), \sum_{k=1}^{\#L(P,W)} \omega^{(k)} = 1\}.$$

$\omega^{(k)}$ is here to refers to the sum of weights of all decision experts associated with the occurrence of probability $P^{(k)}$ under the linguistic term $L^{(k)}$.

Definition 5. [15] The nine segment linguistic evaluation terms are used. In order to increase the flexibility of intermediate evaluation terms, the differentiation is extended when converting them into corresponding interval numbers. The specific conversion scores are shown in Table 1.

All information in Table 1 is represented by the set R.

Table 1. Linguistic evaluation terms conversion table

linguistic evaluation	awful poor	very poor	poor	medium poor	
Terms	(AP)	(VP)	(P)	(MP)	
corresponding interval value	[0,0.1]	[0.1,0.2]	[0.2,0.3]	[0.3,0.45]	
linguistic evaluation	fair	medium good	good	very good	awful good
Terms	(F)	(MG)	(G)	(VG)	(AG)
Corresponding interval value	[0.45,0.55]	[0.55,0.7]	[0.7,0.8]	[0.8,0.9]	[0.9,0.1]

Definition 6. [16] Known set X, and call a PIVHFS on set X as $H_p = \{< x, h_x(p_x) > | x \in X\}$ in whose formula, h_x is the probabilistic interval value hesitant fuzzy element (PIVHFE). In the formula of $h_x(p_x) = \{\gamma_l(p_l) | l = 1,2,\cdots,\#h_x(p_x), \gamma_l \subset [0,1], \sum_{l=1}^{\#h_x(p_x)} p_l = 1\}$, γ_l represents the interval valued degree of membership about $x \in X$ relative to set H_p and $\#h_x(p_x)$ represents the number of γ_l.

Definition 7. Is based on Definition 6, a weighted PIVHFS (weighted probabilistic interval valued hesitant fuzzy sets, WPIVHFS) is defined as $H_p^w = \{< x, h_x(p_x, \omega_x) > | x \in X\}$, in which $h_x(p_x, \omega_x)$ is plus weighted probabilistic interval value hesitant fuzzy element(WPIVHFE). In the formula of $h_x(p_x, \omega_x) = \{\gamma_l(p_l, \omega_l) | l = 1,2,\cdots,\#h_x(p_x, \omega_x), \gamma_l \subset [0,1], \sum_{l=1}^{\#h_x(p_x,\omega_x)} p_l = 1, \sum_{l=1}^{\#h_x(p_x,\omega_x)} \omega_l = 1\}$, γ_l represents the interval value degree of membership about $x \in X$ relative to set H_p^w, and $\#h_x(p_x, \omega_x)$ represents the number of γ_l.

Definition 8. [17] As the non-negative interval number $\tilde{a}$ and $\tilde{b}$, if $\tilde{a} = [a^L, a^U]$, $\tilde{b} = [b^L, b^U]$, $0 < a^L \le a^U$, $0 < b^L \le b^U$, assuming that, $T(\tilde{a} \times \tilde{b}) = \frac{a^L \times a^U}{b^L \times b^U} \times \frac{a^L + a^U}{b^L + b^U}$ is the product type closeness degree of the interval number $\tilde{a}$ and $\tilde{b}$. Stipulation: $\tilde{a} = \tilde{b} = 0$, then $T(\tilde{a} \times \tilde{b}) = 1$; Properties: (1) if $\tilde{a} = \tilde{b}$, then $T(\tilde{a} \times \tilde{b}) = 1$; (2) if $\tilde{a} > \tilde{b}$, then $T(\tilde{a} \times \tilde{b}) > 1$; (3) if $\tilde{a} < \tilde{b}$, then $T(\tilde{a} \times \tilde{b}) < 1$; (4) if $T(\tilde{a} \times \tilde{b}) = t$, then $T(\tilde{b} \times \tilde{a}) = \frac{1}{t}$; (5) if $T(\tilde{a} \times \tilde{b}) > 1$, $T(\tilde{b} \times \tilde{c}) > 1$, then $T(\tilde{a} \times \tilde{c}) > 1 T(\tilde{a} \times \tilde{c}) > 1$.

Definition 9. If $a_i(i = 1, 2, \cdots, n)$ is a set of non-negative real numbers, and $r = 1, 2, \cdots, n$, if

$$MSM^{(r)}(a_1, a_2, \cdots, a_n) = (\sum_{1 \le i_1 < \cdots < i_r \le n} \prod_{j=1}^{r} a_{i_j}/C_n^r)^{\frac{1}{r}}, \tag{1}$$

then (1) is called maclaurin symmetric mean operator, where $i_1, i_2, \cdots, i_r$ is all r tuples of traversal combinations in $r = 1, 2, \cdots, n$. C_n^r is the binomial coefficient. Maclaurin symmetric mean operator has the following operational properties:

(1) If a_i, and $a_i = a \geq 0$, then $MSM^{(r)}(a_1, a_2, \cdots, a_n) = a$.
(2) If a_i, and $0 \leq a_i \leq b_i$, then $MSM^{(r)}(a_1, a_2, \cdots, a_n) \leq MSM^{(r)}(b_1, b_2, \cdots, b_n)$.
(3) If $a_i \geq 0$, we have $Min(a_1, a_2, \cdots, a_n) \leq MSM^{(k)}(a_1, a_2, \cdots, a_n) \leq Max(a_1, a_2, \cdots, a_n)$.

3 The New Measure Paradigm of WPHFS

Definition 2 gives the concept of WPHFS. On the basis of Definition 2, here three-dimensional point coordinates representation of WPHFS is given, such as Definition 10.

Definition 10. A non-empty set X, the binary group $H = \{< x, h_x(p_x) >|x \in X\}$ called three-dimensional point coordinates represents WPHFS in set X, where $h(p, \omega) = h_x(p_x, \omega_x) = \{(\gamma_l, p_l, \omega_l|l = 1, 2, \cdots, |h(p, \omega)|, \sum_{l=1}^{|h(p,\omega)|} p_l = 1, \sum_{l=1}^{|h(p,\omega)|} \omega_l = 1$ called WPHFE, γ_l represents x belongs to the degree of membership about set H, p_l is the probability of γ_l occurrence, and $|h(p, \omega)|$ represents the number of elements in WPHFE.

In order to distinguish it from the WPHFS of Definition 2, the WPHFS below refers to the Definition 10. When the WPHFS is expressed in accordance with the three-dimensional point coordinates, the measure of PHFS in literature [19] is no longer applicable. On the basis of the Definition 10, this paper follows from the perspective of three-dimensional point coordinates, WPHFE's three-dimensional score value model, three-dimensional deviation model, WPHFE's distance model and two WPHFE size comparison rules are established.

Definition 11. The calculation method for defining the three-dimensional score value of WPHFE $h(p, \omega)$ is defined as $\Delta(h(p, \omega))$

$$\Delta(h(p, \omega)) = \frac{\sum_{l=1}^{|h(p,\omega)|} \sqrt{(\gamma_l)^2 + (p_l)^2 + (\omega_l)^2}}{\sqrt{3}|h(p, \omega)|}, \tag{2}$$

$\frac{\sqrt{(\gamma_l)^2+(p_l)^2+(\omega_l)^2}}{\sqrt{3}}$ means: the element in WPHFE $h(p, \omega)$—weighted probabilistic hesitant fuzzy numbers (WPHFN), the ratio of WPHFN to the largest WPHFN $(1, 1, 1)$, in WPHFE, and the average of the sum of the ratio of all WPHFN to the largest WPHFN $(1, 1, 1)$ is taken as the three-dimensional score of WPHFE, which is easily known from Eq. (2) and $\Delta h(p, \omega)$ has the following properties:

(1) $\Delta h(p, \omega)$ about γ_l, p_l, ω_l all monotonous increase;
(2) $\Delta h(p, \omega) \in [0, 1]$.

Definition 12. The calculation method for defining the three-dimensional deviation value of WPHFE $h(p, \omega)$ is defined as $\nabla h(p, \omega)$

$$\nabla(h(p, \omega)) = \begin{cases} \dfrac{\sum_{l=1}^{|h(p,\omega)|} \sum_{l'=1}^{|h(p,\omega)|} \sqrt{(\gamma_l - \gamma_{l'})^2 + (p_l - p_{l'})^2 + (\omega_l - \omega_{l'})^2}}{C_{|h(p,\omega)|}^2} & |h(p, \omega)| \geq 2 \\ 0 & |h(p, \omega)| = 1 \end{cases}, \quad (3)$$

$C_{|h(p,\omega)|}^2 = \frac{|h(p,\omega)|!}{2!(|h(p,\omega)|-2)!}$ is the binomial coefficient in (3), easy to know $\nabla(h(p, \omega)) \in [0, 1]$.

According to the monotonicity of the three-dimensional score value of WPHFE $h(p, \omega)$ in Definition 11, the larger $\Delta h(p, \omega)$, the larger the element-WPHFN in WPHFE($\Delta h(p, \omega)$) is similar to the scoring function of PHFE in literature [18]). The three-dimensional deviation value of WPHFE $\Delta h(p, \omega)$ in Definition 12 reflects the WPHFE's element—WPHFN$(\gamma_l, p_l, \omega_l)$ how far apart they are from each other, the larger $\nabla h(p, \omega)$, the farther away WPHFN $(\gamma_l, p_l, \omega_l)$ is from each other, the more unstable the information within the WPHFE $h(p, \omega)$($\nabla h(p, \omega)$ is similar to the deviation function of PHFE in literature [18]).

In this paper, The WPHFE's three-dimensional score value $\Delta h(p, \omega)$ and the three-dimensional deviation value $\nabla h(p, \omega)$ are used to sort two WPHFEs $h_1(p, \omega)$, $h_2(p, \omega)$ (Definition 13).

Definition 13. $H_1(p, \omega)$ and $h_2(p, \omega)$ are two WPHFEs, according to Definition 11 and Definition 12 has the following comparison rules:

(1) If $\Delta h_1(p, \omega) > \Delta h_2(p, \omega)$, then $h_1(p, \omega) > h_2(p, \omega)$.
(2) If $\Delta h_1(p, \omega) = \Delta h_2(p, \omega)$, then

① When $\nabla h_1(p, \omega) > \nabla h_2(p, \omega)$, there is $h_1(p, \omega) < h_2(p, \omega)$;
② When $\nabla h_1(p, \omega) < \nabla h_2(p, \omega)$, there is $h_1(p, \omega) > h_2(p, \omega)$.

Definition 14. If $h_1(p, \omega)$, $h_2(p, \omega)$, $h_3(p, \omega)$ are three WPHFEs, They are respectively denote as $h_1(p, \omega) = \{(\gamma_{11}, p_{11}, \omega_{11}), (\gamma_{12}, p_{12}, \omega_{12}), \cdots, (\gamma_{1j_1}, p_{1j_1}, \omega_{1j_1})\}$, $h_2(p, \omega) = \{(\gamma_{21}, p_{21}, \omega_{21}), (\gamma_{22}, p_{22}, \omega_{22}), \cdots (\gamma_{2j_2} p_{2j_2}, \omega_{2j_2}), h_3(p, \omega) = \{(\gamma_{31}, p_{31}, \omega_{31}), (\gamma_{32}, p_{32}, \omega_{32}), \cdots, (\gamma_{3j_3}, p_{3j_3}, \omega_{3j_3})\}$. Then the geometric distance between two WPHFEs $h_1(p, \omega)$ and $h_2(p, \omega)$ is defined as

$$D(h_1(p, \omega), h_1(p, \omega)) = \frac{\sum_{j_1=1}^{J_1} \sum_{j_2}^{J_2} \sqrt{(\gamma_{1j_1} - \gamma_{2j_2})^2 + (p_{1j_1} - p_{2j_2})^2 + (\omega_{1j_1} - \omega_{2j_2})^2}}{\sqrt{3}J_1 \cdot J_2}, \quad (4)$$

where J_1, J_2, respectively, is the number of elements in the WPHFE $h_1(p, \omega)$, $h_2(p, \omega)$. We can verify that the geometric distance measure defined in this paper meets the five conditions in literature [19].

In general, distance measure of two PHFEs, requires the two PHFE's elements consistent, If not, you need to follow some rules to add or subtract elements, you also need to sort the elements of the two PHFEs in order from largest to smallest or smallest to largest, and the specific calculation is only to calculate the elements in the corresponding positions of the two PHFEs. Definition 14 in the distance operation of two WPHFEs, all the elements of the two WPHFEs are evaluated one by one, compare with the method in literature [19], and set a new calculation method for the distance of two WPHFEs from another perspective.

4 The WPLTS Decision Problem is Converted to a WPHFS Decision Problem

Definition 15. In multi-attribute group decision problem, decision experts set is $Z = \{z_1, z_2, \cdots, z_t, \cdots, z_T\}$, decision experts' subjective weight, objective weight and comprehensive weight respectively expressed ω'_{z_t}, ω''_{z_t}, ω_{z_t}, where subjective weight is known, objective weight and comprehensive weights to be found, set of plan is $A = \{a_1, a_2, \cdots, a_i, \cdots, a_I\}$, set of attributes is $G = \{g_1, g_2, \cdots, g_j, \cdots, g_J\}$. The external weight, internal weight and comprehensive weight of the attribute are represented by symbols ω'_{g_j}, ω''_{g_j} and ω_{g_j} are unknown, the evaluation information of the i-th plan on the j-th attribute given by the t-th decision expert is represented by linguistic evaluation terms s_{tij}, (belonging to the set R in Definition 5), and the summary linguistic evaluation term information s_{tij} can get the evaluation information of the i-th plan on the j-th attribute as $h_{ij}(s_{ij}, p_{ij})$. Definition of the WPLTS $h_{ij}(s_{ij}, p_{ij})$, $h_{ij}(s_{ij}, p_{ij}) = \{(s_{ij}^{(k)}, p_{ij}^{(k)}, \omega_{ij}^{(k)}) | s_{ij}^{(k)} \in R, \sum_{k=1}^{|h_{ij}(s_{ij}, p_{ij})|} p_{ij}^{(k)} = 1,$ $\sum_{k=1}^{|h_{ij}(s_{ij}, p_{ij})|} \omega_{ij}^{(k)} = 1, \omega_{ij}^{(k)} = \sum_{t \in \{1, 2, \cdots, |s_{ij}^{(k)}|\}} \tilde{\omega}_{z_t}\}$. $|h_{ij}(s_{ij}, p_{ij})|$ represents number of the elements in WPLTS $h_{ij}(s_{ij}, p_{ij})$(number of linguistic evaluation terms), symbol $|s_{ij}^{(k)}| \in \{1, 2, \cdots, T\}$ represents the sum of the weights of all decision experts who endorse the linguistic evaluation term $s_{ij}^{(k)}$, $\tilde{\omega}_{z_t}$ generally refers to the weight of decision-making experts(can be subjective weight, objective weight or comprehensive weight of any one), here in order to keep the consistency with WPLTS below, the WPLTS is also described in the form of three-dimensional point coordinates, the attributes studied in this paper are all benefit categories.

In order to convert the WPLTS in the WPLTS decision problem (Definition 15) to WPIVHFS, we only need to replace the interval values corresponding to the linguistic evaluation terms in Definition 5. The specific conversion process is required, such as Definition 16.

Definition 16. Replaces the linguistic evaluation term in WPLTS $h_{ij}(s_{ij}, p_{ij})$ in Definition 15 with the corresponding number of intervals, $s_{ij}^{(k)} \rightarrow \varphi_{ij}^{(k)} = [\varphi_{ij}^{(k)-}, \varphi_{ij}^{(k)+}], k = 1, 2, \cdots, |h_{ij}(s_{ij}, p_{ij})|$, then WPIVHFE $H_{ij}(\varphi_{ij}, p_{ij})$, where, $H_{ij}(\varphi_{ij}, p_{ij}) = \{(\varphi_{ij}^{(k)}, p_{ij}^{(k)}, \omega_{ij}^{(k)}) | \varphi_{ij}^{(k)} = [\varphi_{ij}^{(k)-}, \varphi_{ij}^{(k)+}] \in R, \sum_{k=1}^{|H_{ij}(\varphi_{ij}, p_{ij})|} p_{ij}^{(k)} = 1, \sum_{k=1}^{|H_{ij}(\varphi_{ij}, p_{ij})|} \omega_{ij}^{(k)} = 1, \omega_{ij}^{(k)} = \sum_{t \in \{1, 2, \cdots, |\varphi_{ij}^{(k)}|\}} \tilde{\omega}_{z_t}\}$, from the transformation process we know that, $|H_{ij}(\varphi_{ij}, p_{ij})| = |h_{ij}(s_{ij}, p_{ij})|$, $|\varphi_{ij}^{(k)}| = |s_{ij}^{(k)}|$.

Since WPIVHFE $H_{ij}(\varphi_{ij}, p_{ij})$ is not easy to calculate in the decision process, the membership of the interval value needs to be processed of $\varphi_{ij}^{(k)}$. The processing result is the number in the interval [0, 1] (membership), and the optimal membership needs to be determined before processing.

Definition 17. For WPIVHFE $H_{ij}(\varphi_{ij}, p_{ij})$, the optimal membership on the jth attribute is defined as

$$\max_{i \in \{1,2,\cdots,I\}} \max_{k \in \{1,2,\cdots,|H_{ij}(\varphi_{ij},p_{ij})|\}} \varphi_{ij}^{(k)} = \tilde{\varphi}_{ij}^{(k)} = [\tilde{\varphi}_{ij}^{(k)-}, \tilde{\varphi}_{ij}^{(k)+}], j = 1, 2, \cdots, J, \quad (5)$$

Definition 18. According to Definition 8, Definition 17, the membership degree $\varphi_{ij}^{(k)}$ in WPIVHFE $H_{ij}(\varphi_{ij}, p_{ij})$ is respectively measured with the optimal membership degree $\tilde{\varphi}_{ij}^{(k)}$, in the form of $T(\varphi_{ij}^{(k)} \times \tilde{\varphi}_{ij}^{(k)}) = \gamma_{ij}^{(k)}$, because $\gamma_{ij}^{(k)} \in [0, 1]$, the WPIVHFE $H_{ij}(\varphi_{ij}, p_{ij})$ is converted to WPHFE. Consider WPHFE $\mu_{ij}(\gamma_{ij}, p_{ij})$, and describe WPHFE $\mu_{ij}(\gamma_{ij}, p_{ij})$ as:

$$\mu_{ij}(\gamma_{ij}, p_{ij}) = \{(\gamma_{ij}^{(k)}, p_{ij}^{(k)}, \omega_{ij}^{(k)}) | \gamma_{ij}^{(k)} = T(\varphi_{ij}^{(k)} \times \tilde{\varphi}_{ij}^{(k)}), \sum_{k=1}^{|\mu_{ij}(\gamma_{ij},p_{ij})|} p_{ij}^{(k)} = 1,$$

$$\sum_{k=1}^{|\mu_{ij}(\gamma_{ij},p_{ij})|} \omega_{ij}^{(k)} = 1, \omega_{ij}^{(k)} = \sum_{t \in \{1,2,\cdots,|\gamma_{ij}^{(k)}|\}} \tilde{\omega}_{z_t}\}. \quad (6)$$

From Definition 15 to Definition 18, we know that, $|\mu_{ij}(\gamma_{ij}, p_{ij})| = |H_{ij}(\varphi_{ij}, p_{ij})| = |h_{ij}(s_{ij}, p_{ij})|$, $|\gamma_{ij}^{(k)}| = |\varphi_{ij}^{(k)}| = |s_{ij}^{(k)}|$.

The WPHFN $(\gamma_{ij}^{(k)}, p_{ij}^{(k)}, \omega_{ij}^{(k)})$ is described in the point coordinate form in (6), so that the WPLTS decision problem is converted to the WPHFS decision problem. Also, the following extended definitions are available from Sect. 2 (Definition 19, Definition 20).

Definition 19. The three-dimensional score value of WPHFE $\mu_{ij}(\gamma_{ij}, p_{ij})$ was defined as $\Delta(\mu_{ij}(\gamma_{ij}, p_{ij}))$,

$$\Delta(\mu_{ij}(\gamma_{ij}, p_{ij})) = \frac{\sum_{k=1}^{|\mu_{ij}(\gamma_{ij},p_{ij})|} \sqrt{(\gamma_{ij}^{(k)})^2 + (p_{ij}^{(k)})^2 + (\omega_{ij}^{(k)})^2}}{\sqrt{3} \cdot |\mu_{ij}(\gamma_{ij}, p_{ij})|}. \quad (7)$$

Definition 20. The three-dimensional deviation of WPHFE $\mu_{ij}(\gamma_{ij}, p_{ij})$ is $\nabla(\mu_{ij}(\gamma_{ij}, p_{ij}))$

$$\nabla(\mu_{ij}(\gamma_{ij}, p_{ij})) = \begin{cases} \dfrac{\sum_{k=1}^{\mu_{ij}(\gamma_{ij},p_{ij})} \sum_{k'=1}^{\mu_{ij}(\gamma_{ij},p_{ij})} \sqrt{(\gamma_{ij}^{(k)} - \gamma_{ij}^{(k')})^2 + (p_{ij}^{(k)} - p_{ij}^{(k')})^2 + (\omega_{ij}^{(k)} - \omega_{ij}^{(k')})^2}}{C_{\mu_{ij}(\gamma_{ij},p_{ij})}^2} & |\mu_{ij}| \geq 2 \\ 0 & |\mu_{ij}| = 1 \end{cases} \quad (8)$$

5 WPHFS Decision Problem Solving Process

This section consists of three parts: (1) calculating the objective weight and comprehensive weight of decision experts; (2) calculating the external weight, internal weight and comprehensive weight of attributes; (3) establishing a decision model under different categories of decision experts and attribute weights.

5.1 Calculate the Weight of the Decision Experts

The subjective weight of the decision expert ω'_{z_t} is known, and the objective weight is now calculating based on the idea of difference maximization, on the basis of Definition 16 to 18. Firstly, let the weight of each decision expert on a scheme in a certain attribute be $\omega^{ij}_{z_t}$, then the calculation method is defined as

$$
\omega^{ij}_{z_t} = \frac{1 - \frac{\left| \gamma_{tij} - \frac{1}{T} \sum_{t=1}^{T} \gamma_{tij} \right|}{\sum_{t=1}^{T} \left| \gamma_{tij} - \frac{1}{T} \sum_{t=1}^{T} \gamma_{tij} \right|}}{\sum_{t=1}^{T} 1 - \frac{\left| \gamma_{tij} - \frac{1}{T} \sum_{t=1}^{T} \gamma_{tij} \right|}{\sum_{t=1}^{T} \left| \gamma_{tij} - \frac{1}{T} \sum_{t=1}^{T} \gamma_{tij} \right|}},
\tag{9}
$$

$(t = 1, 2, \cdots, T)$; $(i = 1, 2, \cdots, I)$; $(j = 1, 2, \cdots, J)$ in (9).

Secondly, let the weight of each decision expert on a scheme on all attributes be $\omega^i_{z_t}$, then the calculation method of $\omega^i_{z_t}$ is defined as

$$
\omega^i_{z_t} = \frac{\sum_{j=1}^{J} \omega^{ij}_{z_t}}{\sum_{t=1}^{T} \sum_{j=1}^{J} \omega^{ij}_{z_t}},
\tag{10}
$$

$(t = 1, 2, \cdots, T)$; $(i = 1, 2, \cdots, I)$; in (10).

Finally, the computational method of the objective weight ω''_{z_t} of decision experts is defined as

$$
\omega''_{z_t} = \frac{\sum_{i=1}^{I} \omega^i_{z_t}}{\sum_{t=1}^{T} \sum_{i=1}^{I} \omega^i_{z_t}}.
\tag{11}
$$

$(t = 1, 2, \cdots, T)$ in (11).

In order to calculate the comprehensive weight of decision experts, this paper selects half of the sum of the subjective and objective weight of decision experts as its comprehensive weight, that is, the comprehensive weight ω_{z_t} of each decision expert is

$$
\omega_{z_t} = \frac{1}{2}(\omega'_{z_t} + \omega''_{z_t}).
\tag{12}
$$

$(t = 1, 2, \cdots, T)$ in (12).

5.2 Determine the Attribute Weight

The common methods to calculate attribute weight include entropy method, multi-objective planning method, game method and difference maximization method. Here, entropy method is used to calculate attribute weight. Due to the calculation of attribute weight need to use the weight of decision experts, and decision experts weight has three (subjective weight, objective weight and comprehensive weight), according to each weight of decision experts can be obtained the attributed of three kinds of weight (external weight, internal weight and comprehensive weight), after the three weights of decision experts, you can choose any one of the weight to calculate the attribute weight. From Definition 16 to Definition 18 as follows:

(1) Calculate the external weights ω'_{gj} of the attributes $(j = 1, 2, \cdots, J)$.

The weight (external weight) is determined according to the overall difference of each scheme. Firstly, calculate the three-dimensional score of WPHFE $h_{ij}(\gamma_{ij}, p_{ij})$ by Formula (7) as $\Delta(h_{ij}(\gamma_{ij}, p_{ij})) = \pi_{ij}$.

Secondly, find the entropy s_j under the attribute g_j, where, $s_j = \frac{-1}{\ln I} \sum_{i=1}^{I} \pi_{ij}$, $(j = 1, 2, \cdots, J)$. Finally, calculate the external of the attribute weight ω'_{gj}, we have

$$\omega'_{gj} = \frac{|1 - s_j|}{\sum_{j'=1}^{J} |1 - s_{j'}|}, (j = 1, 2, \cdots, J). \tag{13}$$

(2) Calculate the internal weight of the attributes ω''_{gj} $(j = 1, 2, \cdots, J)$.

The attribute weight (internal weight) is determined according to the degree of internal score difference given by the decision expert of each plan.

Firstly, from (8) to calculate the three-dimensional difference value of WPHFE $h_{ij}(\gamma_{ij}, p_{ij})$ is $\nabla(h_{ij}(\gamma_{ij}, p_{ij})) = \eta_{ij}$.

Secondly, calculate the three-dimensional difference value ratio of the plan over all the properties is $\tilde{\eta}_{ij}$, where $\tilde{\eta}_{ij} = \frac{\eta_{ij}}{\sum_{j=1}^{J} \eta_{ij}}$, $(i = 1, 2, \cdots, I)$.

Again, find the entropy S_j under the attribute g_j, where, $S_j = \frac{-1}{\ln I} \tilde{\eta}_{ij} \ln \tilde{\eta}_{ij}$, $(j = 1, 2, \cdots, J)$.

Finally, calculate the internal weight of the attribute is ω''_{gj}

$$\omega''_{gj} = \frac{|1 - S_j|}{\sum_{j''=1}^{J} |1 - S_{j''}|}, (j = 1, 2, \cdots, J). \tag{14}$$

(3) Determine the comprehensive weight of the attributes ω_{gj} $(j = 1, 2, \cdots, J)$.

According to the process of calculating the external and internal weights of attributes, the core idea of both is based on the attribute value external score difference and internal score difference to determine the corresponding weight, in view of this, the comprehensive weight of attribute thought is based on the attributes of the external weight and the difference between the internal weight, the greater the separation degree, the attribute in the decision process of the more obvious, therefore, the corresponding weight should be increased. The calculation method is defined as

$$\omega_{gj} = \frac{\left|\omega'_{gj} - \omega''_{gj}\right|}{\sum_{j'=1}^{J}\left|\omega'_{gj'} - \omega''_{gj'}\right|}, (j = 1, 2, \cdots, J). \tag{15}$$

5.3 Decision-Making Steps

When calculating the attribute weight, it is necessary to determine the weight of decision experts first, because decision experts have three weight types (subjective right, objective weight and comprehensive weight), and the weight of each decision expert can be calculated three weights of attributes(external weight, internal weight and comprehensive weight), so involved in the calculation of the properties of the comprehensive value of the attribute weighted has nine kinds of situation, therefore, this paper established a class of using maclaurin symmetrical average operator assembly attribute comprehensive value of nine decision model, and through the specific case of the decision model decision effect of comparative analysis, on the basis of the Definition 16 to 18, the specific decision-making process is as follows:

Step 1 Convert the initial PLTS decision problem into a WPHFS decision problem.
Step 2 Calculate the weight of decision experts (subjective weight, objective weight and comprehensive weight).
Step 3 Determine the weight of attributes (overall weight, individual weight and comprehensive weight).
Step 4 From (7) to calculate the three-dimensional score value $\Delta(h_{ij}(\gamma_{ij}, p_{ij})) = \pi_{ij}$ of WPHFE $h_{ij}(\gamma_{ij}, p_{ij})$, and weight the three-dimensional score value π_{ij}, and obtain the comprehensive three-dimensional score value $\tilde{\pi}_{ij}$, where, $\tilde{\pi}_{ij} = (\pi_{ij})^{\tilde{\omega}_{gj}}$, symbol $\tilde{\omega}_{gj}$ indicates that the overall weight, individual weight and comprehensive weight of attributes can be used in the process of weighting the three-dimensional score value π_{ij}.
Step 5 From (1) to assemble the comprehensive three-dimensional score value $\tilde{\pi}_{ij}$ of each plan, and the assembly results are expressed $F(i, \tilde{\omega}_{gj}, r)$.

$$F(i, \tilde{\omega}_{gj}, r) = MSM^{(r)}(\tilde{\pi}_{i1}, \tilde{\pi}_{i2}, \cdots, \tilde{\pi}_{iJ}) = (\sum_{1 \leq l_1 < \cdots < l_r \leq J} \prod_{i=1}^{r} \tilde{\pi}_{il_j}/C_J^r)^{\frac{1}{r}}, \tag{16}$$

$i = 1, 2, \cdots, I; j = 1, 2, \cdots, J$. The meaning of $F(i, \tilde{\omega}_{gj}, r)$: the comprehensive value of the i-th plan is obtained according to the attribute weight types and parameter r corresponding to the different weight types of decision experts, and the number of comprehensive values obtained $9J$ by each plan is one.

Step 6 In the case of the medium attribute weight type and parameter r being determined in $\tilde{\omega}_{gj}$, the quality of each plan is judged according to the size of $F(i, \tilde{\omega}_{gj}, r)$. From the monotonicity of the Maclaurin operator, if $F(i, \tilde{\omega}_{gj}, r)$ is large, plan of a_i is optimal.
Step 7 The ranking result of $F(i, \tilde{\omega}_{gj}, r)$ is analyzed in the case of expert weight, attribute weight and parameter r change.
Step 8 End.

6   Example Analysis

A well-known journal is preparing to publish a paper working on a hot issue, the editorial department has received 4 related research articles, however, due to layout problems, the editorial department invited 3 review experts to review all 4 papers (decision plan), the paper is reviewed from three dimensions (attributes): innovation (g_1), application (g_2), and readability (g_3). Three evaluation experts expressed $z_t (t = 1, 2, 3)$, their subjective weights respectively are $\omega'_{z_1} = 0.32, \omega'_{z_2} = 0.33, \omega'_{z_3} = 0.35$, objective weight $\omega''_{z_t} (t = 1, 2, 3)$(unknown), comprehensive weight $\omega_{z_t} (t = 1, 2, 3)$ to be asked. Four papers are marked as $a_i (i = 1, 2, 3, 4)$, the external weights of the three review dimensions are marked as $\omega'_{g_j} (j = 1, 2, 3)$(unknown), the internal weights as $\omega''_{g_j} (j = 1, 2, 3)$(unknown), and the comprehensive weights $\omega_{g_j} (j = 1, 2, 3)$ to be determined. The review information given by the review experts is shown in Table 2 (see Table 1 for the linguistic evaluation term information). The four papers are sorted for reference by the editorial department.

Table2. 3 review experts give the paper linguistic terms information table

paper	Innovative (g_1)	application (g_2)	readability (g_3)
a_1	{(MP$|z_1$), (F$|,z_2, z_3$)}	{(F$|,z_1, z_3$), (MG$|z_2$)}	{(VG$|z_1, z_2$), (AG$|z_3$)}
a_2	{(F$|z_1$), (MG$|z_2$), (G$|z_3$)}	{(G$|z_1$), (VG$|,z_2, z_3$)}	{(P$|z_1$), (MP$|,z_2, z_3$)}
a_3	{(F$|,z_1, z_3$), (G$|z_2$)}	{(F$|z_1$), (G$|z_2$), (MG$|z_3$)}	{(MG$|z_1$), (G$|,z_2, z_3$)}
a_4	{(VG$|z_1$),(G$|,z_2, z_3$)}	{(F$|z_1$), (MP$|,z_2, z_3$)}	{(MG$|z_1$), (F$|z_2$), (MP$|z_3$)}

6.1   Sorting Process

It is known that the subjective weight of the three experts is $\omega'_{z_1} = 0.32, \omega'_{z_2} = 0.33, \omega'_{z_3} = 0.35$, according to Formula (9)–(12), the objective weight and comprehensive weight of the three experts are respectively, $\omega''_{z_1} = 0.2971, \omega''_{z_2} = 0.3484, \omega''_{z_3} = 0.3544, \omega_{z_1} = 0.3086, \omega_{z_2} = 0.3392, \omega_{z_3} = 0.3522$, where the objective weight $\sum_{t=1}^{3} \omega''_{z_t} = 0.9999 \neq 1$, 0.0001 is the error value caused by the rounding method in the calculation process.

(1) In the decision-making process, the weight of the evaluation experts adopts the subjective weight, and Table 2 can be converted according to the Definition 15–Definition 18, Table 3 shows the following:

Sections 2.3, 5.2 combined with Table 3 gives the external weight of the review dimension, $\omega'_{g_1} = 0.4857, \omega'_{g_2} = 0.1254, \omega'_{g_3} = 0.3889$, internal weight of the review dimension, $\omega''_{g_1} = 0.325, \omega''_{g_2} = 0.3875, \omega''_{g_3} = 0.2875$, the comprehensive weight of the review dimension according to Formula (15) is $\omega_{g_1} = 0.3066, \omega_{g_2} = 0.5, \omega_{g_3} = 0.1934$. When sorting each paper, different review dimensions and parameters r are selected, and the ranking results are as follows:

Table 3. The review information table of 3 evaluation experts with subjective weight under the point coordinates

paper	innovative (g_1)	application (g_2)	readability (g_3)
a_1	{(0.202 2, 0.666 7, 0.67), (0.082 7, 0.333 3, 0.33)}	{(0.202 2, 0.666 7, 0.67), (0.393 2, 0.333 3, 0.33)}	{(0.715 8, 0.666 7, 0.65), (1, 0.333 3, 0.35)}
a_2	{(0.202 2, 0.333 3, 0.32), (0.393 2, 0.333 3, 0.33), (0.386 3, 0.333 3, 0.35)}	{(0.686 3, 0.333 3, 0.32), (1, 0.666 7, 0.68)}	{(0.017 5, 0.333 3, 0.32), (0.059 2, 0.666 7, 0.68)}
a_3	{(0.202 2, 0.666 7, 0.67), (0.686 3, 0.333 3,0.33)}	{(0.202 2, 0.333 3, 0.32), (0.686 3, 0.333 3, 0.33), (0.393 2, 0.333 3, 0.35)}	{(0.281 4, 0.333 3, 0.32), (0.491 2, 0.666 7, 0.68)}
a_4	{(1, 0.333 3, 0.32), (0.686 3, 0.666 7, 0.68)}	{(0.202 2, 0.333 3, 0.32), (0.082 7, 0.666 7, 0.68)}	{(0.281 4, 0.333 3, 0.32), (0.144 7, 0.333 3, 0.33), (0.059 2, 0.333 3, 0.35)}

① When the review dimension is external weight and the parameters r are 1, 2 and 3, the sorting results are $a_1 \succ a_3 \succ a_4 \succ a_2$.

② When the review dimension is internal weight and the parameters r are 1, 2 and 3, the sorting results are $a_1 \succ a_2 \succ a_3 \succ a_4$.

③ When the review dimension is comprehensive weight and the parameters r are 1, 2 and 3, the ranking results are $a_2 \succ a_1 \succ a_4 \succ a_3$.

(2) In the decision-making process, the weight of the evaluation experts adopts the objective weight, and Table 2 can be converted into the weight by Definition 15–Definition 18, Table 4 is as follows:

Table 4. The review information table of 3 evaluation experts with objective weight under the description of point coordinates

paper	innovative (g_1)	application (g_2)	readability (g_3)
a_1	{(0.202 2, 0.666 7, 0.651 5), (0.082 7, 0.333 3, 0.348 4)}	{(0.202 2, 0.666 7, 0.651 5), (0.393 2, 0.333 3, 0.348 4)}	{(0.715 8, 0.666 7, 0.645 5), (1, 0.333 3, 0.354 4)}
a_2	{(0.202 2, 0.333 3, 0.297 1), (0.393 2, 0.333 3, 0.348 4), (0.386 3, 0.333 3, 0.354 4)}	{(0.686 3, 0.333 3, 0.297 1), (1, 0.666 7, 0.702 8)}	{(0.017 5, 0.333 3, 0.297 1), (0.059 2, 0.666 7, 0.702 8)}
a_3	{(0.202 2, 0.666 7, 0.651 5), (0.686 3, 0.333 3,0.348 4)}	{(0.202 2, 0.333 3, 0.297 1), (0.686 3, 0.333 3, 0.348 4), (0.393 2, 0.333 3, 0.354 4)}	{(0.281 4, 0.333 3, 0.297 1), (0.491 2, 0.666 7, 0.702 8)}
a_4	{(1, 0.333 3, 0.297 1), (0.686 3, 0.666 7, 0.702 8)}	{(0.202 2, 0.333 3, 0.297 1), (0.082 7, 0.666 7, 0.702 8)}	{(0.281 4, 0.333 3, 0.297 1), (0.144 7, 0.333 3, 0.348 4), (0.059 2, 0.333 3, 0.354 4)}

Section 2.3, 5.2 and combined with Table 4, obtained the external weight of the review dimension, $\omega'_{g_1} = 0.5832, \omega'_{g_2} = 0.1177, \omega'_{g_3} = 0.2991$, internal weight of the review dimension, $\omega''_{g_1} = 0.5363, \omega''_{g_1} = 0.5363, \omega''_{g_2} = 0.1956, \omega''_{g_3} = 0.2681$, the comprehensive weight of the review dimension according to Formula (15) is $\omega_{g_1} = 0.301, \omega_{g_2} = 0.5, \omega_{g_3} = 0.199$. When sorting each paper, select different review dimensions and parameter r, and the sorting results are as follows:

① When the review dimension is external weight and the parameters r are 1 and 2, the sorting results are $a_3 \succ a_1 \succ a_4 \succ a_2$. When r is taken as 3, the sorting result is $a_3 \succ a_4 \succ a_1 \succ a_2$.

② When the review dimension is internal weight and the parameters r are 1 and 2, the sorting results are $a_3 \succ a_1 \succ a_4 \succ a_2$. When r is taken as 3, the sorting result is $a_3 \succ a_4 \succ a_1 \succ a_2$.

③ When the review dimension is comprehensive weight and the parameter r is 1 and 2, the sorting results are $a_1 \succ a_2 \succ a_3 \succ a_4$. When r is taken as 3, the sorting result is $a_2 \succ a_1 \succ a_4 \succ a_3$.

(3) In the decision-making process, the weight of the evaluation experts adopts the comprehensive weight, and Table 2 can be converted into it by Definition 15–Definition 18, Table 5 is as follows:

Section 2.3, 5.2 and combined with Table 5, obtained the external weight of the review dimension, $\omega'_{g_1} = 0.493, \omega'_{g_2} = 0.1189, \omega'_{g_3} = 0.3881$, internal weight of the review dimension, $\omega''_{g_1} = 0.5354, \omega''_{g_2} = 0.2009, \omega''_{g_3} = 0.2637$, the comprehensive weight of the review dimension according to Formula (15) is $\omega_{g_1} = 0.1704, \omega_{g_2} = 0.3296, \omega_{g_3} = 0.5$. When sorting paper, different review dimensions and parameters r are selected, and the ranking results are as follows:

Table 5. The review information table of 3 evaluation experts with comprehensive weight under the point coordinates

paper	Innovative (g_1)	application (g_2)	readability (g_3)
a_1	{(0.202 2, 0.666 7, 0.660 8), (0.082 7, 0.333 3, 0.339 2)}	{(0.202 2, 0.666 7, 0.660 8), (0.393 2, 0.333 3, 0.339 2)}	{(0.715 8, 0.666 7, 0.647 8), (1, 0.333 3, 0.352 2)}
a_2	{(0.202 2, 0.333 3, 0.308 6), (0.393 2, 0.333 3, 0.339 2), (0.386 3, 0.333 3, 0.352 2)}	{(0.686 3, 0.333 3, 0.308 6), (1, 0.666 7, 0.691 4)}	{(0.017 5, 0.333 3, 0.308 6), (0.059 2, 0.666 7, 0.691 4)}
a_3	{(0.202 2, 0.666 7, 0.660 8), (0.686 3, 0.333 3,0.339 2)}	{(0.202 2, 0.333 3, 0.308 6), (0.686 3, 0.333 3, 0.339 2), (0.393 2, 0.333 3, 0.352 2)}	{(0.281 4, 0.333 3, 0.308 6), (0.491 2, 0.666 7, 0.691 4)}
a_4	{(1, 0.333 3, 0.308 6), (0.686 3, 0.666 7, 0.691 4)}	{(0.202 2, 0.333 3, 0.308 6), (0.082 7, 0.666 7, 0.691 4)}	{(0.281 4, 0.333 3, 0.308 6), (0.144 7, 0.333 3, 0.339 2), (0.059 2, 0.333 3, 0.352 2)}

① When the review dimension is external weight and the parameters r are 1, 2 and 3, the sorting results are $a_1 \succ a_3 \succ a_4 \succ a_2$.

② When the review dimension is internal weight and the parameters r are 1 and 2, the sorting results are $a_1 \succ a_4 \succ a_3 \succ a_2$. When r is taken as 3, the sorting result is $a_4 \succ a_1 \succ a_3 \succ a_2$.

③ When the review dimension is comprehensive weight and the parameters r are 1, 2 and 3, the ranking results are $a_1 \succ a_2 \succ a_3 \succ a_4$.

6.2 Comparison of the Ranking Results.

To visually show the ranking comparison of the decision process with each paper in Sect. 1, see Table 6.

Table 6. Decision outcomes under the different weight categories

weight categories of review experts (t=1,2,3)	review dimension weight categories (j=1,2,3)		paper ranking results
subjective weights ω'_{z_t}	external weight ω'_{g_j}	r=1,2,3	$a_1 > a_3 > a_4 > a_2$
	internal weight ω''_{g_j}	r=1,2,3	$a_1 > a_2 > a_3 > a_4$
	comprehensive weight ω_{g_j}	r=1,2,3	$a_2 > a_1 > a_4 > a_3$
objective weight ω''_{z_t}	external weight ω'_{g_j}	r=1,2	$a_3 > a_1 > a_4 > a_2$
		r=3	$a_3 > a_4 > a_1 > a_2$
	internal weight ω''_{g_j}	r=1,2	$a_3 > a_1 > a_4 > a_2$
		r=3	$a_3 > a_4 > a_1 > a_2$
	comprehensive weight ω_{g_j}	r=1,2	$a_1 > a_2 > a_3 > a_4$
		r=3	$a_2 > a_1 > a_3 > a_4$
comprehensive weight ω_{z_t}	external weight ω'_{g_j}	r=1,2,3	$a_1 > a_3 > a_4 > a_2$
	internal weight ω''_{g_j}	r=1,2	$a_1 > a_4 > a_3 > a_2$
		r=3	$a_4 > a_1 > a_3 > a_2$
	comprehensive weight ω_{g_j}	r=1,2,3	$a_1 > a_2 > a_3 > a_4$

According to Table 6, the weight of the evaluation experts, the weight of the review dimension and the parameter r all affect the ranking results. According to the different weight categories of evaluation experts and evaluation dimension weight categories, then using different values of the parameter r, four papers are likely to rank first in the ranking process, but considering the ranking in the last paper, only the paper a_1 without this record, that is, during the sorting process, regardless of the perspective of the established decision model, paper a_1 is not the worst, combined with the number of times the paper a_1 was the first, the paper a_1 can be used as a key reference article for the editorial department. The case further illustrates that the decision algorithm in this paper can determine the optimal scheme from multiple perspectives and provides an effective path for scientific sorting of schemes.

7   Conclusion

PLTS allows multiple linguistic terms to describe qualitative problems, easily dealing with complex uncertainty decisions. However, the different linguistic terms in PLTS are given the same importance, which in turn causes it not to more truly reflect those vague information in real problems that must consider the importance of linguistic terms. Therefore, the paper proposed the concept of WPTLS. Considering that WPTLS has higher requirements for integrated operators when solving decision problems, this paper mapped WPTLS to WPHFS, and used WPHFS decision algorithm to indirectly solve the WPTLS decision problem. The following conclusions can be obtained from the application of specific cases:

(1) The WPHFS decision algorithm established from the perspective of three-dimensional point coordinates can achieve the purpose of sorting plan.
(2) It is feasible to solve the WPLTS decision problem into the WPHFS decision problem.
(3) In the decision process, the different weight categories of decision experts and the different weight categories of attributes and parameters will affect the decision, in order to choose a more objective optimization plan, it is necessary to establish the decision model from many angles and make a comprehensive analysis of the ranking results.

Acknowledgements. Recommender: Professor Yi-Quan Zhu from Guangzhou Maritime University in China. Thanks to the scientific research project (2023KTSCX414) of the education bureau of guangdong province for supporting this article.

References

1. Qiong, P., Hai, W., Zeshui, X.: Probabilistic linguistic term sets in multi-attribute group decision making. Inf. Sci. **36**(9), 128–143 (2016)
2. Mao X., Wu, M., Shang, N.: The multi-attribute group decision model based on probabilistic linguistic correlation coefficient. J. Jiangxi Normal Univ. (Nat. Sci.) **42**(03), 267–274 (2018)
3. Rao, Y., Chen, Y., Cai, Z., et al.: Method for multi-attribute decision making considering expert preference approach based on probabilistic hesitation fuzzy entropy. Fire Control Command Control **46**(04), 4–13 (2021)
4. Gao, J., Huang, X., Guo, F., et al.: A probabilistic hesitation fuzzy multi-attribute decision approach based on cumulative prospect theory. Math. Pract. Theor. **51**(10), 45–58 (2021)
5. Wang, X., Zhou, L., Zhu, Y., et al.: Two-side matching decision making method with probabilistic hesitant fuzzy information based on regret theory. Control Decis. **37**(09), 2380–2388 (2022)
6. Luo, H., Wang, Y.: Probabilistic hesitant fuzzy multi-attribute decision-making considering risk preference. Comput. Syst. Appl. **29**(10), 36–43 (2020)
7. Wu, W., Li, Y., Ni, Z., et al.: Probabilistic hesitant fuzzy Maclaurin geometric symmetric average operator and its group decision model. J. Syst. Sci. Math. Sci. **40**(06), 1074–1089 (2020)
8. Luo, H.: Probabilistic hesitant fuzzy multi-attribute decision-making method based on improved distance. J. Wuhan Univ. Technol. (Inf. Manage. Eng.) **42**(03), 239–245 (2020)

9. Su, B., Lu, F., Zhu, F.: Probabilistic dual hesitant fuzzy multi-attribute decision method based on entropy and correlation coefficient. Oper. Res. Manage. Sci. **31**(02), 23–28 (2022)
10. Wang, Z., Fu, M., Wang, P.: Multi-attribute group decision making model based on prospect theory and TOPSIS in a probabilistic hesitating fuzzy environment. Sci. Technol. Eng. **22**(04), 1329–1337 (2022)
11. Zhu, F., Liu, Y., Xu, J., et al.: Probabilistic hesitant fuzzy multi-attribute decision making method based on similarity and improved radar chart. Oper. Res. Manag. Sci. **30**(04), 109–114 (2021)
12. Cao, Q., Liu, X., Zhang, S., et al.: Probabilistic information is completely unknown in the probabilistic of hesitant fuzzy multi-attribute decision method. J. Syst. Sci. Math. Sci. **40**(07), 1242–1256 (2020)
13. Wu, J., Liu, X., Zhang, S., et al.: Probabilistic hesitant fuzzy Bonferroni mean operator and its decision application. Fuzzy Syst. Math. **33**(05), 116–126 (2019)
14. Zhang, S., Xu, Z., He, Y.: Operations and integrations of probabilistic hesitant fuzzy information in decision making. Inf. Fus. **38**(2), 1–11 (2017)
15. Zhu, G.: Research on decision making method based on probabilistic linguistic terms sets considering expert weight. J. Qufu Normal Univ. (Nat. Sci.) **47**(04), 72–80 (2021)
16. He, Y., Xu, Z., Jiang, W.L.: Probabilistic interval reference ordering sets in multi-criteria group decision making. World Sci. Fuzz. Knowl. Based Syst. **25**(2), 189–212 (2021)
17. Zhu, G., Zhuang, L.: Multi-attribute group decision making method based on interval number ranking criteria. J. Guangdong Univ. Petrochem. Technol. **30**(04), 64–69 (2020)
18. Li, J.: Multi-criteria outranking methods with hesitant probabilistic fuzzy sets. Cogn. Comput. **9**(5), 611–625 (2017)
19. Li, D., Zeng, W., Li, J.: New distance and similarity measures on hesitant fuzzy sets and their application in multiple criteria decision making. Eng. Appl. Artif. Intell. **40**(3), 11–16 (2015)

My Humble Opinion on Digital Logistics Theory

Shufeng Wang[1,2](✉)

[1] School of Business Administration, Baiyun University, Guangzhou 510550, Guangdong, China
wangshufeng2015@163.com
[2] Pearl River Delta Regional Logistics Research Center, Baiyun University, Guangzhou 510550, Guangdong, China

Abstract. Digital logistics is based on digital economy, digital technology, machine learning, computational thinking and other theoretical methods. The changeability, complexity and digital industrialization of market demand promote the digital transformation of logistics industry. Digital logistics adopts digital technology to optimize the logistics system for the whole factor and whole process of logistics, so as to realize the digitalization of logistics management process, management means and management technology, improve the supply chain logistics service and efficiency level, and reduce the total factor logistics cost of supply chain, value chain and industry chain. The core of digital logistics is to transform traditional logistics with digital technology. The development of digital logistics not only needs to adopt advanced digital technology, but also needs to have advanced logistics management technology. Digital logistics has been applied to trade logistics, cross-border trade, port and shipping logistics, emergency logistics and other fields and scenarios. Some enterprises build the middle stage of digital logistics and realize the integrated application of digital logistics. In order to promote the development of digital logistics, innovation is needed in management mode, business model and market forecast. Promoting the construction of modern digital logistics system is of milestone significance to the high-quality development of China's logistics industry.

Keywords: Digital Logistics · Digital Technology · Digital Logistics Elements · Data Flow · Digital Flow

1 Theoretical Overview (Development Stage)

The research on digital logistics theory of circulation enterprises in China has mainly gone through five development stages: big logistics theory, logistics system theory, five party logistics theory, logistics efficiency theory and digital logistics theory.

1.1 Material Flow Theory (MFT)

In 2005, Academician Xu Shoubo (2005) proposed the "big logistics theory". The theory of large logistics refers to a complex composed of six elements or six forces of logistics,

© The Author(s), under exclusive license to Springer Nature Singapore Pte Ltd. 2024
B.-Y. Cao et al. (Eds.): ICFIE 2022, LNDECT 207, pp. 268–278, 2024.
https://doi.org/10.1007/978-981-97-2891-6_20

which most effectively applies logistics technology to the national economy and benefits mankind [1]. The six elements refer to: logistics practitioners, logistics labor objects, logistics labor materials, logistics labor environment, logistics labor space, and logistics labor time. Six forces refer to: human, material, financial, transportation, natural and current forces. The theoretical system of big logistics theory includes five aspects: the theory of material flow, the theory of comprehensive logistics, the theory of big logistics engineering, the theory of big logistics science and technology, and the theory of big logistics industry [2]. The main application fields are in production enterprises and circulation enterprises.

1.2 Logistics System Theory (LST)

In 2006, the scholars represented by Professor He Mingke (2006) led the logistics industry to enter the stage of "logistics system theory". The theory of logistics system refers to that in the whole logistics process, six logistics elements, including fluid, carrier, process, flow direction, flow rate and flow, restrict each other to form the whole logistics system. In the professional field of logistics, hardware facilities alone are not enough. The most important is sea, land, air and rail, that is, the connectivity of all transport lines and the use of warehouses [3].

1.3 Fifth Party Logistics Theory (5PLT)

In 2007, Professor Wang Shufeng (2007) proposed the "five party logistics theory". The five party logistics theory refers to having some logistics assets (asset light), providing customers with multiple supply chain management integration services, and having the functional attributes of system integration, process optimization, and resource collaboration. It is a system integrated logistics service provider [4, 5].

The system integrated logistics organization forms efficient and immediate response of the logistics system through linkage mechanism; With the help of e-commerce, Internet of Things and information engineering technology, logistics technology establishes an integrated electronic information network to promote the supply chain system; Logistics operation uses information system to organize, coordinate and implement the entire logistics solution; The logistics service combines the executive members of each interface to optimize the supply chain logistics system. The representative achievements proposed by this theory mainly include enterprise management consulting report [4] and academic papers published in journals [5].

During this period, the five party logistics theory has already had the connotation of document flow between logistics system operation links, and has the preliminary clue of digital logistics theory. It forms the front end of the digital logistics theory system.

1.4 Logistics Efficiency Theory (LET)

In 2012, Professor Wang Shufeng (2012) proposed the "logistics efficiency theory". The theory of logistics efficiency refers to the economic activities that provide the market with logistics services that optimize the allocation of resources, have the logistics functional attributes of system integration, process optimization, link connectivity, and resource collaboration, and achieve the efficiency of regional logistics systems and enterprise logistics systems.

Representative achievements of this theory mainly include provincial and ministerial research report [6] and academic monograph on regional logistics research [7]. In this period, the theory of logistics efficiency has already had the connotation of data flow between logistics system operation links and formed a preliminary system of digital logistics theory.

The rapid optimal allocation of enterprise logistics elements (digital logistics elements) and the rational optimal allocation of social logistics resources (digital logistics resources) are typical application scenarios of this theory.

1.5 Digital Logistics Theory (DLT)

In 2020, the scholars represented by Professor Wang Shufeng (2020) proposed the big development stage of "digital logistics theory". The representative achievements proposed by this theory mainly include academic reports [11] and government decision-making consulting research reports [12].

In this period, the digital logistics theory was clearly put forward. The digital logistics theory, the five party logistics theory and the logistics efficiency theory have an internal logical relationship with the logistics system optimization. It has the connotation of digital flow between logistics system operation links, and has formed a complete theoretical system of digital logistics. The typical application scenario of this theory is to digitize logistics elements, realize data driven logistics business development, guide the collaborative symbiosis of element resources between enterprises and regions, realize the adjustment and optimization of regional economic structure, and realize the economic activities of high-quality logistics services.

Main ideological content. Research on the development of digital logistics industry driven by digital economy, and how to realize digital driven information interaction among logistics enterprises; Research the core framework and evolution pattern of digital logistics, and provide the logistics industry with enterprise information digital services and logistics resources digital elements of the logistics system; On the basis of standards, data acquisition, system links and timeliness, the research focuses on how to get through the whole logistics chain data, how to realize process digitalization, promote the research and optimization of the whole chain digital operation of the logistics system, and improve the logistics efficiency. With the deepening of digitalization, we focus on the theoretical connotation, mechanism and development of "digital logistics", mainly including theoretical research, empirical research and application research. Theoretical model structure, forming a systematic link of "five in one" of business flow, logistics, capital flow, digital flow and information flow. The main application fields are in production enterprises and circulation enterprises.

To sum up, China's digital logistics theory research has mainly gone through five development stages, including big logistics theory, logistics system theory, five party logistics theory, logistics efficiency theory and digital logistics theory. The main development stages of logistics theory research are shown in Table 1.

Table 1. Main development stages of digital logistics theory research.

Development stage	Material Flow Theory, MFT	Logistics System Theory, LST	Fifth Party Logistics Theory, 5PLT	Logistics Efficiency Theory, LET	Digital Logistics Theory, DLT
particular year	2005	2006	2007	2012	2020
representative figure	Shoubo Xu, Academician, Prof.	Mingke He, Prof.	Shufeng Wang, Prof.	Shufeng Wang, Prof.	Shufeng Wang, Prof.

2   The Conceptual Connotation of Digital Logistics Theory

2.1   The Connotation of Digital Logistics

Digital Logistics is based on digital economy (DE), digital technology (DT), machine learning (ML), computational thinking (CT) and other theoretical methods. The core is to transform traditional logistics with digital technology, which requires not only advanced digital technology, but also advanced logistics management technology. This paper constructs the theoretical system of digital logistics based on the theoretical level, technical level and application level. As shown in Fig. 1.

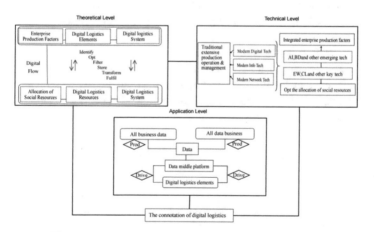

Fig. 1. Theoretical connotation model of digital logistics.

(1) Theoretical level.

   Digital logistics refers to the identification, selection, filtering, storage, conversion and use of digital flows (electronic bill, digital document flow and data flow) between logistics system operation links to guide enterprises to rapidly optimize the allocation of logistics elements (digital logistics elements) and reasonably optimize the allocation of social logistics resources (digital logistics resources), guide enterprises to cooperate with regional factor resources, and realize the adjustment and optimization of regional economic structure, Economic activities to achieve high-quality logistics services. As shown in Fig. 2.

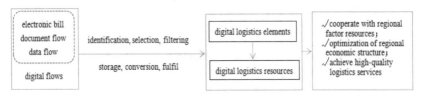

Fig. 2. Digital logistics connotation (theoretical level).

(2) Technical level.

Digital logistics applies modern digital technology, information technology and network technology, including big data, cloud computing, Internet of Things, blockchain, artificial intelligence, 5G communication and other emerging technologies. Among them, digital logistics includes five key technologies, including electronic waybill, machine learning, cloud logistics, automatic driving, and robot process automation (RPA). The rise of digital technology, enterprise competition and market demand are the driving forces for the wide application of digital technology. Under the condition that the traditional extensive operation and management mode has become the bottleneck of enterprise development, it is an effective way to accelerate enterprise transformation by using digital technology and intelligent technology.

AI, including AI overview, search solutions, logic and reasoning, statistical machine learning, application of statistical machine learning algorithms, deep learning, reinforcement learning, AI game, AI development and challenges, algorithm experiments, building AI ecology, and promoting AI specialty and interdisciplinary development. As shown in Fig. 3.

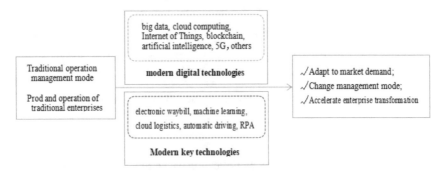

Fig. 3. Digital logistics connotation (technical level).

(3) Application level.

Digital logistics is widely used in new retail, new manufacturing and new logistics. Among them, the representative service markets of the new logistics operation mode mainly include trunk logistics channels (integrated logistics), regional logistics hubs (regional logistics) and urban joint distribution centers (urban logistics). Urban logistics, also known as urban logistics center, is not only the infrastructure of the urban agglomeration, but also the service facilities of the urban agglomeration. It permeates all fields of production and affects all aspects of society. As shown in Fig. 4.

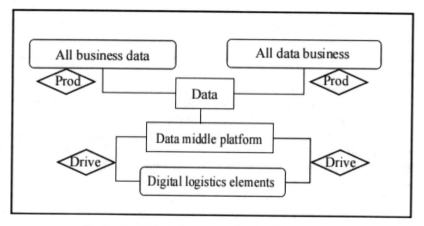

Fig. 4. Digital logistics connotation (application level).

In terms of main application fields, digital logistics takes "all business data" and "all data business" as the guiding principle, carries out intelligent logistics research, and accurately collects the data of the whole logistics supply process; Establish digital and intelligent logistics operation system to provide logistics system guarantee for the development of enterprises.

Digital logistics elements drive the development of logistics business with data. Data middle office is a mechanism for enterprises to use their data, a strategic choice and organizational form, and a mechanism for building a set of continuous data into assets and services. In order to promote the development of digital logistics, innovation is needed in management methods, business models and market forecasts.

2.2 Characteristics of Digital Logistics

(1) Integration of logistics services.

According to the service standards, logistics services include four logistics service levels: functionality, value-added, supply chain and integration. Digital logistics activities include not only all logistics activities generated by internal production activities, but also all logistics activities between external enterprises and individuals as well as between enterprises.

(2) Digital logistics management.

Realize the application of modern technology to logistics management in digitalization, informatization and networking. Among them, digital logistics includes five key technologies, including electronic waybill, machine learning, cloud logistics, automatic driving, and robot process automation (RPA). In the logistics activities, the logistics system operates the document flow, data flow and digital flow between links to conduct digital conversion, and finally realize the electronic bill.

(3) Logistics elements are digitalized.

The people and property that enterprises need to invest in logistics activities constitute the logistics elements of enterprise operation. Through digital technology, the traditional forms of various information resources are converted into binary coded numbers that can be recognized by computers, and the digital logistics elements are formed through calculation, processing, storage, transmission, transmission and restoration, so as to realize the rapid optimal configuration of enterprise logistics elements.

(4) Intelligent logistics resources.

The economic and social development depends on the economic activities of the logistics system. All kinds of people and property that need to be invested constitute the logistics resources for social development. Through intelligent technology, the traditional forms of various social resources are intelligently transformed to form intelligent logistics resources, so as to realize rational and optimal allocation of social logistics resources.

(5) Socialization of logistics resources.

With the rapid development of social economy, diversified changes and demands put forward higher requirements for modern logistics technology and logistics management. In the logistics business, enterprises turn the logistics links they are not good at into professional logistics service providers to achieve social services. The professional logistics provider gathers the advantages of manpower, technology and market, and can adopt more advanced logistics technology and management methods when handling logistics business to help customers realize the rationalization of logistics links and the process of operating efficiency.

3   Application Scenario of Digital Logistics Theory

At present, the main application fields and scenarios of digital logistics theory are: first, industrial Internet platform, integrating the needs of logistics enterprises in the supply chain, and cooperating to meet the needs of logistics enterprises in the service supply chain data service supply chain forecasting and planning (most of them are concentrated in industry leading enterprises, most of them are contract logistics). Second, logistics enterprises focus on warehouse management, WMS cooperation, building intelligent warehouse management under digitalization, and TMS transportation cooperation in supply chain management and control (most industries belong to segmented unicorns or multiple agent operations).

To build a transparent information delivery service in the whole process from warehouse to warehouse, a series of services are still scattered, and there are few one-stop integrated services. The basic mainstream is the outsourcing integration mode, and the effective connection of information is not a link. Third, based on the logistics enterprises' satisfaction of end-to-end demand in the upstream supply chain operation management, they began to jointly implement effective informatization at the end of distribution, to achieve urban intelligent scheduling and service requirements, and to deliver real-time services (excluding takeout services) with accurate algorithms in real time.

(1) Business and trade logistics

Digital logistics theory and application The digital and intelligent capabilities are widely used in all aspects of the logistics industry. Unlike the highly integrated and centralized way of foreign countries, the great development of China's logistics industry over the years depends on industrial coordination, ecological coordination and network coordination. The macro-social coordination from central cities to rural areas is a distributed social network. In the future, logistics services will inevitably move from informatization to digitalization and intelligence.

For example, the globalization of newbie packages is moving towards the globalization of digital supply chains. The modern logistics industry has the complex network characteristics, and more and more "integration" trends among logistics enterprises have gradually formed a new ecology of logistics development in China in recent years. On the one hand, C2C and B2C are developing in the direction of M2C and eventually in the direction of C2M. Data such as consumer preferences are obtained from the consumer side, and the products designed and produced in a timely and rapid manner will be delivered to consumers through digital logistics. On the other hand, with the continuous development of mobile Internet, mobile terminals have become increasingly integrated and digital, from manufacturers to sellers to consumers.

(2) Transportation logistics

Intelligent solutions for digital logistics. Based on blockchain, distributed digital identity, trusted computing, privacy protection, financial risk control and other technologies, solve the problems of logistics transportation and capital authenticity authentication, and realize data cross-validation. Help the upstream and downstream participants in the transportation industry better gain the trust of financial, regulatory agencies and customers. It is applicable to a variety of logistics and supply chain service scenarios, such as road freight, shipping and multimodal transport.

For example, ant chain's powerful map data and algorithm engine make intelligent scheduling more efficient. As the cornerstone of rebuilding trust, blockchain can be used in any scenario that lacks trust. At present, Ant Chain has landed in more than 50 scenarios and has been applied in a wide range of fields to serve the real economy. Ant chain after-sales support team provides multi-type and multi-channel service support to meet the support needs of different customers. Help you put on the chain and use the chain safely. Intelligent scheduling of digital logistics. Ant chain's powerful map data and algorithm engine make intelligent scheduling more efficient.

The main functions include API+SaaS flexible access scheduling algorithm capability, web API supporting multi-scenario computing, SDK for visualization of computing results, multi-constraint and multi-objective, to meet the needs of all industries to reduce costs and improve efficiency, multi-scenario application adaptability from trucks to third-party logistics, large-scale task calculation of fast food and dangerous chemicals, support complex outlets and high order peaks, efficiently calculate tens of thousands of outlets, and stably process multiple orders, Support multiple route planning, truck route planning, bus route planning, motorcycle route planning, electric vehicle route planning, get through Baidu map navigation, strengthen driver distribution control, task distribution driver terminal, real-time navigation location and track monitoring, etc.

(3) Warehousing logistics

Digital logistics elements drive the development of logistics business with data. Data middle office is a mechanism for enterprises to use their data, a strategic choice and organizational form, and a mechanism for building a set of continuous data into assets and services. In order to promote the development of digital logistics, continuous innovation is needed in management methods, business models and market forecasts.

For example, China Southern Power Grid introduced a new management concept of "digital logistics" to implement the integrated management mode of middle and Taiwan. Guangdong Jiangcun Regional Warehouse "Storage, Inspection and Distribution" Integrated Heavy Electrical Materials Intelligent Logistics Demonstration Base of China Southern Power Grid. The base takes the building of "UIM Logistics Middle Platform" as the starting point, Jiangcun Warehouse as the implementation carrier, and through big data, spatial modeling, visual loading, intelligent storage, Internet of Things and other technologies, the middle platform realizes the real interconnection of everything. The demander, supplier, transportation provider and provincial platform operate in series, creating a closed-loop management and control mode with traceable demand, integrated information, intelligent operation and visual process, Realize the goal of standardized, unmanned, visual and intelligent operation of reserved materials.

The results showed that after the application of smart logistics in the middle platform, the input of logistics personnel was reduced by 30%, and the material supply time was shortened from 57.7 days to 10 days now.

4 Conclusion

The digital logistics generated on the basis of technologies and methods such as digital economy, digital technology, machine learning and computational thinking is to guide the rapid optimal allocation of enterprise logistics elements (micro) and the rational optimal allocation of regional social logistics resources (macro) through the identification, selection, filtering, storage, conversion and use of digital flows generated in the operation of the logistics system, so as to form a collaborative symbiosis of logistics elements and resources between enterprises and within regions, And then provide high-quality logistics services for the society. The core of digital logistics is to transform traditional logistics with digital technology and give new impetus to traditional logistics operation mode and management mode. The development of digital logistics needs not only advanced digital technology, but also advanced economic theory, management methods, science and technology.

The rise of digital economy, enterprise competition and market demand are the driving forces for the wide application of digital technology. Under the condition that the traditional extensive operation and management mode has become the bottleneck of enterprise development, digital logistics has become an effective way to accelerate enterprise transformation by using digital technology and intelligent technology.

The use of digital technology can also solve such thorny problems as the lack of scientific means to support inventory forecasting and inventory control, and poor multi system coordination. On the technical level, digital logistics technology comprehensively applies big data, cloud computing, the Internet of Things, blockchain, artificial intelligence, 5G communication and other emerging technologies. In terms of application scenarios, the data middle ground is the key to enable enterprise data to be used, and is to build a mechanism that continuously turns data into assets and services business. In order to promote the development of digital logistics, continuous changes and innovations are needed in management methods, business models and market forecasts.

Acknowledgements. Recommender: Thanks to Han Huang Prof. South China University of Technology, China.

References

1. Xu, S.: Large logistics theory. China's Circ. Econ. **19**(5), 6–9 (2005)
2. Xu, S.: Rediscussion of big logistics. China's Circ. Econ. **21**(10), 9–12 (2007)
3. He, M.: Logistics System Theory. Higher Education Press, Beijing (2006)
4. Wang, S.: Systematic consulting report on enterprise management capacity. Guangzhou Guangbai Logistics Co., Ltd. (2007)
5. Wang, S.: Analysis of the operation mode of the fifth party logistics hub service provider. China Circ. Econ. **29**(06), 36–44 (2015)
6. Wang, S.: Research on the cooperative development strategy of Guangdong, Hong Kong and Macao logistics industry based on the regional growth pole theory. Guangdong Provincial Philosophy and Social Science Planning Leading Group Office (2012)
7. Wang, S.: Regional Logistics Theory and Empirical Research. Science Press, Beijing (2018)
8. Wang, S.: Strategies for building the dynamic mechanism of the regional logistics system around the Pearl River Delta. Logistics Technol. **35**(04), 8–13 (2016)
9. Wang, S.: Research on cross-border e-commerce logistics resource endowment in Guangdong, Hong Kong and Macao—thinking on regional logistics industry layout based on growth pole theory. Bus. Econ. Res. **12**, 78–81 (2016)
10. Wang, S.: Discussion on the construction of regional logistics system in Guangdong, Hong Kong and Macao under the innovation-driven background—from the perspective of value chain resource synergy between the government and logistics enterprises. Sci. Technol. Manage. Res. **36**(16), 90–96 (2016)
11. Wang, S., He, P.: Comprehensive evaluation of logistics competitiveness of Guangzhou Port based on SA-entropy weight method. In: Guangdong Provincial Institute of Operations Research "Innovative Theory and Practice of Logistics Operations Research in the Context of Double Circulation" Academic Conference (2020)
12. Wang, S.: Huangpu District promotes the innovation of modern logistics system. Guangzhou Economic Development Zone, Huangpu District People's Government (2021)

13. Liu, W.: Research on the strategic development of modern logistics enterprises. Bus. Econ. Res. **812**(1), 131–133 (2021)
14. He, L.: Review of the development of China's logistics industry in 2020 and outlook for 2021. China Circ. Econ. **35**(3), 3–8 (2021)
15. Wang, S., He, P., Wu, C.: Digital logistics theory, technical methods and applications - summary of the viewpoints of the digital logistics symposium. China Circ. Econ. **35**(06), 3–16 (2021)
16. Kayikci, Y.: Sustainability impact of digitization in logistics. Procedia Manuf. **21**, 782–789 (2018)
17. Wu, J.: Research on high-quality development of logistics industry in the context of building a new development pattern. Price Monthly **2021**(08), 90–94 (2021)
18. Ouyang, F.: 5G is the path choice to drive the high-quality development of logistics industry. Enterp. Econ. **39**(6), 15–21 (2020)
19. Wang, X., Yang, P.: Design and effect analysis of digital transformation of shipping logistics enterprises. Comput. Eng. Appl. **2021**(8), 1–9 (2021)
20. Gupta, A., Singh, R.K., Gupta, S.: Developing human resource for the digitization of logistics operations: readiness index framework. Int. J. Manpower **43**, 355–379 (2021)
21. Li, N., Qu, F.: Digital supply chain empowers furniture e-commerce logistics "smart" upgrading strategy. Bus. Econ. Res. **822**(11), 107–109 (2021)
22. Lu, Y., Dai, Y., Fan, X.: Application of intelligent logistics distribution system in digital assembly workshop. Manuf. Autom. **39**(6), 12–15 (2017)
23. Bai, J., Guo, H.: Digital picking and distribution logistics system for discrete manufacturing enterprises and its application. Logistics Technol. **8**, 30–33 (2014)
24. Wang, W.: Digital logistics warehouse management system based on radio frequency identification technology. Sci. Technol. Eng. **19**(2), 170–174 (2019)
25. Fan, R.: Analysis of innovative service model of logistics and manufacturing based on collaborative system. Enterp. Econ. **430**(6), 128–132 (2016)

A Dual Referral Optimization Model for Medical Clusters Based on Queuing Theory and Cooperative Game Incentives

Zhiyuan Tong, Yulin Nie, Miaoxia Zhuang, Sitong Liang, Ning Liu, and Caimin Wei[✉]

Department of Mathematics, Shantou University, Shantou 515063, People's Republic of China
{20zytong,cmwei}@stu.edu.cn

Abstract. In this paper, An optimization model for dual referral of medical treatment combination is established via queuing theory and cooperative game incentive mechanism. We establishes a queuing theory model for dual referral in the medical association and analyzes the impact of the referral rate on the revenue and crowding degree of central and community hospitals under the medical association model. In the form of medical alliance, when the referral rate continues to increase, the division of labor between central hospitals and community hospitals becomes more and more clear. Central hospitals focus on high-technology medical treatment such as surgery, while community hospitals focus on basic medical treatment such as rehabilitation, which can maximize the overall benefits. To reduce the overall congestion of the system, the higher the referral rate is not the better, it needs the joint coordination of central and community hospitals. In the initial decision making, the central hospital in the medical consortium can choose a higher referral rate, and at the same time, it should consider the resource limitation of the community hospital and cooperate with the community hospital with more beds. Secondly, through the cooperative game theory, different incentive measures are designed to analyze the situation of simultaneously rewarding central hospitals and community hospitals and only rewarding central hospitals. The latter can achieve a higher referral rate faster with less funds, and the optimal investment capital is about half of the original revenue. Meanwhile, the willingness of community hospitals to cooperate should be considered to achieve a win-win result.

Keywords: Medical cluster · Queuing theory · Cooperation game · Shapley value method

1 Background of the Problem

The construction of integrated medical services started earlier in foreign countries. Different countries have established the index evaluation system which is compatible with their own medical service system [1], with the purpose

B.-Y. Cao et al. (Eds.): ICFIE 2022, LNDECT 207, pp. 279–291, 2024.
https://doi.org/10.1007/978-981-97-2891-6_21

of promoting hierarchical diagnosis and treatment, controlling medical costs, and improving the allocation rate and utilization rate of high-quality medical resources [2]. In recent years, the medical alliance has been included as a key initiative in the new medical reform, and medical institutions at all levels across the country have responded. The sharing of medical equipment under the "medical alliance" model provides a realistic basis for promoting convenient access to medical treatment, reducing the disorderly flow across regions, enhancing the public's sense of access, and achieving a green channel for referrals within the medical alliancen [3]. In practical application, due to the lack of primary medical equipment and technology, central hospitals in the medical alliance need to refer critically ill patients to general hospitals through green channels in a timely manner and return to the original primary hospital for continued treatment and recovery after the condition has been stabilized at the higher level hospital.

Currently, queuing theory methods are widely used in various healthcare operation decisions. Bekke and Bruin [4] explore how time-dependent arrival patterns affect ward bed demand and denial rates by analyzing an $Mt/H/s/s$ queuing model based on a near-infinite server queuing system. Dobson et al. [5] proposed a tandem queuing model with limited server to assess the impact of different work prioritization strategies on patient flow times and institutional service efficiency. Bruin et al. [6] applied queue theory to quantify the impact of a strategy that used the same occupancy criteria for all wards. Some scholars also built a game model, used Shapley value method, and modified Shapley value method to study the benefit distribution of the medical consortium, and studied the benefit distribution within the medical consortium under the optimal social utility [7,8].

2 Description of Model

Through investigation and literature reading, we found that in addition to information docking, medical insurance, and other administrative policy issues, the unfair distribution of benefits between the central and community hospitals is a major obstacle to the implementation of the medical consortium[9]. This paper will design the market and benefit distribution between the central and community hospitals through the queuing theory and cooperative game model, analyze how to promote the dual referral system of medical unions better, and realize the rational allocation of medical resources. The main problem of medical treatment is "easy to turn up and difficult to turn down", so this paper mainly discusses the situation of downward referral. In the short term, the central and community hospitals have "conflicts of interest". Central hospitals have enough medical resources to accommodate a large number of patients in a short period of time, but in the long run, the utilization efficiency of resources is not high[10]. If the resources invested in minor and chronic diseases are used for the treatment and research of major diseases, the long-term interests of the hospitals will be greatly improved. Turn idle hospital beds into more useful resources (e.g. medical research, equipment optimization, $\cdots$) For the whole medical system, from the breadth of expansion to the depth of exploration; For community hospitals, a higher referral rate can increase profits, with which they can develop better, focus on rehabilitation diagnosis and treatment services, fulfill the responsibility that community hospitals should bear, transfer "excess" patients from higher hospitals to community hospitals, and solve the problem of overcrowding in higher hospitals. The following Table 1 is the description of symbols.

Table 1. Description of symbols

Symbol	Instructions
n	Number of hospital beds
n_1	Number of Phase I beds in lower level hospitals
n_2	Number of Phase I beds in higher level hospitals
m	Maximum system capacity
λ	Average rate of arrival
μ	Average service rate
$p_{ref}(t)$	The referral rate of patients from the first stage of higher to lower hospitals at time t
ρ	Load level of the system
p	The probability that the system is idle at the moment
P_{lose}	System loss probability
Q	The relative passing capacity of the system
L_{serve}	Average number of busy service windows
L_q	Average queue for captain
L_s	Average captain
W_s	The average length of time a patient stays in the system
W_q	The average waiting time for patients in the system
B_H	Income from superior Hospitals
B_L	Income of lower hospitals
y_1	The income weight of the higher hospitals
y_2	The income weight of the lower hospitals
R_L	Income from service desks of subordinate hospitals
R_1	Benefits of the first stage unit service desk at the parent hospital
R_2	Benefits of the second stage unit service desk at the parent hospital
$\xi(t)$	In reality, the block factor of resistance encountered at time t
ξ_{max}	The maximum value of blocking factor
c	Unit cost of patient waiting
$v(s)$	The payoff of the subset s
$v(s \backslash i)$	The revenue of subset s excluding hospital i
x_i	i The distribution of the revenue received by the hospital from $v(I)$

Note: The parameters of the superscript or subscript HL_1 are the corresponding parameters of the first stage of the central hospital; The parameters whose superscript or subscript is HL_2 are the corresponding parameters in the second stage of the central hospital. The parameters whose superscript or subscript is LL are the corresponding parameters of community hospitals.

Based on the above background, this paper explores the multi-server hybrid system model $M/M/n/m$ based on queuing theory. Aiming at the rehabilitation stage of chronic diseases, a multi-service window hybrid queuing model $M/M/n/m$ was constructed for central hospitals and community hospitals respectively, with arrival intensity obeying Poisson flow and service time obeying negative exponential distribution. Under the condition of system equilibrium, K-algebraic equations are constructed to obtain the formula expression of each objective parameter, as Fig. 1 follows:

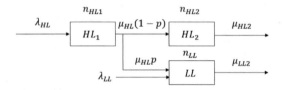

Fig. 1. Multi-service window hybrid queuing model $M/M/n/m$

3   Analysis and Solution of Model

For satisfaction, we use the service utility of patients to describe patient satisfaction, that is, the service utility is higher, the customer satisfaction is higher . The service utility of hospitals at all levels is composed of the overall utility of medical alliances, the treatment price of hospitals at all levels, and the service efficiency of hospitals at all levels:

$$U_i = V - \lambda_i P_i(1 - p) - c\left(\mu_i - (1 - p)\lambda_i\right)^{-1},$$

where $V = V_1 + V_2, V_1 = (1 - p) \cdot V$.

To simplify the calculation, we set the wait time cost to 1. In this section, the community hospital charges are less than the central hospital, so the cost is set to 0. Because the actual number of patients in community hospitals or rural hospitals is small, the customer arrival rate of community hospitals LL is set as 0. In the first diagnosis and treatment stage of the central hospital, the customer does not have the problem of the referral or not. So, for the first diagnosis and treatment, the service utility is only affected by the arrival rate and service rate. Therefore, the service utility of the first diagnosis and treatment of a central hospital is ignored in this section. And the objective function is set as:

$$\max U_{\text{total}} = V - \lambda_1 P_1(1 - p) - (\mu_1 - (1 - p)\lambda_1)^{-1} - (\mu_2 - p\lambda_1)^{-1}.$$

In order to better solve the model, we make the following assumptions:

Hypothesis 1: The feasible domains of hospital service capability are: $2\lambda_1 > \mu_1 > \lambda_1, \mu_2 > \mu_1 - \lambda_1$.

The service capacity of community hospitals should have a lower limit to ensure that patients will choose to be transferred to community hospitals when they choose superior or inferior hospitals.

Hypothesis 2: Patients will definitely choose to seek medical treatment in such a medical system, that is, there will be no situation of not seeking medical treatment when sick.

If $U_1 > U_2$, Patients will choose to stay in a third class hospital for medical treatment. If $U_1 = U_2$, patients can choose either way. The case studied in this section is $U_1(p) \leq U_2(p)$, that is, the service utility of community hospitals is greater than that of higher hospitals, and the referral is encouraged. The model is used to solve the referral rate at the time of maximum customer satisfaction. In this case, when the total service utility V_0 is 5, the referral rate p is 0.4848, and the optimal service utility of patients U is 2.8778, and when the total service utility V_0 is 10, the referral rate p is 0.4934. When the total service utility V_0 is 15, the referral rate p is 0.4958, and the optimal service utility of patients U is 12.882. It can be seen that the change in the total service utility p has a small impact on the referral rate, maintaining at about 0.5. The following figure.2 shows the referral rate of the patient service utility U:

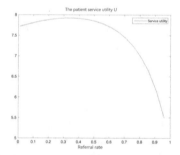

Fig. 2. The referral rate of the patient service utility U

In the case of government intervention, too many referrals lead to an increase in the waiting cost of community hospitals and eventually lead to a rapid decline in customer satisfaction. In addition, for the medical consortium structure with different service effectiveness, the change in the referral rate is not obvious. In the absence of government intervention, the operation effect of the medical consortium is not good.

For the above model, the service utility V of the medical consortium is a fixed constant, which can be used to obtain the optimal referral rate with the highest customer satisfaction under the ideal condition, while the actual service utility will change due to the changes of government incentives. In order to simulate the optimal referral rate affected by government incentives under the actual condition, we set the incentive function $\alpha(p)$:

$$\alpha(p) = 1.31 \cdot \frac{e^p - e^{-p}}{e^p + e^{-p}},$$

where $0 \leq p \leq 1$.

The service utility of medical alliance is $V = \alpha V_0 + V_0/2$, The total government investment is V_0. When $V_0 = 10$, the optimal referral rate is 0.7778, and the patient service efficiency is 8.8991, It can be seen that when the government adds incentives, the referral rate is significantly improved and the patient service utility U is increased. The following Fig. 3 shows the referral rate of the optimized patient service utility U:

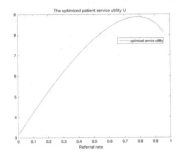

Fig. 3. The referral rate of the optimized patient service utility U

When the referral rate is low, the government will give punishment, and patient satisfaction is not high. In order to ensure the satisfaction of patients, the medical treatment will actively improve the referral rate and improve the service effectiveness of patients. When the service effectiveness of patients reaches the optimal level, the improvement of the referral rate will lead to the crowding of community hospitals, and then lead to the decline of customer satisfaction. Under the government's reasonable incentive measures, the referral rate has been significantly increased, but the lower level hospitals will not be overcrowded, which shows the importance of the government's macro-control.

3.1   Benefits

According to the established objective parameters, the total number of beds in central hospital was set to be fixed in the general mode and the medical association mode. The allocation number and referral rate of beds in the treatment and rehabilitation stages of central hospital were adjusted as the decision variables, and the objective programming model was constructed with the objective function of minimizing waiting cost C and maximizing hospital revenue B.

To simplify the calculation, we quantified the waiting cost as the cost of the hospital to make up for the overcrowding of the queue length by increasing the number of beds, i.e., $C = P_{lose}c$, c is unit cost,

$$\max B = y_1(L_{serve}^{HL1}R_1 + L_{serve}^{HL2}R_2) + y_2L_{serve}^{LL}R_L - C$$
$$s.t. \; n_1 + n_2 = n_{total}$$
$$0 < p_{ref} < 1$$

The following variables are in unit time, and this article sets it to 10 days L_{serve}^{HL1}, L_{serve}^{HL2}, L_{serve}^{LL} are the average number of busy server in the first stage, the second stage and the lower level hospitals at time t, respectively; R_1, R_2, R_L are the income of the server in the first, second and lower-level hospitals respectively; n_1, n_2 are the number of beds in the first and second stage of the central hospital; y_1, y_2 are the income weights of upper and lower hospitals, respectively; p_{ref} is the referral rate. By referring to relevant literature, we obtained a set of reasonable medical conjoint data, which were substituted into the above established multi-service window hybrid queuing model $M/M/n/m$ and multi-objective programming model, and MATLAB software was used to solve the problem. Due to the complexity of queuing theory model parameters, the particle swarm optimization algorithm was mainly adopted for target programming. At this time, the revenue of the upper and lower hospitals as a function of the referral rate is shown in this Fig. 4:

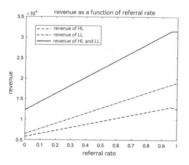

Fig. 4. The revenue of the upper and lower hospitals as a function of the referral rate

The conclusion is as follows: if the total number of beds in central hospital is 50, and the number of beds in community hospital is 20,30,40,50, the solution can be obtained: when the number of beds allocated by central hospital to the first stage is the maximum 48, and the referral rate is 0.9179, the total revenue of central hospital and community hospital is the maximum 30657.4. The number of beds in lower-level hospitals was 11.5564, 15.8346, 20.1129, and 24.3911, respectively, and the utilization rate of beds was 57.8%, 52.8%, 50.3%, 48.8%. It can be seen that in the form of medical alliances when the referral rate continues to increase, the division of labor between central hospitals and community hospitals becomes more and more clear. The central hospital focuses on the high-tech medical stage such as surgery, while the community hospital focuses on the basic medical treatment such as rehabilitation, which can maximize the overall benefits.

The above is an ideal situation, which can simulate the medical alliance without resistance to analyzing the impact of referral rate on the upper and lower hospitals. In order to simulate the resistance encountered by the medical alliance in the real situation, we set the blocking factor $\xi(t)$, meet the conditions $\frac{d\xi}{dt} = (1 - \frac{\xi}{\xi_{max}}\eta)\xi$, ξ_{max} is the maximum value of blocking factor, $\eta = f(B_H, B_L)$, f is a function related to the income of upper and lower hospitals, η_1, η_2 is a constant coefficient:

$$p_{ref}(t+1) = \frac{p_{ref}(t)}{\xi(t)} \tag{1}$$

$$f(B_H, B_L) = \exp(\frac{\eta_1}{B_H}) - \eta_2(\frac{B_H}{B_H - B_L}) \tag{2}$$

The conclusion is as follows: if the total number of beds in central hospital is 50 and the number of beds in community hospital is 20,30,40 and 50, it can be obtained that when the number of beds allocated by central hospital to the first stage is the maximum 48, and the referral rate is 0.33, the total revenue of central hospital and community hospital is the maximum 21050. The number of occupied beds in lower-level hospitals was 7.34, 9.51, 11.68, and 13.85, respectively, and the utilization rate of beds was 36.7%, 31.7%, 29.2%, 27.7%, respectively.

3.2   Congestion

For congestion, we take the wait time as an objective function:

$$\min W_s = W_s^{HL1} + W_s^{HL2} + W_s^{LL}$$

where W_s^{HL1}, W_s^{HL2}, W_s^{LL} are the average waiting time of central hospital in the first stage, central hospital in the second stage, and community hospital respectively.

If the total number of beds in the central hospital and the community hospital is both 50, it can be obtained that when the number of beds in the central hospital in the treatment stage is 33 and the referral rate is 0.7977, the average waiting time of the central hospital and the community hospital is the minimum, which is 2.5187. At this time, the total mean length of stay in the upper and lower hospitals as a function of the referral rate and bed allocation in the higher hospitals is shown below Fig. 5:

Initially, a low referral rate would have very little effect on reducing over-crowding because the central hospital was so overcrowded that referring a small number of patients would not have much impact. However, when the referral rate is between 0.5 and 0.8, the average length of stay decreases rapidly and the marginal benefit is large. It can be seen that to reduce the overall congestion of the system, the higher the referral rate is not the better, but the joint coordination of upper and lower hospitals is needed. In the initial decision making, the

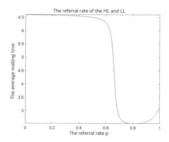

Fig. 5. The total mean length of stay in the upper and lower hospitals as a function of the referral rate and bed allocation in the higher hospitals

central hospital in the medical consortium can choose a higher referral rate, and at the same time, it should consider the resource limitation of the community hospital and cooperate with the community hospital with a larger number of beds.

4   Game of Cooperation

4.1   Shapley Collaborative Response

Assume that there is a leading upper hospital with 50 beds, which is numbered 1, and four subordinate hospitals with beds of 20, 30, 40, and 50, respectively, which are numbered 2, 3, 4, 5. $I = \{1,2,3,4,5\}$ is the set of 5 cooperative hospitals, $s = \{j_1, j_2, \cdots, j_n\}$ is a subset of I.

(1) when there is not a leading upper hospital in the league, we have

$$v(s) = \sum_{k=1}^{n} R_L(j_k) L_{serve}(j_k), n = 1, \cdots, 5. \quad (1 \notin s).$$

(2) when there exists a leading upper hospital in the league, we have

$$v(s) = \sum_{s \in S_i} R(s, p_{ref}) L_{serve}(s, p_{ref}), \quad (1 \in s).$$

The p_{ref} can be calculated by goal-programming revenue model according to the queuing theory above;

(3) The allocation i gets from $v(I)$

$$x_i = \sum_{s \in S_i} w(|s|)[v(s) - v(s \backslash i)], \quad w(|s|) = \frac{(n-s)!(|s|-1)!}{n!}.$$

(4) Part of the results are as the Table 2 follows:

Finally, the revenue distribution ratio of each hospital is as follows $H = 0.673, L_2 = 0.058, L_3 = 0.078, L_4 = 0.090, L_5 = 0.101.$

Table 2. Part of the results

S	H	(1)	(2)	(3)
$v(s)$	16556.4	17053.5	24003.0	29676.7
$v(s\backslash H)$	0	1500	2750	5250
$v(s) - v(s\backslash H)$	16556.4	15553.5	21253.0	24426.7
$|s|$	1	2	3	4
$w(|s|)$	0.20	0.05	0.03	0.05
$w(|s|)[v(s) - v(s\backslash H)]$	3311.28	777.68	708.43	1221.33

4.2   Incentive Measures

(1) Reward institutions were set up to evaluate the cooperation revenue of the upper and lower hospitals. The reward is given according to a set amount of time. The total amount of rewards was K, and the distribution ratio was determined by the Shapley value method. The following figure shows the impact of different K values on referral rates.

With the increase of K, the referral rate roses slowly. Until the government invested 12870, the referral rate reached the maximum of 0.7162, and the total profit is 43528. Under this circumstance, the five hospitals have made 21345.8, 3635.6, 4962.5, 6186.5, and 7397.6, respectively. The utilization rates of beds in subordinate hospitals were 50.6%, 45.6%, 43.1%, and 41.6% as the Fig. 6 follows:

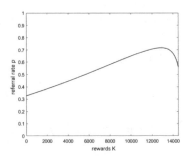

Fig. 6. The impact of different K values on referral rates the impact of different K values on referral rates

According to the queuing theory model, the total revenue of this medical consortium is 21050 without incentive measures, and the invested capital is more than half of it. The growth is slow if the invested capital is small. The reason is that the superior hospital will not pay attention to a small amount of invested capital. However, when the invested capital is too large, there is a downward trend, the reason may be that under a large amount of invested capital, even if the proportion of distribution does not change, central hospitals get more rewards in numerical value, and the gap between them and lower hospitals is more obvious, which increases the blocking factor. Community hospitals feel unfair the allocation, and then, they will be reluctant to cooperate.

(2) Supervisory institutions were set up to evaluate the referral effect of upper and lower hospitals, and only the upper hospitals were given rewards F based on the referral rate as the main evaluation standard. The following Fig. 7 shows the impact of different F values on referral rates:

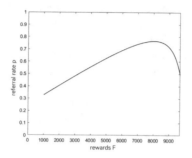

Fig. 7. The impact of different F values on referral rates

With the increase of F, the referral rate rose slowly. Until the government invested 8022, the referral rate reached the maximum of 0.7659, and the total profit is 35640. Under this circumstance, the five hospitals made 19566.3, 2617.7, 3551.5, 4485.4, and 5419.2, respectively. The bed utilization rates of the four community hospitals were 52.4%, 47.4%, 44.9%, and 43.4%. The invested capital is also more than half of the original revenue. The growth rate is slow when the capital is low for the same reason as above. However, when the invested capital is too high, a trend of decline also appears. Combined with the figure above, the reason may be that the dissatisfaction of community hospitals increases the blocking factor.

From the above two incentive measures, it can be concluded that the referral rate can be reached faster with fewer funds t by rewarding the central hospital alone, and the optimal invested capital should be about half of the original revenue. At the same time, the cooperation willingness of the community hospitals should be considered to achieve a win-win result.

5   Conclusions

This paper builds an optimization model for the dual referral of medical treatment combinations based on queuing theory. We analyze the impact of medical treatment on the operation of upper and lower hospitals. Under the form of medical treatment, when the referral rate continues to increase, the division of labor between upper and lower hospitals becomes more and more clear. The central hospital focuses on the difficult medical stage such as surgery, while the lower hospital focuses on the basic medical treatment such as rehabilitation. It maximizes the overall benefit. To reduce the overall congestion of the system, the higher referral rate is not the better. It needs the joint coordination of upper and lower hospitals. In the initial decision making, the central hospital in the medical consortium can choose a higher referral rate, and at the same time, it should consider the resource limitation of the community hospital and cooperate with the community hospital with a larger number of beds.

Design different incentive measures through the cooperative game theory. (1) Set up reward institutions, evaluate the cooperation revenue of upper and lower hospitals, and issue rewards on a half-year cycle with the total amount of rewards being K. (2) To set up supervisory organizations to evaluate the referral effect of upper and lower hospitals, and to award higher hospitals based on the referral rate as the main evaluation criteria; The conclusion is that by rewarding central hospitals alone, higher referral rate can be achieved faster with less invested capital, and the optimal invested capital is about half of the original revenue. Meanwhile, the cooperative intention of the community hospitals should be considered to achieve a win-win result.

Acknowledgements. This research was supported by the Humanities and Social Science Fund of Ministry of Education of China (No. 21YJC630186), the Natural Science Foundation of Guangdong Province (No. 2022A1515012034), the Philosophy and Social Science Planning Project of Guangdong Province (No. GD20CGL19), and the Startup Fund from Shantou University (STF19024).

Recommender: Professor Shengli Lva from School of Science, Yanshan University in China.

References

1. Du, X.L., Gao, H., Li, H.: Comparison and reflection on the construction mode of medical association and hierarchical diagnosis and treatment at home and abroad. China Med. Herald **21**(12), 40–42 (2017)
2. Cheng, Z.Y.: Research on perfecting three-level medical service system in rural areas of China. Anhui University of Finance and Economics (2017)
3. Tian, Y.L.: Discussion on current situation of shared management of medical equipment under the model of "medical consortium". China Med. Devices **36**(9), 156–158 (2021)
4. Bekke, R., de Bruin, A.M.: Time-dependent analysis for refused admissions in clinical wards. Ann. Oper. Res. **178**, 45–65 (2009)
5. Dobson, G., Lee, H.H., Sainathan, A., et al.: A queueing model to evaluate the impact of patient 'batching' on throughput and flow time in a medical teaching facility. Manuf. Serv. Oper. Manag. **14**(4), 584–599 (2012)
6. de Bruin, A.M., Bekker, R., van Zanten, L., et al.: Dimensioning hospital wards using the Erlang loss model. Ann. Oper. Res. **178**, 23–43 (2009)
7. Uz. D., Fang, Z.H., Zhang J.J.: A study on profit allocation model of medical consortium based on Shapley value algorithm. Value Eng. **38**(11), 1–5 (2019)

8. Li, W.M., Zhu, J.H., Liu, S.Y.: Research on benefit distribution of Chinese medical community members based on revisionary Shapley value method. Chin. Hosp. Manag. textbf41(07), 26–29 (2021)
9. Zhong, Y.Y., Chen, J.: Current situation and strategy of two-way referral implementation in integrated delivery system of China. China Med. Herald **14**(6), 154–158 (2017)
10. Ma, M., Wei, C.M., Li, Z.P.: Referral and government investment decision-making study Based on hierarchical medical system. J. Quant. Econ. **37**(04), 130–140 (2020)

Digitally Empowers Yiwen to Promote Barrier-Free Language Communication

Chen Yong-wen[1], Chen Ying-bing[2], Peng Bin[1], and Cao Bing-yuan[3,4(✉)]

[1] Guangdong Provincial Operations Research Society Disabled Maker Research Branch,
Guangdong, China
[2] Jiangxi University of Science and Technology, Ganzhou, China
[3] Editor in Chief of Tsinghua University- IEEE Magazine FIE, Peking, China
j_ifiecao@126.com
[4] School of Mathematics and Information Science, Guangzhou University, Guangzhou, China

Abstract. Yi Wen, invented by us, is a special language that can be used uniformly by people with visual, auditory and other disabilities. It is an important tool for people with disabilities to think, learn and communicate all over the world regardless of nationality or race, and it is also a unique presentation of integrating the diversity of human languages. Yi Wen has four characteristics: easy to write, to spell, to remember and to connect. Yi Wen promotes the communication of disabled people to enter the Internet era. We adopt the digital technology of digital code, set up a special communication system by ourselves, and use the barrier-free language communication of network media to fundamentally solve the world problem of communication between disabled and normal people, and help disabled people start businesses, so as to truly realize the great integration of the people of the world, community of Shared Future for Mankind.

Keywords: digital empowerment · barrier-free language system · coding

1 Introduction

There are 5,651 languages in the world today, in which more than 1,400 independent languages that have not been recognized by people or dying languages. There are more than 200 countries and more than 2,500 nationalities in the world, and different nationalities have their own languages. Thus, the world has formed an intricate language system, which has caused great inconvenience to people all over the world. In order to facilitate people's communication and information sharing around the world, get rid of cultural influence and realize a barrier-free language that can be used all over the world, we invented this language system called Yi Wen. Yi Wen does not change anyone's grammar and culture, but only changes the way of communication to benefit all mankind. At present, in order to enable people all over the world to communicate and share information better, we have adopted a method that people all over the world can recognize and know. The simplified patterns of the sun, the moon, and the stars as well as simple, convenient and universal combination through Arabic numerals code operation would achieve real barrier-free language communication.

2 No Obstacle to the Technical Support Foundation of Language [4]

What is Yi Wen?

Yi Wen is a special language that can be used by disabled people all over the world regardless of nationality or race. It is an important tool for their thinking, learning and communicating, and also a unique presentation of integrating the diversity of human languages. The popularization of Yi Wen is a basic job for the construction of a barrier-free social environment, which will effectively safeguard the language rights of the disabled with an effective means to eliminate communication barriers, solve practical problems and create equal opportunities for survival and development. Doing a good job in the promotion of Yi Wen can greatly improve the educational quality and cultural quality of the disabled, increase their communication ability and social participation level, promote the integration of disabled people with society, promote their equal participation and integrated development, create a barrier-free communication environment of mutual understanding and tolerance in the whole society, and reflect the progress of social civilization.

To realize language communication, a software and hardware system is needed, its foundation meaning digital technology.

Following Comes Specific Technology

Method to edit barrier-free language communication

Technical Field

The technology relates to the field of language communication, in particular, to a barrier-free language communication editing method.

Technical Background

In order to make every Chinese character have a national unified code, in 1980, China promulgated the first national standard for Chinese character coding: GB 2312–80 "Chinese Character Coding Character Set for Information Interchange" [1], which is the basis for the development of Chinese information processing technology in China and the unified standard for all Chinese character systems in China at present. Because the national standard code is a four-digit hexadecimal code, in order to facilitate communication, everyone usually uses a four-digit decimal area code. All national standard Chinese characters and symbols form a 94 × 94 matrix. In this square matrix, each row is called a "zone" and each column a "bit". Therefore, this square matrix actually constitutes a Chinese character set with 94 zones (area codes are 01 to 94 respectively) and 94 bits (bit numbers are 01 to 94 respectively) in each zone. The area code and location number of a Chinese character are simply combined to form the "location code" of the Chinese character. In the location code of Chinese characters, the upper two digits are the area code and the lower two digits are the location number. In the area code, area 01–09 contains 682 special characters, and area 16–87 contains 6763 Chinese characters. Among them, 16–55 areas are first-class Chinese characters (3,755 most commonly used Chinese characters, arranged in alphabetical order), and 56–87 areas are second-class Chinese characters (3,008 Chinese characters, arranged in radical order). According to the related technology, there are many strokes of traditional Chinese characters, which makes it extremely difficult for users to learn and write. How to provide a barrier-free language communication editing method to improve these problems has become an urgent problem for technicians in this field.

Technical Contents

In order to make up for the above shortcomings, this technology provides a barrier-free language communication editing method, aiming at improving the problems existing in the above background technology.

The technology provides a barrier-free language communication editing method, which specifically comprises the following steps [2, 3]:

Step 1: Find out the decimal area code q of the current character.

Step 2: set $Q = abcd$ and generate four $x\,y$ coordinates q1, q2, q3 and Q4 for Q. Where:

$$q1 = (-4\pi + (a - 1) * \frac{\pi}{2}, \cos(X)),$$

$$q2 = (-4\pi + (b - 1) * \frac{\pi}{2}, \cos(X)),$$

$$q3 = ((c + 1) * \frac{\pi}{2}, \cos(X)),$$

$$q4 = (d + 1 * \frac{\pi}{2}\cos(X)).$$

Step 3: the dot matrix is a character with simplified patterns of the sun, the earth, the moon and the stars as the main body;

Step 4, the cosine function takes the left $[-4\,\pi, 0]$ of the interval to form a "W" figure. The cosine function takes the right side of the interval $[\pi, 5\pi]$ to form an "m" graph; "W" graphics + "M" graphics combine to form a nine-digit M + language.

The beneficial effects are as follows:

(1) The application of this barrier-free language communication editing method on the Internet: you can communicate directly in this language; This language can be used as an intermediary to make various languages convertible to each other; Use this language to communicate, transmit and identify information moves on without barriers.
(2) The application of this barrier-free language communication editing method to learning: it greatly reduces the writing difficulty of users, reduces writing strokes, and is easy to write and use; When applied to Braille, it greatly decrease the number of Braille characters, more suitable for deaf-mute people to learn and use, and brings great convenience to the communication between disabled people.
(3) The application of the barrier-free language communication editing method to multi-dimensional code identification or anti-counterfeiting identification: the multi-dimensional code or encrypted anti-counterfeiting identification can be formed by encrypting the language and adding time or space elements, and then polarized light can be generated by laser engraving or 3D printing, which can be used for anti-counterfeiting and traceability.

(4) The barrier-free language communication editing method also has the advantages of "four easy": easy interconnection, writing, spelling and memory, which will effectively solve the problems of numerous strokes of traditional Chinese characters, great difficulties for users to learn and extremely inconvenient writing.

In some specific embodiments, the sun is represented by a big hollow garden "○", the earth is represented by a big solid garden "●", the moon is represented by a small solid garden "•", and the stars are represented by a small hollow garden "°". In the location code of Chinese characters, the area: ten places are represented by the sun "○" and one place is represented by the moon "•".

Specifically as shown in the following Table 1:

Step 5: the cosine function takes the left [-4 π, 0] of the interval to form a "W" figure, the cosine function takes the right [π, 5π] of the interval to form an "M" figure, and the combination of the "W" figure and the "M" figure forms a nine-digit M + language.

The working principle of the barrier-free language communication editing method is as follows: in the specific implementation process, the decimal location code Q of the current character is first found out, for example, the location code of the word "Chen" is Q = 1934, and then four xy coordinates q1, q2, q3 and Q4 are generated for Q;

As shown in Table 2, the four coordinates corresponding to the word "Chen" are:

Table 1. Coordinates and graphic representations of arbitrary words

Coordinate	x	y	Graphical representation
q_1	$-4\pi + (a-1)*\frac{\pi}{2}$	$\cos(x)$	Large solid dot
q_2	$-4\pi + (b-1)*\frac{\pi}{2}$	$\cos(x)$	small solid dot
q_3	$(c+1)*\frac{\pi}{2}$	$\cos(x)$	large hollow dot
q_4	$(d+1)*\frac{\pi}{2}$	$\cos(x)$	small hollow dot

Table 2. Coordinates and graphic representations corresponding to the word "Chen"

coordinate	x	y	Graphical representation
q_1	-4π	1	Large solid dot
q_2	0	1	small solid dot
q_3	2π	1	large hollow dot
q_4	$\frac{5}{2}\pi$	1	small hollow dot

In this process, cosine functions xx(-2 π, 2 π) and y(1,-1) take the maximum value of y, the minimum value of the intersection point of the x axis in sequence along the trajectory of the function image, and the method of connecting three points with one line forms a "W" figure in hexadecimal, forming 1 to 9 coordinate points, namely, 1, 2, 3, 4, 5, 6, 7, 8, 9.

Yi Wen takes the specific communication between disabled people as an example to introduce as follows:

Contributing to the cause of the disabled is one of the original intentions of Chen Yingbing's team to study "Yi Wen". Yi Wen does not change anyone's grammar or culture. "Yi Wen" changed the way of human communication. It enables the deaf and the blind to communicate effectively, and it enables people of all nationalities in the world to communicate effectively.

Yi Wen is different from Braille, where 4–12 dots make up a character, or sign language for the deaf-mutes. Instead, it is a simplified combination of the sun, the moon and the stars that are well known all over the world, and a character is formed by simple 1–4 special dots. This kind of character can be recognized by the whole world, and it is easy to learn and understand, so it is no longer difficult to communicate because of different languages in different countries and regions. The products we developed have four principles: easy to learn, easy to write, easy to remember and easy to use the Internet. You can set the knowledge you want to learn at any time, realize the freedom of learning regardless of time and space, and completely solve the problem of mobility inconvenience for the disabled and the elderly. Yi Wen has strong regularity, high simplicity, uniqueness, exclusiveness and independence from the world.

3   The Great Significance and Development Prospect of Yi Wen and Mo Ma

According to the World Vision Report, at least 2.2 billion people in the world are visually impaired or blind. By 2050, there will be 900 million blind and severely visually impaired patients in the world, and there are 17.31 million blind people in China. There are 120 million elderly people in China who have gradually lost their hearing. According to WHO data, about one third of the people over the age of 65 in the world have moderate or above hearing loss, and about 1.5 billion people around the world have hearing loss. There are 20.45 million deaf people in China, the employment rate of disabled people in China is about 45%, the self-employment rate is 7.7%, and the average employment rate of disabled people in the world is about 44%. On the whole, the employment situation of the disabled is still grim, with the problem affecting the quality of life, and happiness index of disabled people lies in communication tools. Whether it is studying, training or working communication, the proportion of effective studying and communication between people is very low. Governments and social organizations all over the world are trying to expand the employment channels for the disabled and increase the number of disabled people's employment, so as to achieve the balance between economic income and social status of all social strata, integrate the disabled into social production and life, and realize social harmonious development and fairness and justice. However, what is really lacking in all efforts is we need effective tools for learning, training, communication and information sharing.

3.1   Key Points in Communication Between the Blind and the Deaf-Mute

Disabled people are a special group, and account for a large proportion of the global population, which needs the attention and care of the whole society. At present, books for the blind are heavy in paper and waste resources, and there are too many characters, which are easy to be misread or wrongly written, with high error rate, particularly slow learning and low efficiency. Because the blind can't see and the deaf can't speak, they can't communicate effectively each other at all, which is also a difficult problem that the world can't solve at present!

In order to reduce the learning cost of disabled people, improve energy efficiency and solve the problem of effective communication between the blind and the deaf. Chen Yingbing's team invented Yi Wen, which not only enabled the two types of disabled people to communicate with each other, but also enabled them to communicate with normal people, breaking through this world problem.

3.2   The Communication Between the Blind and the Deaf can also Enter the Internet Age

Since human communication has entered the Internet age, paperless office has become the norm. However, the blind with disabilities are still in the primitive era of slash and burn, and they still touch the bumps with their hands to learn knowledge and communicate. In order to make the disabled keep up with the pace of the times quickly and make their work, study and life more convenient, Chen Yingbing's team developed a communication system for the disabled. On the basis of achieving a breakthrough in the world problem-the blind and deaf-mute people communicate with normal people, they upgraded their life and study to the fast lane of the Internet era, and truly integrated them into the big family of society.

3.3   The Programming Adopted by Yi Wen is Mo Code, Divided into Six Sections

3.3.1 Yi wen education and training: start a school with publishing.

3.3.2 Easy-to-read computer (for old, weak, sick and disabled).

3.3.3 Easy-to-read mobile phone (for old, weak, sick and disabled).

3.3.4 Universal language chip,

a. security traceability of various commodities,
b. New computer system;
c. AI cloud service system.

3.3.5 Internet language chip,

a. Various Internet of Things;
b. human-computer interaction;
c. brain-computer interaction.

3.3.6 Information encryption chip, all kinds of information encryption and security.

4   High-Quality and Efficient Communication is the Key to Expanding Knowledge and Improving Skills

It is very important to improve the skills of the disabled themselves. Disabled people have obvious competitive disadvantages and unique development advantages.

At the same time, we use the idea of overall planning to carry out the popularization of Yi Wen project. In order to enable the disabled to study, work and communicate normally, the patentee has developed a series of products for the disabled based on Yi Wen, which have the characteristics of complete functions, convenience and universality. The barrier-free communication between deaf-mute persons and deaf-mute persons can be realized through computers or mobile phones, which can greatly accelerate the ability of disabled people to acquire knowledge and improve their skills, and fundamentally solve the world problems of communication between disabled and normal persons and between disabled people. This product technology keeps unique, the actual operation proves simple and easy to learn, and it is quick on the basis of zero and can be upgraded and iterated.

It is easy to master this technology. Blind and deaf can communicate normally each other, and disabled people and non-disabled people can communicate without barriers. After the product upgrades, people of all countries and nationalities in the world can communicate together without barriers.

5 Expectation

We will apply fuzzy technology and random methods into extending this work to the fields of fuzzy and random, and achieve barrier free communication in human language [5, 6].

6 Conclusion

At present, we have developed a prototype of Yi Wen communication, and our patent has been reported to the Chinese Patent Office, passed by the preliminary examination and is in the process of publicity [4].

At present, it is the era of internet digitalization. Disabled people (including people with different languages) have learned "Yi Wen", and they can acquire rich cultural knowledge and skills as quickly as college students, and step into the Internet era. Anyone who is interested in the industry elite can accumulate cultural knowledge through Yi Wen, and engage in culture, education, lawyers, doctors, writers and even scientific research, so as to truly realize the great integration of the people around world, Community of Shared Future for Mankind.

Acknowledgements. Recommender: Renjie Hu, Associate Professor, Guangdong University of Foreign Studies in China.

References

1. GB2312-1980, Chinese coded character set for information interchange, Beijing: General Administration of National Standards of China (1980)
2. Edited by Cao Bingyuan. Economic Mathematics Tutorial (Part 1), Tianjin Science & Technology Translation & Publishing Corp, Tianjin (1994)

3. Cao, B.: Economic Mathematics Tutorial (Part 2), Linear Programming and Fuzzy Mathematics, Tianjin Science & Technology Translation & Publishing Corp, Tianjin (1994)
4. Chen, Y.: A barrier-free language communication editing method, China PatentNo.: 202111300378.X, China National Intellectual Property Administration, China, Notice of Qualified Preliminary Examination of Invention Patent Application, November 29, 2021
5. Cao, B.: Applied Probability Statistics Tutorial. Science Press, Beijing (2015)
6. Nasseri, S.H., Ebrahimnejad, A., Cao, B.-Y.: Fuzzy Linear Programming. Springer (2019)

Research on the Recommendation Method of Postgraduate Supervisor Based on Natural Intelligence Information Integration

Xiao He[1], Tai-Fu Li[2(✉)], Yu-Yan Li[1], Shao-Lin Zhang[1], Fu-Hong Qing[3], and Qiao Zeng[3]

[1] School of Electrical Engineering, Chongqing University of Science and Technology, Chongqing, China
{2021204040,2020204029}@cqust.edu.cn, liyuyan_mail@163.com
[2] School of Innovation and Entrepreneurship, Chongqing University of Science and Technology, Chongqing, China
litaifumail@qq.com
[3] School of Safety Engineering, Chongqing University of Science and Technology, Chongqing, China

Abstract. In response to the difficulty of choosing a supervisor for prospective postgraduate, this research proposes a method for recommending postgraduate supervisors. The method starts with prospective postgraduate describing their good wishes for selecting a supervisor, initially drawing up the candidate supervisors of the college they apply to, and then using the crowd familiarity level with the candidate supervisors as natural intelligence perceptron, let each natural intelligence perceptron independently rank the candidate supervisors' recommendations, and through the information integration algorithm, obtain the consensus ranking of the candidate supervisors, and prospective postgraduate can select supervisors based on the consensus ranking. To encourage positiveness and professionalism among assessors, the method draws on positive-sum game mechanisms. To encourage evaluators' positiveness and professionalism, this method draws on the positive sum game mechanism to design the recommendation contribution level algorithm, which quantifies the assessor's contribution to the recommendation assessment. And pay the assessor according to the level of contribution, allowing the assessor to be paid negatively. The research also designed an effectiveness feedback algorithm, which assesses the actual effectiveness of prospective postgraduate in selecting a supervisor by the recommendation results, and secondary distribution of the reward to the assessors based on the effectiveness feedback. This research demonstrates the calculation process of the postgraduate supervisor recommendation method using instance.

Keywords: Postgraduate Supervisor Recommendation · Natural Intelligence Information Integration · Positive-Sum Game

B.-Y. Cao et al. (Eds.): ICFIE 2022, LNDECT 207, pp. 300–318, 2024.
https://doi.org/10.1007/978-981-97-2891-6_23

1 Introduction

Since the 21st century, China's postgraduate education has developed rapidly and the scale of postgraduate has been expanding. At present, China's postgraduate training mainly adopts the supervisor system. The supervisor is the first responsible person for postgraduate training, who is responsible for guiding postgraduate on the front of the academic discipline, teaching scientific research methods, and teaching academic standards. Prospective postgraduates select the appropriate supervisor for them, the supervisor can play an important role in postgraduates' studies and life, such as the training of creative ability, the role of exemplary ideological and moral character, the solution of psychological problems, and the guidance of future development direction. However, some prospective postgraduates fail to select an appropriate supervisor for themselves, resulting in strained relations between supervisors and postgraduates, and even extreme cases of supervisors-postgraduates conflict. An important reason for this extreme situation is that when selecting a supervisor, prospective postgraduate lack access to information about the supervisor's personality, research style, and communication style, so it is difficult to judge whether the supervisor is suitable for them, and they choose the "mistake" supervisor with insufficient information, which has laid a hidden danger for future conflicts between supervisors and postgraduates [1]. Therefore, the author of this reteaches hopes to devise a reasonable method to help all prospective postgraduate to choose the appropriate supervisor for themselves.

Many scholars in China have also conducted many relevant kinds of research on the method of selecting postgraduate supervisors. Xu et al. used a system of indicators to derive the satisfaction of postgraduate with each supervisor and the supervisor's satisfaction level with each postgraduate and used the multiplied values of the satisfaction level matrix to compare postgraduate with their supervisors [2]. Wang et al. used the hierarchical analysis method to analyze the matching between postgraduate and their supervisors [3]. However, the method adopted by Xu and Wang has a stronger problem of subjectivity in the assignment of weights to the indicator system and is more complicated to calculate. Xiang et al. address the matching problem of mutual selection between prospective postgraduate and supervisors by using the Gale-Shapley algorithm to match prospective postgraduate and supervisors. However, the method relies more on the information symmetry matching between the two parties, which requires information completeness and is more difficult to implement in reality [4].

According to some real problems in the above methods, this research designs a postgraduate supervisor recommendation method, which invites people familiar with professional supervisors to assess supervisor recommendations for prospective postgraduate. And the group consensus is obtained through a natural intelligent information integration algorithm to find the most appropriate supervisor for prospective postgraduate. The positive-sum game algorithm is also designed to enhance the motivation of the assessor and ensure the fairness of the assessment.

2   Design of the Postgraduate Supervisor Recommendation Method

To improve the accuracy of the supervisor recommendation results, the prospective postgraduate is required to provide an option list of interested supervisors and information about their requirements, such as personality characteristics, academic goals, and interested research directions. Then invite groups of people who are familiar with the professional supervisor to assess the alternative supervisor provided by the graduate student. The people who are familiar with the professional supervisor are seniors of the same school and the same specialty or colleagues and friends of the supervisor.

Postgraduate supervisor recommendation assessment often has problems such as uncertainty and complexity. Facing these problems, using the common linear weighting integration method of integration to obtain group consensus is no longer well adapted to the assessment of complicated objects. Assessors have their own set of judgment indicator systems when recommending postgraduate supervisors, and these indicator systems may have common indicators and individual indicators, and the weight distribution of these indicators is different for different assessors, so a flexible indicator system should be established, and the weight of indicators be established by the assessors themselves. The use of flexible indicators can provide flexible assessment ways for the groups of people involved in the assessment. Therefore, this research replaces the complex artificial intelligence assessment algorithm with a natural person assessor, and also takes into account the uncertainty of natural person assessment, and also forms the assessment method of natural intelligence information integration by integrating the assessment information of the group of natural persons.

Considering the cognitive differences of different assessors, the recommended method for postgraduate supervisors proposed in this research should be able to integrate the group cognition of the assessors, as well as to assess the responsibility and long-term vision of the assessors to participate in the assessment, finally obtaining an objective and accurate recommendation ranking of postgraduate supervisors, which is a topic worthy of research.

For this purpose, three innovative improvements were made in this research: (1) To offset the complexity of postgraduate supervisor recommendation assessment, and to filter out the uncertainty of the level of responsibility and judgment of the assessor group, the Recursion Confidence Level Probability Density Function Weighted Average Algorithm (RCPDFWA) was proposed to obtain each postgraduate supervisor recommendation ranking [5–8, 10]. (2) To prevent arbitrary assessments by assessors and to encourage responsible assessors to participate in assessments, this research proposes a group positive-sum game reward strategy that quantifies the assessors' contribution level and links the level of the contribution to the assessment rewards, as well as allowing some assessors to have negative assessment rewards [9]. (3) To compensate for the consensus mistake of most assessors and to encourage assessors to assess conscientiously and responsibly, prospective postgraduate who finally select a supervisor in compliance with the assessment results provide feedback on their satisfaction level with the chosen supervisor, based on which the assessment compensation of the assessors is secondarily distributed [11, 12].

3 Information Integration Algorithms Based on the RCPDFWA Operator

This research views the postgraduate supervisor recommendation assessment process as a complicated information decision-making process, which requires full information disclosure as a prerequisite for the design of the flexibility indicator system and the display of prospective postgraduate' own needs. At the same time, the accuracy of the postgraduate supervisor recommendation results is greatly influenced by the differences in the assessors' perceptions of the assessed supervisors. Therefore, this research added the assessor's confidence level measure to the assessment. And the weight values that are weighted by the probability density function combined with the assessor's offer of confidence level are used to integrate the shares of their ratings and to calculate the consensus ranking.

Based on the above idea, this research uses the Recursion Confidence Level Probability Density Function Weighted Average (RCPDFWA) operator to achieve the complicated information integration of group assessment results. Assessors are required to provide a ranking of the recommendation level of each supervisor, and also provide a confidence level in the supervisor assessed. The above process is done independently by those who participate in the assessment of postgraduate supervisors' recommendations, and the number of people involved in each assessment varies randomly from 5 or more people. The CPDFWA operator is first used to obtain the ranking of the assessed supervisor through the integrated weighting process of the assessment rankings provided by each participant. Then, the ranking of the assessed supervisors should be iterated until the final ranking is reached.

In this algorithm the number of the assessor is assigned as j, j = 1, 2,…, n; the number of the supervisor is assigned as i, i = 1, 2,…, m. The equation of the information integration algorithm based on the CPDFWA operator is as follows.

$$\mu_i = \frac{\sum_{j=1}^{n} X_{ij}}{n} \tag{1}$$

$$\sigma_i = \sqrt{\frac{\sum_{j=1}^{n} \left(X_{ij} - \mu_i\right)^2}{n}} \tag{2}$$

$$\omega_{ij} = \frac{e^{-\left[\frac{(x_{ij}-\mu_i)^2}{2\sigma_i^2}\right]}}{\sum_{j=1}^{n} e^{-\left[\frac{(x_{ij}-\mu_i)^2}{2\sigma_i^2}\right]}} \tag{3}$$

$$I'_{ij} = \frac{I_{ij}}{\sum_{j=1}^{n} I_{ij}} \tag{4}$$

$$S_i = CPDFWA_i = \sum_{j=1}^{n} n X_{ij} I'_{ij} \omega_{ij} \tag{5}$$

The meanings of the symbols in the above equation are listed below (see **Table 1**).

Table 1. Symbol meaning of information integration algorithm equation based on CPDFWA operator

Symbols	Symbol Meanings
n	The number of assessors involved in the assessment
m	The number of assessed supervisors
X_{ij}	The assessment ranking of the supervisor numbered i by the assessor numbered j
μ_i	The mathematical expectation of the assessment ranking obtained by the supervisor numbered i
σ_i	The standard deviation of the ranking obtained by the supervisor numbered i
ω_{ij}	The normalized value of the weight assigned to the assessment of the supervisor numbered i by the assessor numbered j
I_{ij}	The quantitative value of the confidence level provided by the assessor numbered j to the supervisor numbered i
I'_{ij}	The normalized result of the confidence level provided by the assessor numbered j to the supervisor numbered i
S_i	The score obtained by the supervisor numbered i

Table 2. Table of transformed values of the assessor's confidence level

Transformed Values of the Assessor's Confidence Level					
Confidence Level	Very Confident	Confident	A little Confidence	Not Confident	Very Unconfident
I_{ij}	0.9	0.7	0.5	0.3	0.1

The confidence level provided by the assessors was translated into corresponding numerical values, and the table of values is as follows (see **Table 2**).

After calculating the supervisor recommendation score S_i, the mathematical expectation μ_i is replaced by S_i, and the S_i value is repeated until the error of the S_i value calculated twice adjacent is less than 1%, then the latest calculated value is assigned as S_i. . The final ranking of all supervisors is then obtained (the smaller the score, the higher the ranking).

Through combining the collective wisdom of assessors, we have made a quantitative ranking of postgraduate supervisors' recommendations, which can be used as a reference for the selection of supervisors by prospective postgraduate.

4 Group Positive Sum Game Reward Algorithm

The judgment of the assessor's consensus on the supervisor is integrated through the RCPDFWA operator, it is necessary to use incentives for the assessors to make the assessment results more effective in satisfying the prospective postgraduate's original intentions. This postgraduate supervisor recommendation method rewards assessors according to their contribution level and encourages assessors with good taste and responsibility to participate actively in postgraduate supervisor recommendation assessment. Meanwhile, the group positive sum game reward algorithm contains: (1) The assessment contribution level calculation algorithm. (2) The effectiveness feedback algorithm.

4.1 The Assessment Contribution Level Calculation Algorithm

In the current assessor reward model, assessors are often rewarded in the same amount. This recommendation contribution level calculation method quantifies each assessor's contribution level to the postgraduate supervisor's recommendation assessment and distributes rewards according to the contribution level of each assessor. To further strengthen the motivational effect on the assessors and reduce undesirable phenomena such as bad judgment, low responsibility, and intentionally unfair assessment that affect assessment, this algorithm specifically allows the contribution level of the assessors to be negative.

The assessment contribution level algorithm draws on the game thinking of the prisoner's dilemma, allowing each assessor to play against other unknown assessors, encouraging each assessor to assess the group consensus as the result of its assessment, more rewards are given to assessors close to the consensus result and fewer rewards (or penalizing) are given to assessors far from the consensus result.

The data source for the assessment of the contribution level calculation is the data generated during the integration of information by the RCPDFWA operator. The contribution level is calculated by the following equation.

$$\mu_i' = S_i \tag{6}$$

$$\sigma_i' = \sqrt{\frac{\sum_{j=1}^{n}(X_{ij} - \mu_i')^2}{n}} \tag{7}$$

$$W_{ij} = \frac{1}{\sqrt{2\pi}\sigma_i'} \times e^{-\frac{(X_{ij}-\mu_i')^2}{2\sigma_i'^2}} \tag{8}$$

$$W_{ij}' = \frac{W_{ij}}{\sum_{j=1}^{n} W_{ij}} \tag{9}$$

$$W_{ij}'' = [W_{ij}' + (W_{ij}' - \frac{1}{n})] \cdot l_{ij} \tag{10}$$

$$W_j'' = \frac{\sum_{i=1}^{m} W_{ij}''}{\sum_{j=1}^{n}\sum_{i=1}^{m} W_{ij}''} \tag{11}$$

Table 3. Assessment of the contribution level algorithm equation symbol meaning

Symbols	Symbol Meanings
μ_i'	The mathematical expectation of the ranking obtained by the supervisor numbered i
σ_i'	The standard deviation of the ranking obtained by the supervisor numbered i
W_{ij}	The value of the probability density of the assessment of the supervisor numbered i by the assessor numbered j
W_{ij}'	Normalization of W_{ij}
W_{ij}''	The contribution level of the assessor numbered j to the assessment of the supervisor numbered i
W_j''	The contribution level of the total assessment of all supervisors by the assessor numbered j
N	Budgeted contribution level rewards for postgraduate supervisor assessment
N_j	The contribution level reward assigned to the assessor number j

$$N_j = W_j'' \times N \tag{12}$$

The meanings of the symbols in the above equation are listed below (see **Table 3**).

The assessment contribution level algorithm calculates the contribution level of each assessor and initially assigns rewards according to the contribution level ratio. The algorithm effectively quantifies the assessor's contribution level to the assessment process, which is used as the basis for rewarding, to be as fair and just as possible, and to encourage the participation of assessors with good judgment and responsibility.

4.2 The Effectiveness Feedback Algorithm

The core idea of the algorithm for assessing the contribution level is to encourage the assessors to follow the consensus, but sometimes the consensus of the group may fail, and the truth may be held by the minority. The original intent of this research to propose the postgraduate supervisor recommendation method is to help prospective postgraduate find a suitable supervisor. Whether the prospective postgraduate is satisfied with the most recommended supervisor or not is based on the final feedback from the prospective postgraduate after the supervisor and prospective postgraduate have spent some time together.

To solve the problem of possible mistakes in group consensus, this research then adds the design of an effectiveness feedback algorithm that rewards assessors who are more judgmental and can stick to their judgments. The idea of the algorithm is originated from stock investment: when the assessor selects his recommendation ranking for a supervisor, it is equivalent to buying and selling stocks. If the assessor numbered j ranks the supervisor numbered i higher than the consensus result, it is equivalent to being bullish and buying the stock; if the assessor numbered j ranks the supervisor numbered i lower than the consensus result, it is equivalent to being bearish and selling the stock.

If the number-one ranked supervisor is finally recommended to be the supervisor of the prospective postgraduate, after a period of time, the postgraduate gives five kinds of feedback: "very satisfied", "satisfied", "barely satisfied", "dissatisfied" and "very dissatisfied".

(1) Assuming that the feedback is "very satisfied", the assessor whose rank is higher than the consensus result will be rewarded, and the higher the rank, the more rewarding it will be.
(2) Assuming that the feedback is "very satisfied", the assessor whose rank was lower than the consensus result will be penalized, and the lower the rank, the more penalty it will be.
(3) Assuming that the feedback is "very dissatisfied", the assessor whose rank is lower than the consensus result will be rewarded, and the lower the rank, the more rewarding it will be.
(4) Assuming that the feedback is "very dissatisfied", the assessor whose rank was higher than the consensus result will be penalized, and the higher the rank, the more penalty it will be.

Other cases can be analogous from this, and rewards and penalties can be considered as appropriate. The equation for calculating the effectiveness feedback algorithm is as follows.

$$S_f = \min_{1 \leq i \leq m} \{S_i\} \tag{13}$$

$$T_j = X_{fj} - S_f \tag{14}$$

$$T^+ = \{T_j > 0\} \tag{15}$$

$$T^- = \{T_j < 0\} \tag{16}$$

$$N' = N \times B \tag{17}$$

$$\text{If } T_j \geq 0, \ R_j = -\frac{T_j}{\sum T^+} \times N' \tag{18}$$

$$\text{If } T_j < 0, \ R_j = \frac{T_j}{\sum T^-} \times N' \tag{19}$$

$$N'_j = N_j + R_j \tag{20}$$

The meanings of the symbols in the above equation are listed below (see Table 4).

where B denotes the quantitative satisfaction level of the prospective postgraduate feedback results, and the satisfaction level is translated into a table of corresponding values as follows (see Table 5).

If the feedback result of the prospective postgraduate is a negative assessment, the contribution level reward to be reassigned should be negative numbers, it is considered

Table 4. Effectiveness feedback algorithm equation symbol meaning

Symbols	Symbol Meanings
f	Assessment of the No. 1 ranked supervisor's number
S_f	The score obtained by the assessment of the No. 1 ranked supervisor
X_{fj}	The ranking given by the assessor numbered j to the actual No. 1 ranked supervisor
T_j	The error of the assessment of the actual No. 1 ranked supervisor by the assessor numbered j
$\sum T^+$	The set of all T_j greater than 0
$\sum T^-$	The set of all T_j less than 0
B	The value corresponding to the satisfaction level of the prospective postgraduate feedback results
N'	The contribution level reward that needs to be reassigned after the feedback
R_j	The amount of change in the reward for the amount of contribution of the assessor numbered j after the effect feedback
N'_j	The final contribution level reward obtained by the assessor numbered j in this postgraduate supervisor assessment

Table 5. Corresponding value table of satisfaction level of feedback results

Satisfaction level Quantification Table				
Satisfaction Level	Very Satisfied	Satisfied	Barely Satisfied	Dissatisfied	Very Dissatisfied
B	0.5	0.3	0.1	-0.3	-0.5

as a penalty for the assessor who provided the mistake assessment, so the satisfaction level of the feedback result corresponds to a negative value at this time. The source of the specific values corresponding to the satisfaction level is empirical values obtained by the authors through many experiments.

The effectiveness feedback algorithm calculates the change in reward for each assessor in the secondary distributions, The effectiveness feedback algorithm compensates for the problems caused by the majority of assessors' assessment mistakes, while increasing the motivation of the assessor groups with good taste.

5 Application

During the supervisor selection period of a prospective postgraduate at a university's College of Computer Science and Technology, a prospective postgraduate was provided with a method of recommending a postgraduate supervisor, with the intention of finding the most appropriate supervisor for the prospective postgraduate. The prospective postgraduate first selected 7 supervisors as candidate supervisors in the College of Computer Science and Technology and provided information about his character and his good wishes. Then six seniors of the college were invited to reference this information and to assess and rank the degree of recommendation of the prospective postgraduate with the seven candidate supervisors. The six assessors provided a postgraduate supervisor recommendation assessment ranking table and a supervisor familiarity level table after getting detailed information about the prospective postgraduate. The table is as follows (see Table 6 and Table 7).

Table 6. Ranking of candidate supervisor assessment

	Supervisor 1	Supervisor 2	Supervisor 3	Supervisor 4	Supervisor 5	Supervisor 6	Supervisor 7
Assessor 1	3	6	7	2	1	5	4
Assessor 2	1	6	7	3	4	5	2
Assessor 3	4	5	6	1	3	7	2
Assessor 4	3	5	7	1	2	6	4
Assessor 5	5	4	6	2	1	7	3
Assessor 6	1	6	7	2	3	4	5

Table 7. Level of familiarity of the assessor with the candidate supervisor

	Supervisor 1	Supervisor 2	Supervisor 3	Supervisor 4	Supervisor 5	Supervisor 6	Supervisor 7
Assessor 1	0.7	0.5	0.7	0.9	0.3	0.5	0.7
Assessor 2	0.9	0.3	0.5	0.7	0.9	0.7	0.9
Assessor 3	0.5	0.7	0.9	0.5	0.3	0.5	0.9
Assessor 4	0.3	0.5	0.3	0.9	0.5	0.5	0.7
Assessor 5	0.5	0.3	0.5	0.5	0.9	0.7	0.7
Assessor 6	0.7	0.7	0.7	0.3	0.5	0.5	0.5

Step 1: The assessment ranking table generates the ranking matrix X and the supervisor familiarity level table generates the familiarity level matrix I. Substitute the ranking matrix X into Eq. (1), where n is the number of assessors participating in the postgraduate supervisor recommendation assessment, n = 6. Obtain the expectation of each supervisor's assessment score, $\mu_i (i = 1, 2, 3, 4, 5, 6, 7)$.

$$\mu = \mu_i (i = 1, 2, 3, 4, 5, 6, 7)$$

$$= [2.83, 5.33, 6.67, 1.83, 2.33, 5.67, 3.33]$$

Step 2: Substituting $\mu_i(i = 1, 2, 3, 4, 5, 6, 7)$ and the ranking matrix X into Eq. (2), the standard deviation $\sigma_i(i = 1, 2, 3, 4, 5, 6, 7)$ of the ranking data obtained for each supervisor is obtained.

$$\sigma = \sigma_i(i = 1, 2, 3, 4, 5, 6, 7)$$

$$= [1.46, 0.75, 0.47, 0.69, 1.11, 1.11, 1.11]$$

Step 3: Substituting X, μ, σ, into Eq. (3), the weights $\omega_{ij}(i = 1, 2, 3, 4, 5, 6, 7; j = 1, 2, 3, 4, 5, 6)$ accounted for by the assessment of the supervisor numbered i by the assessor numbered j are calculated, and the weight matrix obtained is table below (see Table 8).

Table 8. Weight of assessors' assessment of candidate supervisor

	Supervisor 1	Supervisor 2	Supervisor 3	Supervisor 4	Supervisor 5	Supervisor 6	Supervisor 7
Assessor 1	0.2509	0.1666	0.2022	0.2363	0.1236	0.2132	0.2132
Assessor 2	0.1151	0.1666	0.2022	0.0576	0.0821	0.2132	0.1236
Assessor 3	0.1837	0.2249	0.0955	0.1167	0.2132	0.1236	0.1236
Assessor 4	0.2509	0.2249	0.2022	0.1167	0.2444	0.2444	0.2132
Assessor 5	0.0843	0.0502	0.0955	0.2363	0.1236	0.1236	0.2444
Assessor 6	0.1151	0.1666	0.2022	0.2363	0.2132	0.0821	0.0821

Step 4: The familiarity matrix I was normalized to obtain the normalized familiarity matrix $\hat{I}_{ij}(i = 1, 2, 3, 4, 5, 6, 7; j = 1, 2, 3, 4, 5, 6)$ provided by the assessor number j to the supervisor number i, according to Eq. (4). The normalized familiarity matrix $\hat{I}$ is presented in the following table (see Table 9).

Table 9. Normalized supervisor familiarity level

	Supervisor 1	Supervisor 2	Supervisor 3	Supervisor 4	Supervisor 5	Supervisor 6	Supervisor 7
Assessor 1	0.1944	0.1667	0.1944	0.2368	0.0882	0.1471	0.1591
Assessor 2	0.2500	0.1000	0.1389	0.1842	0.2647	0.2059	0.2045
Assessor 3	0.1389	0.2333	0.2500	0.1316	0.0882	0.1471	0.2045
Assessor 4	0.0833	0.1667	0.0833	0.2368	0.1471	0.1471	0.1591

(continued)

Table 9. (*continued*)

	Supervisor 1	Supervisor 2	Supervisor 3	Supervisor 4	Supervisor 5	Supervisor 6	Supervisor 7
Assessor 5	0.1389	0.1000	0.1389	0.1316	0.2647	0.2059	0.1591
Assessor 6	0.1944	0.2333	0.1944	0.0789	0.1471	0.1471	0.1136

Step 5: The final assessment score $S_i (i = 1, 2, 3, 4, 5, 6, 7)$ for each supervisor was calculated according to Eq. (5). Each value in the equation is given before this step, and the recommended scores of the supervisor participating in this assessment can be obtained directly.

$$S = S_i (i = 1, 2, 3, 4, 5, 6, 7)$$

$$= [2.52, 5.82, 6.53, 1.72, 2.12, 5.67, 3.21]$$

Step 6: Replacing the expectation μ in step 1 with S, steps 1 to 5 were repeated and the cycle was repeated until the error of S values calculated twice adjacent to each other was less than 1%. The final scores of the selected supervisors were calculated after a total of four iterations, and the error of the S values of the two adjacent calculations was kept within 1%. The table of the results of the iterative calculation is as follows (see Table 10).

Table 10. Iterative values of candidate supervisor scores

	Supervisor 1	Supervisor 2	Supervisor 3	Supervisor 4	Supervisor 5	Supervisor 6	Supervisor 7
Performing 1 iteration	2.5250	5.8193	6.5281	1.7177	2.1174	5.6726	3.2143
Performing 2 iteration	2.8850	5.0726	6.6566	2.0078	2.9457	5.2009	3.3649
Performing 3 iteration	2.8956	5.0746	6.6578	2.0033	2.9469	5.1961	3.3710

The errors of the supervisor scores for 2 iterations and the supervisor scores for 3 iterations are shown in the following table (see Table 11).

From the above table, it can be seen that the error of the S values of two adjacent calculations is controlled within 1%, so the value of 3 iterations is the final score of all supervisors. The line chart for the change in supervisors' scores is shown below (see Fig. 1).

Table 11. The errors of the supervisor scores for 2 iterations and the supervisor scores for 3 iterations

	Supervisor 1	Supervisor 2	Supervisor 3	Supervisor 4	Supervisor 5	Supervisor 6	Supervisor 7
Error	−0.0037	−0.0004	−0.0002	0.0022	−0.0004	0.0009	−0.0018

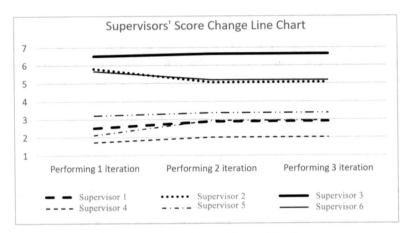

Fig. 1. Supervisors' Score Change Line Chart

The final scores were ranked and the prospective postgraduate was ranked in terms of fit to the supervisor in the following table (see Table 12).

Table 12. Final score and ranking of candidate supervisors

	Final score	Ranking
Supervisor 4	2.0033	1st
Supervisor 1	2.8956	2st
Supervisor 5	2.9469	3st
Supervisor 7	3.3710	4st
Supervisor 2	5.0746	5st
Supervisor 6	5.1961	6st
Supervisor 3	6.6578	7st

According to the results obtained from the RCPDFWA algorithm, the supervisor number 4 had the highest recommendation.

Step 7: The total amount of reward offered to the assessors who participated in the postgraduate supervisor assessment was set at $200. The reward was also assigned to the six assessors, and the algorithm used was the assessment contribution level calculation algorithm. From Eq. (6), the value of the mathematical expectation $\mu\prime$ of the assessed ranking obtained by the supervisor numbered i is equal to the final assessment score $S_i (i = 1, 2, 3, 4, 5, 6, 7)$ for each supervisor.

$$\mu\prime = \mu_i\prime (i = 1, 2, 3, 4, 5, 6, 7)$$

$$= (2.90, 5.07, 6.66, 2.00, 2.95, 5.20, 3.37)$$

Because n, X, $\mu\prime$ are known, the standard deviation $\sigma\prime$ of the assessed ranking obtained by the supervisor with number i can be obtained from Eq. (7).

$$\sigma\prime = \sigma_i\prime (i = 1, 2, 3, 4, 5, 6, 7)$$

$$= [1.46, 0.79, 0.47, 0.71, 1.26, 1.20, 1.11]$$

Step 8: The normalized probability density matrix $W\prime$ for each postgraduate supervisor's recommendation assessment was calculated. Based on Eqs. (8) and (9), the normalized probability density matrix $W'_{ij} (i = 1, 2, 3, 4, 5, 6, 7; j = 1, 2, 3, 4, 5, 6)$ of the postgraduate supervisor's recommendation evaluation is shown in the following table (see Table 13).

Table 13. Normalized probability density matrix for recommendation assessment of candidate supervisors

	Supervisor 1	Supervisor 2	Supervisor 3	Supervisor 4	Supervisor 5	Supervisor 6	Supervisor 7
Assessor 1	0.2514	0.1291	0.2007	0.2437	0.0751	0.2448	0.2174
Assessor 2	0.1090	0.1291	0.2007	0.0904	0.1736	0.2448	0.1186
Assessor 3	0.1896	0.2556	0.0987	0.0893	0.2454	0.0804	0.1186
Assessor 4	0.2514	0.2556	0.2007	0.0893	0.1855	0.1984	0.2174
Assessor 5	0.0897	0.1016	0.0987	0.2437	0.0751	0.0804	0.2416
Assessor 6	0.1090	0.1291	0.2007	0.2437	0.2454	0.1512	0.0864

Step 9: Based on Eq. (10), the assessment contribution level $W''_{ij} (i = 1, 2, 3, 4, 5, 6, 7; j = 1, 2, 3, 4, 5, 6)$ of the assessor numbered j to the supervisor numbered i was calculated. Where n is the number of assessors equal to 6 and I is the familiarity matrix. The matrix W'' of the contribution level of each assessor to the assessment of the supervisor is as follows (see Table 14).

Each assessor's total contribution level $W''_j (j = 1, 2, 3, 4, 5, 6)$ is then obtained based on W'', based on Eq. (11).

$$W'' = W''_j (j = 1, 2, 3, 4, 5, 6)$$

Table 14. Contribution level of the assessors to the assessment of the supervisors

	Supervisor 1	Supervisor 2	Supervisor 3	Supervisor 4	Supervisor 5	Supervisor 6	Supervisor 7
Assessor 1	0.2353	0.0457	0.1643	0.2886	-0.0050	0.1615	0.1877
Assessor 2	0.0461	0.0274	0.1173	0.0099	0.1625	0.2261	0.0634
Assessor 3	0.1063	0.2412	0.0276	0.0059	0.0972	-0.0029	0.0634
Assessor 4	0.1008	0.1723	0.0704	0.0107	0.1022	0.1150	0.1877
Assessor 5	0.0063	0.0109	0.0153	0.1604	-0.0149	-0.0041	0.2216
Assessor 6	0.0359	0.0640	0.1643	0.0962	0.1620	0.0678	0.0031

$$= [0.27, 0.16, 0.13, 0.19, 0.10, 0.15]^T$$

Step 10: After obtaining the total contribution level of each assessor, the reward is assigned in proportion to the contribution level. The assessment reward for each assessor can be obtained according to Eq. (12). The assessors were rewarded for their assessments as follows (see Table 15).

Table 15. Assessor reward

Assessor reward	
Assessor 1	$53.67
Assessor 2	$32.50
Assessor 3	$26.82
Assessor 4	$37.79
Assessor 5	$19.69
Assessor 6	$29.54

According to the assessment contribution level algorithm, we calculate the contribution level of each assessor in this assessment, and use it as the basis to assign the assessment reward. While achieving the differentiation of assessment rewards, it also shows the fairness of the assignment method, which can better inspire the assessors to improve their ability to judge and assess carefully.

Step 11: If the prospective postgraduate finally selects the supervisor with the highest recommendation, he will provide satisfaction feedback about the supervisor after a period of time. For example, the feedback result is "Satisfied". The assessment feedback algorithm provides a secondary distribution of the assessment reward to the participants of the assessment.

The assessment feedback algorithm calculates the difference $T_j (j = 1, 2, 3, 4, 5, 6)$ between the assessor's ranking of the supervisor numbered 4 and the actual ranking of the supervisor numbered 4, based on Eq. (14). We know that $S_f = 2.0033$, where $X_j(j = 1, 2, 3, 4, 5, 6)$ is the ranking provided by each assessor for the supervisor numbered 4.

$$X_j(j = 1, 2, 3, 4, 5, 6) = [2, 3, 1, 1, 2, 2]^T$$

$$T = T_j(j = 1, 2, 3, 4, 5, 6)$$

$$= [0.00, 1.00, -1.00, -1.00, 0.00, 0.00]^T$$

Create a vector T^+, T^+ as the set of elements in T greater than or equal to 0; create a vector T^-, T^- as the set of elements in T less than 0. And sum the elements in T^+ and T^- individually to obtain T^+ and T^-.

The result of the calculation is $\sum T^+ = 1.00$; $\sum T^- = -2.02$.

Step 12: According to Eq. (17), Eq. (18) and Eq. (19) calculate the amount of change $R_j(j = 1, 2, 3, 4, 5, 6)$ in reward for the assessor numbered j after the secondary assignment. Where N' is the assessment reward for participating in the secondary assignment, N is the reward for this assessment set to \$200, and B is the conversion value of satisfaction level. Calculated $N'=60$. $R_j(j = 1, 2, 3, 4, 5, 6)$ is the amount of change in reward for assessor number j after the secondary assignments.

$$R = R_j(j = 1, 2, 3, 4, 5, 6)$$

$$= [0, -60, 30, 30, 0, 0]^T$$

Based on Eq. (20), the final reward obtained by each assessor in the assessment of that postgraduate supervisor after passing the satisfaction level feedback from the prospective postgraduate was calculated. The final reward table for each assessor's assessment is as follows (see Table 16).

Table 16. Assessor reward after secondary feedback

	Contribution reward before secondary feedback	Contribution reward change amount	Final contribution reward
Assessor 1	$53.67	$0.00	$53.67
Assessor 2	$32.50	-$60.00	-$27.50
Assessor 3	$26.82	$30.00	$56.82
Assessor 4	$37.79	$30.00	$67.79
Assessor 5	$19.69	$0.00	$19.69
Assessor 6	$29.54	$0.00	$29.54

Through the above example, the quantitative algorithm of assessment contribution is used, making the assessment reward linked to the contribution level, which reasonably makes the differentiated assessment reward; apply assessment feedback algorithms to make the assessor responsible for the forward value of the assessment results, let the feedback results of prospective postgraduate adjust the reward of assessor. Assessors who are discerning and insist on careful assessment are encouraged to participate in the supervisor recommendation assessment.

In order to demonstrate the superiority of the supervisor selection method used in this paper over the traditional method, the other 6 prospective postgraduates from the same college were allowed to select their supervisor using the method described in this paper, making a total of 7 prospective postgraduates using the method in this paper as a group. 9 prospective postgraduates who used the traditional method to select the supervisor were made to serve as a comparison group. After a period of time, by asking these postgraduates about their satisfaction level with their current supervisor, it is clear that the average satisfaction level of the group of postgraduates who used the method described in this paper was higher than that of the group of postgraduates who used the traditional method. The experiment proves the superiority of the proposed method in this paper. The specific experimental data are listed in the following table (see Table 17).

Table 17. Feedback form for postgraduates' satisfaction level with their supervisors

Feedback form for postgraduates' satisfaction level with their supervisors			
Postgraduate groups who used the methods in this paper		Postgraduate group who used traditional methods	
Evaluators	Satisfaction level score	Evaluators	Satisfaction level score
Postgraduate 1	3	Postgraduate 8	4
Postgraduate 2	5	Postgraduate 9	5
Postgraduate 3	5	Postgraduate 10	4
Postgraduate 4	4	Postgraduate 11	3
Postgraduate 5	5	Postgraduate 12	3
Postgraduate 6	5	Postgraduate 13	4
Postgraduate 7	4	Postgraduate 14	1
		Postgraduate 15	5
		Postgraduate 16	2
Average score	4.43	Average score	3.44

Note: The satisfaction score range is 0 to 5, 0 is considered very unsatisfactory, 5 is considered very satisfactory, and the postgraduate feedback score is a whole number

6   Conclusion

The natural intelligence information integration-based postgraduate supervisor recommendation method proposed in this research has two major advantages over the traditional postgraduate supervisor recommendation method:

(1) Traditional postgraduate supervisor recommendation methods often use linear weighting with fixed weights, but the proportion of weights assigned to indicators tends to be different for each assessor, and the use of a fixed-weight model will affect the full use of the assessor's natural intelligence. The traditional methods tend to assign the same weights to the assessors and do not consider the level differences among the assessors. In this research, the consensus of the group assessor is obtained by using the calculation method of flexible indicators and nonlinear weighting. While fully using the natural intelligence of the assessors, taking into consideration that there are differences in the assessment levels of the assessors and that there are differences in the familiarity of the assessors with different supervisors, the algorithm automatically integrates the assessment weights of the assessors, so that the final supervisor recommendation results can be obtained relatively accurately.

(2) In postgraduate supervisor recommendations, the assessor's positiveness directly affects the accuracy of the recommendation result. However, the traditional method of postgraduate supervisor recommendation ignores the aspect of enhancing the assessor's positiveness or simply calls on the assessor to participate in the assessment seriously. In this research, the postgraduate supervisor recommendation method has specific and actionable methods to enhance assessors' positiveness. By quantifying the assessor's contribution level to the assessment process to not-average assigning the assessor's reward, and setting up the prospective postgraduate assessment feedback method to reassign the assessor's reward. All of the above methods improve the assessor's positiveness and ultimately enhance the accuracy of the postgraduate supervisor's recommendation results.

Acknowledgment. This work was supported by the Research funding project of Chongqing University of Science and Technology (ckrc2019021); Science and Technology Research Program of Chongqing Municipal Education Commission (Grant No: KJQN202001536); Chongqing Natural Science Foundation General Project (cstc2021jcyj-msxmX0833).

Recommender: Associate Professor Yang Jie, Chongqing Industry & Trade Polytechnic in China.

References

1. Xi, C.: Design and implementation of two-way selection management and assisted decision-making system for prospective postgraduate and supervisors. Jiangxi University of Finance and Economics, Jiangxi China (2022)
2. Haohua, X., Zhiming, D., et al.: Optimization model for postgraduate admissions. J. Math. Pract. Theor. **7**, 115–119 (2005)
3. Hongxia, W., Xilin, Z., Liping, Z.: Mathematical modeling of postgraduate admissions problems. J. Taiyuan Univ. Technol. **153**(05), 467–470 (2007)
4. Bing, X. Wenjun, L.: Optimal match between prospective postgraduate and supervisors for two-way selection. Future Develop. **40**(04), 91–94 (2016)
5. Taifu, L., Xinyao, H., Wucan, J., et al.: A confidence-based nonlinear weighted integration operator for a system of assigning performance, CN202010059945.6. 2020–01–19
6. Taifu, L. Xinyao, H. Wucan, J., et al.: A blockchain-based decentralized business performance appraisal method, CN202010059941.8. 2020–01–19

7. Taifu, L., Xinyao, H., Yin, D., et al.: A blockchain-based business performance reward allocation system, CN202010057239.8. 2020–01–19
8. Xinyao, H., Wucan, J., Taifu, L., et al.: A DApp system for performance assignment based on confidence level weighted integration operator, CN202010059956.4. 2020–01–19
9. Taifu, L., Shaolin, Z., Xiao, H., et al.: A method for calculating the contribution level in the assessment process, CN202210142192.4.2020–02–16
10. Taifu, L., Shaolin, Z., Xiao, H., et al.: A blockchain mutual evaluation reward management method, system and storage device thereof, CN202111257503.3.2021–10–27
11. Taifu, L., Xiao, H., Shaolin, Z., Fuhong, Q.: Design of judging system of Internet+ innovation and entrepreneurship competition based on positive-sum game of expert group Operations Research Management and Fuzzy Mathematics, **10**(1), 7–25 (2022)
12. Taifu, L., Shaolin, Z., Xiao, H., et al.: An enterprise performance pay assignment method with mutual shareholding for everyone. CN202111255559.5.3.2021–10–27

The Qudratic Programming
with Max-Min Fuzzy Relation Equations
Constraint

Xue-Gang Zhou[1] and YongBin OuYang[2]([✉])

[1] School of Financial Mathematics and Statistics, Guangdong, University of Finance,
Guangzhou 510521, Guangdong, China
[2] Guangzhou Huali Science and Technology Vocational College, Guangzhou 511325,
Guangdong, China
15011973027@163.com

Abstract. In this paper, a new solution method of quadratic programming with max-min fuzzy relation equation constraint is putting forward. Firstly, the optimal solution to some special fuzzy relation quadratic programming is researched. Secondly, some rules are presented to simplify the original programming. Finally, the new algorithm is given based on the simplified programming and the branch and bound method, the new algorithm does not need to find all feasible minimal solutions. Some numerical examples are given to illustrate the feasibility and effectiveness of the presented new algorithm.

Keywords: Quadratic programming · Fuzzy relation equation · Fuzzy relation quadratic programming · Optimal solution

MSC: 90C70 · 90C20

1 Introduction

Fuzzy relation equation (FRE) plays an important role in theory of fuzzy mathematics and its applications, the solution set of a fuzzy relation equation has been investigated by many researchers [1–10]. The optimization problem with the fuzzy relation equations constraint was first proposed to the literature [11], and since then, it gradually becomes a hot topic in the fuzzy optimization field. In recent years, the optimization problem with the fuzzy relation equations constraint produced several different research directions, including fuzzy relation linear programming [12–16], fuzzy relation multi-objective programming [17,18], fuzzy relation quadratic programming [19,20], fuzzy relation geometric programming [21,22], fuzzy relation nonlinear programming [23,24], and so on. As is well known, Quadratic Programming (QP) can be viewed as a generalization of linear programming problems. It arises from a wide variety of scientific and engineering applications including regression analysis and function approximation [25], signal processing and image restoration [26], Pattern Recognition [27], Inventory management [28], portfolio selection [29], engineering design [30], and so on. In addition, some nonlinear programming problems with high difficulty degrees often can be solved by the sequential quadratic programming (SQP) method. Yang et al. [19] have given an algorithm for solving the global optimal solution of fuzzy relation Quadratic Programming, the main method is to look for the maximum solution and all minimal solutions of fuzzy relation equations by conservative path method, and then some sub-problems of classic quadratic programming with box constraints can be constructed, the global optimal solution of fuzzy relation quadratic programming can be obtained by solving every sub-problems of classic quadratic programming. This method is limited by the size of the fuzzy relation equation. Under normal circumstances, finding all the minimal solutions are a NP-hard problem with large-scale fuzzy relation equations.

We consider the quadratic programming problem with the fuzzy relation equation constraint. This problem can be formulated as follows:

$$\min \ f(x) = \frac{1}{2} x^T Q x + c^T x \tag{1}$$
$$\text{s.t. } A \circ x = b,$$

where $Q = (q_{ij})_{n \times n}$ is a n order real symmetric matrix, $A = (a_{kj})_{m \times n}$ is a $m \times n$ fuzzy matrix, $x = (x_1, \cdots, x_n)^T$, $b = (b_1, \cdots, b_m)^T$, $c = (c_1, \cdots, c_n)^T$, $M = \{1, 2, \cdots, m\}$, $N = \{1, 2, \cdots, n\}$, " $\circ$ " is $\vee - \wedge$ Operator,it is the most widely used one in fuzzy set theory. Without loss of generality,$1 \geqslant b_1 \geqslant b_2 \geqslant \cdots \geqslant b_m > 0$, otherwise the fuzzy relation equation $A \circ x = b$ can be adjusted to satisfy the above condition

Due to the importance of quadratic programming and fuzzy relational equations with the max-min composition in theory and applications, we are motivated to propose a fuzzy relational quadratic programming with the max-min composition. Therefore, it is important to solve the quadratic programming problem with fuzzy relational inequality constraint on the max-min composition. Since the objective function is nonlinear and the feasible domain consisting of a maximum solution and a finite number of minimal solutions is nonconvex, it is a global optimization problem. But it cannot be solved by classical nonlinear optimization algorithms. So, several sufficient conditions for determining the optimal solution to the original plan are given, and some results of reducing the size of the original plan are also given. Using these conditions, some methods to simplify the problem are proposed. Then the simplified problem is transformed into a traditional quadratic programming problem. A global optimization algorithm is proposed based on the branch and bound method, the global optimal solution to fuzzy relation quadratic programming can be obtained by solving some simplifying quadratic programming, the algorithm does not need to solve all minimal solutions of fuzzy relation equations.

This paper is organized as follows: In Sect. 2, the quadratic programming problem with fuzzy relation equation constraint is presented, and some definitions and properties of fuzzy relation equations and their feasible solution sets are formulated. In Sect. 3, some sufficient conditions are proposed to determine the optimal solution to the original programming. Some results are also given to reduce the size of the original programming. In Sect. 4, calculation procedures are suggested to simplifying the reduced original programming, and an algorithm is designed to solve this simplified programming. Some numerical examples are presented to illustrate the effectiveness of the algorithm. Finally, conclusions are presented in Sect. 5.

2   Fuzzy Relation Equation with ∨ − ∧ Operator

In this section, some concepts and properties of fuzzy relation equation will be introduced based on the following form.

$$A \circ x = b, 0 \leqslant x_j \leqslant 1, j \in N. \tag{2}$$

Here, " $\circ$ " represents $\vee - \wedge$ operator[6–8].
   Let $X(A, b)$ is the solution set of fuzzy relation Eq. (2).

Definition 2.1. [9] If there exists a solution to Eq. (2), it is called compatible.
   Suppose $X(A, b) = \{(x_1, x_2, \cdots, x_n)^T \in R^n | A \circ x = b, 0 \leq x_j \leq 1\}$ is a solution set of Eq. (2). We can define $x^1 \leq x^2 \Leftrightarrow x_j^1 \leq x_j^2$ $(1 \leq j \leq n), \forall x^1, x^2 \in X(A, b)$. So, " $\leq$ " is a partial order relation on $X(A, b)$.

Definition 2.2. [11] If $\exists \hat{x} \in X(A, b)$, such that $x \leq \hat{x}$, $\forall x \in X(A, b)$, then $\hat{x}$ is called maximum solution to Eq. (2). If $\exists \check{x} \in X(A, b)$, such that $\check{x} \leq x$, $\forall x \in X(A, b)$, then $\check{x}$ is called a minimum solution to Eq. (2). And if $\exists \check{x} \in X(A, b)$, when $x \leq \check{x}$, then $x = \check{x}$, $\check{x}$ is called a minimal solution to Eq. (2).
   Let

$$\hat{x}_j = \bigwedge_{1 \leq i \leq m} (b_i | b_i < a_{ij}) \; (1 \leq j \leq n), \tag{3}$$

suppose that $\{\wedge \emptyset = 1\}$.
   If $\hat{x} = (\hat{x}_1, \hat{x}_2, \cdots, \hat{x}_n)^T$ is a solution to Eq. (2), we can easily prove that $\hat{x}$ must be a maximal solution to one. For a maximal solution to Eq. (2), we have the following lemma.

Lemma 2.1. [11] $A \circ x = b$ is compatible if and only if there exists a maximal solution $\hat{x}$ to Eq. (2).
   If a minimal solution exists on Eq. (2), then solution set of (2) can be easily confirmed. However, that is not necessarily the case. The minimal solution does not often exist on Eq. (2). Even under the situation of $X(A, b) \neq \emptyset$, We can not find an effective method to confirm whether $X(A, b)$ has a minimum solution at present, which makes solving Eq. (2) more complicated. In order to simplify and reduce the complexity of the problem, the paper always assumes that every solution to Eq. (2) contains a minimum solution that is less than or equal to it, and the number of minimum solution is finite. If we denote all minimum solutions to

Eq. (2) by $\check{X}(A,b)$, then solution set of Eq. (2) can be denoted as follows.

$$X(A,b) = \bigcup_{\check{x} \in \check{X}(A,b)} \{x|\check{x} \le x \le \hat{x}, \ x \in [0,1]^n\}. \tag{4}$$

We can clearly see by Formula (4), solution set structure of Eq. (2) can be ascertained by $\check{X}(A,b)$, solving $\check{X}(A,b)$ means $X(A,b)$ is known. Up to now, the most effective ways to solve $\check{X}(A,b)$ is called conservative path method.

Let $N_k = \{j \in N| \min\{(a_{kj}, \hat{x}_j\} = b_k\}(k \in M), M_j = \{k \in M|a_{kj} \wedge \hat{x}_j = b_k\}(j \in N)$ and $\Lambda = N_1 \times N_2 \times \cdots \times N_m$. The vector $q = (q_1, q_2, \cdots, q_m) \in \Lambda$ if and only if $q_k \in N_k, \forall k \in M$. For all $q \in \Lambda$, we calculated the index set

$$M_q^j = \{k \in M \mid q_k = j\}, j \in N. \tag{5}$$

Defined functions $F : \Lambda \longrightarrow R^n$

$$F_j(q) = \begin{cases} \max_{k \in M_q^j} b_k & \text{if } M_q^j \ne \emptyset, \\ 0 & \text{if } M_q^j = \emptyset, \end{cases} \qquad \forall j \in N. \tag{6}$$

The vector $q \in \Lambda$ is called conservative path (C path) to Eq. (2). Let the set of all conservative paths of Eq. (2) is CP. It is easy to obtain the following theorem[6, 27]:

Theorem 2.1. [11, 14] Let $X(A,b) \ne \emptyset$. Then

(1) if $q \in \Lambda$, then $F(q) \in X(A,b)$;
(2) for all $x \in X(A,b)$, exist $p \in \Lambda$ have $F(p) \le x$.

Definition 2.3. [14] The vector $q \in \Lambda$ is called fuzzy relation equation path (FRE path), If it satisfies

$$q_k \begin{cases} \in N_1, & \text{if } k = 1, \\ \in N_k, & \text{if } N_k \bigcap \{q_1, q_2, \cdots, q_{k-1}\} = \emptyset; \quad k \in M, \\ = 0, & \text{otherwise.} \end{cases} \tag{7}$$

Let the sets of all FRE path are $FREP$, It is obviously having $FREP \subseteq CP$.

Definition 2.4. [14] Let $q \in CP$. The $x^q = (x_1^q, x_2^q, \cdots, x_m^q)^T = F(q)$ is called quasi-minimal solution the corresponding C path q, q is called C path the corresponding solution x^q, where $x_j^q = F_j(q)(j \in N)$.

Theorem 2.2. [14] IF $X(A,b) \ne \emptyset$, then $X^* = \{x \in X \mid x^q \le x \le \hat{x}, q \in CP\} = \{x \in X \mid x^q \le x \le \hat{x}, q \in FREP\}$.

Theorem 2.3. [14] Suppose $\check{x}$ is a minimal solution to fuzzy relation Eq. (2), then there is a FRE path $q \in FREP$ to satisfy $\check{x} = x^q$. Further, If fuzzy relation Eq. (2) satisfies $d_1 > d_2 > \cdots > d_m$, then for each FRE path q, $x^q = F(q)$ is a minimal solution of Eq. (2).

From Theorem 2.3, In order to obtain all minimal solutions to Eq. (2), We just need to find all FRE paths of Eq. (2). If $k_1 \ge k_2$ and $N_{k_1} \supseteq N_{k_2}$, then deleting N_{k_1} from the Λ does not affect the minimal solution sets of Eq. (2).

3   Main Results and Algorithm

This section first analyzes optimal solution to Model (1) under some special circumstances. Then seeks to some simplifying rules of Model (1) under general circumstances. Finally, the new algorithm for solve global optimum solution to Model (1) is constructed based on branch and bound method [26, 31–33].

For all $i_0 \in N$, Suppose $N_{i_0}^+ = \{j \in N | q_{i_0 j} \geqslant 0\}, N_{i_0}^- = N \backslash N_{i_0}^+$.

Lemma 3.1. If there exists some $i_0 \in N$ satisfies $c_{i_0} \leqslant 0, q_{i_0 i_0} \leqslant 0$, and $c_{i_0} + 0.5 q_{i_0 i_0} \hat{x}_{i_0} + \sum\limits_{j \in N_{i_0}^+} q_{i_0 j} < 0$, then, any optimal solution x^*, we can get $x_{i_0}^* = \hat{x}_{i_0}$.

Proof. Let x^* is optimal solution to Model (1) and $x_{i_0}^* < \hat{x}_{i_0}$, suppose $\bar{x} = (\bar{x}_1, \cdots, \bar{x}_n)$ satisfies

$$\bar{x}_i = \begin{cases} x_i^*, & \text{If } i \neq i_0, \\ \hat{x}_i, & \text{If } i = i_0. \end{cases}$$

Obviously, $0 \leqslant x^* \leqslant \bar{x} \leqslant \hat{x}$, that is, for all $i \in N$, have $0 \leqslant x_i^* \leqslant \bar{x}_i \leqslant \hat{x}_i$. Since $c_{i_0} \leqslant 0, q_{i_0 i_0} \leqslant 0$, and $c_{i_0} + 0.5 q_{i_0 i_0} \hat{x}_{i_0} + \sum\limits_{j \in N_{i_0}^+} q_{i_0 j} < 0$, the following inequality holds:

$$\begin{aligned} f(x^*) &= \tfrac{1}{2} x^{*T} Q x^* + c x^* \\ &= \tfrac{1}{2} \sum_{i=1, i \neq i_0}^{n} \sum_{j=1, j \neq i_0}^{n} q_{ij} x_i^* x_j^* + \sum_{j \in N_{i_0}^-, j \neq i_0} q_{i_0, j} x_{i_0}^* x_j^* + \sum_{i=1, i \neq i_0}^{n} c_i x_i^* \\ &\quad + \Big(\sum_{j \in N_{i_0}^+, j \neq i_0} q_{i_0, j} x_j^* + \tfrac{1}{2} q_{i_0 i_0} x_{i_0}^* + c_{i_0} \Big) x_{i_0}^* \\ &> \tfrac{1}{2} \sum_{i=1, i \neq i_0}^{n} \sum_{j=1, j \neq i_0}^{n} q_{ij} x_i^* x_j^* + \sum_{j \in N_{i_0}^-, j \neq i_0} q_{i_0, j} \hat{x}_{i_0} x_j^* + \sum_{i=1, i \neq i_0}^{n} c_i x_i^* \\ &\quad + \Big(\sum_{j \in N_{i_0}^+, j \neq i_0} q_{i_0, j} x_j^* + \tfrac{1}{2} q_{i_0 i_0} \hat{x}_{i_0} + c_{i_0} \Big) \hat{x}_{i_0}. \end{aligned}$$

It is contradiction to the assumption that x^* is optimal solution to Model (1). Thus the optimal solution x^* must be $x_{i_0}^* = \hat{x}_{i_0}$.

By Lemma 3.1 the following some conclusions can be directly obtained.

Corollary 3.1. If there exists some $i_0 \in N$ satisfies $c_{i_0} \leqslant 0$, and for all $j \in N, q_{i_0 j} \leqslant 0$ and $c_{i_0}, q_{i_0 i_0}$ at least one is not 0, then, for any optimal solution x^*, we have $x_{i_0}^* = \hat{x}_{i_0}$.

Corollary 3.2. If $c_i \leqslant 0, q_{ii} \leqslant 0$, for all $i, j \in N$, and $c_i + 0.5 q_{ii} \hat{x}_i + \sum\limits_{j \in N_i^+} q_{ij} < 0$, then maximal solution $\hat{x}$ of the Eq. (2) is optimal one of Model (1).

Corollary 3.3. If $c_i \leqslant 0, q_{ij} \leqslant 0$, for all $i, j \in N$, then maximal solution $\hat{x}$ of the Eq. (2) is optimal one of Model (1).

Lemma 3.2. If there exists some $i_0 \in N$ satisfies $c_{i_0} \geqslant 0, q_{i_0 i_0} \geqslant 0$, and $c_{i_0} + \sum\limits_{j \in N_{i_0}^-} q_{i_0 j} > 0$, then there exists a minimal solution $\breve{x}$ satisfies $x_{i_0}^* = \breve{x}_{i_0}$.

Proof. Since x^* is a feasible solution to Model (1), then must have a minimal solution $\breve{x}$ satisfy $0 \leqslant \breve{x} \leqslant x^* \leqslant \hat{x}$, that is to say, for all $i \in N, 0 \leqslant \breve{x}_i \leqslant x_i^* \leqslant \hat{x}_i$. To prove the conclusion, only need to prove $\breve{x}_{i_0} = x_{i_0}^*$. When $0 \leqslant \breve{x}_{i_0} < x_{i_0}^*$, since $c_{i_0} \geqslant 0, q_{i_0 i_0} \geqslant 0$, and $c_{i_0} + \sum\limits_{j \in N_{i_0}^-} q_{i_0 j} > 0$, then we have

$$
\begin{aligned}
f(x^*) &= \tfrac{1}{2} x^{*T} Q x^* + c x^* \\
&= \tfrac{1}{2} \sum_{i=1, i \neq i_0}^{n} \sum_{j=1, j \neq i_0}^{n} q_{ij} x_i^* x_j^* + \sum_{j \in N_{i_0}^+, j \neq i_0} q_{i_0,j} x_{i_0}^* x_j^* + \sum_{i=1, i \neq i_0}^{n} c_i x_i^* \\
&\quad + \big(\sum_{j \in N_{i_0}^-, j \neq i_0} q_{i_0,j} x_j^* + \tfrac{1}{2} q_{i_0 i_0} x_{i_0}^* + c_{i_0} \big) x_{i_0}^* \\
&> \tfrac{1}{2} \sum_{i=1, i \neq i_0}^{n} \sum_{j=1, j \neq i_0}^{n} q_{ij} x_i^* x_j^* + \sum_{j \in N_{i_0}^+, j \neq i_0} q_{i_0,j} \breve{x} x_j^* + \sum_{i=1, i \neq i_0}^{n} c_i x_i^* \\
&\quad + \big(\sum_{j \in N_{i_0}^-, j \neq i_0} q_{i_0,j} x_j^* + \tfrac{1}{2} q_{i_0 i_0} \breve{x}_{i_0} + c_{i_0} \big) \breve{x}_{i_0}.
\end{aligned}
$$

It is contradiction to the assumption that x^* is optimal solution to Model (1). Therefore, there exists a minimal solution $\breve{x}$ satisfies $x_{i_0}^* = \breve{x}_{i_0}$.

By Lemma 3.2 the following conclusions can be directly obtained.

Corollary 3.4. Let x^* is an optimal solution yo Model (1). If there $i_0 \in N$ satisfies $c_{i_0} \geqslant 0$, and for all $j \in N$, $q_{i_0 j} \geqslant 0$ and $c_{i_0}, q_{i_0 i_0}$ at least one is not 0, then there must be a minimal solution $\breve{x}$ that satisfies $x_{i_0}^* = \breve{x}_{i_0}$.

Corollary 3.5. If $c_i \geqslant 0, q_{ii} \geqslant 0$, for all $i, j \in N$, and $c_i + \sum\limits_{j \in N_i^-} q_{ij} > 0$, then there exists a minimal solution $\breve{x}$ of Eq. (2) that it is optimal solution to Model (1).

Corollary 3.6. If $c_i \geqslant 0, q_{ij} \geqslant 0$, for all $i, j \in N$, then there exists a minimal solution $\breve{x}$ of Eq. (2) that it is optimal solution to Model (1).

Lemma 3.3. If there exists some $i_0 \in N$ satisfied:

(1) $M_{i_0} = \emptyset$;
(2) $c_{i_0} \geqslant 0, q_{i_0 i_0} \geqslant 0$, and $c_{i_0} + \sum\limits_{j \in N_{i_0}^-} q_{i_0 j} > 0$,

then, for any optimal solution x^* of Model (1), we have $x_{i_0}^* = 0$.

Proof. Suppose an optimal solution to Model (1) is x^*. It follows from Lemma 3.2 and Theorem 2.2 that we can know that must exist a minimal solution $\breve{x}$ to satisfy $x_{i_0}^* = \breve{x}_{i_0}$. So we only need to prove $\breve{x}_{i_0} = 0$. Since $M_{i_0} = \emptyset$, then for all

$k \in M$, must have $i_0 \notin N_k$, that is to say, all FRE path p has $p_k \neq i_0$, for all $k \in M$. And according to Theorem 2.3, for any one minimal solution $\breve{x}$. There must be an FRE path p that satisfies $\breve{x} = F(p) = x^p$. According to (6), there must have $\breve{x}_{i_0} = F_{i_0}(p) = x^p_{i_0} = 0$.

Corollary 3.7. If there exists some $i_0 \in N$ satisfy:

(1) $M_{i_0} = \emptyset$;
(2) $c_{i_0} \geq 0$, for all $j \in N$, $q_{i_0 j} \geq 0$;
(3) $c_{i_0}, q_{i_0 i_0}$ at least one is not 0,

then, for any one optimal solution x^*, $x^*_{i_0} = 0$ holds.

Lemma 3.4. If there exists some $k_0 \in M, j_0 \in N$ satisfies:

(1) $N_{k_0} = \{j_0\}$,
(2) for all $k < k_0$, $j_0 \notin N_k$,
(3) $c_{j_0} \geq 0, q_{j_0 j_0} \geq 0$, and $c_{j_0} + \sum_{j \in N^-_{j_0}} q_{j_0 j} > 0$,

then, for any one optimal solution x^*, it is $x^*_{j_0} = b_{k_0}$.

Proof. Since $N_{k_0} = \{j_0\}$, and for all $k < k_0$, $j_0 \notin N_k$, then any one FRE path $p = (p_1, \cdots, p_m)$ of Eq. (2) must satisfy $p_{k_0} = j_0$ and $p_k \neq j_0 (k < k_0)$, then all minimal solution $\breve{x}$ must have $\breve{x}_{j_0} = b_{k_0}$. By the Condition (3) and Lemma 3.2, for any one optimal solution x^*, we have $x^*_{j_0} = b_{k_0}$.

Corollary 3.8. If there exists some $k_0 \in M, j_0 \in N$ satisfies:

(1) $N_{k_0} = \{j_0\}$,
(2) for all $k < k_0$, $j_0 \notin N_k$,
(3) $c_{j_0} \geq 0$, for all $j \in N$ have $q_{j_0 j} \geq 0$, and $c_{j_0}, q_{j_0 j_0}$ at least one is not 0,

then, for any one optimal solution x^*, we can see $x^*_{j_0} = b_{k_0}$.

Corollary 3.9. If there exists some $i \in N$ satisfies:

(1) $M_i \neq \emptyset$, and for all $l \in N \backslash \{i\}$ have $M_i \cap M_l = \emptyset$,
(2) $c_i \geq 0, q_{ii} \geq 0$, and $c_i + \sum_{j \in N^-_i} q_{ij} > 0$,

then, for any one optimal solution x^*, it is $x^*_i = \max\{b_k | i \in M_i\}$.

Proof. Suppose any one optimal solution to Model (1) is x^*, By Lemma 3.2, there exists a minimal solution $\breve{x}$ satisfy $x^*_i = \breve{x}_i$. Let $k_1, \cdots, k_r \in M_i$ and $k_1 < k_2 < \cdots < k_r$, since for all $l \in N \backslash \{i\}$ have $M_i \cap M_l = \emptyset$, then have $N_{k_1} = N_{k_2} = \cdots = N_{k_l} = \{i\}$, $i \notin N_k(k \in M, k \notin \{k_1, \cdots, k_r\})$. By the definition of FRE path, for any one FRE path $p = (p_1, \cdots, p_m)$ satisfy $p_{k_1} = i$ and $p_k \neq i(k \neq k_1)$, but according to Theorem 2.3, for any one minimal solution $\breve{x}$ must exist FRE path p satisfy $\breve{x} = F(p) = x^p$, By Formula (5), there $\breve{x}_i = F_i(p) = x^p_i = b_{k_1} = \max\{b_k | k \in M_i\}$.

By Corollary 3.5 and 3.9, the following corollary can easily be obtained:

Corollary 3.10. If Model (1) satisfies:

(1) for all $i \in N$ have $c_i \geqslant 0, q_{ii} \geqslant 0$, and $c_i + \sum\limits_{j \in N_i^-} q_{ij} > 0$;

(2) for all $k, l \in M$ have $N_k \cap N_l = \emptyset$,

then Model (1) must have only one optimal solution $x^* = (x_1^*, \cdots, x_n^*)^T$, and

$$x_j^* = \begin{cases} \max\{b_k | k \in M_j\}, & \text{If } M_j \neq \emptyset, \\ 0, & \text{If } M_j = \emptyset, \end{cases} \quad j \in N.$$

Theorem 3.1. If there exists some $j_1 \in N$ satisfies,

(1) $\bigcup\limits_{j=1, j \neq j_1}^{n} M_j \subset M_{j_1}$ and $1 \in M_{j_1} - \bigcup\limits_{j=1, j \neq j_1}^{n} M_j$,

(2) for all $i \in N$ have $c_i \geqslant 0, q_{ii} \geqslant 0$, and $c_i + \sum\limits_{j \in N_i^-} q_{ij} > 0$;

then Model (1) must have only one optimal solution $x^* = (x_1^*, \cdots, x_n^*)^T$, and

$$x_j^* = \begin{cases} b_1, & \text{If } j = j_1, \\ 0, & \text{If } j \neq j_1, \end{cases} \quad j \in N.$$

Proof. When $X(A, b) \neq \emptyset$, there $M_{j_1} = M$, by the condition, otherwise, exist $k \in M$ satisfy $k \notin M_{j_1}$, so have $k \notin M_j(\forall j \in N)$, that is for all $j \in N, a_{kj} \wedge \hat{x}_j = b_k$ can not be satisfied, this is equivalent to K-th equation of $A \circ x = b$ has no solution, this is a contradiction to $X(A, b) \neq \emptyset$. So for all $k \in M$ have $a_{kj} \wedge \hat{x}_j = b_k$, that is to say $x = (0, \cdots, 0, \hat{x}_j, 0, \cdots, 0)^T$ is one solution of $A \circ x = b$. And because $1 \in M_{j_1} - \bigcup\limits_{j=1, j \neq j_1}^{n} M_j$, then have $N_1 = \{j_1\}$. We can see from the proof of Corollary 3.9, any one minimal solution $\breve{x}$ of Eq. (2) must have $\breve{x}_{j_1} = \hat{x}_{j_1} = b_1$, and for all $i \in N$, $c_i \geqslant 0, q_{ii} \geqslant 0$, $c_i + \sum\limits_{j \in N_i^-} q_{ij} > 0$, then optimal solution x^* satisfy $x_{j_1}^* = b_1$. So $(0, \cdots, 0, b_1, 0, \cdots, 0)^T$ is only one optimal solution to Model (1).

We can obtain the following theorem based on the above discussion,

Theorem 3.2. If there exists some $j_1, \cdots, j_r \in N$ satisfies,

(1) for all $l \in \{1, 2, \cdots, r\}$ have $\bigcup\limits_{j=1, j \notin \{j_1, \cdots, j_r\}}^{n} M_j \subset, M_{j_l}$ and $M_{k_l} = M$;

(2) let $i_1, \cdots, i_l \in I_k - \bigcup\limits_{j=1, j \neq k}^{n} I_j$ and $i_1 < \cdots < i_l$,

(3) for all $i, j \in N$ have $c_i \geqslant 0, q_{ij} \geqslant 0$, and $c_i + \sum\limits_{j \in N_i^-} q_{ij} > 0$;

the Model (1) have only one optimal solution $x^* = (x_1^*, \cdots, x_n^*)^T$, and

$$x_j^* = \begin{cases} b_{i_1}, & \text{If } j = k, \\ 0, & \text{If } j \neq k, \end{cases} \quad j \in N.$$

Now we propose several rules to simplify the Model (1) based on the above discussion.

Rule 3.1. Suppose that $N^0 = \{j \in N | c_j \leqslant 0, q_{jj} \leqslant 0, c_j + 0.5q_{jj}\hat{x}_j + \sum_{i \in N_j^+} q_{ij} < 0\}$. For all $j \in N^0$, let $x_j^* = \hat{x}_j$, and remove the j-th column of matrix A. Suppose that $M^0 = \{k \in M | a_{kj} \wedge \hat{x}_j = b_k, \forall j \in N^0\}$, if $k \in M^0$, remove k-th row of matrix A and k-th component of vector b. Model (1) reduces to the following questions:

$$\min \ f(x) = \frac{1}{2} \sum_{i \in N \setminus N^0} \sum_{j \in N \setminus N^0} q_{ij} x_i x_j + \sum_{i \in N \setminus N^0} c_i x_i$$
$$+ \frac{1}{2} \sum_{i \in N^0} \sum_{j \in N \setminus N^0} q_{ij} \hat{x}_i x_j + \frac{1}{2} \sum_{i \in N \setminus N^0} \sum_{j \in N^0} q_{ij} x_i \hat{x}_j + \alpha \quad (8)$$

s.t. $A' \circ x = b'$,

where A', b' is a matrix and vector to be removed rows and columns by Rule 3.1,

$$\alpha = \frac{1}{2} \sum_{i \in N^0} \sum_{j \in N^0} q_{ij} \hat{x}_i \hat{x}_j + \sum_{i \in N^0} c_i \hat{x}_i.$$

Rule 3.2. If $k_1 \geqslant k_2$ and $N_{k_1} \supseteq N_{k_2}$, then removing N_{k_1} does not affect the minimal solution sets of (8) in Λ.

Rule 3.3. Let $N^1 = \{j \in N | M_j = \emptyset, c_j \geqslant 0, q_{jj} \geqslant 0, c_j + \sum_{i \in N_j^-} q_{ij} > 0\}$. For all $j \in N^1$, let $x_j^* = 0$, remove the j-th column of matrix A, that is to remove all j in N_k.

Rule 3.4. Let $N^2 = \{j \in N | \exists k_0 \in M, N_{k_0} = \{j_0\}, \forall k < k_0, j_0 \notin N_k, c_{j_0} \geqslant 0, q_{j_0 j_0} \geqslant 0, c_{j_0} + \sum_{j \in N_{j_0}^-} q_{j_0 j} > 0\}$. For all $j \in N^2$, let $x_j^* = b_{k_0}$, remove the j-th column of matrix A, that is to remove all j in N_k.

By using Rule 3.1–3.4, Model (1) can be reduced to the following optimization question:

$$\min \ f(x) = \frac{1}{2} \sum_{i \in N'} \sum_{j \in N'} q_{ij} x_i x_j + \sum_{i \in N'} c_i x_i + \frac{1}{2} \sum_{i \in N^0} \sum_{j \in N'} q_{ij} \hat{x}_i x_j$$
$$+ \frac{1}{2} \sum_{i \in N'} \sum_{j \in N^0} q_{ij} x_i \hat{x}_j + \frac{1}{2} \sum_{i \in N^2} \sum_{j \in N'} q_{ij} x_i^* x_j + \frac{1}{2} \sum_{i \in N'} \sum_{j \in N^2} q_{ij} x_i x_j^* + \alpha' \quad (9)$$

s.t. $A'' \circ x = b''$,

where A'', b'' is a matrix and vector to be removed rows and columns by rule 3.1, 3.3 and 3.4, $N' = N \backslash N^0 \backslash N^1 \backslash N^2$

$$\alpha' = \frac{1}{2} \sum_{i \in N^0} \sum_{j \in N^0} q_{ij} \hat{x}_i \hat{x}_j + \sum_{i \in N^0} c_i \hat{x}_i + \frac{1}{2} \sum_{i \in N^2} \sum_{j \in N^2} q_{ij} x_i^* x_j^* + \sum_{i \in N^2} c_i x_i^*.$$

Suppose that the sets of all FRE paths are $FREP$ in Model (9), and $|FREP| = h$. In order to obtain the optimal solution to Model (9), the following h convex quadratic programming must be solved:

$$\min \ f(x)$$
$$\text{s.t. } x_j^q \leqslant x_j \leqslant \hat{x}_j, \quad j \in N', \tag{10}$$

where $q \in FREP$. Suppose that the optimal solution to Model (10) is $x^l (l = 1, 2, \cdots, h)$, then the optimal solution to Model (1) is x^*, and $f(x^*) = \min_{l=1,2,\cdots,h} \{f(x^l)\}$. In order ms with Suppose that the all FRI path set of the problem (10) is FRIP, and —FRIP— = h. In order to solving problem (10), we must solve the following h quadratic programming problems with interval constraint by numerical algorithms in Bazaraa et al. [26].

If the condition of Corollary 3.5 is satisfied, and suppose that the sets of all FRE paths of Model (1) are $FREP$, and $|FREP| = h$, then we can just solve the following problem:

$$\min \ f(x)$$
$$\text{s.t. } x = x^q, \quad q \in FREP. \tag{11}$$

We can solve the problem (11) by using branch and bound method [31–33].

According to the above-presented results, we develop an algorithm for searching an approximate optimal solution to Model (1).

Step 3.1. Calculate the maximum solution $\hat{x}$ of $\vee - \wedge$ fuzzy relation equation based on (3), if $A \circ \hat{x} = b$ is satisfied, then goes to step, otherwise, there is no feasible solution to the original problem, stop.

Step 3.2. Determine whether the condition of Corollary 3.2 is satisfied, if it is satisfied, then the maximum solution $\hat{x}$ is optimal one of Model (1), stop, otherwise, go to step 3.

Step 3.3. Determine whether the condition of Corollary 3.5 is satisfied, if it is satisfied, then use FRE path and branch and bound method for solving Model (11), the optimal solution to Model (1) can be obtained, stop, otherwise, go to step 3.4.

Step 3.4. Simplify the Model (1) based on the Rule 3.1–3.4, the Model (9) can be obtained.

Step 3.5. Solving all solution to Model (10) based on solver technology of convex quadratic programming problems, the optimal solution to Model (1) can be obtained.

4   Numerical Examples

Example 4.1. We consider the following fuzzy relation quadratic programming.

$$\min \ f(x) = \frac{1}{2}x^T Q x + cx$$
$$\text{s.t. } A \circ x = b,$$

(12)

where $c = (-2.415, -1.947, -1.772, -3.522, -1.762, -3.4429)^T$, $b = (0.7, 0.6, 0.5, 0.2, 0.2)^T$,

$$Q = \begin{bmatrix} -1.2786 & -1.1147 & 0.1382 & 0.3666 & 0.4695 & 0.9516 \\ -1.1147 & -1.0097 & 0.1495 & 0.9153 & 0.6918 & -1.0907 \\ 0.1382 & 0.1495 & -1.2473 & 0.4678 & -0.9782 & 0.2769 \\ 0.3666 & 0.9153 & 0.4678 & -0.6719 & -1.2666 & 0.8538 \\ 0.4695 & 0.6918 & -0.9782 & -1.2666 & -0.2561 & 0.3359 \\ 0.9516 & -1.0907 & 0.2769 & 0.8538 & 0.3359 & -0.9978 \end{bmatrix},$$

$$A = \begin{bmatrix} 0.4 & 0.7 & 1 & 0.7 & 0.8 & 0.9 \\ 0.5 & 0.8 & 0.7 & 0.3 & 0.6 & 0.6 \\ 0.4 & 0.6 & 0.3 & 0.5 & 0.2 & 0.9 \\ 0.2 & 0.7 & 0.2 & 0.1 & 0.2 & 0.4 \\ 0.1 & 0.2 & 0.1 & 0.1 & 0.2 & 0.1 \end{bmatrix}.$$

Solution. Step 1 Calculate the maximum solution $\hat{x}$ of Model (12) based on (3), $\hat{x} = (1, 0.2, 0.6, 1, 0.7, 0.2)^T$ can be gotten, obviously, $A \circ \hat{x} = b$ is satisfied, go to step 2.

Step 2. Obviously, for all $i = 1, 2, 3, 4, 5, 6$, $c_i \leqslant 0$, $q_{ii} \leqslant 0$, $c_i + 0.5 q_{ii} \hat{x}_i + \sum_{j \in N_i^+}^{n} q_{ij} < 0$. By Corollary 3.2, the optimal solution to Model (12) is $\hat{x}$, the optimal value is $f(\hat{x}) = -11.4489$.

Example 4.2. We consider the following fuzzy relation quadratic programming.

$$\min \ f(x) = \frac{1}{2}x^T Q x + cx$$
$$\text{s.t. } A \circ x = b,$$

(13)

where
$c = (1.9820, 2.2920, 3.3907, 1.9876, 2.2624)^T$, $b = (0.8, 0.8, 0.8, 0.5, 0.5, 0.4)^T$,

$$Q = \begin{bmatrix} 0.6376 & -0.8483 & 1.0157 & -0.1710 & -0.5250 \\ -0.8483 & 0.0584 & 1.8577 & 1.4607 & -0.9772 \\ 1.0157 & 1.8577 & 0.9177 & -0.9262 & 1.0936 \\ -0.1710 & 1.4607 & -0.9262 & 1.2481 & 1.3583 \\ -0.5250 & -0.9772 & 1.0936 & 1.3583 & 1.7703 \end{bmatrix},$$

$$A = \begin{bmatrix} 0.9 & 0.8 & 0.6 & 0.3 & 0.9 \\ 0.8 & 0.7 & 0.8 & 1 & 0.8 \\ 0.6 & 0.9 & 0.8 & 0.9 & 0.5 \\ 0.4 & 0.2 & 0.5 & 0.6 & 0.2 \\ 0.3 & 0.3 & 0.5 & 0.2 & 0.1 \\ 0.4 & 0.1 & 0.2 & 0.3 & 0.5 \end{bmatrix}.$$

Solution. Step 1 Calculate the maximum solution $\hat{x}$ of Model (13) based on (3), the $\hat{x} = (0.8, 0.8, 1.0, 0.5, 0.4)^T$. Obviously, $A \circ \hat{x} = b$ is satisfied, go to step 2.

Step 2 the conditions of Corollary 3.2 is not satisfied, go to Step 3.

Step 3 Obviously, for all $i = 1, 2, 3, 4, 5$ have $c_i \geqslant 0, q_{ii} \geqslant 0, c_i + \sum\limits_{j \in N_i^-}^{n} q_{ij} > 0$. One of the minimal solution to Model (13)is an optimal solution to (13). We will find the optimal solution of Model (13) based on the following FRE path and branch and bound method. For all $k = 1, 2, 3, 4, 5, 6$ calculate the index set N_k,

$$N_1 = \{1, 2\}, N_2 = \{1, 3\}, N_3 = \{2, 3\}, N_4 = \{3, 4\}, N_5 = \{3\}, N_6 = \{1, 5\}.$$

Figure 1 is the process of branch and bound algorithm, by the algorithm, the optimal solution to Model (13) is $x^* = (0.8, 0, 0.8, 0, 0, 0)$, and optimal value is $f(x^*) = 5.4459$.

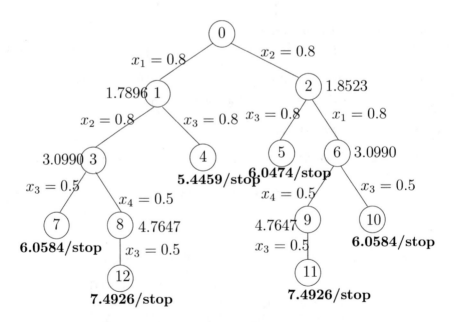

Fig. 1. Illustrates for the process of the branch and bound algorithm

Note 1. Comparison to the Algorithm of Ref. [19]. In Ref. [19], they need to use the FRE path to calculate all minimum solutions of the fuzzy relational equation and the corresponding Quadratic programming problems. However, using our algorithm, there is no need to find the minimal solution and Quadratic programming.

Example 4.3. [19] We consider the following fuzzy relation quadratic programming.

$$\min \ f(x) = \frac{1}{2}x^T Q x + cx$$
$$\text{s.t. } A \circ x = b, \tag{14}$$

where $c = (-0.8217, 0.2341, 1.4725, 0.9535, -1.3194, 1.3396, -0.3306)^T$,
$b = (0.8, 0.5, 0.5, 0.5, 0.5, 0.3)^T$,

$$Q = \begin{bmatrix} -0.4326 & -0.0376 & -0.1364 & -1.3362 & -1.4410 & 0.6686 & -1.0565 \\ -0.0376 & 0.3273 & 0.1139 & 0.7143 & 0.5711 & 1.1908 & 1.4151 \\ -0.1364 & 0.1139 & 1.0668 & 1.6236 & -0.3999 & -1.2025 & -0.8051 \\ -1.3362 & 0.7143 & 1.6236 & -0.6918 & 0.6900 & -0.0198 & 0.5287 \\ -1.4410 & 0.5711 & -0.3999 & 0.6900 & 0.8156 & -0.1567 & 0.2193 \\ 0.6686 & 1.1908 & -1.2025 & -0.0198 & -0.1567 & -1.6041 & -0.9219 \\ -1.0565 & 1.4151 & -0.8051 & 0.5287 & 0.2193 & -0.9219 & -2.1707 \end{bmatrix},$$

$$A = \begin{bmatrix} 0.9 & 0.7 & 0.9 & 0.6 & 0.9 & 0.8 & 0.9 \\ 0.1 & 0.5 & 0.2 & 0.5 & 0.3 & 0.4 & 0.5 \\ 0.5 & 0.2 & 0.1 & 0.8 & 0.3 & 0.4 & 0.4 \\ 0.5 & 0.5 & 0.1 & 0.3 & 0.4 & 0.9 & 0.5 \\ 0.5 & 0.2 & 0.1 & 0.5 & 0.4 & 0.5 & 0.5 \\ 0.2 & 0.3 & 0.2 & 0.1 & 0.3 & 0.1 & 0.3 \end{bmatrix}.$$

Solution. Step 1 Noting the example has condition $b = d$, calculate the maximum solution $\hat{x}$ of Model (14) based on (3), the $\hat{x} = (0.8, 1.0, 0.8, 0.5, 0.8, 0.5, 0.8)^T$, obviously, $A \circ \hat{x} = b$ is satisfied, go to step 2.

Step 2. The conditions of Corollary 3.2 are not satisfied, go to Step 3.

Step 3. The conditions of Corollary 3.5 are not satisfied, go to Step 4.

Step 4 Since $N^0 = \{1\}, M^0 = \{1, 3, 4, 5\}$, according to Rule 3.1, let $x_1 = 0.8$, remove 1-th column and 1, 3, 4, 5 rows of matrix A, remove 1, 3, 4, 5 components of vector b, the Model (14) can be simplified to the following form:

$$\min \ f(x) = \frac{1}{2}\sum_{i=2}^{7}\sum_{j=2}^{7} q_{ij}x_i x_j + \sum_{i=2}^{7} c_i x_i + \sum_{j=2}^{7} q_{1j}\hat{x}_1 x_j + \alpha$$
$$\text{s.t. } A' \circ x = b', \tag{15}$$

where $b' = (0.8, 0.4)^T$,

$$A = \begin{array}{c c} j\backslash i & \begin{array}{cccccc} 2 & 3 & 4 & 5 & 6 & 7 \end{array} \\ \begin{array}{c} 2 \\ 6 \end{array} & \begin{pmatrix} 0.5 & 0.2 & 0.5 & 0.3 & 0.4 & 0.5 \\ 0.3 & 0.2 & 0.1 & 0.3 & 0.1 & 0.3 \end{pmatrix} \end{array}.$$

when $k = 2, 6$, have $N_2 = \{2, 4, 7\}$, $N_6 = \{2, 5, 7\}$. Obviously, Rule 3.2–3.4 can not be satisfied, and FRE path sets are $FREP = \{q_1, q_2, q_3, q_4, q_5\}$, where

$$q_1 = (2, 0), q_1 = (4, 2), q_3 = (4, 5), q_4 = (4, 7), q_5 = (7, 0),$$

the minimal solution is

$$x^{q_1} = (0.5, 0, 0, 0, 0, 0), \quad x^{q_2} = (0.3, 0, 0.5, 0, 0, 0), \quad x^{q_3} = (0, 0, 0.5, 0.3, 0, 0)$$
$$x^{q_4} = (0, 0, 0.5, 0, 0, 0.3), \quad x^{q_5} = (0, 0, 0, 0, 0, 0.5).$$

Step 5 Solving the following five convex quadratic programming:

$$\min f(x) = \frac{1}{2} \sum_{i=2}^{7} \sum_{j=2}^{7} q_{ij} x_i x_j + \sum_{i=2}^{7} c_i x_i + \sum_{j=2}^{7} q_{1j} \hat{x}_1 x_j + \alpha \tag{16}$$

s.t. $x_j^{q_l} \leqslant x_j \leqslant \hat{x}_j$, $j = 2, 3, 4, 5, 6, 7$,

where $l = 1, 2, 3, 4, 5$. The following Table 1 is the optimal solution and optimal value for any l,

Table 1. The optimal solution and optimal value for any l.

l	x^*	$f(x^*)$
1	(0.8,0.5,0.0,0.0,0.8,0.0,0.8)	-3.07
2	(0.8,0.3,0.0,0.5,0.8,0.0,0.8)	-3.00
3	(0.8,0,0.0,0.5,0.8,0.0,0.8)	-3.66
4	(0.8,0,0.0,0.5,0.8,0.0,0.8)	-3.66
5	(0.8,0,0.0,0,0.8,0.0,0.8)	-4.01

Thus, the optimal solution to Model (12) is $x^* = (0.8, 0, 0.0, 0, 0.8, 0.0, 0.8)$, and the optimal value is $f(x^*) = -4.01$.

Note 2. Comparison to the Algorithm of Ref. [19]. In Ref. [19], they need to use the FRE path to calculate all 23 minimum solutions of the 7-dimensional fuzzy relational equation and solve 23 Quadratic programming problems. However, using our algorithm, we only need to use the conservative path to calculate five minimal solutions of the 6-dimensional fuzzy relational equation and solve five Quadratic programming problems.

In summary, the main advantages of this article are: (1) There is no need to solve all minimal solutions. (2) Only a small amount of quadratic programming needs to be solved.

5   Conclusion

We have presented a new algorithm based on FRE paths and branch and bound method of solving fuzzy relation quadratic programming, the algorithm has abandoned a rather large amount of dots that have nothing to do with optimal solution to feasible region, because it does not need to solve all minimal solutions to fuzzy relation equations, the efficiency of new algorithm has been improved. Numerical examples have proved that the algorithm can smoothly reach the optimal point when the variable scale of Model (2) are not very large. However, when the size of Model (2) is a huge large, how to effectively solve Model (1) is still a problem to be studied.

Acknowledgements. This work was supported in part by the National Natural Science Foundation of China under Grant 12271132, Grant 61877014, in part by the Natural Science Foundation of Guangdong Province under Grant 2022A1515011460.

Recommender: ProfessorXiaopeng Yang, School of Mathematics and Statistics, Hanshan Normal University in China.

References

1. Sanchez, E.: Resolution of composite fuzzy relation equation. Inform. Control. **30**, 38–48 (1976)
2. Pedrycz, W.: On generalized fuzzy relational equations and their applications. Soft. Comput. **107**, 520–536 (1985)
3. Chen, L., Wang, P.P.: Fuzzy relation equations (I): the general and specialized solving algorithms. Soft. Comput. **6**, 428–435 (2002)
4. Chen, L., Wang, P.P.: Fuzzy relation equations (II): the branch-point-solutions and the categorized minimal solutions. Soft. Comput. **11**, 33–40 (2007)
5. Li, P., Fang, S.C.: A survey on fuzzy relational equations, Part I: classification and solvability. Fuzzy Optim. Decis. Making **8**, 179–229 (2009)
6. Nola, A.D., Sessa, S., Pedrycz, W., Sanchez, E.: Fuzzy Relation Equations and Their Applications to Knowledge Engineering. Kluwer Academic Publishers, Dordrecht, Boston/London (1989)
7. Baets, B.D.: Analytical solution methods for fuzzy relation equations. In: Dubois, D., Prade, H. (eds.) The Handbook of Fuzzy Sets Series, vol. 1, pp. 291–340. Kluwer Academic Publishers, Dordrecht (2000)
8. Stamou, G.B., Tzafestas, S.G.: Fuzzy relation equations and fuzzy inference systems: an inside approach. IEEE Trans. Syst., Man Cybern.-Part B **29**, 694–702 (1999)
9. Perfilieva, I., Tonis, A.: Compatibility of systems of fuzzy relation equations. J. General Syst. **29**, 511–528 (2000)
10. Peeva, K.: Resolution of fuzzy relational equations: method, algorithm and software with applications. Inf. Sci. **234**, 44–6 (2013)
11. Wang, P.Z., Zhang, D.Z., Sanchez, E., Lee, E.S.: Latticized linear programming and fuzzy relation inequalities. J. Math. Anal. Appl. **159**(1), 72–87 (1991)
12. Fang, S.C., Li, G.: Solving fuzzy relations with a linear objective function. Fuzzy Sets Syst. **103**, 107–113 (1999)
13. Khorram, E., Ghodousian, A.: Linear objective function optimization with fuzzy relation equation constraints regarding max-av composition. Appl. Math. Comput. **173**, 872–886 (2006)
14. Guo, F.F., Pang, L.P., Meng, D., Xia, Z.Q.: An algorithm for solving optimization problems with fuzzy relational inequality constraints. Inf. Sci. **252**, 20–31 (2013)

15. Ghodousian, A., Falahatkar, S.: Linear optimization constrained by fuzzy inequalities defined by Max-Min averaging operator. J. Algorithms Comput. **52**(2), 13–28 (2020)

16. Ghodousian, A.: Optimization of linear problems subjected to the intersection of two fuzzy relational inequalities defined by Dubois-Prade family of t-norms. Inf. Sci. **503**, 291–306 (2019)

17. Loetamonphong, J., Fang, S.C., Young, R.E.: Multi-objective optimization problems with fuzzy relation equation constraints. Fuzzy Sets Syst. **127**, 141–164 (2002)

18. Guu, S.-M., Wu, Y.-K., Lee, E.S.: Multi-objective optimization with a max-t-norm fuzzy relational equation constraint. Comput. Math. Appl. **61**, 1559–1566 (2011)

19. Yang, J.H., Cao, B.Y., Lv, J.: The global optimal solutions for fuzzy relation quadratic programming. Fuzzy Syst. Math. **27**(6), 154–161 (2013)

20. Molai, A.A.: The quadratic programming problem with fuzzy relation inequality constraint. Comput. Ind. Eng. **66**(1), 256–263 (2012)

21. Yang, J.H., Cao, B.Y.: Monomial geometric programming with fuzzy relation equation constraints. Fuzzy Optim. Decis. Making **6**(4), 337–349 (2007)

22. Zhou, X.G., Ahat, R.: Geometric programming problem with single-term exponents subject to max-product fuzzy relational equations. Math. Comput. Model. **53**, 55–62 (2011)

23. Lu, J., Fang, S.C.: Solving nonlinear optimization problems with fuzzy relation equations constraints. Fuzzy Sets Syst. **119**, 1–20 (2001)

24. Ghodousian, A., Naeeimib, M., Babalhavaeji, A.: Nonlinear optimization problem subjected to fuzzy relational equations defined by Dubois-Prade family of t-norms. Comput. Ind. Eng. **119**, 167–180 (2018)

25. Boyd, S., Vandenberghe, L.: Convex Optimization. Cambridge University Press (2004)

26. Bazaraa, M.S., Sherali, H.D., Shetty, C.M.: Nonlinear Programming: Theory and Algorithms, 2nd edn. Wiley, New York (1993)

27. Yildiz, O.T.: Quadratic programming for class ordering in rule induction. Pattern Recogn. Lett. **54**, 63–68 (2015)

28. Abdel-Malek, L.L., Areeratchakul, N.: A quadratic programming approach to the multi-product newsvendor problem with side constraints. Eur. J. Oper. Res. **176**, 855–861 (2007)

29. Zhang, W.G., Nie, Z.K.: On admissible efficient portfolio selection policy. Appl. Math. Comput. **169**, 608–623 (2005)

30. Peterse, J.A.M., Bodson, M.: Constrained quadratic programming techniques for control allocation. IEEE Trans. Control Syst. Technol. **14**, 91–98 (2006)

31. Horst, R., Tuy, H.: Global Optimization: Deterministic Approaches. Springer, Berlin (1993)

32. Hoai An, L.T., Tao, P.D.: A branch and bound method via d.c. optimization algorithms and ellipsoidal technique for box constrained nonconvex quadratic problems. J. Global Optim. **13**(2), 171–206 (1998)

33. Buchheim, C., Caprara, A., Lodi, A.: An effective branch-and-bound algorithm for convex quadratic integer programming. Math. Programm. **135**(1), 369–395 (2012)

Feature Selection Algorithm for Multi-label Classification Based on Graph Operations

Qianyao Tang, Fuyi Wei$^{(\boxtimes)}$, Zhihong Liu, Hang Zhang, Ying Guo, Peiwei Su, and Dongxin Li

Department of Mathematics 701, College of Mathematics and Information, South China Agricultural University, Guangzhou 510642, China
weifuyi@scau.edu.cn

Abstract. Referring to the method of Guo Yankui [1] and others in the field of single classification, this paper proposes a correlation attribute based on graph operation. Multi-label classification and selection algorithm. The algorithm takes the correlation between labels and attributes and between attributes as the weights of bipartite graphs and complete graphs respectively, sets thresholds for graph operation, constructs maximal connected subgraphs, and finally obtains the optimal attribute subset. This subset can effectively contain the information in the original data set. In this paper, the commonly used multi-label classification data sets are selected for experiments. Experiments show that this algorithm can effectively reduce the data dimension, reduce redundant information and improve the classification rate in multi-label data sets.

Keywords: Graph operation · Correlation attribute · Maximal connected subgraph

1 Introduction

Effective operation and management of big data is a great challenge for data science at present. Selecting feature as the preprocessing step of machine learning can effectively reduce the dimension, remove irrelevant data and improve the speed and accuracy of learning model. With the advent of the era of big data, it is necessary to preprocess the feature values containing a large amount of data in the fields of text classification, image retrieval, data mining, etc., and the huge data scale brings a severe test to the original feature recognition algorithm. Therefore, it is very important to establish a fast and effective feature selection model. In order to reduce the dimension and improve the classification performance, researchers put forward multi-label feature selection methods, which mainly include filtering-based feature selection method, packaging-based feature selection method and embedded-based feature selection method.

Filter feature selection method usually considers some feature properties separately. In view of the fact that ReliefF algorithm can't remove redundancy, Yan Tao et al.

F. Wei—Professor, research direction: intelligent network and graph theory.

© The Author(s), under exclusive license to Springer Nature Singapore Pte Ltd. 2024
B.-Y. Cao et al. (Eds.): ICFIE 2022, LNDECT 207, pp. 335–342, 2024.
https://doi.org/10.1007/978-981-97-2891-6_25

[2] proposed a feature selection algorithm based on ReliefF sorting filter considering the relationship between genes, which can effectively screen and eliminate redundant genes. Xu Yao et al. [3] proposed an improved MRMR algorithm based on group strategy (MRMRE), aiming at the defect that MRMR algorithm will ignore the local correlation between feature groups. A large number of experiments on image and gene sequence data sets in UCI machine learning day database show that compared with MRMR algorithm, the proposed algorithm has higher result stability and classification accuracy. Compared with the feature selection method based on filtering, the feature selection method based on packaging can get higher classification accuracy. Zhang ML et al. [4] put forward a multi-label naive Bayes algorithm, which combines naive Bayes classifier with principal component analysis (PCA) and genetic algorithm (GA). The former is used to eliminate irrelevant and redundant features, while the latter is used to select the most appropriate feature subset and classify. This method greatly improves the performance of the experiment. Jiang et al. [5] put forward a method of FSkNN, which uses fuzzy similarity (FSN) and proximity algorithm (kNN) to classify multi-label texts. Experiments show that this method is feasible and effective. Aiming at the characteristics of low classification accuracy of filtering feature selection method and low efficiency of packaging feature selection method, embedded feature selection method can select and classify features at the same time. At present, sparse regularization is widely used, but Chen Hong [6] considers the geometric structure of feature popularity into the research of multi-label feature selection, and proposes a sparse regularization multi-label feature selection algorithm based on correlation entropy and feature popularity learning. Ge Lei et al. [7] put forward the multi-label embedded feature selection method (MEFS), which adopts the backward sequence search method, effectively improves the accuracy of multi-label learning, and this method is also superior to some popular multi-label dimensionality reduction methods.

In single classification, Guo Yankui and others took a different approach and proposed an attribute selection algorithm based on maximal connected subgraph correlation. Experiments show that the attribute selection algorithm can obviously improve the classification accuracy in the data preprocessing process of the classification algorithm. In this paper, a multi-label classification and selection algorithm of correlation attribute based on graph operation is proposed, and the threshold is set to operate the series graph, and finally the optimal attribute subset is obtained. This algorithm can effectively improve the classification accuracy of multi-label algorithm, aiming at providing a better solution to the multi-label classification problem.

2 Model Building

In multi-label feature selection algorithm, information entropy is the core concept. According to the information entropy formula, this model leads to a series of formulas such as correlation degree. The related concepts and derivation process of correlation degree are introduced below.

2.1 Information Entropy Formula

Information entropy is an index to measure the uncertainty of information. The lower the information entropy, the lower the uncertainty of the information contained in the set.

Let set X be the label attribute set, Y the feature attribute set. $X = \{x1, x2, ..., xm\}$, $Y = \{y1, y2......, yn\}$. The information entropy [8] of X and Y are respectively

$$H(X) = -\sum_{i=1}^{m} p(x_i) \log p(x_i) \tag{1}$$

$$H(Y) = -\sum_{i=1}^{n} p(y_i) \log p(y_i) \tag{2}$$

2.2 Conditional Entropy Formula

Conditional Entropy [8] of Set Y under the given condition of Set X

$$H(Y|X) = -\sum_{i=1}^{m}\sum_{j=1}^{n} p(x_iy_j) \log p(y_j|x_i) \tag{3}$$

The formula indicates the uncertainty of information in set Y under the condition that set X appears.

2.3 Derivation of Correlation Formula

According to the above formula, the dependence of Y on X Sim (Y | X) [9] is defined as

$$Sim(Y|X) \equiv \frac{H(Y) - H(Y|X)}{H(Y)} \tag{4}$$

Exchange X and Y to get the dependence of X on Y:

$$Sim(X|Y) \equiv \frac{H(X) - H(X|Y)}{H(X)} \tag{5}$$

Since the dependencies are mutual, define the following equation:

$$H(X|Y) = H(X|Y) = Sim(X, Y) \tag{6}$$

Sim(X,Y) is defined as the interdependence of X and Y, that is, the correlation between them.

$$Sim(X, Y) = 2\left[\frac{H(Y)+H(X)-H(X,Y)}{H(X)+H(Y)}\right] \tag{7}$$

3 Algorithm Description

In this algorithm, firstly, the weighted bipartite graph is constructed by taking label attributes and feature attributes as vertices, and their correlation degree is weight. Then, a series of graph operations are carried out, and finally the feature attributes corresponding to the remaining isolated points form the optimal attribute subset.

Step 1: Define the X label attribute set $X = \{v1, v2, ...vn\}$, and Y is the feature attribute set $Y = \{u1, u2, ...un\}$. Set the correlation degree between label attributes and feature attributes as the weight and calculate the correlation degree between each label and each feature, and record it as $(vi, vj) = aij$. Get the fully weighted bipartite graph $B0(X, Y)$.

Step 2: Set the correlation threshold between label attributes and feature attributes as α,if $aij < \alpha$, then delete aij. After deletion, if there is an isolated point, set this kind of feature attribute set as $Z = \{uk, uk + 1, ..um\}$, and remove the elements in Z from the Y set to get the set $Y1 = Y/Z$, so as to filter the relationship with low correlation between feature attributes and tag attributes. The final bipartite graph $B1(X, Y)$ is a complete graph.

Step 3: Take any two elements in the Y1 set as vertices, take the correlation between the two vertices as the weight, and record them as $(ui,uj) = cij$. $G(Y1)$ is a complete graph. Set the correlation threshold β and delete the edges of $cij < \beta$ in $G(Y1)$ to get the undirected graph $G2(X, Y1)$. The larger the bij is, the information contained in the two features can represent each other. The purpose of this step is to obtain the maximal connected subgraph with a certain degree of correlation between the characteristic attributes, namely vertices.

Step 4: For each maximal connected subgraph in $G2(Y1) = G2(X, Y1)/X$, if the number of vertices with the largest degree is not unique, the vertex with the smallest correlation between each label and feature and the edges connected to it will be deleted first, until only isolated points left in the graph, that is, the feature attributes left after filtering. The purpose of this step is to delete redundant information between feature attributes and ensure that as many feature attributes as possible are obtained.

4 Experiment

In this paper, the proposed algorithm uses MLkNN classification algorithm to compare and analyze the classification accuracy and other indicators.

4.1 Experimental Setting

In the experiment, the parameter setting range of the algorithm is $\alpha = 0.1$, $\beta \in [0.8, 1]$. To test the performance of the algorithm, this paper uses four multi-label data sets, Emotions, Flags, Scene and Yeast, and the experimental data comes from the multi-label learning library Mulan. Table 1 is an introduction to the data set.

Table 1. Introduction of experimental data set.

Number	Name	Features	Labels	Instances
1	emotions	72	6	593
2	flags	19	7	194
3	scene	294	6	2407
4	yeast	103	14	2417

In this paper, three commonly used evaluation indexes are used to evaluate the performance of the algorithm, which are average accuracy, Hamming loss and ranking loss. The range of the above indexes is [0, 1], in which the higher the average accuracy, the better the performance of the algorithm, while the Hamming loss and sorting loss are opposite. The specific definitions of the three indicators are as follows:

average precision [10]:

$$AP = \frac{1}{p} \sum_{i=1}^{p} \frac{1}{|yi|} \sum_{jk \in yi} \frac{|\{rank(xi,lj) \leq rank(xi,lk), lj \in yi\}|}{rank(xi,lk)} \tag{8}$$

Hamming loss(HL) [10]:

$$HL = \frac{1}{pC} \sum_{i=1}^{p} |y_i \prime \Delta y_i| \tag{9}$$

where Δ represents the symmetry difference between two sets.

ranking loss(RL) [10]:

$$RL = \frac{1}{p} \sum_{i=1}^{p} \frac{|\{(l_j,l_k)|f_j(x_i),(l_j,l_k) \in y_i \times \overline{y_i}\}|}{|y_i||\overline{y_i}|} \tag{10}$$

In addition, all experiments were classified by MLkNN, and k was set to 11 for 5 times cross-validation.

4.2 Experimental Results and Analysis

In the experimental process, the threshold sum in the algorithm has certain influence on the speed and accuracy of the algorithm: the threshold value α affects the edges connected by feature attributes and label attributes, and the larger the threshold value α is, the more edges will be deleted, possibly losing some important attributes; The smaller the threshold α is, the fewer edges are deleted, and the higher the redundancy of feature attributes. The threshold value β affects the connecting edges between feature attributes, and the threshold value β affects which features will be selected. Therefore, many experiments are needed to determine the threshold.

In order to evaluate and analyze the performance of the algorithm, α and β of the following data sets are determined through many experiments, and the average accuracy, Hamming loss and ranking loss are used for evaluation. The average accuracy of this correlation attribute algorithm is generally higher than that before data cleaning, and the Hamming loss and sorting loss are lower than that before data cleaning. It shows that this correlation attribute algorithm can screen features well, delete redundant attributes to the maximum extent, and improve the feature selection speed of multi-tags. On the whole, the correlation attribute algorithm has its merits, and the algorithm has simple steps and is easy to popularize. However, the algorithm still has some shortcomings, such as manually determining the threshold and deleting features excessively.

The following experimental data are the results when the average accuracy is the best, including the thresholds α of emotions, flags, scene and yeast are 0.1, the thresholds β are 0.9, 0.8, 0.8 and 0.98 respectively, and the number of selected features is 18, 11, 98 and 80 respectively.

Table 2. Average Accuracy of different multi-label feature selection algorithms.

Name	Correlation attribute algorithm	Uncleaned data
emotions	0.6566	0.6299
flags	0.7893	0.7675
scene	0.8144	0.7092
yeast	0.6913	0.6974

It can be seen from Table 2 that in some data sets, the correlation attribute algorithm can obviously improve the average accuracy of the classifier, and the accuracy of some data decreases slightly, but the dimension of the data is effectively reduced.

Table 3. Hamming Loss of different multi-label feature selection algorithms.

Name	Correlation attribute algorithm	Uncleaned data
emotions	0.2705	0.2725
flags	0.2567	0.3261
scene	0.0326	0.0906
yeast	0.2074	0.2068

The smaller the Hamming loss, the better the result of the classification algorithm. The experimental data in Table 3 reflect that multi-label feature algorithm can effectively reduce Hamming loss.

The experimental data in Table 4 show that the multi-label algorithm can reduce the dimensions of the data set, while not increasing the sorting loss excessively.

Table 4. Sorting Loss of different multi-label feature selection algorithms.

Name	Correlation attribute algorithm	Uncleaned data
emotions	0.5818	0.6118
flags	0.4379	0.5286
scene	0.1856	0.2876
yeast	0.4883	0.4430

5 Conclusion and Prospect

Based on graph theory, this paper transforms the problem of multi-classification feature selection in high-dimensional big data into a series of graph operations, and finally obtains the optimal attribute subset. This subset can effectively contain the information in the original data set. Experiments show that this algorithm can effectively reduce the data dimension, reduce redundant information and improve the classification rate in multi-label data sets. This method can be applied to unbalanced multi-label scenes to broaden the application scope of the model.

Acknowledgments. This project is supported by "Intelligent Interactive System for Identification and Tracking of Accident Vehicles", a college students' science and technology innovation project of Guangdong Province in 2022. (Number S202210564109).

Recommender: Professor Degui Yang, South China Agricultural University in China.

References

1. Guo, Y., Hu, J, Xu, C., Xu, W.: Attribute selection algorithm using correlation based on maximal connected subgraph. Software **05**, 69–72 (2014)
2. Tao, Y.: Application and research of filter ranking feature selection method in leukemia typing. (Master's thesis, Harbin Institute of Technology) (2020). https://kns.cnki.net/KCMS/detail/detail.aspx?dbname=CMFD 202201&filename=1021899313.nh
3. Xu, Y., Hu, X., Li, P.: Filter feature selection algorithm based on group mechanism. Application Research of Computers (05),1322–1326 (2016)
4. Min-Ling, Z., Peajosé, M., Victor, R.: Feature selection for multi-label naive bayes classification. Information Sciences Informatics and Computer Science, Intelligent Systems, Applications: An International Journal (2009)
5. Jiang, J.Y., Tsai, S.C., Lee, S.J.: Fsknn: multi-label text categorization based on fuzzy similarity and k nearest neighbors. Expert Syst. Appl. **39**(3), 2813–2821 (2012)
6. Hong, C.: Research on embedded multi-label feature selection algorithm (Master's thesis, Xi 'an Polytechnic University) (2019). https://kns.cnki.net/KCMS/detail/detail.aspx?dbname=CMFD202001&filename=1019954870.nh
7. Lei, G., Li, G.Z., You, M.Y.: Embedded feature selection for multi-label learning. J. Nanjing Univ. (Natural Sciences) **45**(5), 671–676 (2009)
8. Peng, H., Long, F., Ding, C.: Feature selection based on mutual information criteria of max-dependency, max-relevance, and min-redundancy. IEEE Trans. Pattern Anal. Mach. Intell. **27**(8), 1226–1238 (2005)

9. Yu, L., Liu, H.: Feature selection for high-dimensional data: a fast correlation-based filter solution. In: International Conference on Machine Learning. AAAI Press (2003)
10. Huang, R., Wu, Z.: Multi-label feature selection with dependence maximization and sparse regression. Comput. Eng. Design (07), 1898–1904 (2022). https://doi.org/10.16208/j.iss n1000-7024.2022.07.013

A Comprehensive Evaluation Method for Tourism Value Co-creation Based on Fuzzy Analysis

Wu Nan[1], Hu ChuXiong[2], and Wang Xiaoyu[3(✉)]

[1] School of International Cruise Yacht, Guangzhou Maritime University, Guangzhou 510000, China
[2] Macau University of Science and Technology, Macau 999087, China
[3] School of Management, Ji'nan University, Guangzhou 510632, China
244190054@qq.com

Abstract. With the development of the experience economy, tourists actively participate in the process of product production and service delivery to obtain experience value. From the perspective of tourist perception, this paper constructs a "Motivation-Behavior-Outcome" mechanism of tourist value co-creation using factor analysis, regression analysis, and fuzzy evaluation methods, based on the data collected in a recent survey. This paper has concluded that tourism enterprises should stimulate the motivation of tourist value co-creation, promote tourist value co-creation behavior, and enhance tourist experience value, for the purpose of achieving win-win cooperation between supply and demand, and promote the development of tourism industry.

Keywords: Fuzzy Comprehensive Evaluation · Tourism · Value Co-Creation · Tourist Participation · Perceived Value

1 Introduction

With the development of the experience economy and internet technology, tourists have more convenient access to information and more choices. Tourists invest time, money, energy, and other resources to participate in the whole process of tourism product production and service transmission, and jointly create value with tourism enterprises to obtain high-quality tourism experience. Therefore, the way of value co-creation between tourists and tourism enterprises is one of the hot topics in tour-ism marketing research. This paper aims to discover the mechanism of tourist value co-creation from the perspective of tourists' perception. Specifically, based on "Motivation-Behavior-Outcome" theory, this paper employs a series of factor analysis, regression analysis, and fuzzy evaluation methods to construct a model to explain the intrinsic mechanism of tourist value co-creation.

B.-Y. Cao et al. (Eds.): ICFIE 2022, LNDECT 207, pp. 343–357, 2024.
https://doi.org/10.1007/978-981-97-2891-6_26

2   Research Status at Home and Abroad

It is different from the traditional model of producers creating value alone, value co-creation theory believes that value is created by the cooperation between enterprises and their customers, and the interaction is the basis of value creation (Normann, Ramirez, 1993) [1], highlighting the core position of customers. At present, scholars have focused on the motivation, behavior, content, and results of tourists' participa-tion in value co-creation, such as Chouki & Peter (2013) explored a new tourism experience network framework, demonstrating how tourists and tourism enterprises can co-create experiences in the process of interaction [2]. Elaine & Hyelin (2016) analyzed how the co-creation of travel experiences affect traveler behavior, satisfaction, well-being, and loyalty [3]. Gao Shijun (2009) studied how to manage and co-create value in the Web 2.0 travel website community [4], and Li Lijuan (2012) explored the mechanism of tourist participation in tourism value co-creation from the perspective of antecedent factors, processes and results [5]. Jian Zhaoquan (2015) studied Ctrip's method of building an efficient service value network for travel websites [6]. However, there is little litera-ture to study the value co-creation mechanism based on tourists' perception. Therefore, this paper tastes from the perception of tourists, explores the price co-creation mech-anism, fees, achieves win-win cooperation between supply and demand, promotes the development of the tourism industry, improves the resources of tourists, co-creates insti-tutions, considers discretion, re-search combination, win-win cooperation, promotes the development of the tourism industry, improves tourists' resources, co-creates ingenuity, considers discretion, and studies a substance.

3   Statistical Analysis Methods, Research Data Sources, and Indicator Selection

3.1   Statistical Analysis Methods

This paper will use factor analysis to extract the questionnaire variables into new vari-ables, regression analysis to study the intrinsic correlation of new variables, and fuzzy evaluation to analyze the significance of new variables.

Factor analysis methods. The purpose of factor analysis is to seek the essence of variables, extract a few key variables, explain complex problems, and analyze the moti-vation variables, behavior variables, and value variables of the "Motivation-Behavior-Outcome" mechanism of tourist value co-creation.

Regression analysis methods. The purpose of regression analysis is to find the con-nection between variables and use the stepwise entry method to analyze the regres-sion relationship between the motivation, behavior, and results of the tourist value co-creation mechanism.

Fuzzy evaluation methods. The fuzzy evaluation method is to vaguely evaluate tourists' motivation, behavior, and results by establishing fuzzy similarity relation-ships.

3.2   Data Acquisition and Processing

The data were collected from February 15 to February 22, 2022, and tourists were randomly selected through an online platform for the questionnaire survey. While 145 questionnaires were distributed, 132 questionnaires were recovered, and 9 invalid questionnaires were eliminated using the Z test. As a result of data cleaning, 123 valid questionnaires were usable.

3.3 Construction of Indicator System

This paper assumes that motivation significantly affects behavior, and behavior significantly affects the outcome. This paper explores the mechanism of "Motivation-Behavior-Outcome" of tourist value co-creation, where motivation is the independent variable, behavior is the mediating variable, and the outcome is the dependent variable. The questionnaire uses the Likert 5-point scale to compile variable items, and a score of 1–5 represents the attitude of tourists to state their views: 5 means strong agreement, 4 means somewhat agree, 3 means average, 2 means disagree, and 1 means strongly disagree.

Definition and measurement of tourists' value co-creation motivation. Tourism is an ideal way to balance the needs of diversity. At present, scholars use satisfaction theory to study customer motivation, tourists' material needs, spiritual needs, and self-worth needs, corresponding to economic benefit motivation, emotional motivation, and social motivation. Measured by the X_i variable, $i = 1 \dots 7$.

Economic motivation. Kelly et al. (1990) argue that lower prices drive consumer engagement [7], helping consumers save on purchase price, time cost, risk cost, etc. The economic motive is to meet the basic practical needs of tourists, save economic costs and obtain material rewards. Questionnaire Table 1 is measured by $X_7 X_8$.

Emotional motivation. Rodie & Kleine (2000) pointed out that psychological pleasure and satisfaction in event interactions are the main motivations for customer participation [8]. Emotional motivation is the psychological need of tourists to experience friendship and belonging in group interaction and to satisfy a sense of pleasure. Questionnaire Table 1 is measured by $X_4 X_5 X_6$.

Social motivation. Naman (2002) concluded that customers expect to connect with other visitors, seek social recognition, and improve self-esteem through interactive engagement [9]. Social motivation is for tourists to exert themselves and gain a sense of accomplishment. Questionnaire Table 1 is measured by $X_1 X_2 X_3$.

Definition measurement of tourists' value co-creation and behavior. Yi and Gong (2013) argue that in the service industry, the necessary behaviors of customer value co-creation are composed of information search, information sharing, responsible behavior, and interpersonal interaction, and customer value co-creation is composed of feedback, advocacy, helpfulness, and tolerance [12]. This paper defines the behavior of tourist value co-creation as information sharing, responsibility behavior, interpersonal interaction, feedback, and advocacy. Measured by the Y_i variable, $i = 1 \dots 15$.

Information sharing. Busser and Shugla (2018) argued that the contribution of customer knowledge, experience, and skills promoted value co-creation [13]. Information sharing refers to tourists feeding back their needs to tourism enterprises, enterprises providing products and services that meet personalized needs, and tourists assisting in solving tourism problems by sharing tourism knowledge and experience. Questionnaire Table 2 was measured by $Y_1 Y_2 Y_3$.

Responsible acts. Bettencourt (1997) pointed out that customers and employees work together and follow the rules to create value with employees [14]. In tourism activities, tourists need to collaborate with employees to complete tourism activities and realize value. Questionnaire Table 2 was measured by $Y_4Y_5Y_6Y_7$.

Interpersonal interaction. Interpersonal interaction is the interpersonal relationship between customers and employees (Ennew, Binks, 1999) [15] and between tourists (Zhang Tianwen, 2015) [16]. Interpersonal interaction in this article refers to the interaction between tourists and employees and other visitors. Questionnaire Table 2 was measured by Y_8Y_9.

Feedback. Thuy (2015) believed that feedback is that customers voluntarily and proactively provide information to employees in the course of service to promote service improvement [17]. The feedback behavior in this article is that tourists evaluate tourism products and make suggestions to tourism enterprises. Questionnaire Table 2 was measured by $Y_{10}Y_{11}Y_{12}$.

Advocacy. Bettencourt (1997) pointed out that advocacy was the recommendation of products by customers to relatives and friends, which was an expression of customer loyalty to the enterprise and created additional value for the enterprises [14]. The advocacy behavior in this article is that tourists share travel experiences and recommend or encourage relatives and friends to purchase travel products. Questionnaire Table 2 was measured by $Y_{13}Y_{14}Y_{15}$.

Definition measurement of tourists value co-creation results. From the perspective of tourist perception, the participation result of the research was customer-perceived value, and customer-perceived value is the connotation of the result of co-creating value between enterprises and customers [5]. Sweeney and Soutar (2001) developed a four-dimensional model of user-perceived value including quality value, sentiment value, social value, and price value [18]. This paper argues that the result of tourist value co-creation participation is to improve tourists' perceived value, including functional value, emotional value, social value, and economic value, measured by Z_i variable, $i =$ 1...11. Functional value refers to tourism products to meet the needs of tourists for food, accommodation, travel, shopping, entertainment, broaden their horizons and increase their knowledge, questionnaire Table 3 was measured by $Z_1Z_2Z_3$; Emotional value refers to tourism products to meet tourists' intrinsic psychological needs such as pleasure, sense of achievement, and aesthetic experience, questionnaire Table 3 was measured by $Z_4Z_5Z_6$; Social value refers to tourism products to meet tourists' needs for socialization and self-esteem, questionnaire Table 3 was measured by $Z_7Z_8Z_9$; Economic value refers to tourists' perception of the cost of tourism products, including purchase price, time and effort costs, questionnaire Table 3 was measured by $Z_{10}Z_{11}$.

4 Empirical Analysis of Tourism Value Co-creation Mechanism Based on Factor Analysis

4.1 Analysis of the Demographic Characteristics of the Sample

63.6% of female tourists and 36.4% of men; 61.4% of tourists aged 20–35 and 28.8% of tourists aged 35–55, mainly young and middle-aged visitors. 75.7% had a bachelor's degree or above; 40.2% were employees of enterprises and institutions, and 12.1% were freelancers; Annual income below 100,000 yuan accounted for 49.2%; Access to travel information: Fliggy and other online travel service platforms accounted for 22.8%, Mafengwo and other travel guide websites accounted for 19.6%, WeChat channels accounted for 17.2%, and family and friends' referrals accounted for 11.1%; Relatives and friends accounted for 48.5% and 45.5% respectively.

4.2 Factor Analysis and Validity Test of Samples

In Table 1, Table 2, and Table 3, using the principal component analysis method of SPSS22.0 and the rotation method of Caesar's normalized maximum variance method, factor analysis, and validity tests were performed on 33 questionnaire variables of the visitor value co-creation mechanism, $Alpha = 0.945$, variance maximization rotation.

X_i extracts the motivation variable U_1; Y_i extracts the behavior variable U_2; Z_i extracts the result variable U_3. According to the content of the questionnaire, this article names the U_{ij}, as shown in Tables 1, 2, and 3. The largest common factor of eigenvalue is the variable with the greatest influence U_{ij}, the core motivation is social motivation U_{11}, the key behavior is interactive feedback U_{21}, and the result is mainly social value U_{31}.

$$U_{\text{Motivation}} = \begin{pmatrix} .575 & .786 & .795 & 0 & 0 & 0 & 0 \\ 0 & 0 & 0 & .839 & .851 & 0 & 0 \\ 0 & 0 & 0 & 0 & 0 & .546 & .923 \end{pmatrix}$$

$$U_{\text{Behavior}} = \begin{pmatrix} .659 & 0 & 0 & 0 & 0 & 0 & .703 & .663 & .631 & .690 & .765 & .636 & .586 & 0 & 0 \\ 0 & 0 & 0 & .658 & .725 & .763 & 0 & 0 & 0 & 0 & 0 & 0 & 0 & 0 & 0 \\ 0 & 0 & 0 & 0 & 0 & 0 & 0 & 0 & 0 & 0 & 0 & 0 & 0 & .767 & .761 \\ 0 & .772 & .616 & 0 & 0 & 0 & 0 & 0 & 0 & 0 & 0 & 0 & 0 & 0 & 0 \end{pmatrix}$$

$$U_{\text{Outcome}} = \begin{pmatrix} 0 & 0 & 0 & 0 & 0 & .732 & .767 & .733 & .666 & 0 & 0 \\ 0 & .759 & .853 & .728 & .557 & 0 & 0 & 0 & 0 & 0 & 0 \\ 0 & 0 & 0 & 0 & 0 & 0 & 0 & 0 & 0 & .822 & .800 \\ .902 & 0 & 0 & 0 & 0 & 0 & 0 & 0 & 0 & 0 & 0 \end{pmatrix}$$

4.3 Fuzzy Analysis of Tourist Value Co-creation Mechanism

Regression Analysis of the Correlation of the Visitor value Co-creation Mechanism of the Sample. Using SPSS22.0 stepwise entry regression method, the relationship between multiple linear regression of tourist value co-creation mechanism was analyzed. Each regression model has a p-value of 0.000, and the regression relationship is significant. The collinear statistical VIF value of each regression model is much less than 10, which excludes the problem of collinearity between variables, and the regression results are valid.

Table 1. Results of exploratory factor analysis of tourist "Motivation"

Tourist "Motivation" items	Factor loading		
	U_{11} Social motivation	U_{12} Emotional motivation	U_{13} Economic motivation
X_3 Participate in tourism activities allows me to have more contact with other tourists and make more friends	.795		
X_2 Participate in travel activities enhances my friendly relationship with staff	.786		
X_1 Participate in tourism activities gives me a sense of engagement	.575		
X_5 Get excited by interacting with travel activities		.851	
X_4 Participate in tourism activities and interact to get a sense of leisure pleasure		.839	
X_7 Participate in tourist activities can save on the cost of playing			.923
X_6 Participate in interactive tourism activities to get material rewards such as souvenirs and raffle prizes			.546
Eigenvalue	2.155	1.929	1.259
Cumulative variance explanatory value (%)	30.785	58.347	76.335
KMO = 0. 820; Bartlett's spherical test = 330.950, p = 0.000			

Table 2. Results of exploratory factor analysis of tourist "Behaviors"

Tourist "Behavior" items	Factor loading			
	U_{21} Interactive feedback	U_{22} Responsible Conduct	U_{23} Advocacy	U_{24} Information sharing
Y_{11} I would like to comment on issues arising from tourism products and services	.765			
Y_7 In the interaction of tourism activities, I cooperate with other tourists	.703			
Y_{10} I would like to rate the products and services I experienced	.690			
Y_8 I talk to my employees and exchange feelings	.663			
Y_1 Tourist information and activities are detailed and easily accessible	.659			
Y_{12} I would like to come up with new ideas for tourism development, such as adding new activities or services	.636			
Y_9 I interact with other tourists	.631			
Y_{13} I share my travel itinerary and experience with my family and friends	.586			

(*continued*)

Table 2. (*continued*)

Tourist "Behavior" items	Factor loading			
	U_{21} Interactive feedback	U_{22} Responsible Conduct	U_{23} Advocacy	U_{24} Information sharing
Y_6 I work with my staff in the interaction of travel activities		.763		
Y_5 I treat employees politely		.725		
Y_4 During the interaction of travel activities, I follow the arrangements of the staff		.658		
Y_{15} I encourage friends and family to buy travel products			.767	
Y_{14} I recommend travel products to my family and friends			.761	
Y_2 I am willing to present my needs to my employees and get personalized products and services				.772
Y_3 I am happy to contribute my experience and knowledge to the interaction of travel activities				.616

(*continued*)

Table 2. (*continued*)

Tourist "Behavior" items	Factor loading			
	U_{21} Interactive feedback	U_{22} Responsible Conduct	U_{23} Advocacy	U_{24} Information sharing
Eigenvalue	3.948	2.597	2.274	1.453
Cumulative variance explanatory value (%)	26.322	43.637	58.799	68.482

KMO = 0.873; Bartlett's spherical test = 906.864, p = 0.000

Table 3. Results of exploratory factor analysis of tourist " Outcomes"

Tourist "Outcome" items	Factor loading			
	U_{31} Social value	U_{32} Intellectual emotional value	U_{33} Economic value	U_{34} Functional value
Z_7 I make more friends by interacting with travel activities	.767			
Z_8 I get a sense of engagement by interacting with my employees	.733			
Z_6 I love interacting with travel activities	.732			
Z_9 I became more confident during the tour	.666			
Z_5 It's interesting to interact with tourist activities		.557		
Z_3 I have broadened my horizons by interacting with travel activities		.853		

(continued)

Table 3. (*continued*)

Tourist "Outcome" items	Factor loading			
	U_{31} Social value	U_{32} Intellectual emotional value	U_{33} Economic value	U_{34} Functional value
Z_2 I learned new knowledge and skills through participating in tourism activities		.759		
Z_4 It's fun to interact with tourist activities		.728		
Z_{10} Individual travel products are moderately priced			.822	
Z_{11} Overall tourism products are good value for money			.800	
Z_1 After expressing my needs to my employees, the products and services they provide (such as food, accommodation, travel, shopping, and entertainment) are more in line with my requirements				.902
Eigenvalue	3.112	2.624	1.893	1.529
Cumulative variance explanatory value (%)	25.935	47.802	63.576	76.317
KMO = 0.862; Bartlett's spherical test = 828.325, p = 0.000				

KMO is greater than 0.8, Bartlett spherical test is greater than 300, $p = 0.000$, to verify the convergence and discriminant validity of the scale. In this paper, 11 common factor U_{ij} with feature values greater than 1 are extracted from the $X_i Y_i Z_i$ variable to obtain the transpose factor load matrix.

Table 4. Multiple linear regression analysis results of tourist value co-creation mechanism

Motivation			Behavior	Outcome			
U_{11} Social motivation	U_{12} Emotional motivation	U_{13} Economic motivation	Normalize regression coefficients	U_{31} Social value	U_{32} Intellectual emotional value	U_{33} Ec-onomic value	U_{34} Functional value
.259	—	.259	U_{21} Interactive feedback	.384	—	—	.417
.253	—	-.178	U_{22} Responsible Conduct	—	.340	—	—
—	.224	.200	U_{23} Advocacy	.183	.227	—	.169
.255	—	—	U_{24} Information sharing	.206	—	—	.243

Note: The value is the standardized regression coefficient of the significant regression at the 0.05 level, and the no coefficient value is the insignificant regression at the 0.05 level

In Table 4, the standardized regression coefficients reflect the significance effect, the coefficients indicate there is a significant regression effect between the two variables, and no coefficients indicate there is no significant regression effect between the two variables. Write the following matrix of correlation coefficients:

$$C_{\text{Motivation}-\text{Behavior}} = \begin{pmatrix} .259 & .253 & 0 & .255 \\ 0 & 0 & .224 & 0 \\ .259 & -.178 & .200 & 0 \end{pmatrix}$$

and

$$C_{\text{Behavior}-\text{Outcome}} = \begin{pmatrix} .384 & 0 & 0 & .417 \\ 0 & .340 & 0 & 0 \\ .183 & .227 & 0 & .169 \\ .206 & 0 & 0 & .243 \end{pmatrix}$$

Fuzzy Comprehensive Evaluation Based on Samples. The values of each row of the above matrix are taken absolute and normalized to obtain the "Motivation-Behavior" evaluation matrix:

$$R_{\text{Motivation}-\text{Behavior}} = \begin{pmatrix} .3377 & .3299 & 0 & .3325 \\ 0 & 0 & 1 & 0 \\ .4066 & .2794 & .3140 & 0 \end{pmatrix}$$

With the "Behavior-Outcome" evaluation matrix:
From the value 0, the value co-creation behavior has nothing to do with economic value.

$$R_{\text{Behavior}-\text{Outcome}} = \begin{pmatrix} .4794 & 0 & 0 & .5206 \\ 0 & 1 & 0 & 0 \\ .3161 & .3921 & 0 & .2919 \\ .4588 & 0 & 0 & .5412 \end{pmatrix}$$

Get the "Motivation-Outcome" evaluation matrix:

From the value 0, the value co-creation motivation has nothing to do with economic value.

$$R_{\text{Motivation}-\text{Outcome}} = \begin{pmatrix} .3144 & .3299 & 0 & .3557 \\ .3161 & .3921 & 0 & .2919 \\ .2942 & .4025 & 0 & .3033 \end{pmatrix}$$

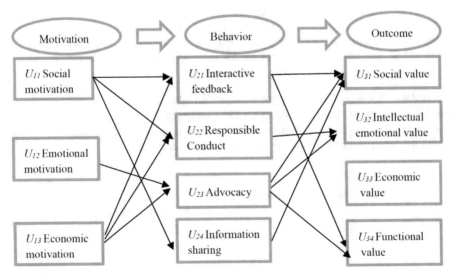

Fig. 1. Multiple linear regression analysis of tourist value co-creation mechanism

The motivation of tourist value co-creation significantly affects the value co-creation behavior, the value co-creation behavior significantly affects the value co-creation outcome, and the "Motivation-Behavior-Outcome" mechanism of tourist value co-creation is established (see Fig. 1). The key variables with the greatest in-fluence of motivation, behavior and outcome are social motivation, interactive feed-back, and social value, the core motivation is social motivation, the key behavior is interactive feedback, and the result is mainly social value. The path dependence of key variables was explored, and the social motivation of visitor value co-creation (core motivation) significantly affected interactive feedback (key behavior), and interactive feedback (key behavior) significantly affected social value (primary out-come).

5 Conclusions and Suggestions

The previous sections have constructed a mechanism for the co-creation of tourist value from the perspective of tourists, such that tourism enterprises are enabled to stimulate the motivation of tourists for value co-creation, to promote value co-creation behavior among tourists, and to enhance the perceived value of tourists. It can be concluded that the key of the proposed mechanism is stimulating the social motivation of tourists' value co-creation, which promotes interactive feedback behavior of tourists' value co-creation, and enhances tourists' perceived social value, intellectual and emotional value and functional value. Therefore, tourism enterprises are suggested to adopt the following measures to promote these values respectively.

5.1 Stimulate Social Motivation for Social Value

Tourism Enterprises Should Propagate the Cultural Connotation of Tourism Products in Multiple Ways. Including disseminating tourism leisure, cultural entertainment, and social value, for the purpose of stimulating the social motivation of tourists, promoting the co-creation of all values in the behavior process, and meeting the social value of tourists' identity and status. Such that a solid foundation for the high-quality development of the tourism industry could be formed.

Meanwhile, Tourism Enterprises Should Recognize and Advocate Tourists' Responsible Behaviors. Encouraging them to follow the itinerary of tourism enterprise employees, cooperating with employees and fellow tourists. As a result, a variety of immersive tourism activities can be jointly experienced by the tourists, which achieves the goal of promoting social value by meeting tourists' sense of being respected, self-utility, self-affirmation, and self-realization.

5.2 Stimulate Emotional Motivation for Emotional Value

Tourism Enterprises Should Stimulate Tourists' Emotional Motivation. Encourage information interaction with tourists, and enhance the emotional value. Tourist enterprises should provide opportunities and space for tourists who travel with relatives and friends, to enhance family communication, which in turn deepens the emotional value. On the other hand, Tourist enterprises should encourage interaction among tourists, enterprise employees, and other tourists, by guiding tourists to participate in tourism activities, enabling them to experience pleasant friendship in group interactions, promoting the breadth of their interpersonal interactions, and as a result, expanding their emotional value.

Tourism Enterprises Should Also Design New, Interesting, Educational Leisure Activities. Focusing on the theme of tourism products. In addition to the traditional music, song and dance, magic, and other ornamental performance activities, interaction and shared experience among tourists with local cultural characteristics should be deeply explored. These new and interesting entertainment activities and unparalleled integration of science and technology would provide tailor-made products for tourists. These measures stimulate emotional motivation, encourage responsible behaviors and emotional communication, which make tourists happy and satisfied, realizing their anticipation of emotional value.

5.3 Stimulate Functional Motivation for Functional Value

Tourism Enterprises Should Design Reasonable Entertaining Experience Activities. Including and not limited to catering, accommodation, travel, and shopping. So that tourists can enjoy high-quality tourism products and services, and promote tourists' information interaction and emotional communication. By careful design of product itineraries, plans of catering, arrangement of special theme activities, service quality would be greatly improved.

Tourism Enterprises can Also Provide Functional Value. By means of displaying intangible experiences in tangible forms to price-sensitive tourists, such as photos and souvenirs, giving them material rewards such as credits and gifts, or properly propagate the cost performance of the tour.

5.4 Dilute the Motivation for Economic Value

Although the economic interest motive of tourists' value co-creation can promote value co-creation behaviors, neither the value co-creation motivation nor the behaviors can result in significant increase of economic value. Perhaps tourists' participation in the process of value co-creation needs to invest a lot of time, energy and money costs, which is contrary to tourists' pursuit of economic value. Therefore, enterprises should weaken the motivation for economic interests, dilute economic value, and should not overly pursue cost performance.

The survey of this paper had been conducted from February 15 to February 22 2022, during which the tourism consumption delayed by the pandemic might be fully released, at an expected positive growth rate of the economy abundant with enthusiastic and confident consumers. Therefore, the above survey results have only shown people's tourism consumption behaviors in an economic growth cycle. Future research could focus on tourism consumption behaviors under an economic downturn.

Acknowledgements. Recommender: Professor Jian-Xin Li, School of Mathematics and Statistics, Guangdong University of Technology in China.

References

1. Richard, N., Ramirez, R.: From value chain to value constellation: designing interactive strategy. Harv. Bus. Rev. **71**(4), 65–77 (1993)
2. Sfandla, C., Björk, P.: Tourism experience network: co-creation of experiences in interactive processes. Int. J. Tourism Res. **15**(5) (2013)
3. Mathis, E.F., Kim, H.: The effect of co-creation experience on outcome Variable. Ann. Tour. Res. **9**(57), 62–75 (2016)
4. Gao, S.: Research on Community Management and Value Co-creation of Web 2.0 Travel Websites: A Case Study of Travel Websites Along Kikushima. Taipei: College of Business and Management, Chengchi University, 1–15 (2009)
5. Li, L.: Research on co-creation of tourism experience value. Beijing Forestry University (2012)

6. Jian, Z., Xiao, X.: Service innovation and value co-creation in network environment: a case study of ctrip. J. Ind. Eng. Eng. Manage. **29**(01), 20–29 (2015)
7. Kelley, S.W., James, H., Donnelly Jr. J., Skinner, S.: Customer participation in service production and delivery. J. Retailing 1990(3), 315–335
8. Rodie, A.R., Kleine, S.S.: Customer participation in services production and delivery [M]. Sage Publications, Thousand Oaks, CA (2000)
9. Nambisan, S.: Designing virtual customer environments for new product development: toward a theory. Acad. Manag. Rev.Manag. Rev. **27**(3), 392–413 (2002)
10. Dabhholkar, P.A.: How to improve perceived service quality by improving customer participation in developments in marketing science. In: Dunlap, B.J. (ed.) Academy of Marketing Science, Cullowhee, pp. 483–487 (1990)
11. Wu, C.H.-J., Liang, R.-D.: Effect of experiential value on customer satisfaction with service encounters in luxury-hotel restaurants. Int. J. Hosp. Manag.Manag. **28**, 586–593 (2009)
12. Yi, Y., Gong, T.: Customer value co-creation behavior: scale development and validation. J. Bus. Res. **66**(9), 1279–1284 (2013)
13. Busser, J.A., Shulga, L.V.: Co-created value: multidimensional scale and nomological network. Tour. Manage. **65**, 69–86 (2018)
14. Bettencourt, L.A.: Customer voluntary performance: customer as partners in service delivery. J. Retail. **73**(3), 383–406 (1997)
15. Ennew, C.T., Binks, M.R.: Impact of participative service relationships on quality, satisfaction and retention: an exploratory study. J. Bus. Res. **46**(2), 121–132 (1999)
16. Zhang, T.: Research on cruise tourism happiness based on rooted theory. Shanghai University of Engineering Science (2015)
17. Thuy, P.N.: Customer participation to co-create value in human transformative services: a study of higher education and health care services. Service Business, pp. 1–26 (2015)
18. Sweeney, J.C., Soutar, G.N.: Consumer perceived value: the development of a multiple item scale. J. Retail. **77**(2), 203–220 (2001)
19. Bing-Yuan, C.: Optimal Models and Methods with Fuzzy Quantity. Springer (2010)

U-type Tube Vibration Based Fuel Density Portable Testing Instrument Research

Fuhong Qin[1], Yuyan Li[2(✉)], Taifu Li[1], Xianguo Wang[3], Hongchao Zhao[2], and Qiang He[1]

[1] School of Safety Engineering, Chongqing Institute of Science and Technology, Chongqing, China

[2] School of Electrical Engineering, Chongqing Institute of Science and Technology, Chongqing, China
1070339600@qq.com

[3] Chongqing Torchheart Intelligent Technology Research Institute, Chongqing, China

Abstract. To overcome the shortcomings of traditional engine fuel testing equipment such as poor portability, poor stability, and expensive, a portable fuel density testing instrument was designed and developed. The physical anti-Jamming technology and temperature compensation technology are used to simplify the tedious experimental process used in the traditional method, and the impact of temperature changes on the measurement accuracy of the testing instrument is analyzed and explained, and the U-type tube vibration technology is used to address the problem of low measurement accuracy of the current fuel density measurement method, and the final real-time density measurement accuracy reaches $0.0001 g/cm^3$ and the measurement range is $(0.6000–2.000)\ g/cm^3$. The results show that the instrument can meet the demand of fuel density measurement with high accuracy, strong stability, and portability through the core components of double U-type tube and temperature compensation technology.

Keywords: U-type Tube Vibration Technology · Fuel Density · Temperature Compensation · Density Detection · Portability · Instrument

1 Introduction

Density is one of the important indicators to characterize the quality of fuel and fuel products, density tester is a physical analysis instrument to measure the density of fluids, mineral fuel of fuel, whose density data can reflect its combustibility and volatility [1]. With the development of science and technology, the requirements for instrumentation accuracy are getting higher and higher. The military industry requires very high and fuel density characteristics of engine fuel density indirectly affects the performance and life of the engine, which in turn creates uncertain safety hazards.

Currently, many scholars have conducted a series of studies on fuel density detection technology. Yang, D. et al. [2] used two indicators of the transmission time difference and phase of ultrasonic waves to calculate the density of oil and designed a monitoring device, but its structure is complicated and the measurement effect is not very high. Guo, Y. et al. [3] proposed a method to measure the density of fuel using the relationship of ultrasonic wave, temperature, and speed of sound, and the error of fuel density measurement was about 0.1%, although the relationship of temperature was considered, the effect after temperature compensation was not studied. Li K. et al. [4] proposed a combination of ZigBee wireless network and ultrasonic time difference density measurement method to solve the detection of multi-channel density using a CC2530 chip as the core controller, which improved the density measurement accuracy and made the error less than 0.5%. However, the wireless network is susceptible to environmental factors, which leads to unstable results. He, B. et al. [5] designed a resonant density logging method and detection circuit through the function between the resonant frequency of the tuning fork sensor and the density of the fuel dependent on the tuning fork, which enabled the measurement accuracy to reach 0.1%, but the measurement method did not take into account the influence of relevant variables such as temperature. Quemet, A. et al. [6, 7] investigated uranium from generated radioactive effluent, simplicity, uncertainty estimation, and detection limit for uranium concentration detection and found that the detection limit and uncertainty of ID-TIMS is lower than that of KED. The radioactivity method is fast, and if not used correctly it can cause harm to humans and the environment and is less applicable. Cheng, X. [7] used three different vibration modes of fluid flow through a pipe to establish a vibration mathematical model and designed a MIMR-XZ6 pipeline liquid density meter the error is about 1%, which meets the industrial demand but does not consider the existence of a certain correlation between liquid density and uncertain parameters such as temperature. Yang, F. et al. [8] came up with a temperature compensation method for this monitoring system to improve the measurement accuracy of the image-based roadbed settlement monitoring system.

In recent years, in the field related to oil density detection, the commonly used method techniques include regression method [9], Monte Carlo and other methods [10], and neural network method [11, 12] for direct or indirect density measurement. After a literature survey, it is found that the current trend is to transition from traditional methods to new methods of measurement with artificial intelligence technology, which has the advantages of high accuracy, high automation and portable lines. Secondly, the new components also have a great impact on their measurement accuracy [13]. Translated with www.DeepL.com/Translator (free version). Literature [14–16] studied gamma ray as a key method for rapid fuel density measurement, and the results showed that the average relative error of 3.2% for the total flow, 4.3% for the gas flow, 11.5% for the fuel flow and 7.8% for the water flow in the experimental study of horizontal fuel-gas-water flow by gamma ray. Sleiti, A. et al. [17] proposed a new GOR study method with wide applicability in order to establish a reliable gas-surge detection model and simulation to accurately predict the PVT properties of gas, and a comprehensive evaluation of 63 empirical correlations of four gas properties. Based on the evaluation results, the most accurate gas viscosity and density correlations for high temperature and high pressure (HTHP) conditions were recommended, and the results showed a minimum average absolute relative error (AARE) of 3.50% and a maximum error of 4.45%.Jiao [18] used a commercial vibrating tube densitometer (VTD) calibrated with a reliable physical model to regress calibration data (vibration period, temperature, and pressure) linearly and nonlinearly into a physics-based calibration model to obtain the calibration parameters. The results show that the relative deviation of the measured density from the density calculated using the GERG-2008 equation of state is between −0.5% and 0.4%, and that the accuracy of the vibration method is highest when applied to liquid density detection.

In summary, the traditional fuel density detection equipment has low detection accuracy, large size, and long detection period, which cannot meet the needs of modern fast and high-precision detection. Therefore, many domestic and foreign scholars have launched decades-long research on density detection methods, anti-Jamming strategies, and how to optimize detection accuracy.

Therefore, in this paper, we design a new double U-type glass tube device for vibration frequency acquisition, enhance frequency feature extraction, use a polynomial temperature compensation model to compensate for density difference in detection results, and finally propose a model of U-type vibration fuel density portable detection instrument incorporating a machine learning algorithm for temperature compensation, which is verified to be more effective. This instrument has the features of high detection accuracy, portable type, and a high degree of automation to enhance the automation of fuel density measurement and continuously improve the efficiency of fuel metering and fuel supply.

2 U-type Tube Fuel Density Measurement Principle

Figure 1 shows the working principle of the U-type tube vibration, which consists of a double U-type tube, a magnet, and a base. After the liquid to be measured enters and fills the U-type tube, it vibrates together with the U-type tube, the equivalent mass of the system changes, and the resonant frequency also changes [19]. Since this resonant frequency has a definite relationship with the density of the liquid, the density of the liquid to be measured can be found by determining the resonant frequency of the system [20].

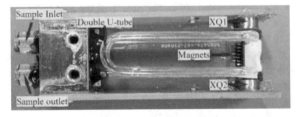

Fig. 1. U-type vibrating tube working principle diagram

Calculate the air density at the test temperature from Eq. (1).

$$d_a = 0.01293 \times \frac{273.15}{t} \times \frac{p}{101.3} \tag{1}$$

Density (D) and relative density (D/D) can be calculated from Eq. (2) and Eq. (3).

$$D = d_w + K_1\left(T_s^2 - T_w^2\right)$$ (2)

$$D/D = 1 + K_2 \times \left(T_s^2 - T_w^2\right)$$ (3)

Using the period of vibration of air and water and the corresponding density values, K1 and K2 are calculated by Eq. (4) and Eq. (5).

$$K_1 = \frac{d_w - d_a}{T_w^2 - T_a^2}$$ (4)

$$K_2 = \frac{1.0000 - d_a}{T_w^2 - T_a^2}$$ (5)

where: t is the test temperature, K; P is the pressure at t temperature; K_1 is the instrument constant for the density test; K_2 is the instrument constant for the relative density test; T_w is the vibration period of water in the U-type tube, s; Tw is the vibration period when the U-type tube is filled with sample, unit s; d_w is the density of water at the test temperature, g/ml; D is the density, g/cm^3; D/D is the relative density.

The liquid density calculation in this paper is based on the densities of water and air at the calibration temperature, and by measuring the frequency Ts of fuel vibration in the U-type tube and finally using the above-mentioned relationship, the density of fuel can be accurately calculated. It can be seen that the ambient temperature is an uncertain key variable and it is impossible to be at an absolute constant temperature, so the temperature compensation appears necessary to be studied.

3 Design of U-type Vibration Based Fuel Density Detection Device

3.1 Hardware Design

The densitometer instrument is divided into a vibration module and a signal processing module. Firstly, the vibration module is to generate the vibration frequency through the U-type tube, and secondly, the obtained signal is passed to the data processing module, and the STM32 microcontroller records the resonant frequency of the U-type tube by pulse counting [21], and the excitation signal is amplified by power and then the electromagnet generates the excitation force to supplement the energy loss of the U-type tube during the vibration to make it vibrate stably, and finally the liquid density value is calculated and displayed on the resistive touch screen [22]. The overall design of the instrument is shown in Fig. 2.

3.2 Resonant Circuit and Frequency Acquisition Circuit

After the density detector is powered on, the potential difference on coil XQ2 is generated through LMOUT2, and electromagnetic induction is generated using an electromagnet. At this time, on the other side of the U-type tube, coil XQ1 is cut off, which results in a

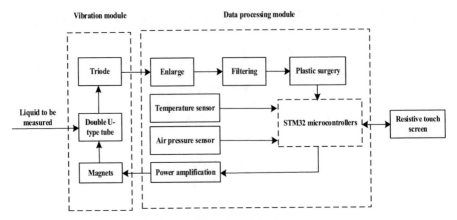

Fig. 2. Overall design of the instrument

voltage change in coil XQ1, which is then passed back to coil XQ2 after an amplification circuit, at which time the whole circuit forms a closed loop and the U-type vibrating tube is quickly stabilized. Using STM32 as the core microcontroller, the vibration frequency signal of the U-type tube is collected through the triode PL. The resonant circuit is shown in Fig. 3.

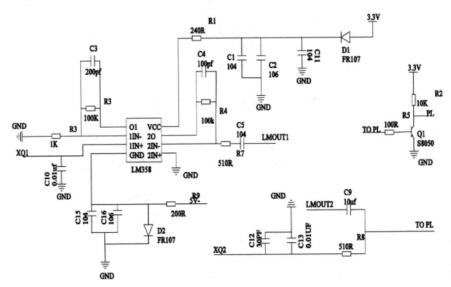

Fig. 3. Resonant circuit and frequency acquisition circuit diagram

3.3 Software Design

The system software is written in C programming language through keil5 compiler [23], and the system program mainly includes a frequency detection program, timing interrupt program, Bluetooth transmission program, resistive touch screen TJC3224T024_011R display program and microcontroller minimum system program, etc. After the system is powered on, it first performs a reset operation and then sends a control command to the microcontroller STM32 through the SPI bus to generate an excitation signal to excite the U-type vibrating tube. The feedback circuit is then used to collect the resonant frequencies of different liquid samples, and the temperature sensor MAX 31865 and the air pressure sensor HP203B are connected to the microcontroller to measure the sample data at a certain temperature. Finally, the density of the liquid to be measured is calculated and analyzed inside the microcontroller. The density of the liquid to be measured is calculated and analyzed inside the microcontroller. The density calculation results are sent to the host computer using Bluetooth wireless transmission module, which is connected to the resistive touch screen TJC3224T024_011R for display. The program flow is shown in Fig. 4.

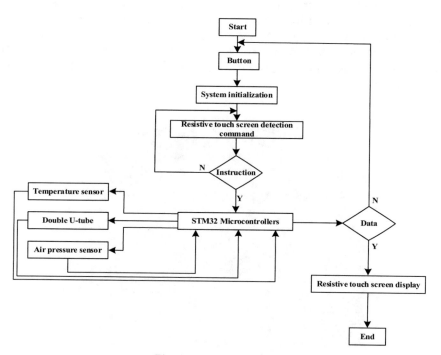

Fig. 4. Program flow chart

4 Experimental Data Analysis

To verify the accuracy of the instrument's test results, a comprehensive test analysis of the instrument is required. The imported high-precision fuel density testing instrument DMA35 is used, and the data measured by this instrument is used as the standard value, whose density measurement range is between 0 and 3.0000 g/cm³, and the measurement accuracy is 0.0010 g/cm³ [24]. The density of the sample fuel in the experiment must be within the measurement range of this instrument, and the standard value of the parameters such as the density of the sample fuel is measured by this standard instrument firstly, and then the measured value is compared with the standard value by the intelligent density detector developed in this paper. The density of the sample fuel must be within the measurement range of the instrument.

4.1   Relationship Between Fuel Vibration Frequency, Density and Temperature

Two prototypes, as shown in Fig. 5, were developed by combining hardware and software, and experimental tests were conducted to explore the relationship between vibration frequency, density, and temperature, and to propose effective instrument calibration measures. Firstly, the DMA35 instrument and No.1 prototype were used to test 0# diesel fuel at different temperatures, and the test results are shown in Table 1. The standard value of 0# fuel density 0.8266 g/cm3 in Table 1 is obtained by the actual measurement of the standard instrument DMA35, while the experimental data of real-time density at different temperatures are collected by the developed prototype, and the standard error value is the difference between the prototype and the standard value after the temperature compensation.

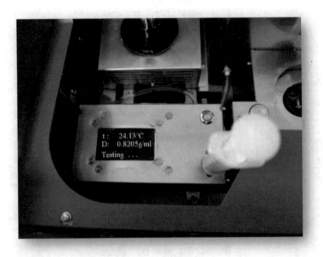

Fig. 5. Density testing prototype

Secondly, compared with the direct fitting method for ultrasonic density measurement in the literature [3], the residual sum of squares from 31 sets of sampling points were compared and found that the residual square of the U-type vibration method (0.001667) was better than the residual square of the function fitting method (0.001669), indicating that the U-type vibration technique is more effective for sample fuel density measurement. In this paper, the U-type vibration was used to analyze the experimental data with

Table 1. 0# diesel high temperature test (standard value: 0.8266 g/cm3)

Fuel temperature	Changes	Real-time density	Corrected density	Standard Error	Fuel temperature	Changes	Real-time density	Corrected density	Standard Error
4.0 °C	4.2	0.8375	0.8268	0.0002	22.6 °C	26.3	0.8219	0.8264	−0.0002
	4.4	0.8373	0.8267	0.0001		26.4	0.8219	0.8264	−0.0002
	4.2	0.8375	0.8267	0.0001		26.5	0.8219	0.8265	−0.0001
	4.3	0.8375	0.8266	0.0000		26.5	0.8218	0.8264	−0.0002
9.1 °C	9.6	0.8336	0.8265	−0.0001	26.3 °C	26.3	0.8219	0.8264	−0.0002
	9.6	0.8336	0.8264	−0.0002		26.4	0.8219	0.8264	−0.0002
	9.6	0.8335	0.8264	−0.0002		26.5	0.8219	0.8265	−0.0001
	9.4	0.8335	0.8263	−0.0003		26.5	0.8218	0.8264	−0.0002
13.6 °C	13.7	0.8309	0.8267	0.0001	32.5 °C	32.1	0.8177	0.8264	−0.0002
	13.9	0.8309	0.8267	0.0001		32.4	0.8176	0.8264	−0.0002
	14	0.8307	0.8266	0.0000		32.4	0.8174	0.8263	−0.0003
	13.9	0.8307	0.8265	−0.0001		32.3	0.8174	0.8265	−0.0001
18.5 °C	18.7	0.8275	0.8266	0.0000	37.3 °C	36.8	0.8148	0.8267	0.0001
	18.8	0.8273	0.8265	−0.0001		63.9	0.8146	0.8267	0.0001
	18.9	0.8274	0.8266	0.0000		37	0.8146	0.8267	0.0001
	18.9	0.8272	0.8265	−0.0001		37.1	0.8148	0.8269	0.0003

31 sets of sampling points, which were plotted by Matlab 2018 [25] in Fig. 6. Where the horizontal coordinate represents the temperature and the vertical coordinate represents the sample fuel vibration frequency, and it can be seen that the sample fuel vibration frequency increases with increasing temperature.

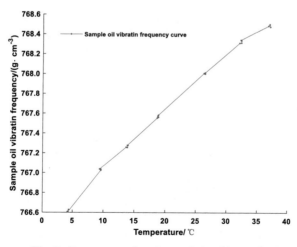

Fig. 6. Temperature-frequency relationship graph

The correlation experiments conducted by SPSSPRO software yielded that the correlation coefficient value between temperature and real time density showed significance, which was -0.9340, and the correlation coefficients of the remaining variables were also less than 0, implying that there was a negative correlation between temperature and real time density. At the same time, the correlation coefficient value is close to 0. There is no significant and correlation between temperature and standard density, calibration value and 2 terms.

It is important to study the method to overcome the correlation between temperature and real-time density according to the practical needs. Therefore, it is necessary to study the method to overcome the correlation between temperature and real time density according to the practical needs, so that it reaches the density value at a certain temperature of calibration is a constant value. In this paper, the non-linear relationship between real time density and temperature, as shown in Fig. 7, can be corrected by temperature compensation to obtain the standard density of the fuel.

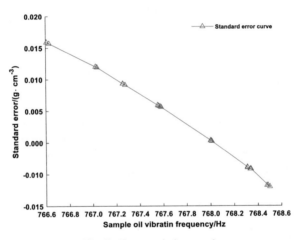

Fig. 7. Error variation graph

4.2 Polynomial-Based Temperature Compensation Technique

In this paper, a polynomial fit is used for temperature correction, and the goodness of fit R2 is 0.9999. Figure 8 shows that the real-time density measured by Prototype 1 can almost coincide well with the calibration value after the polynomial temperature compensation technique. The density values calculated by this algorithm are fluctuating above and below the calibration value at different temperatures as can be seen from Table 1 and Table 3. In addition, at the fuel temperature of 37.7 °C, it is obvious that it is a false detection, after the outlier value is rejected. A study was conducted to explore the relationship between the difference between the real-time density and the standard density and the frequency.

The test was conducted using prototypes No. 1 and No. 2, with the fuel temperature calibrated was calibrated at 17.7 °C and the calibration value was 0.8266 g/cm3. Since

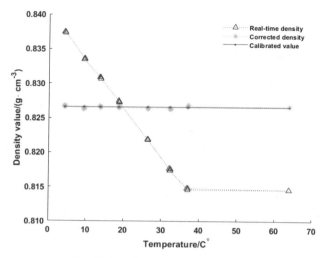

Fig. 8. Density-temperature test chart

the vibration will produce certain heat change, which leads to the sample being tested temperature change, and the U-type vibration technique in the temperature has a greater impact on the detection results, therefore, the polynomial is used to correct the real-time density, through the difference between the calibration value and the corrected density is formulated as the standard error in this paper. The final statistical results are shown in Table 2.

The temperature in Table 2 is the real-time temperature of the fuel sample, measured by the temperature sensor MAX 31865, and the real-time density is obtained from the measurement of this instrument. The maximum sample density difference of 0.0002 g/cm3 between different samples of corrected density can be analyzed, and the result statistics are shown in the Table 3. The test results of each sample are the same, that is, they are reproducible and meet the practical needs. Meanwhile, the algorithm involved in the instrument developed in this paper does not distinguish the sample fuel, as long as it belongs to the fuel that can calculate the standard density value. So the same use of these 1 and 2 good samples to test 3# jet diesel, the data as shown in Table 4. Similarly, the maximum sample density difference of 3# jet fuel is 0.0002 g/cm3 by analyzing Table 4, and the density error analysis is shown in Table 5.

The above values between temperature and real-time density in Table 1 show that a temperature change of 0.1 °C will have an impact on the four decimal places of density value measurement, which must be controlled within the national standard error of liquid density measurement to be meaningful for the study. The research test shows that the polynomial temperature compensation technology based on the instrument can achieve better results through several experiments and the density detection accuracy of many different fuels are high, the fuel density detection accuracy can reach 0.01%.

Table 2. 0# diesel fuel test comparison data

0# Diesel(Fuel temperature:17.7 °C Calibrated value:0.8266 g/cm³)

Machine No. 1				Machine No. 2			
Temperature	Real-time density	Corrected density	Standard Error	Temperature	Real-time density	Corrected density	Standard Error
17.4	0.8284	0.8267	0.0001	17.6	0.8484	0.8267	0.0001
17.6	0.8283	0.8267	0.0001	17.7	0.8484	0.8268	0.0002
17.6	0.8282	0.8266	0.0000	17.7	0.8283	0.8267	0.0001
17.7	0.8282	0.8266	0.0000	17.7	0.8484	0.8268	0.0002
17.8	0.8282	0.8267	0.0001	17.8	0.8484	0.8269	0.0003
17.8	0.8281	0.8266	0.0000	17.8	0.8484	0.8269	0.0003
17.8	0.8282	0.8267	0.0001	17.8	0.8283	0.8268	0.0002
17.8	0.8282	0.8267	0.0001	17.9	0.8283	0.8268	0.0002
17.8	0.8282	0.8267	0.0001	17.9	0.8484	0.8269	0.0003
17.9	0.8282	0.8268	0.0002	17.8	0.8282	0.8268	0.0002

Table 3. 0# diesel density error analysis

Maximum prototype density difference of standard density :0.0002 g/cm³

	Machine No. 1				Machine No. 2			
Maximum value	17.9	0.8282	0.8268	0.0002	17.9	0.8484	0.8269	0.0003
Minimum value	17.6	0.8281	0.8266	0.0000	17.8	0.8282	0.8267	0.0001
Maximum Difference	0.3	0.0001	0.0002	0.0002	0.1	0.0202	0.0002	0.0002
Average value	17.8	0.8282	0.8267	0.0001	17.8	0.8383	0.8268	0.0002

4.3 Decision Tree-Based Temperature Compensation Technique

Machine learning is a very powerful and popular prediction or classification method, which has been widely used in medical diagnosis, big data forecasting, etc. The decision tree can be constructed using the frequency density error dataset in this paper, and by several common machine learning methods, firstly, setting the ratio of a training set to the validation set as 7:3, and secondly, using the 10-fold cross-check method to compare the results, as shown in Table 6. The results show that the decision tree for temperature error prediction works best, and since the size of the decision tree is independent of the size of the data set, this algorithm is used as a correction for temperature deviation.

Table 4. 3# jet fuel test comparison data

3# Jet fuel(Fuel temperature:17.7 °C Calibrated value:0.7985)

Machine No. 1				Machine No. 2			
Temperature	Real-time density	Corrected density	Standard Error	Temperature	Real-time density	Corrected density	Standard Error
17.5	0.8004	0.7986	0.0001	17.6	0.8003	0.7986	0.0001
17.6	0.8003	0.7986	0.0001	17.7	0.8003	0.7987	0.0002
17.6	0.8004	0.7987	0.0002	17.7	0.8002	0.7986	0.0001
17.7	0.8003	0.7986	0.0001	17.8	0.8002	0.7986	0.0001
17.7	0.8003	0.7986	0.0001	17.8	0.8002	0.7986	0.0001
17.7	0.8003	0.7986	0.0001	17.9	0.8003	0.7988	0.0003
17.7	0.8003	0.7987	0.0002	17.9	0.8003	0.7988	0.0003
17.7	0.8003	0.7987	0.0002	17.9	0.8002	0.7987	0.0002
17.8	0.8003	0.7987	0.0002	17.9	0.8002	0.7988	0.0003
17.8	0.8002	0.7986	0.0001	17.9	0.8001	0.7986	0.0001

Table 5. 3# jet fuel density error analysis

Maximum prototype density difference of standard density:0.0002 g/cm^3

	Machine No. 1				Machine No. 2			
Maximum value	17.8	0.8004	0.7987	0.0002	17.9	0.8003	0.7988	0.0003
Minimum value	17.5	0.8002	0.7986	0.0001	17.6	0.8001	0.7986	0.0001
Maximum Difference	0.3	0.0002	0.0001	0.0001	0.3	0.0002	0.0002	0.0002
Average value	17.7	0.8003	0.7986	0.0001	17.8	0.8002	0.7987	0.0002

Table 6. Comparison of machine learning algorithms

Machine Learning Algorithms	Average error	Mean standard deviation
SVM	0.0047%	0.0026%
LASS	0.0050%	0.0025%
EN	0.0050%	0.0025%
CART	0.0001%	0.0000%

In data set D, the input space is divided into M cells. Each cell has an output value Cm, the regression tree model can be expressed as

$$f(x) = \sum_{m=1}^{M} C_m I(x \in R_m) \tag{6}$$

$$D = \{(x_1, y_1), (x_2, y_2), \ldots, (x_n, y_n)\} \tag{7}$$

$$M = \{R_1, R_2, \ldots, R_n\} \tag{8}$$

$$C_m = ave(y_i | x_i \in R_m) \tag{9}$$

where I is the indicator function, if X belongs to the division region Rm, it is 1, otherwise, it is 0. Using the squared error and minimum to solve the optimal value of each cell, the absolute error with the calibrated density value after using the decision tree CART for temperature compensation and polynomial temperature compensation is shown in Fig. 9. The absolute error after polynomial temperature compensation fluctuates around 0.0001 g/cm^3, but the effect of temperature compensation based on the decision tree algorithm is more stable and only one absolute error is 0.0002 g/cm^3. Therefore, this method is more effective and the average error is smaller.

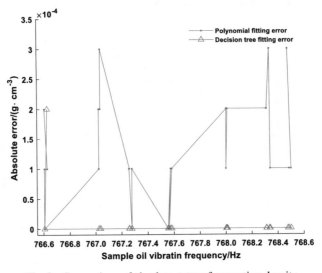

Fig. 9. Comparison of absolute error of correction density

5 Conclusion

This paper researched a new portable density testing instrument with double U-typed tubes, completed the hardware design and software design of the instrument device, and conducted experimental research on the designed instrument. The relationship between fuel density, frequency, and temperature is obtained through data processing, and the temperature compensation algorithm is used to correct the density difference and integrate the machine learning algorithm for temperature compensation, which overcomes the shortcomings of the traditional density measurement method such as complicated operation and large error. The experimental results show that it can meet the requirements of high precision, strong stability, and portable measurement of fuel products. Meanwhile, the density measurement of fuel products is an important research problem for indirect mass flow measurement methods, and the measurement accuracy of the instrument reaches the national fuel delivery standard of 0.35% mass flow meter requirement [26]. The instrument in this paper has been successfully applied in a military enterprise, which provides some reference and reference for the next research on related instrumentation devices and has good application prospects.

Acknowledgments. This work was supported by the Research funding project of Chongqing University of Science and Technology (ckrc2019021); Science and Technology Research Program of Chongqing Municipal Education Commission (Grant No: KJQN202001536); Chongqing Natural Science Foundation General Project(cstc2021jcyj-msxmX0833); Chongqing University of Science and Technology Science and Technology Innovation Project (YKJCX2120715).

Recommender: Associate Professor Yang Jie, Chongqing Industry & Trade Polytechnic in China.

References

1. Wang, X., Zhang, L., Zhang, R., et al.: Study on density variation characteristics of high temperature fuel in combustion chamber and control method. Exp. Fluid Mech. **35**(3), 83–87 (2021)
2. Yang, D., Zhou, B., Xu, C., et al.: Online calibration method for pneumatic conveying capacitive laminar imaging. Chin. J. Electr. Eng. **30**(14), 31–35 (2010)
3. Guo, Y., Zhang, X., Meng, F., et al.: Research on online fuel density detection device based on ultrasonic wave velocity. Instrum. Technol. Sens. **9**, 106–109 (2018)
4. Li, K., Cui, Y., Yang, W.: Design of distributed density detection system based on ZigBee and TDC-2. Instrum. Technol. Sens. **11**, 73–75 95 (2017)
5. He, B., Wei, Y., Yu, H., et al.: Resonant tuning fork density logging method and detection circuit design. Instrum. Technol. Sens. **5**, 119–123 (2020)
6. Quemet, A., Ruas, A., Esbelin, E., et al.: Development and comparison of two high accuracy methods for uranium concentration in nuclear fuel: ID-TIMS and K-edge densitometry. J. Radioanal. Nucl. Chem. **321**(3), 997–1004 (2019)
7. Cheng, X.: Development of resonant density meter and its application in tailings backfilling. Adv. Civ. Eng. **2021**, 1–7(2021)
8. Yang, F.: Temperature compensation method for image-based roadbed settlement monitoring system. J. Instrum. 1–9 (2022)
9. Akhmadiyarov, A., Marczak, W., Petrov, A.P., et al.: Measurements of density at elevated pressure – a vibrating-tube densimeter calibration, uncertainty assessment, and validation of the results. J. Mol. Liq. **336**, 116196 (2021)
10. Berlizov, A.N., Sharikov, D.A., et al.: A quantitative Monte Carlo modelling of the uranium and plutonium X-ray fluorescence (XRF) response from a hybrid K-edge/K-XRF densitomete. Nucl. Inst. Methods Phys. Res. **615**(1), 127–135 (2010)

11. Abdulaziz, s.: Application of artificial intelligence and gamma Attenuation__Techniques for Predicting Gas–Fuel–Water Volume Fraction in__Annular regime of three-phase flow independent of Fuel__Pipeline's Scale Layer. Mathematics **2021**, 1–14 (2021)

12. Roshani, M., Phana, G., Farajc, R., et al.: Proposing a gamma radiation based intelligent system for simultaneous analyzing and detecting type and amount of petroleum by-products. Nucl. Eng. Technol. **53**(4), 1277–1283 (2020)

13. Makarov, D.M., Egorov, G.I., et al.: Density of water - 2-pyrrolidone mixture a new vibrating tube densimeter from (278.15–323.15) K and up to 70 MPa. J. Mol. Liq. **335**, 116113 (2020)

14. Yahaya, D.B., Joseph, X.F.R., Aliyu, M.A., et al.: Characteristics of horizontal gas-liquid two-phase flow measurement in a medium-size d pipe using gamma densitometry. Sci. Afr. **10**, e00550 (2020)

15. Sungyeop, J., Yewon, K., Seunghoon, P.: Material mixture solution analysis by a hybrid L-edge/L-XRF densitometer. Appl. Radiat. Isot. **146**, 1–4 (2019)

16. Ma, Y., Li, C., Pan, Y., et al.: A flow rate measurement method for horizontal fuel-gas-water three-phase flows based on Venturi meter, blind tee, and gamma-ray attenuation. Flow Meas. Instrum. **80**, 101965 (2021)

17. Sleiti, A.K., Al-Ammari, W.A., et al.: Comprehensive assessment and evaluation of correlations for gas-fuel ratio, fuel formation volume factor, gas viscosity, and gas density utilized in gas kick detection. J. Petrol. Sci. Eng. **207**, 109135 (2021)

18. Jiao, F., Al Ghafri, S.Z.S., et al.: Extended calibration of a vibrating tube densimeter and new reference density data for a methane-propane mixture at temperatures from (203 to 423) K and pressures to 35 MPa. J. Mol. Liq. **310**, 113219 (2020)

19. Feng, S., Du, H., Wang, H.Z.: Design and implementation of resonant liquid density meter. Sens. Microsys. **38**(10), 92–95 (2019)

20. GBT 29617–2013 Test method for determining the density, relative density and API specific gravity of liquids by digital densitometer (2013)

21. Guo, Z., Yuan, J.: Design of STM32-based intelligent control system for pulsed lasers. J. Quantum Electron. **36**(2), 161–167 (2019)

22. Lv, X., Zhang, B., Liu, Z.J., et al.: STM32's frequency acquisition method for real-time operating systems. Electron. Technol. Softw. Eng. **21**, 36–38 (2021)

23. Wang, F., Du, X.: Design of integrated screwdriver controller based on keil software. Software **40**(9), 75–80 (2019)

24. Parmar, P., Lopez-Villalobos, N., Tobin, J.T., et al.: Effect of temperature on raw whole milk density and its potential impact on milk payment in the dairy industry. Int. J. Food Sci. Technol. **56**(5), 2415–3242 (2020)

25. Li, J., Mao, Y.: Leakage simulation analysis of large LNG storage tanks based on MATLAB. Refinery Technol. Eng. **50**(4), 50–54 (2020)

26. Su, Y., Guo, W., Sheng, J.: Flow Measurement and Testing (2nd edition). China Metrology Press, Beijing (2007)

The Research on System Structure Weights of Feedback, Scale, Inclusion and Their Utility Functions

Liting Zeng, Jianbo Guan[✉], and Yunshi Fong

Guangzhou College of Technology and Business, Guangzhou, People's Republic of China
787420515@qq.com

Abstract. The diversity and complexity of system structure determines to study the utility function of system structure weights, and the relationship between the system structure and the utility function. The feedback, scale, and inclusion are the three basic system structures, which characteristics are similar to those of the previously studied series and parallel. They all consist of subsystems which can be regarded as elements of a system, if the subsystem is regarded as the new parent system, the relationship between the subsystem and its next level subsystem is similar to that between the subsystem and its parent system. In this paper, we have established utility function of system structure weights of feedback, scale and inclusion, and tried to take the green innovation system of manufacturing industry as an example.

Keywords: Systems Structure Weights · Utility Function · Green Innovation System in Manufacturing Industry

1 Introduction

A number of interconnected, interacting elements with certain functions in organism, which is called a system. There are two conditions must be satisfied: first, the set X of objects contains at least two different objects; second, objects which refers to system elements are grouped in a certain way. In reality, systems can be found everywhere, such as ecosystems in nature, meteorological systems, and also economic systems in human economic life. Understanding the characteristics of systems are beneficial to people's study of systems, and helps to optimize systems and make decision. The purpose of complexity research is to carry out cross-scientific and cross-domain research. It is believed that the complexity of things is developed from simplicity and is generated in the process of appropriate environment. It is believed that there are some general laws in the rule system that control the behavior of these systems.

System structures refers to the whole of the ways and orders in which the elements in the system structure interact with each other, they have two characteristics: relative stability and absolute variability [2]. Relative stability means that the system tends to maintain a certain state. The absolute variability of system is caused by the variability of

© The Author(s), under exclusive license to Springer Nature Singapore Pte Ltd. 2024
B.-Y. Cao et al. (Eds.): ICFIE 2022, LNDECT 207, pp. 373–382, 2024.
https://doi.org/10.1007/978-981-97-2891-6_28

the environment, with a gradual spiral of change. The system environment refers to all transactions which are related to the system. The system environment is ubiquitous. For example, the connection between children and their parents, and the social connection between people around them. Such changes in the system environment can lead to changes in the whole system. The structure of system can determine the function of system, the system environment affects the function of system output, and the system reacts to the environment. To sum up, in order to further study the system error and system optimization, it is necessary to study the system structures. This paper takes the green innovation system in manufacturing industry as an example, to discuss three basic system structures of feedback, scale, inclusions, and their utility function of system structure weights.

2 Feedback Structure

Feedback is a form of interaction between systems and its environment. The output of system becomes part of the input during the interaction, which in turn acting on the system itself, and affecting the output of system. According to the influence of feedback, this structure can be divided into positive and negative. The former enhances the output of system; the latter weakens the output of system, and the study of negative feedback is the core issue of cybernetics. Taking the evolution of manufacturing innovation system as an example, the system is composed of state subsystem and process subsystem, which can produce positive and negative feedback mechanism with the evolution of green innovation system in manufacturing industry. The feedback that causes the system to further deviate from the original state is positive feedback, which can promote the continuous evolution of green innovation system in manufacturing industry from low level to high level. The feedback that promotes the system to keep developing to the original state is negative feedback, which will hinder the evolution of the green innovation system of manufacturing industry. Under the interaction of positive and negative feedback, the evolution process of green innovation system in manufacturing industry shows obvious dynamics and complexity (Fig. 1) [1]. Feedback is a form of interaction between systems and its environment. The output of system becomes part of the input during the interaction, which in turn acting on the system itself, and affecting the output of system. According to the influence of feedback, this structure can be divided into positive and negative. The former enhances the output of system; the latter weakens the output of system, and the study of negative feedback is the core issue of cybernetics. Taking the evolution of manufacturing innovation system as an example, the system is composed of state subsystem and process subsystem, which can produce positive and negative feedback mechanism with the evolution of green innovation system in manufacturing industry. The feedback that causes the system to further deviate from the original state is positive feedback, which can promote the continuous evolution of green innovation system in manufacturing industry from low level to high level. The feedback that promotes the system to keep developing to the original state is negative feedback, which will hinder the evolution of the green innovation system of manufacturing industry. Under the interaction of positive and negative feedback, the evolution process of green innovation system in manufacturing industry shows obvious dynamics and complexity [8].

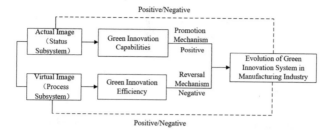

Fig. 1. System Architecture Diagram of Feedback System Structure

The utility function of feedback structure system is described as FSG ($s_1, s_2 \ldots\ldots s_n$), $n \geq 2$, and the utility in the feedback structure system is directed, such as $\vec{a}, \vec{b}$. Thus, the utility function $FSG(s_1, s_2 \ldots\ldots s_n)$ of feedback system weight is $FSG_y(s_i, s_j) = G_f(s_i, s_j) = (w_{ij} + \Delta w_{ij})G_y(s_i) \wedge G_y(s_j)$ or $(w_{ij} + \Delta w_{ij}) \wedge G_y(s_i).G_w(s_j)$.

Among them, $G_w(s_i)$ and $G_w(s_j)$ are functions of subsystem s_i and subsystem s_j respectively, w_{ij} is the directed structure weight of subsystem s_i and subsystem s_j.

$$FSG_y(s_1, s_2 \ldots\ldots s_n) = G_y(s_1, s_2 \ldots\ldots s_n)$$
$$= (w_{12} + \Delta w_{12})G_y(s_1) \wedge w_{23}G_y(s_2) \wedge \ldots\ldots \wedge w_{ij}G_y(s_j) \ldots\ldots$$
$$\wedge w_{n1}G_y(s_1)$$

Or $w_{12}(G_y(s_1) + \Delta G_y(s_1)) \wedge w_{23}G_y(s_2) \wedge \ldots\ldots \wedge w_{ij}G_y(s_j) \wedge \ldots\ldots \wedge w_{n1}G_y(s_1)$

Or $w_{12}(G_y(s_1) + \Delta G_y(s_1)) + w_{23}G_y(s_2) + \ldots\ldots + w_{ij}G_y(s_j) + \ldots\ldots + w_{n1}G_y(s_1)$

Or $((w_{12} + \Delta w_{12}) \wedge w_{23} \wedge \ldots\ldots \wedge w_{ij} \wedge \ldots\ldots \wedge w_{n1} \wedge (G_y(s_1) + G_y(s_2) + \ldots\ldots + G_y(s_j) + \ldots\ldots + G_y(s_1))$

Or $(w_{12} + \Delta w_{12})G_y(s_1).w_{23}Gy(s_2) \ldots\ldots w_{ij}G_y(s_j) \ldots\ldots w_{n1}G_y(s_1)$ or other types.

As $w_{12} = w_{23} = \ldots\ldots = w_{n-1,n} = w$, $FSG_y(s_1, s_2 \ldots\ldots s_n) = G_y(s_1, s_2 \ldots\ldots s_n) = (w + \Delta w)G_y(s_1) \wedge G_y(s_2 \ldots\ldots \wedge G_y(s_j) \wedge \ldots\ldots \wedge G_y(s_n) \wedge G_y(s_1)$.

Or $(w + \Delta w)G_y(s_1) \vee G_y(s_2) \vee \ldots\ldots \vee G_y(s_j) \vee \ldots\ldots \vee G_y(s_n) \vee G_y(s_1)$.

Or $w(G_y(s_1) + \Delta G_y(s_1)) \wedge G_y(s_2) \wedge \ldots\ldots \wedge G_y(s_j \wedge \ldots\ldots \wedge G_y(s_1)$

Or $w(G_y(s_1) + \Delta G_y(s_1)) + G_y(s_2) + \cdots\ldots + G_y(s_j) + \cdots\ldots + G_y(s_1)$

Or $(w + \Delta w)G_y(s_1).G_y(s_2).\ldots\ldots G_y(s_j).\ldots\ldots G_y(s_n).G_y(s_1)$

Or $(w + \Delta w)G_y(s_1) + G_y(s_2) + \ldots\ldots + G_y(s_j) + \ldots\ldots + G_y(s_n) + G_y(s_1)$ or other types.

As $w_{12} = w_{23} = \ldots\ldots = w_{n-1,n} = w = 1$,

$FSG_y(s_1, s_2 \ldots\ldots s_n) = G_y(s_1, s_2 \ldots\ldots s_n) = (1 + \Delta w)G_y(s_1) \wedge G_y(s_2) \wedge \ldots\ldots \wedge G_y(s_j) \wedge \ldots\ldots \wedge G_y(s_1)$.

Or $(1 + \Delta w)G_y(s_1) + G_y(s_2) + \ldots\ldots + G_y(s_j) + \ldots\ldots + G_y(s_1)$.

Or $(G_y(s_1) + \Delta G_y(s_1)) \wedge G_y(s_2) \wedge \ldots\ldots \wedge G_y(s_j) \wedge \ldots\ldots \wedge G_y(s_1)$.

Or $(G_y(s_1) + \Delta G_y(s_1) + G_y(s_2) + \ldots\ldots + G_y(s_j) + \ldots\ldots + G_y(s_1)$.

Or $(1 + \Delta w)(G_y(s_1).G_y(s_2).\ldots\ldots G_y(s_j).\ldots\ldots G_y(s_1)$ or other types.

As $w_{12}, w_{23}, \ldots\ldots, w_{n-1,n},$ one of them equals 0, $FSG_y(s_1, s_2 \ldots\ldots s_n) = G_y(s_1, s_2 \ldots\ldots s_n) = 0$.

Among them, $G_w(s_1)$, $G_w(s_2)$ ......, $G_w(s_n)$ are the functions of subsystems s_1, s_2 ......s_n respectively, w_{ij} is the directed structure weight of subsystem s_i and subsystem s_j.

3   Scale Structure

The arrangement of elements in the system structure makes the relationship (information flow or logistics, etc.) divided into n outflow through one element, $n \geq 2$, or n outflow flows into the same element at the same time, and the branches are not connected with each other. If this relationship is unidirectional, it is called the directed expansion structure, and if this relationship is bidirectional, it is called an undirected expansion structure [3].

3.1   Expansion

The connection and interaction between the elements with the contraction, the function of the starting elements of the system structure generally affects the function of the ending element. Taking Green Innovation Efficiency in Manufacturing Industry as an example (Fig. 2) [8]:

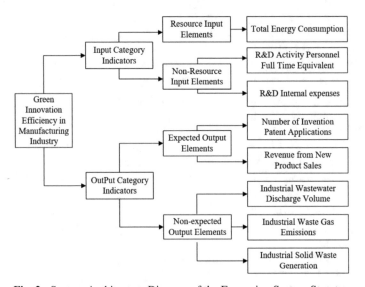

Fig. 2.   System Architecture Diagram of the Expansion System Structure

3.2   Contraction

The following system contains subsystems with contraction between elements and multiple expansion subsystems. Then taking the manufacturing green innovation capability

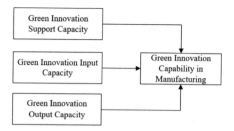

Fig. 3. System Architecture Diagram of the Contraction System Structure

system as an example. The function of any element at the beginning of the contraction system structure will generally affect the function of the end element (Fig. 3) [8].

The utility function $KSG(s_1, s_2 \ldots s_n)$ of scale structure system weights is $KSG_y(s_i, s_j) = G_f(s_i, s_j) = w_{ij}G_y(s_i) \vee G_y(s_j)$ or $w_{ij}G_y(s_i) + G_w(s_j)$ $G_w(s_i)$ and $G_w(s_j)$ are the functions of subsystems s_i and s_j respectively, and w_{ij} are the directed structure weights of subsystems s_i and s_j.

$$KSG_y(s_1, s_2 \ldots \ldots s_n) = G_y(s_1, s_2 \ldots \ldots s_n)$$
$$= w_{12}G_y(s_1) \vee w_{23}G_y(s_2) \vee \ldots \ldots \vee w_{(n-1)n} G_n(s_n)$$
Or $w_{12}G_y(s_1) + w_{23}G_y(s_2) + \ldots \ldots + w_{(n-1),n}G_n(s_n)$.Or other types.
As $w_{12} = w_{23} = \ldots \ldots = w_{n-1,n} = w$, $KSG_y(s_1, s_2 \ldots \ldots s_n) = G_y(s_1, s_2 \ldots \ldots s_n) =$
$w(G_y(s_1) \vee G_y(s_2) \vee \ldots \ldots \vee G_n(s_n))$
Or $w(G_y(s_1) + G_y(s_2) + \ldots \ldots + G_n(s_n))$.Or other types.
As $w_{12} = w_{23} = \ldots \ldots = w_{n-1,n} = w = 1, KSG_y(s_1, s_2 \ldots \ldots s_n) = G_y(s_1, s_2 \ldots \ldots s_n) = G_y(s_1) \vee$
$G_y(s_2) \vee \ldots \ldots \vee G_n(s_n)$
Or $G_y(s_1) + G_y(s_2) + \ldots \ldots + G_n(s_n)$. Or other types.
As $w_{12} = w_{23} = \ldots \ldots = w_{n-1,n} = w = 0$, $KSG_y(s_1, s_2 \ldots \ldots s_n) = G_y(s_1, s_2 \ldots \ldots s_n) = 0$

Among them, $G_w(s_1)$, $G_w(s_2) \ldots \ldots$, $G_w(s_n)$ are the functions of subsystems s_1, s_2 $\ldots \ldots s_n$ respectively, w_{ij} is the directed structure weight of subsystems s_i and s_j.

4   Inclusion Structure

Inclusion is an inclusive relationship. If X contains Y, the meaning of Y is contained in X. The function of any element other than the first or last element in the system structure will not directly affect the function of other elements, but it will affect the function of system. The function of inclusion structure system weight, if each level is regarded as a subsystem, it is a series system, but only the first or last element of the series system is connected with or directly affects other elements of the system [4].

If each level is regarded as a subsystem, the arrangement of elements in the system structure makes the relationship (information flow or logistics, etc.) all pass through each element, and the structure has the sequential relationship of elements [7]. Only the first or last element of the series system is connected with other elements of the system. In addition, the function of any element in the system structure other than the first or last element will not directly affect the function of other elements, but it will affect the function of system. If this relationship is unidirectional, it is called directed inclusion structure, and is bidirectional, it is called undirected inclusion structure [5]. In the case of green innovation systems in manufacturing, there are embedded organic links and interactions between the elements (Fig. 4) [8].

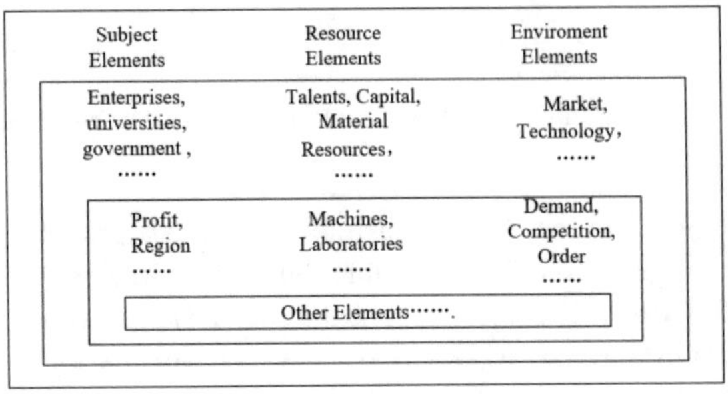

Subject Elements	Resource Elements	Enviroment Elements
Enterprises, universities, government , ...	Talents, Capital, Material Resources, ...	Market, Technology, ...
Profit, Region	Machines, Laboratories	Demand, Competition, Order ......
Other Elements·······.		

Fig. 4. System Architecture Diagram of Inclusion System Structure

If each level of the inclusion structure is regarded as a subsystem, which is a series system CSG (s_1, s_2... s_n), the utility function of inclusion structure system weight is.

4.1 Undirected Inclusion Structure System

The utility function YSG(s_1, s_2 ......s_n) of undirected inclusion structure system weight is

$$YSG_w(s_i, s_j) = G_w(s_i, s_j) = (G_w(s_i) \wedge G_w(s_j))w_{ij}$$
$$YSG_y(s_1, s_2 \ldots \ldots s_n) = G_y(s_1, s_2 \ldots \ldots s_n)$$
$$= w_{12}G_w(s_1) \wedge w_{23}G_w(s_2) \wedge \ldots \ldots \wedge w_{ij}G_w(s_j) \wedge \ldots \ldots$$
$$\wedge w_{n-1,n}G_y(s_{n-1}) \wedge G_y(s_n)$$

Or $(w_{12} \wedge w_{23} \wedge \ldots \ldots \wedge w_{ij} \wedge \ldots \ldots \wedge w_{n-1,n})(G_w(s_1) + G_w(s_2) + \ldots \ldots + w_{ij}G_w(s_j) + \ldots \ldots + G_y(s_{n-1}) + G_y(s_n))$

Or $(w_{12} \wedge w_{23} \wedge \ldots \ldots \wedge w_{ij} \wedge \ldots \ldots \wedge w_{n-1,n})(G_w(s_1) \vee Gw(s2) \vee \ldots \ldots \vee wijGw(sj) \vee \ldots \ldots \vee Gy(sn - 1) \vee Gy(sn))$

Or $w_{12}G_w(s_1). w_{23}G_w(s_2). \ldots \ldots \ldots w_{ij}G_w(s_j). \ldots \ldots w_{n-1,n}G_w(s_{n-1}). G_y(s_n)$. （The weights in the expression can also be right-to-left）.Or other types.

As $w_{12}{=}w_{23}{=} \cdots\cdots{=}w_{n-1,n}{=}w$, $YSG_y(s_1, s_2 \ldots \ldots s_n) = G_y(s_1, s_2 \ldots \ldots s_n) = w(G_w(s_1) \wedge G_w(s_2) \wedge G_w(s_3) \wedge \ldots \ldots \wedge G_w(s_j). \ldots \ldots \wedge G_y(s_n))$,

Or $w(G_w(s_1) + G_w(s_2) + G_w(s_3) + \ldots \ldots + G_w(s_j). \ldots \ldots + G_y(s_n))$,

Or $w(G_w(s_1) \vee G_w(s_2) \vee G_w(s_3) \vee \ldots \ldots \vee G_w(s_j). \ldots \ldots \vee G_y(s_n))$,

Or $wG_w(s_1). G_w(s_2)G_w(s_3) \ldots \ldots G_w(s_j) \ldots \ldots G_y(s_n)$.Or other types.

As $w_{12}{=}w_{23}{=} \cdots\cdots{=}w_{n-1,n}{=}w{=}1$, $G_y(s_1) \wedge G_y(s_2) \ldots \ldots) \wedge G_y(s_j) \wedge \ldots \ldots \wedge G_y(s_n)$,

Or $G_y(s_1) + G_y(s_2) \ldots \ldots + G_y(s_j) + \ldots \ldots + G_y(s_n)$,

Or $G_y(s_1) \vee G_y(s_2) \vee \ldots \ldots \vee G_y(s_j) \vee \ldots \ldots \vee G_y(s_n)$,

Or $G_y(s_1)). G_y(s_2) \ldots \ldots G_y(s_j). \ldots \ldots G_y(s_n)$ .Or other types.

As w_{12} , w_{23} , $\cdots\cdots$, $w_{n-1,n}$, one of them is equal to zero，

$YSG_y(s_1, s_2 \ldots \ldots s_n) = G_y(s_1, s_2 \ldots \ldots s_n) = 0$.Or other types.

Among them, $G_w(s_1), G_w(s_2)$ ......,$G_w(s_n)$ are the functions of subsystems s_1, s_2 ......s_n respectively, w_{ij} is the directed structural weight of subsystems s_i and s_j.

4.2 Directed Inclusion Structure System

The utility function $CSG(s_1, s_2 ......s_n)$ of directed inclusion structure system weight is
$$YSG_y(s_i, s_j) = G_y(s_i, s_j) = (w_{ij}G_y(s_i)) \wedge G_y(s_j) \text{ or } (w_{ij}G_y(s_i)).G_y(s_j).$$
Among them, $G_w(s_i)$ and $G_w(s_j)$ are the function of subsystems s_j and s_j respectively, and w_{ij} is the directed structural weights of subsystems s_i and s_j.

$$YSG_y(s_1, s_2 ......s_n) = G_y(s_1, s_2 ......s_n)$$
$$= w_{12}G_y(s_1) \wedge w_{23}G_y(s_2) \wedge ......\wedge w_{ij}G_y(s_j) \wedge ......\wedge w_{(n-1)n}G_y(s_n).$$
Or $w_{12}G_y(s_1) \vee w_{23}G_y(s_2) \vee ......\vee w_{ij}G_y(s_j) \vee ......\vee w_{(n-1)n}G_y(s_n).$
or $w_{12}G_y(s_1).w_{23}G_y(s_2). ......w_{ij}G_y(s_j). ......w_{(n-1)n}G_y(s_n).$

or

$w_{12}G_y(s_1) + w_{23}G_y(s_2) + ... + w_{ij}G_y(s_j) + ... + w_{(n-1)n}G_y(s_n). \text{ Or other types.}$

As $w_{12}=w_{23}= \cdots\cdots=w_{n-1, n}=w$, $YSG_y(s_1, s_2 ......s_n) = G_y(s_1, s_2 ......s_n) =$
$w(G_y(s_1) \wedge G_y(s_2) \wedge ......\wedge G_y(s_j) \wedge ......\wedge G_y(s_n)).$
Or $w(G_y(s_1) + G_y(s_2) + ... + G_y(s_j) + ... + G_y(s_n)).$
Or $w(G_y(s_1)).G_y(s_2) ......G_y(s_j). ...... G_y(s_n)). \text{ Or other types.}$

As $w_{12}=w_{23}= \cdots\cdots=w_{n-1, n}=w=1$, $G_y(s_1) \wedge G_y(s_2) ......\wedge G_y(s_j) \wedge ......\wedge G_y(s_n).$
Or $G_y(s_1) + G_y(s_2) ...... + G_y(s_j) + ... + G_y(s_n).$
Or $G_y(s_1) \vee G_y(s_2) \vee ......\vee G_y(s_j) \vee ......\vee G_y(s_n).$
Or $G_y(s_1).G_y(s_2) ......).G_y(s_j). ...... G_y(s_n) . \text{ Or other types.}$

As w_{12} , w_{23} , $\cdots\cdots$, $w_{n-1, n}$, one of them is equal to zero,
$$YSG_y(s_1, s_2 ......s_n) = G_y(s_1, s_2 ......s_n) = 0.$$

Among them, $G_w(s_1), G_w(s_2)$ ......,$G_w(s_n)$ are the function of subsystems s_1, s_2 ......and s_n respectively, w_{ij} is the directed structure weight of subsystem s_i and s_j.

5 Establishing Utility Function of System S Structure Weight

In the previous study we discussed the basic system structure of series and parallel type, and in this study, we discuss the basic system structure of feedback, scale and inclusion. They all consist of subsistems which can be regarded as elements of a system, if the subsystem is regarded as the new parent system, the relationship between the subsystem and its next level subsystem is similar to that between the subsystem and its parent system [6].

Since it has not been proven that "the structure of any system can be decomposed into five basic structures", we divide the system into types below:

①series; ② parallel; ③feedback; ④ scale (a. expansion type; b. contraction type); ⑤ inclusion; ⑥other basic types. Since the basic system structure ①, ② and ④ can be combined into several special cases of m to n. That is, ① is the case of m = n = 1; ② is the case of m = n ≥ 2; ④ of a) is the case of m = 1 and n ≥ 2; ④ of b) is the case of m ≥ 2 and n = 1.

Thus, the basic structure can be divided into:

①m-to-n type; ②feedback type; ③ inclusion type; and ④ other basic types.

In the basic structure, since as system S, its subsystems can be considered as elements, we label them all as elements in the basic structure.

As shown in the Fig. 5 [1], system S can be split into the following two subsystems S_1 and S_2, S_1 can be split into tandem subsystem S_{11} and subsystem S_{12}, while subsystem S_{12} can be split into elemental subsystem S_{121} and two subsystems at the next level of expansion subsystem S_{122}. In the process of splitting, we continue to split into subsystems until it is the basic structure.

Subsystem S_1 and subsystem S_2 is series structure.

Figure 1 Subsystem S_{121} and subsystem S_{122} is series structure.

At this point, we start building the utility function of the system structure weight from the most basic one, and build them level by level from the basic structure until the systems structures weights utility functions are completely established.

Establishing the utility function for the system S structure weight. Starting from the most basic level, from the previous discussion, it is assumed that there are

$$G_y(S_{122}) = G_y(c, d, e, n) = w_{nc}G_w(c) + w_{nd}G_w(d) + w_{ne}G_w(e),$$
$$G_y(s_{12}) = G_y(m, s_{122}) = w_{m, s_{122}}G_w(m) \wedge G_w(s_{122}),$$
$$G_y(s_{11}) = w_{ab}G_w(a) \wedge G_w(b),$$
$$G_y(s_1) = ws_{11, s_{12}}G_w(s_{11}) \wedge G_w(s_{12}),$$
$$G_y(s_2) = w_{fg}G_w(g) + w_{fh}G_w(h) + w_{fk}G_w(k). \text{ Substitute to get}$$

$$\begin{aligned}
G_y(s) &= ws_{1, s_2}G_w(s1) \wedge G_w(s_2) \\
&= ws_{1, s_2}(ws_{11, s_{12}}G_w(s_{11}) \wedge G_w(s_{12})) \wedge (w_{fg}G_w(g) + w_{fh}G_w(h) \\
&\quad + w_{fk}G_w(k)) \\
&= ws_{1, s_2}(ws_{11, s_{12}}(w_{ab}G_w(a) \wedge G_w(b)) \wedge (w_{m, s_{122}}G_w(m) \wedge G_w(s_{122})) \wedge (w_{fg}G_w(g) + \\
&\quad w_{fh}G_w(h) + w_{fk}G_w(k)) \\
&= ws_{1, s_2}(ws_{11, s_{12}}(w_{ab}G_w(a) \wedge G_w(b)) \wedge (w_{m, s_{122}}G_w(m) \wedge (w_{nc}G_w(c) + \\
&\quad w_{nd}G_w(d) + w_{ne}G_w(e)) \wedge (w_{fg}G_w(g) + w_{fh}G_w(h) + w_{fk}G_w(k)).
\end{aligned}$$

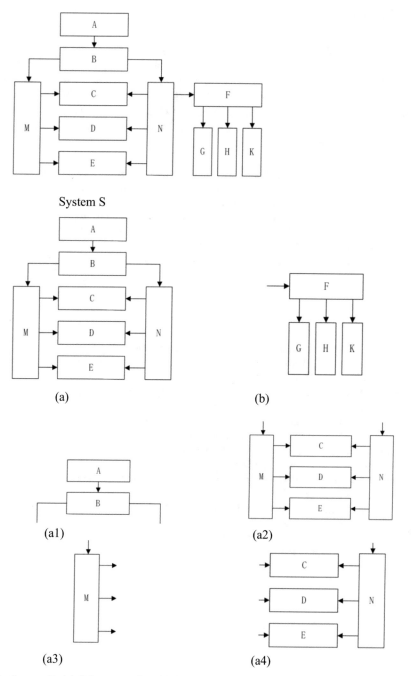

System S

(a) Subsystems S_1

(b) (Expansion) Subsystems S_2

(a1)

(a2)

(a3)

(a4)

Fig. 5. System S, (a) Subsystems S_1, (b) (Expansion) Subsystems S_2, (a1) (series) Subsystem S_{11}, (a2) Subsystem S_{12}, (a3) (Element) Subsystem S_{121}, (a4) (Expansion) Subsystem S_{122}

6 Conclusion

The system environment affects the function of the system output, and the system reacts on the environment. If the system provides positive and beneficial output to the environment, it is called function. If the system provides negative output, the harmful output is pollution. The environment provides energy to the system, and the ability of the environment to absorb or assimilate the substances discharged by the system is called sink capacity. In this way, a cycle is formed with sustainable development. This is the basis of system science for sustainable development.

Acknowledgments. The project is supported in part by The Characteristic Innovation Projects of Regular Universities in Guangdong Province: Research on the Innovative Driving Path of Integration between Enterprises and Universities in Application-oriented Undergraduate Universities in the New Era(2022WTSCX138); in part by Social Science Planning Project of Foshan City: Research on the Trade Potential between Foshan and RCEP's High-end Manufacturing Industry under the Goal of High-quality Development(2022-QN014).

Recommender: Professor Guo Kaizhong from Guangdong University of Technology in China.

References

1. Guo, K.Z., Liu, S.Y.: Fundamentals of Error Theory. Springer, (9) (2018)
2. Guo, K.Z., Liu, S.Y.: Error Systems: Concepts, Theory and Applications. Springer (2020). https://doi.org/10.1007/978-3-030-40760-5
3. An, Q.X., Meng, F.Y., Sheng, A., Chen, X.H.: A new approach for fair efficiency decomposition in two-stage structure system. Oper. Res. Int. Journal **1**, 257–272 (2018)
4. Xiong H.: Analysis of Error System Structure Dynamic Structure and Static Structure. In: International Conference on Fuzzy Information & Engineering (2019)
5. Patelli, E, Feng, G., Coolen, F.P.A., Coolen-Maturi, T.: Simulation methods for system reliability using the survival signature. Reliab. Eng. Syst. Saf. **167**, 327–337 (2017)
6. Liao L.P., Guo, K.Z.: Study on destruction transformation of error system structure. Fuzzy Sets Oper. Res. 293–291 (2017)
7. Kosterin, A.V., Skvortsov, E.V.: Analytical formula for integral whose kernel containing error function. Lobachevskii J. Math. **37**, 266–267 (2016)
8. Feng, Z.J., Ming, Q.: Research on the Evaluation and Policy System of Coordinated Development of Green Innovation System in Chinese Manufacturing Industry. Economic Science Press, pp. 69–70 (2020)

Screening for Late-Onset Fetal Growth Restriction in Antepartum Fetal Monitoring Using Deep Forest and SHAP

Jianhong Huo[1], Guohua Li[1], Chongwen Li[1], Xia Li[2], Guiqing Liu[3], Qinqun Chen[1], Jialu Li[1], Yuexing Hao[4], and Hang Wei[5(✉)]

[1] School of Medical Information Engineering, Guangzhou University of Chinese Medicine, Guangzhou, China
[2] The Second Affiliated Hospital, Guangzhou Medical University, Guangzhou, China
[3] The First Affiliated Hospital, Guangzhou University of Chinese Medicine, Guangzhou, China
[4] Department of Human Centered Design, Cornell University, Ithaca, NY, USA
[5] School of Medical Information Engineering, Guangzhou University of Chinese Medicine, Guangzhou, China
crwei@gzucm.edu.cn

Abstract. Fetal growth restriction (FGR) is the second leading cause of perinatal death, of which late-onset fetal growth restriction (LFGR) accounts for 70%–80% and has a low detection rate. Cardiotocography (CTG) is a routine tool for antepartum fetal monitoring, continuously recording fetal heart rate (FHR) to assess the development of the fetus. Therefore, in this paper, we proposed a hybrid DF-SHAP model that screens LFGR in routine CTG monitoring using deep forest (DF) and Shapley Additive Explanation (SHAP). Firstly, principal component analysis, spearman correlation analysis and logistic regression analysis were implemented to explore significant FHR features for LFGR. After data preprocessing, deep forest multi-granularity scanning was introduced to probe the connection among the features. Then the cascade forest phase, which was designed to integrate random forest, extra trees, logistic regression and extreme gradient boosting as the basic classifiers, iteratively generated new layers and finally got the best performance model. Finally, SHAP was introduced to enhance the interpretability of DF and to interpret the impact of each feature on the predicted value. The experimental results showed that the proposed DF-SHAP model outperformed the state-of-the-art LFGR recognition models using CTG data and had good interpretability. This indicates that the DF-SHAP model is feasible for screening LFGR in antepartum fetal monitoring.

Keywords: Cardiotocography · Late-onset fetal growth restriction · Deep forest · Shapley Additive Explanation(SHAP)

1 Introduction

Fetal growth restriction (FGR) is a complication of pregnancy with high incidence, complex etiology, limited treatment and prognosis. Moreover, FGR is the

B.-Y. Cao et al. (Eds.): ICFIE 2022, LNDECT 207, pp. 383–394, 2024.
https://doi.org/10.1007/978-981-97-2891-6_29

second leading cause of perinatal death. Clinically, there are two main phenotypes of FGR which are early and late FGR based on before or after diagnosis 32–34 weeks' gestation [7]. It is shown that late FGR (LFGR) accounts for 70–80% of FGR incidences and at least 50% of unexplained intrauterine stillbirths result from late-onset or full-term FGR [4].

Hence, some researchers have been focusing on how to establish prediction models for screening LFGR in the antepartum period. Crovetto F et al. [1] conducted logistic regression model to predict LFGR based on maternal characteristics, mean arterial pressure (MAP), uterine artery (UtA) Doppler and placental growth factor (PlGF), etc. and the detection rate was 65.8%. Zheng H et al. [2] established a multivariate screening model, achieving a 52.6% detection rate of LFGR by using logistic regression with maternal characteristics, the second-trimester head circumference (HC/AC) and estimated fetal weight (EFW). The researches have achieved prediction models to screen fetuses at high risk of LFGR for supporting assistance in clinical diagnosis. Nevertheless, the detection of LFGR is still unsatisfactory, between 52.6% and 65.8% [1, 2].

Cardiotocography (CTG) plays an important role in assessing fetal condition, by continuously recording fetal heart rate (FHR) and uterine contraction. The changes of FHR are monitored and the feedback on uterine contraction are made dynamically, which achieve the purpose of evaluating the fetal intrauterine condition in time [6]. Nowadays, researchers have introduced CTG data with machine learning models to identify late-onset growth-restricted fetuses to provide effective assistance to physicians. Signorini M G et al. [8] extracted 12 FHR features from CTG records and obtained the best method random forest with an average classification accuracy of 0.911. Pini N et al. [5] established radial basis function support vector machine (RBF-SVM) model combined with quantitative features extracted from FHR signals, and the classification accuracy was 0.93. Above researches have shown the feasibility to introduce CTG data for prediction LFGR, but the interpretability of model hasn't been further studied yet.

In this work, we proposed a hybrid model that screening for LFGR in routine CTG monitoring using deep forest (DF) and Shapley Additive Explanation (SHAP). Deep forest multi-granularity scanning and cascade forest were designed to build Deep Forest framework. SHAP interpreted the complex structure of deep forest and the classification result in the model to better address the complexity of CTG explanation. The results demonstrate the proposed DF-SHAP model has enhanced performance for screening LFGR and improved the model interpretability.

The rest of the paper is organized as follows: In Sect. 2, materials and methods are described in the present study. The experimental results, the performance of the model and the interpretation in the model are concluded in Sect. 3. Then, the conclusions and future work are summarized in Sect. 4.

2   Materials and Methods

2.1   Data

The dataset studied in this paper are publicly available CTG cases collected by the University of Azienda Ospedaliera Federico II [5]. It consists of 31 features (Table 1) extracted from 262 records of FHR, of which 160 and 102 are healthy and LFGR sample, respectively. Each case contains classical FHR features and their nonlinear features.

Table 1. Features information.

NO.	feature	Description	NO.	feature	Description
1	GA	Gestational Age	17	SampEn	Samping Entropy
2	MA	Maternal Age	18	LCZ_BIN_0	Lempel Ziv Complexity_binary
3	Sex	Male=1; Female=2	19	LCZ_TER_0	Lempel Ziv Complexity_ternary
4	Mean FHR	Mean Fetal heart rate	20	AC_T1_s2	Average Acceleration, T=1, s=2
5	Std FHR	Standard Fetal heart rate	21	DC_T1_s2	Deceleration Capacities, T=1, s=2
6	DELTA	Delta	22	DR_T1_s2	Deceleration Reserve, T=1, s=2
7	II	Interval Index	23	AC_T5_s5	Average Acceleration, T=5, s=5
8	STV	Short Term Variability	24	DC_T5_s5	Deceleration Capacities, T=5, s=5
9	LTI	Long Term Irregularity	25	DR_T5_s5	Deceleration Reserve, T=5, s=5
10	ACC_L	Large accelerations	26	AC_T9_s9	Average Acceleration, T=9, s=9
11	ACC_S	Small accelerations	27	DC_T9_s9	Deceleration Capacities, T=9, s=9
12	CONTR	Contractions	28	DR_T9_s9	Deceleration Reserve, T=9, s=9
13	LF	Low frequency	29	AC_T40_s1	Average Acceleration, T=40, s=1
14	MF	Movement frequency	30	DC_T40_s1	Deceleration Capacities, T=40, s=1
15	HF	High frequency	31	DR_T40_s1	Deceleration Reserve, T=40, s=1
16	ApEn	Approximate Entropy	State	Clinical Diagnosis	HEALTHY=1; LFGR=2

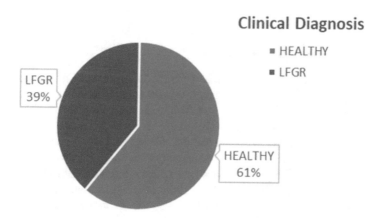

Fig. 1. Comparison of the number of HEALTHY and LFGR.

Figure 1 demonstrates the percentages of fetuses with healthy and LFGR are 61% and 39%, respectively, indicating a relatively balanced data distribution.

Due to strict data collection and exclusion of currently known interference factors, the dataset is free of missing values and outliers. Furthermore, the relation between fetal status and features was observed by principal component analysis. According to the results (Fig. 2), there was a large degree of crossover between healthy and LFGR, which indicated that it was extremely easy to misclassify healthy and LFGR in antepartum fetal monitoring.

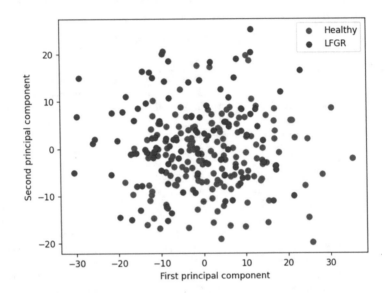

Fig. 2. Principal component analysis 2D distribution graph of the CTG dataset.

2.2   Feature Selection for LFGR

Feature selection can relieve dimension, screen out important features and reduce the difficulty of learning tasks to improve effectively the classification performance of the model. With 0.2 as the feature importance threshold, principal component analysis (PCA), logistic regression analysis (LRA) and spearman correlation analysis were used to obtain the influencing features for screening LFGR. The central idea of principal component analysis is mapping N-dimensional features to K-dimension (KN), which are new orthogonal features reconstructed on the basis of the original features. In logistic regression analysis, the following formula (formula (1)) was used to find the significant features for LFGR.

$$L = \prod_{i=1}^{n} P_i^{Y_i} (1 - P_i)^{1-Y_i} \tag{1}$$

P_i represents the probability that the observation object of case i is LFGR. If the result is LFGR, Y_i is set to 1, otherwise, Y_i is set to 0.

Spearman correlation coefficient represents the correlation between rank variables. The following formula (formula (2)) was used to calculate the coefficients for LFGR feature screening.

$$\rho = \frac{\sum (x_i - \bar{x})(y_i - \bar{y})}{\sqrt{\sum (x_i - \bar{x})^2 (y_i - \bar{y})^2}} \tag{2}$$

where x_i and y_i are the rank of the two variables sorted by size respectively. $\bar{x}$ and $\bar{y}$ are mean values.

2.3 Deep Forest for Screening LFGR

Deep Forest is a stacked forest model proposed by Zhou [9], which is an integration of the traditional forest model in terms of breadth and depth. For LFGR recognition, the overall structure of the whole forest was multi-granularity scanning followed by cascading forest, finally the final predictions were obtained (Fig. 3).

Multi-granularity Scanning. Multi-granularity scanning not only finds out the sequence relation between each feature, but also enhances learning ability of classification for LFGR. FHR features were scanned and re-represented as new feature by sliding windows of multiple dimensions to prepare for the cascade forest. The principle of this phase was shown in Fig. 3, sliding windows of length 2 and 3 were set for multi-granularity scanning, and multiple different feature vectors were generated as the input of the cascade forest.

Cascade Forest. In order to reflect the diversity of the structure, each layer is composed of several ensemble learning classifiers. In our study, each layer of the cascading forest contained random forest, extra trees, logistic regression and extreme gradient boosting (Fig. 3). In this structure, the classifier was trained by using input features. Each classifier conducted independently supervised learning and then outputted a probability vector of length 2 as a new feature. The new features were merged with the original input feature to constitute a new input for the next layer of feature generation. After each new feature generation, the new feature set was trained in this layer, whose effect was tested on the validation set with the four classifiers. When the classification performance was no longer improved significantly after iterations, there would be no new layer generated (Fig. 3)

2.4 SHAP for Model Interpretation

Inspired by cooperative game theory, SHAP constructs an additive explanatory model in which all features have an impact on outcome and shap value is the contribution value of each [3]. Interpretation model of deep forest for LFGR is

$$y_i = y_{base} + f(x_{i1}) + f(x_{i2}) + \cdots + f(x_{ij}) \tag{3}$$

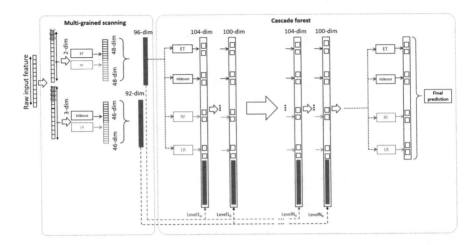

Fig. 3. Model structure of multigranular cascade forest for screening LFGR.

y_{base} represents the mean of the target variable for all the samples and $f(x_{ij})$ represents shap value of the j feature of the i sample.

The shap value of x_i in deep forest can be obtained from the following equation:

$$\varphi(x_i) = \sum_{S \subseteq N_{\{i\}}} \frac{|S|!(N - |S| - 1)!}{N!}(f(S \cup \{i\}) - f(S)) \tag{4}$$

where S is a subset of the features used in the deep forest, $N_{\{i\}}$ represents all set of features excluding feature x_i, N is the number of features, $f(S)$ is the prediction for feature values in set S and $f(S\{i\})$ is the prediction for feature values in set S and x_i.

2.5   Evaluation Indicators

To verify the testing effect of DF screening for LFGR in antepartum fetal monitoring, Accuracy, Precision, Recall, Specificity, the value of F1, Average F1, Kappa, MCC and AUC were used for evaluation in this study.

$$Accuracy = \frac{TP + TN}{TP + TN + FP + FN} \tag{5}$$

$$Precision = \frac{TP}{TP + FP} \tag{6}$$

$$Recall = \frac{TP}{TP + FN} \tag{7}$$

$$Specificity = \frac{TN}{TN + FP} \tag{8}$$

$$F1 - score = \frac{2}{\frac{1}{Precision} + \frac{1}{Recall}} \tag{9}$$

$$Kappa = \frac{Accuracy - P_c}{1 - P_c} \tag{10}$$

$$MCC = \frac{TP * TN - FP * FN}{\sqrt{(TP + FP) * (TP + FN) * (TN + FP) * (TN + FN)}} \tag{11}$$

3   Results and Discussion

3.1   Important Features for LFGR

In PCA, three principal components whose cumulative contribution rate reached more than 50% were selected, and twenty-three features were screened out eventually (Fig. 4). In LRA, there were six features that met the requirements. In

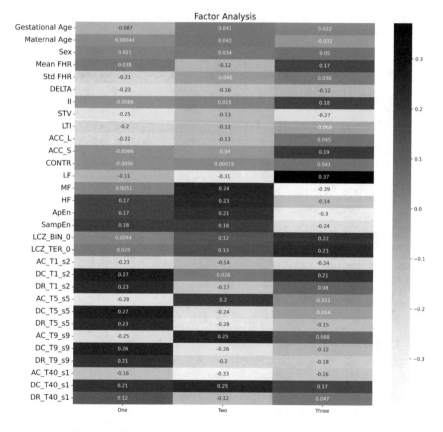

Fig. 4. The feature importance obtained by the PCA.

spearman correlation analysis, the correlation between features and the importance of features were analyzed. Finally, the terminal feature set including seventeen features for screening LFGR was determined by aggregating the features found by the three methods.

3.2   Classification Performance of DF-Based Model

Randomly selecting 20% of the dataset as the test set, the model was validated 10 times and the average value was taken as evaluation indicator. The following models were all implemented using programming language Python3.10 and the compilers PyCharm Community Edition 2021.3.2. The deep forest model made a certain contribution in predicting fetal wellbeing. The average F1 value was 94.8393(93.3829,96.2157), AUC value was 94.4625 (93.1051,95.8199), specificity and sensitivity were 0.9750 (0.9545,0.9955) and 0.9143(0.8874,0.9412), respectively. Compared with the other twelve machine learning models, the proposed DF model has higher performance. The prediction performance to screen LFGR of the comparison model is shown in Table 2 and Fig. 5. Compared with these models, each evaluation index of deep forest model was promoted. It could be intuitively concluded that DF solved the problem of screening LFGR in antepartum fetal monitoring effectively.

Table 2. Performance comparison of different machine learning models.

Evaluation Indicators		Precision (%)	Recall(%)	Specificity(%)	F1 value(%)	Average F1(%)	Kappa(%)	MCC(%)	Accuracy(%)	AUC value
LOGISTIC	HEALTH	81.14	76.34	87.40	78.33	82.04	64.21	64.63	83.02	0.9069
	LFGR	84.43	87.40	76.34	85.75					
NB	HEALTH	71.35	73.37	82.92	72.11	77.99	56.08	56.38	79.87	0.8402
	LFGR	85.13	82.92	73.37	83.87					
KNN	HEALTH	92.23	63.41	97.01	74.96	82.01	64.63	67.08	84.90	0.8915
	LFGR	82.36	97.01	63.41	89.06					
BP	HEALTH	81.14	67.58	90.87	73.63	80.12	60.52	61.25	82.39	0.9140
	LFGR	82.83	90.87	67.58	86.61					
SVM	HEALTH	84.55	65.64	92.95	73.82	80.55	61.44	62.59	83.02	0.9265
	LFGR	82.33	92.95	65.64	87.29					
DT	HEALTH	57.28	64.17	72.14	59.71	67.10	34.98	35.90	69.18	0.6926
	LFGR	78.25	72.14	64.17	74.49					
RF	HEALTH	81.57	63.03	91.24	70.39	77.90	56.29	58.02	80.50	0.8683
	LFGR	80.66	91.24	63.03	85.42					
GBDT	HEALTH	73.28	79.92	83.12	76.28	80.68	61.45	61.77	81.76	0.8840
	LFGR	87.28	83.12	79.92	85.07					
XGBoost	HEALTH	77.71	72.31	86.83	74.48	80.00	60.13	60.63	81.76	0.9018
	LFGR	84.53	86.83	72.31	85.51					
LightGBM	HEALTH	74.08	77.99	83.95	75.88	80.52	61.09	61.27	81.76	0.8738
	LFGR	86.53	83.95	77.99	85.16					
Catboost	HEALTH	73.65	76.89	83.95	75.07	79.90	59.85	60.05	81.13	0.8884
	LFGR	85.65	83.95	76.89	84.72					
Adaboost	HEALTH	76.07	73.14	85.99	74.31	79.63	59.37	59.67	81.13	0.8748
	LFGR	84.18	85.99	73.14	84.96					
DF [this paper]	HEALTH	96.70	91.43	97.50	93.99	94.83	89.25	89.50	95.09	0.9446
	LFGR	85.32	97.50	91.43	95.85					

Note: The bolded black numbers in the table represents the index data of the models with better results in the model comparison.

Compared with existing LFGR recognition models using CTG data [5], the proposed DF model removing sex from features has advantages in accuracy, specificity and AUC value, which can better screen for LFGR in antepartum fetal

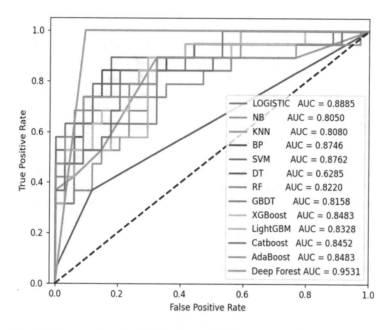

Fig. 5. Comparison for ROC curves of different classification models.

Table 3. Comparison of results of existing LFGR recognition models using CTG data.

	Model		
Method	RBF-SVM[5]	RBF-SVM(sex -removed)[5]	DF[this paper]
Specificity	83.95%	79.05%	**97.50%**
Sensitivity	**92.87%**	92.47%	91.43%
AUC	0.9277	0.9208	**0.9462**

monitoring. The proposed model increased by 18.45% and 2.54% in the specificity and AUC, respectively (Table 3). In clinical diagnosis, the higher the specificity is, the lower the misdiagnosis rate is. Correct classification of healthy class could avoid effectively excessive treatment or intervention and even cesarean. The results of comparison show that the misclassification rate of deep forest was the lowest and deep forest has promising screening ability in antepartum fetal monitoring.

3.3 SHAP Based Interpretability

The proposed DF was a stack of different kinds of classifiers in width and depth with certain complexity. Therein SHAP was implemented to interpret the deep forest model for LFGR and study the impact of input features furthermore.

Figure 6 shows the SHAP values of each feature. Gestational age, MF, LF and other features have a significant impact on the deep forest model, among which the most important feature is gestational age. With the increase of gestational age, the fetus is more likely to be healthy in the same conditions. MF and LF also have significant effects on fetal status. With the increase of the two feature values, the risk of LFGR will increase. Features of frequency domain are highly correlated with LFGR, and the larger the frequency domain characteristic value, the higher the risk of fetal LFGR.

Four features with significant influence on the deep forest model, namely gestational age, MF, LF and LCZ_TER_0, were selected to draw the dependency diagram respectively. Figure 7 shows the feature dependency diagram of important features. It can be seen that with the increase of gestational age and LCZ_TER_0, SHAP values decreased, indicating a reverse relationship with LFGR. On the contrary, the increase of MF and LF values showed a positive relationship with LFGR. As LCZ_TER_0 and gestational age increasing, the risk of LFGR reduced, indicating that both of them had reverse effects on LFGR. With the increase of LF, the CONTR increases first and then decreases, so the interactive relationship of them is complicated.

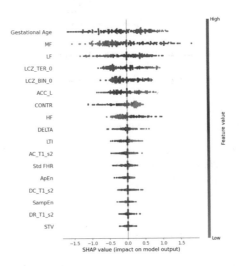

Fig. 6. The SHAP values of features interpreted by SHAP.

In addition, SHAP can be applied to analyze the individual influencing factors. Figure 8 and 9 respectively show the SHAP features contribution diagrams of cases predicted to be healthy and LFGR by the proposed DF model. In Fig. 8, the features in blue represent the supporting features predicted to be healthy, while the features in red represent the interfering features. It was forecasted as healthy mainly because of smaller MF, larger ACC_L, older Gestational Age, and larger LCZ_TER_0 and LCZ_BIN_0 values. Likewise, in Fig. 9, the red features are the supporting factors for LFGR, and the blue features are the interfering factors for LFGR. The case was predicted as LFGR due to the larger MF value and the smaller LCZ_TER_0 and LCZ_BIN_0 values.

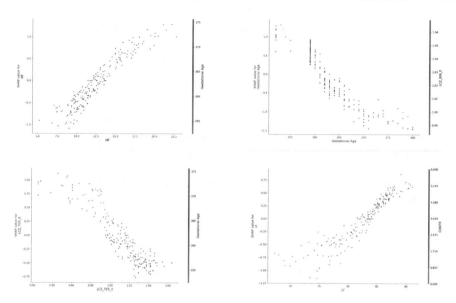

Fig. 7. The dependency diagram of four features interpreted by SHAP.

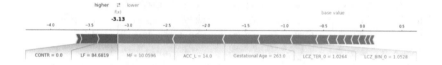

Fig. 8. The healthy pregnant woman interpreted by SHAP.

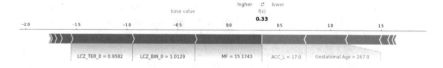

Fig. 9. The LFGR pregnant woman interpreted by SHAP.

4   Conclusion

In this paper, the principal component analysis, logistic regression analysis and spearman correlation analysis were employed to screen out the significant features for LFGR; then DF model was established and has showed its excellent performance compared with the state-of-the-art models in screening for LFGR using CTG data. Furthermore, SHAP interpreted the complex structure of DF and the classification result in the model. In summary, the proposed DF-SHAP model has crucial practical significance for clinical LFGR screening in antepartum fetal monitoring.

Nonetheless, the dataset for the prediction model establishing in this paper was derived from retrospective case control study, so the application scenario in clinic has some limitation. At the same time, the model interpretability can be further improved, such as mechanisms for diagnose LFGR, to provide strong support for medicine workers.

Acknowledgement. This work is supported by Natural Science Foundation of China No.61976052 and No.71804031, Medical Scientific Research Foundation of Guangdong Province No. A2019428 and National Undergraduate Innovation and Venture Training Project No. 202210572005.

Recommender: Associate Professor Bo Xu, Guangdong University of Finance and Economics in China.

References

1. Crovetto, F., et al.: First-trimester screening with specific algorithms for early- and late-onset fetal growth restriction. Ultrasound Obstet. Gynecol. **48**(3), 340–348 (2016)
2. Feng, Y., Zheng, H., Fang, D., Mei, S., Zhong, W., Zhang, G.: Prediction of late-onset fetal growth restriction using a combined first- and second-trimester screening model. J. Gynecol. Obstet. Hum. Reprod. **51**(2), 102273 (2022)
3. Lundberg, S.M., Lee, S.I.: A unified approach to interpreting model predictions. In: Advances in Neural Information Processing Systems, vol. 30 (2017)
4. Oros, D., Figueras, F., Cruz-Martinez, R., Meler, E., Gratacos, E.: Longitudinal changes in uterine, umbilical and fetal cerebral doppler indices in late-onset small-for-gestational age fetuses. Ultrasound Obstet. Gynecol. **37**(2), 191–195 (2011)
5. Pini, N., et al.: A machine learning approach to monitor the emergence of late intrauterine growth restriction. Front. Artif. Intell. **4**, 622616 (2021)
6. Qi, T.I., Sun, Y.L.: Clinical analysis of prenatal stressless fetal heart monitor combined with electronic fetal heart monitor. Chin. J. Fam. Plann. Gynecotokology **11**(1), 73–76 (2019)
7. Savchev, S., et al.: Evaluation of an optimal gestational age cut-off for the definition of early- and late-onset fetal growth restriction. Fetal Diagn. Ther. **36**(2), 99–105 (2014)
8. Signorini, M.G., Pini, N., Malovini, A., Bellazzi, R., Magenes, G.: Integrating machine learning techniques and physiology based heart rate features for antepartum fetal monitoring. Comput. Methods Programs Biomed. **185**, 105015 (2020)
9. Zhou, Z.H., Feng, J.: Deep forest: towards an alternative to deep neural networks. In: IJCAI, pp. 3553–3559 (2017)

The Solution Closest to a Given Vector in the System of Fuzzy Relation Inequalities

Miaoxia Chen[1,2], Abdul Samad Shibghatullah[2], and Xiaopeng Yang[2,3(✉)]

[1] Department of Education, Hanshan Normal University, Chaozhou 521041, China
[2] Institute of Computer Science and Digital Innovation, UCSI University, 56000 Kuala Lumpur, Malaysia
happyyangxp@163.com
[3] School of Mathematics and Statistics, Hanshan Normal University, Chaozhou 521041, China

Abstract. In recent years, the fuzzy relation inequalities, with addition-min or max-min composition, were introduced for describing the P2P network system. A solution of the fuzzy relation inequalities corresponds to a feasible scheme in the P2P network system. In this work, we aim to find a solution of the max-min fuzzy relation inequalities, which is closest to a given vector. Here, the given vector is not a solution. We propose an novel method for obtaining the closest solution. A numerical example is given to illustrate our proposed method.

Keywords: Fuzzy relation inequality · Max-min composition · P2P network system · Closest solution

1 Introduction

The fuzzy relation equation (FRE), since introduced by E. Sanchez in 1976 [1], has played an important role in fuzzy mathematics and fuzzy systems. It is induced by the fuzzy matrix and the fuzzy composed operations. Its mathematical form is as

$$(\alpha_{i1} \wedge y_1) \vee (\alpha_{i2} \wedge y_2) \vee \cdots \vee (\alpha_{in} \wedge y_n) = \beta_i, \quad i = 1, 2, \cdots, m, \tag{1}$$

where α_{ij} and β_i are some given constants. In general, when the system is consistent (i.e., has at least a solution), it has only one maximum solution and finitely many minimal solutions. Its complete solution set could be simply and perfectly characterized by the unique maximum solution and all the minimal solutions. It should be mentioned here that the solution set should be non-convex when there are more than one minimal solutions. There were several approaches for solving the FRE system [1–6].

The fuzzy relation inequality (FRI) system corresponding to system (1) could be represented by

$$\underline{\beta}_i \leq (\alpha_{i1} \wedge y_1) \vee (\alpha_{i2} \wedge y_2) \vee \cdots \vee (\alpha_{in} \wedge y_n) \leq \overline{\beta}_i, \quad i = 1, 2, \cdots, m. \tag{2}$$

B.-Y. Cao et al. (Eds.): ICFIE 2022, LNDECT 207, pp. 395–404, 2024.
https://doi.org/10.1007/978-981-97-2891-6_30

In fact, the inequality system (2) could be solved by the conservative path approach [7] or the FRI path approach [8]. Moreover, the resolution and the structure of solution set to system (2) are similar to those to system (1). As a consequence, the above equality system (1) and the inequality system (2) are usually collectively referred to as fuzzy relation system (FRS), with max-min composition.

In [9–11], the max-min FRS was introduced to characterize the server-to-client system. Recently, the max-min FRS was further developed to describe the peer-to-peer network system [12–15]. For an FRS, either refer to system (1) or refer to system (2), is said to be consistent, if it has at least a solution. Otherwise, it is said to be inconsistent. Both consistent and inconsistent FRSs have been investigated.

For an inconsistent FRS, one would be interested in the approximate solution(s) for the system [16–19]. To define the approximate solution, the concept of distance between two given fuzzy vectors should be determined firstly. In [20], the authors listed three kinds of commonly used formulaes for the distance, i.e., the L_1 norm distance, the L_2 norm distance and the L_∞ norm distance. The the L_∞ norm distance was introduced in [16,17], while the L_1 one was applied in [18,19]. Correspondingly, different definitions of the distance induce distinguishing resolution methods for the approximate solution.

As pointed out in [17], when considering the requirement of the total download speed for each terminal, the peer-to-peer network system could be reduced into the addition-min inequalities as [21–24]

$$\alpha_{i1} \wedge y_1 + \alpha_{i2} \wedge y_2 + \cdots + \alpha_{in} \wedge y_n \geq \beta_i, \quad i = 1, 2, \cdots, m. \tag{3}$$

while considering the biggest (highest) download speed for each terminal, the peer-to-peer network system could be reduced into the following max-min inequalities,

$$(\alpha_{i1} \wedge y_1) \vee (\alpha_{i2} \wedge y_2) \vee \cdots \vee (\alpha_{in} \wedge y_n) \geq \beta_i, \quad i = 1, 2, \cdots, m. \tag{4}$$

Here, all the parameters and variables $\alpha_{ij}, \beta_i, y_j$ are restricted to the unit interval $[0, 1]$. The abbreviated form of system (4) could be written as

$$A \circ y \geq b, \tag{5}$$

where $A = (\alpha_{ij})_{m \times n}$, $b = (\beta_i)_{m \times 1}$, $y = (y_j)_{n \times 1}$. In fact, when system (4) is consistent, any its solution represents a feasible flow scheme in the peer-to-peer network system.

Now, suppose system (4) is consistent and there is a given flow scheme, denoted by the vector $v = (v_1, v_2, \cdots, v_n)$. Assume that v is not feasible in system (4), i.e., v is not a solution to (4). Note that the sign of inequality in system (4) is "$\geq$". As a result, since v is "not big enough", v is not feasible in (4). Moreover, since (4) is consistent, we know that when a fuzzy vector is bigger enough, it would be a solution of (4). In fact, $(1, 1, \cdots, 1)$ is the maximum solution of (4) as presented in [17]. Hence, we are interested in searching a solution y^* of system (4), which is closest to the given vector v, and bigger than or equal to v (i.e., $y^* \geq v$). Here, we adopt the L_∞ norm distance to characterized the concept of "closest". As a consequence, the solution closest to the given vector v is indeed the optimal solution of the following optimization problem,

$$\min \quad z(y) = (y_1 - v_1) \vee (y_2 - v_2) \vee \cdots \vee (y_n - v_n)$$

$$\text{s.t.} \quad \begin{cases} (\alpha_{11} \wedge y_1) \vee (\alpha_{12} \wedge y_2) \vee \cdots \vee (\alpha_{1n} \wedge y_n) \geq \beta_1, \\ (\alpha_{21} \wedge y_1) \vee (\alpha_{22} \wedge y_2) \vee \cdots \vee (\alpha_{2n} \wedge y_n) \geq \beta_2, \\ \cdots\cdots\cdots\cdots\cdots\cdots\cdots\cdots\cdots\cdots\cdots\cdots\cdots\cdots\cdots\cdots\cdots\cdots \\ (\alpha_{m1} \wedge y_1) \vee (\alpha_{m2} \wedge y_2) \vee \cdots \vee (\alpha_{mn} \wedge y_n) \geq \beta_m, \\ y \geq v. \end{cases} \quad (6)$$

In this work, we aim to find an optimal solution of the above problem (6). In Sect. 2 we propose a feasible approach for finding the optimal solution of problem (6). Furthermore, a numerical example is provided in Sect. 3 to illustrate our proposed approach. Section 4 is simple conclusion.

2   Solving the Solution Closest to a Given Vector in System (4)

In what follows, we always assume that system (4) is consistent, i.e., it has at least a solution. Denote its solution set by

$$\mathcal{S}^{A,b} = \{y \in [0,1]^n | A \circ y \geq b\}. \quad (7)$$

Then we have $\mathcal{S}^{A,b} \neq \emptyset$. Moreover, we also assume that

$$v = (v_1, v_2, \cdots, v_n) \in [0,1]^n, \quad (8)$$

is a given vector with $v \notin \mathcal{S}^{A,b}$. That is to say, v is a given fuzzy vector, but not a solution of system (4). In addition, we denote

$$I = \{1, \cdots, m\}, \quad J = \{1, \cdots, n\}.$$

Theorem 1. $\mathcal{S}^{A,b} \neq \emptyset$ if and only if $\bigvee_{j \in J} \alpha_{ij} \geq \beta_i$ holds for any $i \in I$.

It is indicated in Theorem 1 that when system (4) is consistent, then for each $i \in I$, there exists a $j_i \in J$ such that $\alpha_{ij_i} \geq \beta_i$. Thus, for the consistent system (4), it should be

$$\{j \in J | \alpha_{ij} \geq \beta_i\} \neq \emptyset.$$

Since $v \notin \mathcal{S}^{A,b}$, there exists some $i \in I$ such that

$$(\alpha_{i1} \wedge v_1) \vee (\alpha_{i2} \wedge v_2) \vee \cdots \vee (\alpha_{in} \wedge v_n) < \beta_i \quad (9)$$

Hence, if we denote

$$I^< = \{i \in I | (\alpha_{i1} \wedge v_1) \vee (\alpha_{i2} \wedge v_2) \vee \cdots \vee (\alpha_{in} \wedge v_n) < \beta_i\}, \quad (10)$$

it is obvious that $I^< \neq \emptyset$. For any $i \in I^<$, we further denote

$$J_i = \{j \in J | \alpha_{ij} \geq \beta_i\}, \tag{11}$$

and

$$j_i^* = \arg\max_{j \in J_i}\{v_j\}. \tag{12}$$

Until now, for each $i \in I^<$, we find a relevant index set J_i and a relevant index j_i^*. Based on all our obtained indices $\{j_i^* | i \in I^<\}$, next we construct a series of vectors, denoted by $y^1, y^2, \cdots, y^m$.

Let $y^0 = v$. For $i = 1, 2, \cdots, m$, let $y^i = (y_1^i, y_2^i, \cdots, y_n^i)$, where

$$y_j^i = \begin{cases} \beta_i \vee y_j^{i-1}, & j = j_i^*, i \in I^<, \\ y_j^{i-1}, & j \neq j_i^*, i \in I^<, \\ y_j^{i-1}, & i \notin I^<. \end{cases} \tag{13}$$

In the following we will prove that the vector y^m is exactly an optimal solution of problem (6).

Proposition 1. $y^0 \leq y^1 \leq y^2 \leq \cdots \leq y^m$.

Proof. Directly obtained by (13). □

Proposition 2. *For arbitrary $j \in J$, we have*

(i) *if $\{k \in I^< | j_k^* = j\} \neq \emptyset$, then*

$$y_j^m \geq \beta_i, \quad \forall i \in \{k \in I^< | j_k^* = j\},$$

(ii) *if $\{k \in I^< | j_k^* = j\} = \emptyset$, then $y_j^m = v_j$.*

Proof. (i) Suppose $\{k \in I^< | j_k^* = j\} \neq \emptyset$. Take an arbitrary $i \in \{k \in I^< | j_k^* = j\} \neq \emptyset$. Then it is clear that $j_i^* = j$. By (13), we have $y_j^i = \beta_i \vee y_j^{i-1} \geq \beta_i$. Considering Proposition 1, we have $y^m \geq y^i$. As a result, $y_j^m \geq y_j^i \geq \beta_i$.

(ii) Suppose $\{k \in I^< | j_k^* = j\} = \emptyset$. Obviously, for any $i \in I^<$, it holds that $j_i^* \neq j$. Thus we have $y_j^i = y_j^{i-1}$. On the other hand, for any $i \notin I^<$, it follows from (13) that $y_j^i = y_j^{i-1}$. As a consequence, we have

$$y_j^i = y_j^{i-1}, \quad \forall i \in I.$$

This indicates $y_j^m = y_j^{m-1} = \cdots = y_j^1 = y_j^0 = v_j$. □

Corollary 1. *For arbitrary $i \in I^<$, it holds that $y_{j_i^*}^m \geq \beta_i$.*

Proof. Denote $j' = j_i^*$. It is clear that $i \in \{k \in I^< | j_k^* = j\} \neq \emptyset$. According to Proposition 2, we have $y_{j'}^m \geq \beta_i$, i.e., $y_{j_i^*}^m \geq \beta_i$. □

Proposition 3. *For arbitrary $j^0 \in J$, if $\{k \in I^< | j_k^* = j^0\} \neq \emptyset$, then there exists $i^* \in \{k \in I^< | j_k^* = j^0\}$, such that $y_{j^0}^m = \beta_{i^*}$.*

Proof. By (10), we have

$$(\alpha_{i1} \wedge v_1) \vee (\alpha_{i2} \wedge v_2) \vee \cdots \vee (\alpha_{in} \wedge v_n) < \beta_i, \quad \forall i \in I^<, \tag{14}$$

i.e.,

$$\alpha_{ij} \wedge v_j < \beta_i, \quad \forall i \in I^<, j \in J. \tag{15}$$

By (11), we have

$$\alpha_{ij} \geq \beta_i, \quad \forall i \in I^<, j \in J_i. \tag{16}$$

Combining (15) and (16), we have

$$v_j < \beta_i, \quad \forall i \in I^<, j \in J_i. \tag{17}$$

For any $i \in \{k \in I^< | j_k^* = j^0\}$, it is clear that $i \in I^<$ and $j^0 = j_i^* \in J_i$. Following (17), we have

$$v_{j^0} < \beta_i, \quad \forall i \in \{k \in I^< | j_k^* = j^0\}. \tag{18}$$

This indicates

$$v_{j^0} < \bigvee_{i \in \{k \in I^< | j_k^* = j^0\}} \beta_i, \tag{19}$$

i.e.,

$$y_{j^0}^0 < \bigvee_{i \in \{k \in I^< | j_k^* = j^0\}} \beta_i. \tag{20}$$

Observing (13), for $j^0 \in J$, it holds that

$$y_{j^0}^i = \begin{cases} \beta_i \vee y_{j^0}^{i-1}, & i \in \{k \in I^< | j_k^* = j^0\}, \\ y_{j^0}^{i-1}, & i \notin \{k \in I^< | j_k^* = j^0\}, \end{cases} \quad i = 1, 2, \cdots, m. \tag{21}$$

Note that $\{k \in I^< | j_k^* = j^0\} \neq \emptyset$. By (21) we find

$$y_{j^0}^m = y_{j^0}^0 \vee \left(\bigvee_{i \in \{k \in I^< | j_k^* = j^0\}} \beta_i \right). \tag{22}$$

Considering (20), we further get

$$y_{j^0}^m = \bigvee_{i \in \{k \in I^< | j_k^* = j^0\}} \beta_i. \tag{23}$$

Therefore, there exists $i^* \in \{k \in I^< | j_k^* = j^0\}$ such that $y_{j^0}^m = \beta_{i^*}$. □

Theorem 2. y^m *is a solution of system* (4), *i.e.,* $y^m \in \mathcal{S}^{A,b}$.

Proof. It follows from Proposition 1 that $y^m \geq y^0 = v$. Take arbitrary $i \in I$.
Case 1. If $i \notin I^<$, then by (10),

$$(\alpha_{i1} \wedge v_1) \vee (\alpha_{i2} \wedge v_2) \vee \cdots \vee (\alpha_{in} \wedge v_n) \geq \beta_i. \tag{24}$$

Since $y^m \geq v$, we have

$$(\alpha_{i1} \wedge y_1^m) \vee (\alpha_{i2} \wedge y_2^m) \vee \cdots \vee (\alpha_{in} \wedge y_n^m)$$
$$\geq (\alpha_{i1} \wedge v_1) \vee (\alpha_{i2} \wedge v_2) \vee \cdots \vee (\alpha_{in} \wedge v_n) \qquad (25)$$
$$\geq \beta_i.$$

Case 2. If $i \in I^<$, then we could find J_i and j_i^*, according to (11) and (12), respectively. Denote $j' = j_i^* \in J_i$. It is clear that

$$i \in \{k \in I^< | j_k^* = j'\} \neq \emptyset.$$

Hence, by Proposition 2 (i), we have $y_{j'}^m \geq \beta_i$. On the other hand, $j' \in J_i$ implies that $\alpha_{ij'} \geq \beta_i$, by (11). As a result,

$$\alpha_{ij'} \wedge y_{j'}^m \geq \beta_i \wedge \beta_i = \beta_i. \qquad (26)$$

Thus,

$$(\alpha_{i1} \wedge y_1^m) \vee (\alpha_{i2} \wedge y_2^m) \vee \cdots \vee (\alpha_{in} \wedge y_n^m) \geq \alpha_{ij'} \wedge y_{j'}^m \geq \beta_i. \qquad (27)$$

Cases 1 and 2 contribute to

$$(\alpha_{i1} \wedge y_1^m) \vee (\alpha_{i2} \wedge y_2^m) \vee \cdots \vee (\alpha_{in} \wedge y_n^m) \geq \alpha_{ij'} \wedge y_{j'}^m \geq \beta_i, \quad \forall i \in I. \qquad (28)$$

Hence we have $y^m \in \mathcal{S}^{A,b}$. □

Theorem 3. *The vector y^m obtained by (13) is exactly an optimal solution of problem (6).*

Proof. Let $x = (x_1, x_2, \cdots, x_n)$ be a feasible solution for problem (6). Then we have $x \geq v$ and

$$(\alpha_{i1} \wedge x_1) \vee (\alpha_{i2} \wedge x_2) \vee \cdots \vee (\alpha_{in} \wedge x_n) \geq \beta_i, \quad \forall i \in I. \qquad (29)$$

$x \geq v$ indicates

$$z(x) = (x_1 - v_1) \vee (x_2 - v_2) \vee \cdots \vee (x_n - v_n) \geq 0. \qquad (30)$$

According to Proposition 1 and Theorem 2, we know that y^m is feasible in problem (6). To prove it is further an optimal solution to (6), we just need to examine $z(y^m) \leq z(x)$. Take arbitrary $j^0 \in J$.

(i) If $\{k \in I^< | j_k^* = j^0\} = \emptyset$, then by Proposition 2 (ii), $y_{j^0}^m = v_{j^0}$. Hence,

$$y_{j^0}^m - v_{j^0} = 0 \leq z(x). \qquad (31)$$

(ii) If $\{k \in I^< | j_k^* = j^0\} \neq \emptyset$, then by Proposition 3, there exists $i^* \in \{k \in I^< | j_k^* = j^0\}$, such that $j_{i^*}^* = j^0$ and $y_{j^0}^m = \beta_{i^*}$. Note that $j_{i^*}^* = \arg \max_{j \in J_{i^*}} \{v_j\}$, by (12).

We gave

$$v_{j^0} = v_{j_{i*}^*} \geq v_{j'}, \quad \forall j' \in J_{i*}. \tag{32}$$

Furthermore,

$$y_{j^0}^m - v_{j^0} = \beta_{i*} - v_{j^0} \leq \beta_{i*} - v_{j'}, \quad \forall j' \in J_{i*}. \tag{33}$$

On the other hand, by (29), it is clear that

$$(\alpha_{i*1} \wedge x_1) \vee (\alpha_{i*2} \wedge x_2) \vee \cdots \vee (\alpha_{i*n} \wedge x_n) \geq \beta_{i*}. \tag{34}$$

There exists $j'' \in J$ such that

$$\alpha_{i*j''} \wedge x_{j''} \geq \beta_{i*}. \tag{35}$$

This indicates

$$\alpha_{i*j''} \geq \alpha_{i*j''} \wedge x_{j''} \geq \beta_{i*}. \tag{36}$$

and

$$x_{j''} \geq \alpha_{i*j''} \wedge x_{j''} \geq \beta_{i*}. \tag{37}$$

By (11) and (36), we have $j'' \in J_{i*}$. Thus, Inequality (33) indicates

$$y_{j^0}^m - v_{j^0} \leq \beta_{i*} - v_{j''}. \tag{38}$$

Considering (37) and (38), we further get

$$y_{j^0}^m - v_{j^0} \leq x_{j''} - v_{j''} \leq (x_1 - v_1) \vee (x_2 - v_2) \vee \cdots \vee (x_n - v_n) = z(x). \tag{39}$$

Cases (i) and (ii) contribute to $y_{j^0}^m - v_{j^0} \leq z(x)$ for all $j^0 \in J$. So we have

$$z(y^m) = (y_1^m - v_1) \vee (y_2^m - v_2) \vee \cdots \vee (y_n^m - v_n) \leq z(x). \tag{40}$$

As a result, y^m is an optimal solution of problem (6).   □

3   Numerical Example

Example 1. Consider the following max-min fuzzy relation inequalities,

$$\begin{cases} (0.8 \wedge y_1) \vee (0.7 \wedge y_2) \vee (0.5 \wedge y_3) \vee (0.3 \wedge y_4) \vee (0.2 \wedge y_5) \geq 0.5, \\ (0.6 \wedge y_1) \vee (0.8 \wedge y_2) \vee (0.4 \wedge y_3) \vee (0.5 \wedge y_4) \vee (0.6 \wedge y_5) \geq 0.6, \\ (0.5 \wedge y_1) \vee (0.3 \wedge y_2) \vee (0.6 \wedge y_3) \vee (0.5 \wedge y_4) \vee (0.3 \wedge y_5) \geq 0.4, \\ (0.2 \wedge y_1) \vee (0.5 \wedge y_2) \vee (0.8 \wedge y_3) \vee (0.7 \wedge y_4) \vee (0.4 \wedge y_5) \geq 0.5, \\ (0.3 \wedge y_1) \vee (0.4 \wedge y_2) \vee (0.9 \wedge y_3) \vee (0.8 \wedge y_4) \vee (0.7 \wedge y_5) \geq 0.6. \end{cases} \tag{41}$$

Denote the matrix form of (41) by $\bar{A} \circ y \geq \bar{b}$. Let

$$v = (0.3, 0.5, 0.2, 0.2, 0.4)$$

be a given vector. Since

$$
\begin{cases}
(0.8 \wedge 0.3) \vee (0.7 \wedge 0.5) \vee (0.5 \wedge 0.2) \vee (0.3 \wedge 0.2) \vee (0.2 \wedge 0.4) = 0.5, \\
(0.6 \wedge 0.3) \vee (0.8 \wedge 0.5) \vee (0.4 \wedge 0.2) \vee (0.5 \wedge 0.2) \vee (0.6 \wedge 0.4) = 0.5 < 0.6, \\
(0.5 \wedge 0.3) \vee (0.3 \wedge 0.5) \vee (0.6 \wedge 0.2) \vee (0.5 \wedge 0.2) \vee (0.3 \wedge 0.4) = 0.3 < 0.4, \\
(0.2 \wedge 0.3) \vee (0.5 \wedge 0.5) \vee (0.8 \wedge 0.2) \vee (0.7 \wedge 0.2) \vee (0.4 \wedge 0.4) = 0.5, \\
(0.3 \wedge 0.3) \vee (0.4 \wedge 0.5) \vee (0.9 \wedge 0.2) \vee (0.8 \wedge 0.2) \vee (0.7 \wedge 0.4) = 0.4 \geq 0.6,
\end{cases}
\tag{42}
$$

it is clear that $v = (0.3, 0.5, 0.2, 0.2, 0.4)$ is not a solution of system (41). Next we try to find solution of system (41), which is closest to the given vector v. That is to find an optimal solution of the following problem,

$$
\begin{aligned}
\min \quad & z(y) = (y_1 - v_1) \vee (y_2 - v_2) \vee (y_3 - v_3) \vee (y_4 - v_4) \vee (y_5 - v_5) \\
\text{s.t.} \quad & \bar{A} \circ y \geq \bar{b}, \quad y \geq v.
\end{aligned}
\tag{43}
$$

Solution: According to (10), we find that

$$
I^< = \{2, 3, 5\}.
\tag{44}
$$

Furthermore, according to (11), for $i = 2, 3, 5$, we have

$$
\begin{cases}
J_2 = \{j \in J | \alpha_{2j} \geq \beta_2\} = \{1, 2, 5\}, \\
J_3 = \{j \in J | \alpha_{3j} \geq \beta_3\} = \{1, 3, 4\}, \\
J_5 = \{j \in J | \alpha_{5j} \geq \beta_5\} = \{3, 4, 5\}.
\end{cases}
\tag{45}
$$

Following (12), for $i = 2, 3, 5$, we have

$$
\begin{cases}
j_2^* = \arg\max_{j \in J_2}\{v_j\} = \arg\max\{v_1, v_2, v_5\} = 2, \\
j_3^* = \arg\max_{j \in J_3}\{v_j\} = \arg\max\{v_1, v_3, v_4\} = 1, \\
j_5^* = \arg\max_{j \in J_5}\{v_j\} = \arg\max\{v_3, v_4, v_5\} = 5.
\end{cases}
\tag{46}
$$

Since $1 \notin I^< = \{2, 3, 5\}$, we have

$$
y^1 = y^0 = v = (0.3, 0.5, 0.2, 0.2, 0.4).
\tag{47}
$$

For $i = 2$, since $j_2^* = 2$, by (13) we have $y_2^2 = \beta_2 = 0.6$ and $y_j^2 = y_j^1$ for $j \neq 2$. That is

$$
y^2 = (0.3, 0.6, 0.2, 0.2, 0.4).
\tag{48}
$$

For $i = 3$, since $j_3^* = 1$, by (13) we have $y_1^3 = \beta_3 = 0.4$ and $y_j^3 = y_j^2$ for $j \neq 1$. That is

$$
y^3 = (0.4, 0.6, 0.2, 0.2, 0.4).
\tag{49}
$$

Since $4 \notin I^< = \{2, 3, 5\}$, we have

$$
y^4 = y^3 = (0.4, 0.6, 0.2, 0.2, 0.4).
\tag{50}
$$

For $i = 5$, since $j_5^* = 5$, by (13) we have $y_5^5 = \beta_5 = 0.6$ and $y_j^5 = y_j^4$ for $j \neq 5$. That is

$$y^5 = (0.4, 0.6, 0.2, 0.2, 0.6). \tag{51}$$

y^5 is the optimal solution of problem (43).   □

4   Conclusion

In the existing works, both the consistent and inconsistent fuzzy relation inequalities system were investigated. For the consistent system, most scholars focused on finding the complete solution set, while for the inconsistent system, one usually attempted to find some approximate solution. In this work, we consider a consistent system consisting of some max-min fuzzy relation inequalities. Suppose there is a given vector, denoted by v. We are interested in finding the so-called closest solution in system (4), which is closest to v. In the future, we would further try to investigate the closest solution for some other types of fuzzy relation systems

Acknowledgement. Supported by the National Natural Science Foundation of China (12271132, 61877014) and the funds provided by Department of Education of Guangdong Province (2022A1515011460, 2021ZDJS044, QD202211, PNB2103).

Recommender: Associate Professor Jianjun Qiu, School of Mathematics and Statistics, Lingnan Normal University in China.

References

1. Sanchez, E.: Resolution of composite fuzzy relation equations. Inf. Control **30**, 38–48 (1976)
2. Shieh, B.S.: Solutions of fuzzy relation equations based on continuous t-norms. Inf. Sci. **177**(19), 4208–4215 (2007)
3. Li, P.K., Fang, S.C.: On the resolution and optimization of a system of fuzzy relational equations with sup-t composition. Fuzzy Optim. Decis. Making **7**, 169–214 (2008)
4. Wang, S., Li, H.: Column stacking approach to resolution of systems of fuzzy relational inequalities. J. Franklin Inst. **356**, 3314–3332 (2019)
5. Sun, F., Qu, X.B.: Resolution of fuzzy relation equations with increasing operations over complete lattices. Inf. Sci. **570**, 451–467 (2021)
6. De Baets, B.: Analytical solution methods for fuzzy relational equations. In: Dubois, D., Prade, H. (eds.) Fundamentals of Fuzzy Sets. The Handbooks of Fuzzy Sets Series, vol. 1, pp. 291–340. Kluwer Academic Publishers, Dordrecht (2000)
7. Wang, P.Z., Zhang, D.Z., Sanchez, E., Lee, E.S.: Latticized linear programming and fuzzy relation inequalities. J. Math. Anal. Appl. **159**(1), 72–87 (1991)
8. Guo, F.F., Pang, L.P., Meng, D., Xia, Z.Q.: An algorithm for solving optimization problems with fuzzy relational inequality constraints. Inf. Sci. **252**, 20–31 (2013)
9. Lee, H.C., Guu, S.M.: On the optimal three-tier multimedia dtreaming dervices. Fuzzy Optim. Decis. Making **2**, 31–39 (2003)

10. Lin, H., Yang, X.: Optimal strong solution of the weighted minimax problem with fuzzy relation equation constraints. IEEE Access **6**, 27593–27603 (2018)
11. Xiao, G., Hayat, K., Yang, X.: Evaluation and its derived classifcation in a server-to-client architecture based on the fuzzy relation inequality. Fuzzy Optim. Decis. Making **22**, 213–245 (2023)
12. Zhong, Y., Xiao, G., Yang, X.: Fuzzy relation lexicographic programming for modelling P2P file sharing system. Soft. Comput. **23**, 3605–3614 (2019)
13. Ma, Y., Yang, X., Cao, B.: Fuzzy-relation-based lexicographic minimum solution to the P2P network system. IEEE Access **8**, 195447–195458 (2020)
14. Chen, Y., Liu, X., Zhang, L.: Interval solution to fuzzy relation inequality with application in P2P educational information resource sharing systems. IEEE Access **9**, 96166–96175 (2021)
15. Zhou, X., Yang, X., Cao, B.: Posynomial geometric programming problem subject to max-min fuzzy relation equations. Inf. Sci. **328**, 15–25 (2016)
16. Xiao, G., Zhu, T., Chen, Y., Yang, X.: Linear searching method for solving approximate solution to system of max-min fuzzy relation equations with application in the instructional information resources allocation. IEEE Access **7**, 65019–65028 (2019)
17. Yang, X.: Evaluation model and approximate solution to inconsistent max-min fuzzy relation inequalities in P2P file sharing system. Complexity **2019**, 6901818 (2019)
18. Wu, Y.K., Lur, Y.Y., Kuo, H.C., Wen, C.F.: An analytical method to compute the approximate inverses of a fuzzy matrix with max-product composition. IEEE Trans. Fuzzy Syst. **30**(7), 2337–2346 (2022)
19. Wu, Y.K., Lur, Y.Y., Wen, C.F., Lee, S.J.: Analytical method for solving max-min inverse fuzzy relation. Fuzzy Sets Syst. **440**, 21–41 (2022)
20. Wen, C.F., Wu, Y.K., Li, Z.: Algebraic formulae for solving systems of max-min inverse fuzzy relational equations. Inf. Sci. **622**, 1162–1183 (2023)
21. Li, J.X., Yang, S.J.: Fuzzy relation equalities about the data transmission mechanism in bittorrent-like peer-to-peer file sharing systems. In: Proceedings of the 2012 9th International Conference on Fuzzy Systems and Knowledge Discovery, FSKD 2012, pp. 452–456 (2012)
22. Yang, X.P., Lin, H.T., Zhou, X.G., Cao, B.Y.: Addition-min fuzzy relation inequalities with application in bittorrent-like peer-to-peer file sharing system. Fuzzy Sets Syst. **343**, 126–140 (2018)
23. Guu, S.M., Wu, Y.K.: A linear programming approach for minimizing a linear function subject to fuzzy relational inequalities with addition-min composition. IEEE Trans. Fuzzy Syst. **25**(4), 985–992 (2017)
24. Li, M., Wang, X.: Remarks on minimal solutions of fuzzy relation inequalities with addition-min composition. Fuzzy Sets Syst. **410**, 19–26 (2021)

Fuzzy Analysis Model of Financial System -- Application in Credit Risk of Commercial Banks

Yin Tian-hui[1,2] and Cao Bing-yuan[2,3(✉)]

[1] CFETS Financial Data Co., Ltd., Shanghai, China
[2] School of Mathematics and Information Science, Guangzhou University, Guangzhou, China
j_ifiecao@126.com
[3] Tsinghua University- IEEE Magazine FIE, Peking, China

Abstract. In financial systems, many concepts or attributes are often unclear, and in economic operations, there are also many factors that are beyond human control and difficult to define. However, these uncertain factors play a non-negligible role in ensuring the smooth operation of a system. In this paper, we analyze the financial system from the perspective of fuzzy mathematics. In the use of online survey data, mainly adopts the fuzzy comprehensive evaluation method, combined with multi-level analysis method and expert investigation method, with reference to satisfaction, feasibility, and other indicators, to judge and statistically analyze the selected financial system and establish a reasonable fuzzy evaluation model. This paper selects the credit status assessment of enterprises when commercial banks conduct credit business as the object and establishes an index system by analyzing the financial data indicators of enterprises, the previous financial status of the enterprises, and the data quality of credit reference system to propose a new method for the evaluation of the financial system and provide a scientific basis for the improvement and optimization of the financial system.

Keywords: Fuzzy Mathematics · Comprehensive Evaluation · Enterprise Credit · Risk Management

1 Introduction

1.1 Background and Significance of the Study

Due to the unclear boundary of credit risk, the financial institutions' measurement of credit risk is easily overlooked in the fuzzy nature of credit risk assessment when converting the measurement into corporate financial indicators. Therefore, credit analysis and assessment techniques in China are still at the stage of traditional ratio analysis [5]. Based on this, this paper establishes a fuzzy mathematical evaluation model to evaluate commercial bank's measurement of credit risk, and at the same time establishes a credit risk evaluation system through a dynamic quantitative credit risk management technology that integrates multiple technologies, and puts forward relevant suggestions. In order to make a correct evaluation of the financial system, put forward a new method to study the financial system, and then improve and optimize the financial system, select

© The Author(s), under exclusive license to Springer Nature Singapore Pte Ltd. 2024
B.-Y. Cao et al. (Eds.): ICFIE 2022, LNDECT 207, pp. 405–421, 2024.
https://doi.org/10.1007/978-981-97-2891-6_31

the commercial bank's credit risk evaluation system for enterprises, that is, the enterprise credit reference system, use the fuzzy integral evaluation method, referring to satisfaction, feasibility, etc., establish a reasonable fuzzy comprehensive evaluation model, and provide a scientific basis for optimizing the financial system through statistical analysis.

Currently in China, there have been researched achievements in this field by Wang Ji-kui [1], Yu Xiu-yan, and Hu Ku-jin [2], starting from the λ fuzzy measure and Choquet fuzzy integral; Xu Jing [3] used the fuzzy mathematics comprehensive evaluation method for analysis; Zhou Zhen-hua [4] based on fuzzy integral integrated support vector machine, and in accordance with the classification results of sub-support vector machines, proposed a new method based on fuzzy integral integrated support vector machine, and assigning different weights to each sub-support vector machine through fuzzy integration for integration. Internationally, there have been models based on statistical discriminant methods, which involve discriminant analysis, classification, and establishment of new discriminant models based on the category of historical samples. The main methods used are multivariate discriminant analysis and logistic analysis models, which have the advantage of obvious interpretability. However, the drawback is that the preconditions are too strict, and there is no definite expression of the relationship between risk factors and their consequences caused. In China, researches on the fuzzy comprehensive evaluation method, gray relational projection model, cloud center of the gravity model, BP neural network model, and rough set theory are relatively mature, but they all take the mutual independence of each index as the precondition, without considering the interaction between subjective evaluation indicators and other indicators, and the indicators are often not strictly independent.

1.2   Research Methodology and Theoretical Significance

By analyzing the financial statements of the enterprise, a comprehensive evaluation of its financial performance is carried out. According to the different impact of each indicator, as well as the completeness of enterprise information, and reliability, availability, compliance with business rules, data integrity, detail, stability, and timeliness of credit system data, a comparative judgment matrix is constructed to calculate the weights of each indicator, and adopt the analytic hierarchy process (AHP) evaluation method to determine the effective index through scientific methods and establish an accurate quantitative model to solve the problem of credit evaluation issues; By combining the preferences of different models within the fuzzy comprehensive evaluation method and conducting calculations and analysis, the comprehensive score is obtained to find out the optimal evaluation result [6]. As for the modified evaluation model, it can be combined with various methods, and the concepts of fuzzy measure and fuzzy integral can be introduced.

The main methods are as follows:

1) Comparative analysis method: Using the fuzzy comprehensive evaluation method, and according to the financial statements of the enterprise, analysis and evaluation of the enterprise are conducted based on the asset management capability, profitability, and debt repayment capacity. Commercial banks will not lend to companies with poor operating conditions.

2) Literature research method: According to the requirements of the topic, use various channels to search for literature related to the selected topic, and grasp the issues to be studied comprehensively and correctly.

3) Qualitative analysis method: Based on intuition and experience, analyze the selected enterprises' past and present continuation status and recent credit status.

4) Quantitative analysis method: Based on the collected data, establish mathematical models, and use the mathematical model to calculate the comprehensive scores of enterprise credit in the credit reference system, so as to conduct scientific analysis on the data.

This paper utilizes knowledge of fuzzy mathematics to establish an evaluation model to evaluate the financial system scientifically and provides a scientific solution to the evaluation problem of the financial system so that the problems such as credit evaluation, performance evaluation, and other blurred boundaries can be scientifically solved.

2 Literature Review

2.1 Literature Review on Relevant Research in China

There are many scholars in China who have devoted their efforts to studying the application of fuzzy mathematics in the credit system of commercial banks and they have achieved significant research results. These research achievements have played a significant role in the financial risk prevention of commercial banks.

(1) Xu Jing [3] objectively views the important influence of potential risks on the decisions of financial information providers and users. By introducing fuzzy concepts and establishing a set of influential factors causing financial risks, significant factors are identified through theoretical reasoning and found out the factors that have a greater impact through theoretical demonstrations. This paper expounds the evaluation results by using fuzzy mathematics, and discloses the judgment basis of various indicators in the financial statements, so as to reduce the damage to the interests of enterprises by accounting manipulation of data, help auditors to make correct evaluations, and ultimately achieve the purpose of preventing risks.

(2) Xie Yu, Wang Ya-lin, Lin Hong-jin, and Yan Miao-miao [5] evaluated the credit and default risk process of commercial loans. They believed that factors such as the operational status and future development trends of enterprises are often overlooked in credit rating. In practice, the economic-related data usually exhibit a certain degree of fuzziness. The use of fuzzy methods in the credit rating of enterprises can improve the rating process and obtain more reasonable credit rating methods. By utilizing fuzzy integration, combining it with the financial conditions of the enterprise and the subjective factors of data sources, the hierarchical structure is formed through classification from top to bottom to find out subordinating degree function and calculate the final results based on the different weights of each data indicator.

(3) Yu Wei [6], in "An Empirical Analysis of Credit Data Quality Evaluation for Commercial Banks Based on Fuzzy Analytic Hierarchy Process," pointed out the quality of credit data and the subjectivity of credit rating processes have a very important impact on the rationality of credit. Through empirical analysis, this study analyzed

the problem of credit evaluation for lending customers and established an evaluation matrix. Based on this, it considered issues such as uncertainty in individual factors and data deficiencies by making reasonable modifications. In terms of credit data quality evaluation, it proposed a new perspective for credit risk evaluation. However, the analysis does not integrate the detailed economic situation of customers.

(4) Wang Ji-kui [1] introduces λ fuzzy measures and Choquet fuzzy integrals, based on the fact that the fuzzy integral lower the requirements of various indicators and does not require the assumption of mutual independence among the indicators, proposed an easily implementable algorithm which transforms the comprehensive evaluation system into a search tree, continuously analyzes the sub-levels and performs fuzzy integration to obtain the final result. A detailed structural plan is given by the model tree and combined with JAVA programming to implement.

2.2 Defects and Improvements of Domestic Research

Domestic scholars have adopted various methods in credit risk evaluation. At present, the more advanced method is to use the relevant knowledge of fuzzy mathematics for modelings, such as the use of fuzzy comprehensive evaluation, the introduction of fuzzy integration, and even the establishment of the BP neural network model. On the one hand, some scholars ignore the mutual independence or non-linearity among data indicators in the research process, some scholars only focus on the research of subjective factors, while others have not combined enterprise financial conditions with subjective factors and data quality deficiencies, which leads to one-sided research results. On the other hand, the People's Bank of China has transformed and upgraded the original bank credit registration and consultation system, and replaced it with the enterprise credit reference system and personal credit reference system. The relevant systems can automatically collect data and summarize the credit information of customers from various commercial banks in China. However, prior permission is required to use it, which makes it more difficult to obtain and use credit data. Generally, it is not frequent for enterprises and individuals to inquire about their own credit data through the People's Bank of China and its affiliated branches. As a result, it's difficult to find hidden errors in the credit system, and errors in credit data have a certain impact on evaluating the credit status of individuals and enterprises. Therefore, when commercial banks use credit data to avoid credit risks, they will also be affected accordingly.

Based on the above, only by comprehensively analyzing the credit status of the enterprise (or individual) in combination with the financial status of the enterprise (or individual), the data quality of the credit reference system, and the subjective factors in the implementation process of credit evaluation, only then the optimal credit management of commercial banks can be achieved.

3 Fuzzy Comprehensive Evaluation of Commercial Bank Credit Risk Management

3.1 Fuzzy Comprehensive Evaluation Method

3.1.1 Concept

The fuzzy comprehensive evaluation method is a comprehensive evaluation method based on fuzzy mathematics. It transforms qualitative evaluations of research subjects into quantitative evaluations by the theory of membership degrees in fuzzy mathematics. By employing relevant theories in fuzzy mathematics, the method makes a comprehensive evaluation of the research object according to the characteristics that the research object is restricted by multiple factors and the quantitative definition of factors is fuzzy.[7].

3.1.2 Basic Ideas and Fuzzy Sets [7]

Basic Idea: Replace belonging or non-belonging with the degree of belonging.

Definition of Fuzzy Subset: For any mapping from the domain of discourse U to the closed interval $[0, 1]$, $\tilde{A} : U \rightarrow [0, 1]$, if for any $u \in U$, $u \xrightarrow{\tilde{A}} \tilde{A}(u)$, $\tilde{A}(u) \in [0, 1]$, then $\tilde{A}$ is called a fuzzy subset of U, $\tilde{A}(u)$ is called the membership function of u, also denoted as $\mu_{\tilde{A}}(u)$, and is referred to as the degree of membership of u to $\tilde{A}$.

Usually, without misunderstanding, the fuzzy subset $\tilde{A}$ and its membership function $\tilde{A}(u)$ are not distinguished, and the fuzzy subset is referred to as a fuzzy set.

Representation Methods of Fuzzy Sets:

1. Zadeh's Representation:

If the set X is a finite set, assuming the domain of discourse $X = \{x_1, x_2, \cdots x_n\}$, then the fuzzy set is represented as:

$$\tilde{A} = \frac{\tilde{A}(x_1)}{x_1} + \frac{\tilde{A}(x_2)}{x_2} + \cdots + \frac{\tilde{A}(x_n)}{x_n} = \sum_{i=1}^{n} \frac{\tilde{A}(x_i)}{x_i}.$$

Among them, $\sum \frac{\tilde{A}(x_i)}{x_i}$ represent the membership degrees $\tilde{A}(x_i)$ of the point x_i to the fuzzy set.

If the set X is an infinite set, then a fuzzy set over X is represented as $\tilde{A} = \int_{x \in X} \frac{\tilde{A}(x)}{x}$.

Among them, $\int$ represents the meaning of the infinite logical sum, $\frac{\tilde{A}(x)}{x}$ represents the membership degree $\tilde{A}(x)$ of the point x to the fuzzy set.

2. When the domain of discourse X is a finite set,

(1) Representation based on definition:

$$\tilde{A} = \left\{ \left(\tilde{A}(x_1), x_1\right), \left(\tilde{A}(x_2), x_2\right), \cdots, \left(\tilde{A}(x_n), x_n\right) \right\}$$

(2) Representation in vector form:

$$\tilde{A} = \left(\tilde{A}(x_1), \tilde{A}(x_2), \cdots, \tilde{A}(x_n)\right)$$

Note that X and $\emptyset$ can be regarded as fuzzy sets of X. When the membership functions $\tilde{A}(x) \equiv 1$ and $\tilde{A}(x) \equiv 0$, $\tilde{A}$ represent the universal set X and the empty set $\emptyset$, respectively.

3.1.3 Fuzzy Comprehensive Evaluation Steps [7]

The general evaluation steps are as follows:

(1) Identify the evaluation objects and establish the factor set.
(2) Determine the comment set of the indicators and establish the indicator system.
(3) Give weights and determine the weight set.
(4) Make single-factor evaluations.
(5) Perform comprehensive evaluation and analysis.

Applicable fuzzy comprehensive evaluation models:
There are four types of comprehensive evaluation methods.
$M(\wedge, \vee), M(\cdot, \vee), M(\cdot, \oplus), M(\cdot, +)$: "dominant factor determination type", "dominant factor prominence type", "weighted average type", and the last one is actually an ordinary matrix multiplication operation but it belongs to this category as well.

Note: When there are many factors in the evaluation, it is appropriate to adopt the multi-level comprehensive evaluation method.

The applicable evaluation steps for credit risk discussed in this paper are as follows:

3.1.3.1 Determine the Factor Domain of the Evaluation Object
Assuming that there are M evaluation index objects in the evaluation system, then $X = \{x_1, x_2, \cdots \cdots, x_m\}$.

3.1.3.2 Determine the Domain of Comments Grades
$Y = \{y_1, y_2, \cdots \cdots, y_n\}$, that is, a set of grades. The factors may be obtained, and each grade corresponds to a fuzzy subset. Fuzzy evaluation is a fuzzy set on Y.

3.1.3.3 Establish Fuzzy Relationship Matrix
After constructing the fuzzy subsets of the hierarchical structure, each individual factor $x_i (i = 1, 2, \cdots, m)$ in the evaluation system is quantified one by one. Now, assume that each factor $x_i (i = 1, 2, \cdots, m)$ has a fuzzy evaluation $R_i = \{r_{i1}, r_{i2}, \cdots, r_{in}\} \in F(X \times Y)$ (assumed $F(X)$ to be the whole of the upper fuzzy set on X). For m factors, there are m fuzzy evaluations $R_1, R_2, \cdots, R_m$. Therefore, the membership degree of each single factor in the hierarchical fuzzy subset of the evaluation system is determined $(R|x_i)$ first, and then the fuzzy relationship matrix is obtained:

$$R = \begin{pmatrix} R_1 \\ R_2 \\ \vdots \\ R_m \end{pmatrix} = \begin{bmatrix} R|x_1 \\ R|x_2 \\ \cdots \\ R|x_m \end{bmatrix} = \begin{bmatrix} r_{11} & r_{12} & \cdots & r_{1n} \\ r_{21} & r_{22} & \cdots & r_{2n} \\ \cdots & \cdots & \cdots & \cdots \\ r_{m1} & r_{m2} & \cdots & r_{mn} \end{bmatrix}_{m \times n}.$$

This matrix is called the single-factor evaluation matrix.

The element r_{ij} in the i-th row and j-th column of the matrix R, represents the membership degree of a certain evaluation system from the perspective of indicator factor x_i in the y_j-th hierarchical fuzzy subset.

The degree of performance of an evaluation system in a particular factor x_i is reflected by the fuzzy vector $(R|x_i) = (r_{i1}, r_{i2}, \cdots \cdots, r_{im})$, while in other evaluation methods it is mostly reflected by the actual value of an indicator. Thus, it can be observed that the fuzzy comprehensive evaluation method requires more information in the application process to carry out evaluation comprehensively and comprehensively.

3.1.3.4 Give Weights and Determine the Weight Vector of Evaluation Factors

In fuzzy comprehensive evaluation, the weight vector for evaluation factors is determined as follow: $A = (a_1, a_2, \cdots\cdots, a_m)$. The element a_i in the weight vector A represents the membership degree of factors to the fuzzy subsets. Considering that there are large number of system indicators evaluated in this paper, weights are given to determine the importance of each factor based on indicator characteristics. Furthermore, the weight coefficients are determined to normalize the data for the convenient calculation and come to $\sum_{i=1}^{m} a_i = 1, a_i \geq 0, i = 1, 2, \cdots, m$.

3.1.3.5 Obtain the Fuzzy Comprehensive Evaluation Result Vector

By synthesizing the factor weights A and the single-factor fuzzy evaluation of the evaluation system to be evaluated R, the fuzzy comprehensive evaluation results of the evaluation system is obtained, which is denoted as vector B, that is:

$$B = (b_1, b_2, \cdots\cdots, b_m) = A \circ R = (a_1, a_2, \cdots\cdots, a_m) \begin{bmatrix} r_{11} & r_{12} & \cdots & r_{1n} \\ r_{21} & r_{22} & \cdots & r_{2n} \\ \cdots & \cdots & \cdots & \cdots \\ r_{m1} & r_{m2} & \cdots & r_{mn} \end{bmatrix}.$$

Here, b_j is obtained by performing operations between the j-th element of A and the j-th column of R, which represents the degree y_j of membership of the evaluated object to the fuzzy subset of the grade as a whole; "$\circ$" represents the synthesis operation, that is, the synthesis operator. Under the generalized fuzzy operation, the elements of B are $b_j = (a_1 * r_{1j}) \overset{+}{\underset{*}{}} (a_2 * r_{2j}) \overset{+}{\underset{*}{}} \cdots \overset{+}{\underset{*}{}} (a_m * r_{mj})(1 \leq j \leq n)$ respectively, denoted as $M \begin{pmatrix} \cdot & + \\ * & * \end{pmatrix}$, $\overset{\cdot}{\underset{*}{}}$ is denoted as the "AND" operation under generalized fuzziness, and $\overset{+}{\underset{*}{}}$ is denoted as the "OR" operation under generalized fuzziness.

Model 1 Replace "$\overset{\cdot}{\underset{*}{}}$" with "$\wedge$", and "$\overset{+}{\underset{*}{}}$" with "$\vee$", that is model $M(\wedge, \vee)$, then $b_j = \overset{m}{\underset{i=1}{\vee}}(a_i \wedge r_{ij})$, and then $b_j = \max\{\min(a_1, r_{1j}), \cdots, \min(a_m, r_{mj})\}$, where $\wedge$ and $\vee$ represent the minimum and maximum operations, respectively.

Model 2 Replace "$\overset{\cdot}{\underset{*}{}}$" with "$\cdot$", and "$\overset{+}{\underset{*}{}}$" with "$\vee$", that is model $M(\cdot, \vee)$, then $b_j = \overset{m}{\underset{i=1}{\vee}}(a_i r_{ij})$, and then $b_j = \max\{a_1 r_{1j}, \cdots, a_m r_{mj}\}$, where "$\cdot$" represents real number multiplication.

Model 3 Replace "$\overset{\cdot}{*}$" with "·", and "$\overset{+}{*}$" with "$\oplus$", that is model $M(\cdot, \oplus)$, then

$$b_j = \min\left\{1, \sum_{i=1}^{m} a_i r_{ij}\right\},$$ where $a \oplus r = \min(1, a + r)$ is the summation with upper bound 1.

Model 4 Replace "$*$" with "·", and "$\overset{+}{*}$" with "+", that is model $M(\cdot, +)$, then

$$b_j = \sum_{i=1}^{m} a_i \cdot r_{ij}\ [7].$$

The four operators come into being four different evaluation results, and the characteristics of the four models are shown in follow Table 1:

Table 1. Different Evaluation Effects Generated by Different Operators

Characteristics	Operators			
	$M(\wedge, \vee)$	$M(\cdot, \vee)$	$M(\cdot, \oplus)$	$M(\cdot, +)$
Type	Dominant Factor Determination Type	Dominant Factor Prominence Type	Weighted Average Type	Weighted Average Type
Role of Weights	Not Obvious	Obvious	Obvious	Not Obvious
Utilization of R	Insufficient	Insufficient	Sufficient	Sufficient
Degree of Comprehensive	Weak	Weak	Strong	Strong

3.1.3.6 Analysis of Fuzzy Evaluation Vectors

Generally, three methods are commonly used: Firstly, the principle of maximum membership $M = \max(B_1, B_2, \cdots, B_m)$, which is also the most commonly used method. However, sometimes it may result in the error phenomenon in the loss of important information and lead to evaluation results that are inconsistent with common sense. Secondly,

the weighted average method $x^* = \dfrac{\sum\limits_{i=1}^{m} a_i \cdot r_i^k}{\sum\limits_{i=1}^{m} r_i^k}$ can be adopted to rank multiple factors to be

evaluated based on their hierarchical positions. Thirdly, fuzzy vector singularization.

3.2 Establishment of Indicator System

3.2.1 Principles for Constructing an Indicator System

Banks have various criteria for classifying risks [8]. In the classification of bank risks, the principles of accuracy, comprehensiveness, and marketability are initially followed. Secondly, it should facilitate management personnel, investors, and financial institutions to intuitively and comprehensively understand and grasp the information about bank risks;

and again, other methods of risk classification should be considered. Commercial bank risks generally include credit risk, exchange rate risk, interest rate risk, market risk, environmental risk, liquidity risk, etc. Among these, credit risk is the primary risk in commercial banks. Credit risk is also known as default risk. As the name implies, in the case of diversified business of commercial banks, it refers to the possibility that the debtor backed by bank credit fails to repay the debt in a timely and sufficient manner due to various reasons, such as poor management, in a situation of diversified business operations [3]. This means that they fail to repay the principal and interest in full and on time according to the contract, resulting in potential losses for the bank or counterparty. This article specifically selects the aspect of credit assessment of enterprises when commercial banks grant loans.

3.2.2 Selection of Indicator

When commercial banks grant loans to enterprises, since their primary source of profit comes from loans, they prefer to look for customers with strong capital, strong repayment ability and high creditworthiness. When conducting the credit authorization, commercial banks evaluate the credit of the enterprise based on the information of the loan applicant; Simultaneously, they investigate the legality and profitability of the loan, determine the degree of risk of the loan and issue a credit analysis report. Afterwards, the approvers review and evaluate the loan qualification of the lender based on the analysis report, re-evaluate the risk degree of the loan, and approve it within their authorized powers or submit it to the superior for approval. After the loan is approved, the bank negotiates with the lender on the loan amount, loan term and interest rate. Ultimately, a loan contract is signed and loan instructions is issued to complete the loan application and disbursement.

When applying for a loan, the commercial bank initiates a formal analysis of the enterprise's credit status. Through the analysis of the financial statements of the enterprise, a preliminary evaluation of the enterprise's finances is carried out. According to the different influence of various indicators, together with the completeness of enterprise information and the credibility, availability, compliance of business rules, data integrity, detail, stability and timeliness of the credit system data, a comparative judgment matrix is constructed to calculate the weights of each indicator.

Regarding the financial status of the enterprise, Alexander. Wall has proposed the concept of creditworthiness index, which evaluates the credit level of the enterprise by combining seven financial ratio indicators, known as Wall's weighted evaluation method. In evaluating enterprise performance in China, an indicator system is commonly used which includes four aspects: enterprise profitability, asset operation, solvency, and development capacity [9]. The eight basic indicators in the enterprise performance evaluation index system include return on equity, return on total assets, total asset turnover, turnover of current assets, asset liability ratio, interest earned multiples, sales growth rate and capital accumulation rate are the core of the evaluation index system. The previous credit history record of an enterprise is a valuable reference indicator for commercial banks when reviewing loans. The credit management system used by the Credit Reference c = Center of the People's Bank of China, that is, the data quality of the current credit reference system has a certain effect on evaluating the credit status of enterprises. It has a relatively less impact on the current financial status of enterprises and previous credit records, but cannot ignored.

Using software to draw a diagram of the credit rating indicator system, see Fig. 1.

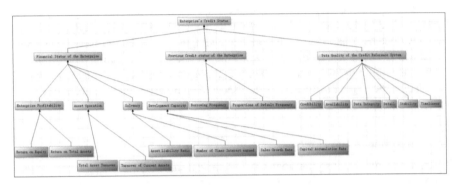

Fig. 1. Credit Evaluation Index System

3.2.3 Establishment of the Evaluation Index System

Assuming that the credit status evaluation system factor set is X, including three aspects, namely, the financial status of the enterprise, the previous credit status of the enterprise, and the data quality of the credit reference system, then $X = \{x_1, x_2, x_3\}$. Each aspect includes different indicators. According to the figure above, the enterprise profitability, asset operation, solvency, and development capacity are represented by x_{11}, x_{12}, x_{13}, and x_{14}, respectively. The indicators such as return on equity, return on total assets, total asset turnover, turnover of current assets, asset liability ratio, number of times interest earned, sales growth rate and capital accumulation rate are represented by x_{111}, x_{112}, x_{121}, x_{122}, x_{131}, x_{132}, x_{141}, and x_{142}, respectively. The borrowing frequency and the proportions of default frequency are denoted as x_{21}, x_{22}. The credibility, availability, data integrity, detail, stability and timeliness of the credit system data are represented by x_{31}, x_{32}, x_{33}, x_{34}, x_{35}, x_{36}. To sum up, the factor set of the evaluation system is $X = \{x_{111}, x_{112}, x_{121}, x_{122}, x_{131}, x_{132}, x_{141}, x_{142}, x_{21}, x_{22}, x_{31}, x_{32}, x_{33}, x_{34}, x_{35}, x_{36}\}$.

The evaluation level set of the index evaluation system is represented by Y, $Y = \{y_1, y_2, y_3, y_4, y_5\}$, representing the credit audit levels are in order as follows: "Safe Loan", "Eligible for Loan", "Loan with Risk", "Loan with High Risk - Further Authorization Required", and "Not Eligible for Loan".

Afterwards, when the established evaluation system is applied to commercial banks for credit business, a credit review on the enterprise is conducted based on their actual financial condition, previous credit history, and the effectiveness of the current credit reference system. The weights are determined according to the in above. Different banks may have different preferences, and weights for certain indicators are also different during the review process. According to the actual situation, the weight vector is found and the evaluation analysis is carried out through calculation.

It is a remarkable fact that the original data, which are selected as indicators for the financial condition and previous credit history of the enterprise, are not exactly the same and special processing is required.

3.3 Empirical Analysis

An enterprise was selected for analysis based on its financial statements (for research purposes only, so the name of the enterprise is not mentioned). The results are presented in Table 2:

Table 2. Financial Analysis Data for a Certain Enterprise in 2012

Financial Analysis Data for a Certain enterprise in 2012			
Solvency		Development Capability	
Current Ratio	1.3962	Growth Rate of Main Business Income	–
Quick Ratio	0.2857	Growth Rate of Operating Profit	33.3043
Cash Ratio	0.2013	Growth Rate of Gross Profit	33.306
Debt-to-Equity Ratio	57.7	Growth Rate of Net Profit	–
Equity-to-Asset Ratio	16.8493	Growth Rate of Net Assets	20.4988
Equity-to-Debt Ratio	21.5145	Growth Rate of Current Assets	28.3489
Equity Multiplier	5.935	Growth Rate of Fixed Assets	1.0273
Long-term debt to working capital ratio	0.3501	Growth Rate of Total Assets	27.8835
Long-term Debt Ratio	9.72	Growth Rate of Diluted Earnings per Share	30.3995
Interest Coverage Ratio	13.1134	Growth Rate of Net Assets per Share	20.332
Equity-to-Fixed Assets Ratio	3958.77	Growth Rate of Net Operating Cash Flow per Share	10.0649
Fixed Assets-to-Long-term Debt Ratio	3958.77	Three-Year Arithmetic Average Return on Equity	18.1005
Tangible Net Worth-to-Debt Ratio	493.6176		
Liquidation Value Ratio	127.4612		
Debt Coverage Ratio	16.63	Capital Structure	
Cash Flow Ratio	1.43	Equity-to-Asset Ratio	16.8493
Net Tangible Worth Assets per Share	–	Long-term Debt Ratio	9.72
Operating Funds per Share	9.362	Equity-to-Fixed Assets Ratio	3958.77

(*continued*)

Table 2. (*continued*)

Financial Analysis Data for a Certain enterprise in 2012

Solvency		Development Capability	
Total Debt /EBITDA	333.03	Debt-to-Equity and Owner's Equity Ratio	464.804
		Long-term Assets to Long-term Funds Ratio	15.9235
		Capitalization Ratio	36.59
		Capital Fixation Ratio	25.112
		Fixed Asset Proportion	0.4256
Profitability		Net Operating Cash Flow-to-Sales Ratio	3.61
Net Rate of Return on Total Assets	3.7188		
Return On Invested Capital	4.891		
Cost -to-Profit Ratio	29.2548	Operating Efficiency	
Operating Profit Ratio	20.378	Operating Cycle	1281.2
Cost of Main Business Ratio	63.4445	Inventory Turnover Days	1275.27
Net Profit Margin on Sales	12.1719	Accounts Receivable Turnover Days	5.9374
Return on Total Assets	6.7583	Current Asset Turnover Days	1126.65
Gross Profit Ratio on Sales	36.5555	Total Asset Turnover Days	1178.3
Proportion of Three Expenses	6.4019	Inventory Turnover Rate	0.2823
Operating Expense Ratio	2.964	Accounts Receivable Turnover Rate	60.6323
Management Expense Ratio	2.6963	Current Asset Turnover Rate	0.3195
Financial Expense Ratio	0.7416	Fixed Asset Turnover Rate	64.2845
Non-core Business Proportion	0.2712	Total Asset Turnover Rate	0.3055
Operating Profit Proportion	99.73	Net Asset Turnover Rate	1.3752
Earnings Before Interest, Taxes, Depreciation, and Amortization	–	Equity Turnover Rate	1.7658
Earnings Before Interest & Tax (EBIT)	2.0744	Working Capital Turnover Rate	–
EBITDA/ (Income from Main Business)	22.2974	Year-on-year Growth Rate of Inventory	–
Debt-to-Asset Ratio	78.3163	Year-on-year Growth Rate of Accounts Receivable	–

The fuzzy input vectors are $A' = (0.6, 0.35, 0.05)$. Because the selected evaluation system in this article involves multiple factors, and it is divided into several layers, Therefore, single-factor evaluations are provided first. The fuzzy evaluation matrix R cannot be given directly. R needs to be calculated first. In terms of enterprise financial status, $A_1 = (0.25, 0.13, 0.09, 0.09, 0.12, 0.08, 0.12, 0.12)$, for enterprise previous credit status, $A_2 = (0.2, 0.8)$, for credit reference system data quality, $A_3 = (0.25, 0.2, 0.1, 0.1, 0.1, 0.25)$.

Therefore, the fuzzy vector and the fuzzy evaluation matrix of 16 indicators are

$$A = \begin{bmatrix} A_1'A_1 \\ A_2'A_2 \\ A_3'A_3 \end{bmatrix} = \begin{pmatrix} 0.15 \\ 0.078 \\ 0.054 \\ 0.054 \\ 0.072 \\ 0.048 \\ 0.072 \\ 0.072 \\ 0.07 \\ 0.28 \\ 0.0125 \\ 0.01 \\ 0.005 \\ 0.005 \\ 0.005 \\ 0.0125 \end{pmatrix}, R = \begin{pmatrix} 0.35 & 0.45 & 0.1 & 0.1 & 0 \\ 0.45 & 0.3 & 0.2 & 0 & 0 \\ 0.1 & 0.25 & 0.4 & 0.2 & 0.05 \\ 0.1 & 0.3 & 0.4 & 0.15 & 0.05 \\ 0.55 & 0.3 & 0.1 & 0.05 & 0 \\ 0.5 & 0.35 & 0.1 & 0.05 & 0 \\ 0.3 & 0.4 & 0.1 & 0.15 & 0.05 \\ 0.4 & 0.35 & 0.15 & 0.05 & 0.05 \\ 0.5 & 0.25 & 0.15 & 0.1 & 0 \\ 0.6 & 0.35 & 0.05 & 0 & 0 \\ 0.5 & 0.35 & 0.05 & 0.05 & 0 \\ 0.5 & 0.25 & 0.15 & 0.05 & 0.05 \\ 0.2 & 0.4 & 0.2 & 0.1 & 0.1 \\ 0.3 & 0.35 & 0.1 & 0.1 & 0.15 \\ 0.4 & 0.45 & 0.05 & 0.05 & 0.05 \\ 0.4 & 0.25 & 0.2 & 0.1 & 0.05 \end{pmatrix}.$$

According to the model calculation, $R^T \circ A = B = (0.3889\ 0.2838\ 0.1347\ 0.0649\ 0.0152)$.

3.4 Model Improvement

In the mentioned method, the model nature of some indicators is weakened. Although the results obtained during the evaluation are more reliable, the determination of the weight indicators is often accompanied by uncertainty.

First of all, in order to simplify calculations and research in this paper, this study selects fewer financial condition indicators for the sake of simplicity, which actually does not reflect the financial situation of the enterprise comprehensively. Secondly, the weight of the evaluation factors, $A = (a_1, a_2, \cdots\cdots, a_m)$, some weights are not given reasonably. There are many indicators associated with enterprise credit, and in this case, it is more appropriate to choose analytic hierarchy process (AHP). By establishing the hierarchical structure of the problem, constructing a matrix for comparison and judgement, calculating the weights in the judgment matrix, conducting a consistency check to calculate the combination weight of elements at each level, and finally a comprehensive score is made for the evaluation index system through calculation. In decision-making problems, the conclusions derived from the analytic hierarchy process are often more reliable.

A single model usually cannot perfectly solve the problem of blurred boundaries in reality. In practice, if we want to comprehensively evaluate the credit status of an enterprise, we should adopt multiple methods and multiple perspectives for comprehensive evaluation. Combining the fuzzy comprehensive evaluation method, AHP, and expert opinions through surveys can significantly improve the accuracy of the evaluation.

4 Qualitative Analysis of Credit Risk Management

4.1 Credit Risk and Credit Products

As companies face credit risk, commercial banks appeared to the public with a new identity: commercial banks undertake, transfer and disperse or redistribute financial risks. They give birth to the derivatives market, and commercial banks, especially investment banks, have become the driving force behind the scenes. Since the 1980s, with the rapid development of the securities industry, credit products have become a common occurrence in developed economic markets, resulting in a enormous bond market. Credit derivatives have increased the possibility of comprehensive risk coverage. Commercial banks transfer credit risk back to the market and become hedging agents in financial market. They convert assets through diversified investments, real-time monitoring, and active risk management.

The purpose of risk management is to avoid large-scale economic losses, and control the stability of bank profits by commercial banks. However, in the case of banks with weak capital structures, only financing measures can be taken. Owing to information asymmetry and conflicts of interest between managers and investors, the proxy costs are relatively large, resulting in high financing costs and significant risks. In credit approval business, commercial banks have competitive advantages over the market. Commercial banks have the priority right to collect credit information. Compared with investors, they have resource and information advantages. When the debt supervisor and the authorized regulatory body adhere to the optimal behavioral pattern, the bank firmly maintains its information advantage.[10].

4.2 Default Loss Rate

In the process of credit risk analysis, loan repayment rate and default loss rate are critical pieces of information, which are the adverse consequences during the credit risk assessment process.

When analyzing the creditworthiness of an enterprise, the probability of default reflects the creditworthiness of the evaluated enterprise very well, while the loan repayment rate is mostly defined at the institutional level.

Basel II emphasizes the importance of loss given default. The loan repayment rate is influenced by measurable events and various unquantifiable factors, such as the capacity of creditors [10].

In the process of approval for credit business, commercial banks must conduct careful examination and review of the company. After gaining a comprehensive understanding of the enterprise's economic conditions and operations, continuous monitoring should be implemented to mitigate risks actively. The debt structure of an enterprise can affects its credit risk. Negligent supervision of credit can lead to failure to detect poor operating conditions of the enterprise in a timely manner, further increasing credit risk.

Loss given default is a major obstacle in credit risk measurement. It is equally important as default rate but often be overlooked. The measurement in this area lacks accuracy and some factors are hardly to quantify, which leads to the underestimation of credit risk at the portfolio level.

4.3 Credit Rating

Purpose of rating an enterprise is to give a standard credit rating based on certain data, and to provide the credit status of the credit business applicant, which is a significant task of credit risk management.

Rating agencies play an active role in the bond market. Rating agencies the returns of investors in the bond market, and provide reliable indicators after eliminating regional and industry disparities. Objective and transparent credit ratings from rating agencies are playing an increasingly important role in the financial market. The credit rating results represent professional opinions given by rating agencies regarding the creditworthiness of debtors in specific aspects. Different credit rating agencies give different ratings, and the indicators referenced in the rating are also not exactly the same. For example, the world-renowned credit rating agencies such as Standard & Poor's and Moody's, Standard & Poor's mainly refers to the possibility of bond issuer default, while Moody's reflects the expected loss on specific debt, which is the probability of default multiplied by the loss. Credit ratings are divided into long-term debt products and short-term debt products. Credit ratings include two categories, investment grade and non-investment grade.

During the credit rating process, different agencies have their own rating criteria. However, they generally consist of two parts: enterprise history and quantitative analysis, in which enterprise history includes enterprise competitiveness, enterprise management, quality of enterprise policies, business foundation, regulatory behavior, market, operation and cost control.

Commercial banks can also assess the credit risk of enterprises using credit rating methodologies. However, credit rating in the banking differs from that in the securities market. The credit risk of enterprises is affected by the different industries in which they operate and by market development much less than in the securities market. The credit rating of an enterprise has a significant impact on its orderly operations, especially for large listed companies, as the stock price may fluctuate drastically following a credit rating revision, which will lead to significant on the enterprise's economy.

The establishment of credit risk evaluation aims to better manage the credit work. Both domestic and international commercial banks basically conduct comprehensive management of commercial banks themselves, with special management of various types of risks and risk avoidance through a combination of improving the bank's profitability. At this stage, China adopts the analysis theory based on customer credit, emphasizes the analysis of management and financial quality, and determines whether the credit business is approved or not and the amount of credit through the measurement of risks. The preceding content of this article also addresses these aspects. However, it is easy to ignore the possibility of realizing the project target income used by credit when measuring risks. The credit analysis process often follows traditional models. Although the current credit reference system by the central bank has been greatly facilitated the credit evaluation, if artificial intelligence, prediction, and decision-making can be combined and multi-model can be used to analyze credit status, the reliability of credit analysis will be greatly improved.

4.4 Risk Management

The establishment of credit risk evaluation model aims to reasonably and scientifically evaluate the credit status of the enterprise, provide convenience for commercial banks' credit authorization, to avoid risks and to reduce opportunity costs.

In the actual use of credit risk evaluation model, each factor is evaluated and scored. For indicators with low scores, banks must pay special attention when reviewing them; As for indicators with low scores, banks should pay special attention when reviewing. While enterprises pay attention to their credit status, they should deeply analyze the real reasons for reduction of their credit, so that they can plug loopholes, take preventive measures, and improve creditworthiness, enable enterprises to operate permanently and healthily in the long term.

5 Conclusion

Credit evaluation is mainly based on the financial status and historical information of the enterprise by commercial banks or relevant evaluation agencies, and evaluates the solvency, future development trend and trustworthiness of the enterprise according to certain standards. Commercial banks select enterprises with high credit ratings for loans, which greatly reduces lending risks. At the same time, enterprises can reduce opportunity costs, optimize the use of funds, and improve competitiveness. From the evaluation and analysis of enterprise credit risk when commercial banks conduct credit business in this paper, it can be seen that introducing fuzzy mathematics theory into the evaluation of financial system is a relatively scientific new method. The focus point in the process of credit risk evaluation is the establishment of evaluation system indicators. When evaluating the credit of an enterprise, both financial and non-financial indicators must be considered. Non-financial indicators should include the nature of the company itself, past historical information, and mechanical errors that cannot be ignored in the process of data collection which can only be mitigated by improvements in the means of data collection, but cannot be completely eliminated. In the evaluation process, the data units of many indicators are different. The indicators with different units are normalized or dimensionless to facilitate the implementation of subsequent steps and facilitate the comparison of indicators. Only a model cannot perfectly solve the problem of fuzzy evaluation. Compared with other models, the model after introducing fuzzy mathematics is more scientific.

All in all, introducing fuzzy mathematical theory [11] into the evaluation system of financial system allows for the rational establishment of evaluation factor sets and comment sets through fuzzy comprehensive evaluation. It enables the judgment of indicators that are difficult to classify, explains the importance of each indicator in the entire system evaluation from the perspective of membership degrees, and resolves the situations where subordination is difficult to distinguish. Therefore, it is feasible to introduce the evaluation of fuzzy numbers on the basis of the fuzzy comprehensive evaluation method, and it is superior to the evaluation of classical models and other evaluation methods.

Acknowledgements. Recommender: Renjie Hu, Associate Professor, Guangdong University of Foreign Studies in China.

References

1. Ji-kui, W.: Design of customer satisfaction comprehensive evaluation algorithm based on fuzzy integration. J. Gansu Sci. **21**(1), 101–104 (2009)
2. Xiu-yan, Y., Ke-jin, H.: Empirical study on evaluation of enterprise core IT capability based on modified fuzzy integration evaluation method. Sci. Technol. Manage. Res. **7**, 164 (2009)
3. Jing, X.: Application of fuzzy mathematics in bank risk prevention. Econ. Res. Guid. **11**, 050 (2011)
4. Zhen-hua, Z.: Research on Credit Risk Evaluation Model of Commercial Banks Based on Fuzzy Integration Integrated Support Vector Machine. Harbin Institute of Technology (2006)
5. Xie, Y., Ya-lin, W., Hong-jin, L., et al.: Research on enterprise credit rating method based on fuzzy integration. Chin. Soft Sci. Mag. **9**, 145–149 (2004)
6. Wei, Y.: Empirical analysis of commercial bank credit data quality evaluation based on fuzzy analytic hierarchy process. Shanghai Financ. **3**, 102–105 (2011)
7. Bing-Yuan, C.: Optimal Models and Methods with Fuzzy Quantity. Springer (2010). https://doi.org/10.1007/978-3-642-10712-2
8. Jiang-ping, P.: Risk Management: Theory and System of Commercial Bank Risk Management, pp. 16–17. Southwestern University of Finance and Economics Press (2001)
9. Jing: Basic Accounting Tutorial, vol. 3, pp. 261–262. Lixin Accounting Publishing House (2008)
10. De Servigny, A., Renault, O., Yong-quan, P., et al.: Credit Risk: Measuring and Managing Credit Risk. China Machine Press (2012). 22–23, 33–34, 108–110
11. Lin, H.R., Cao, B.Y., Liao, Y.Z.: Fuzzy Sets Theory Preliminary. Springer International Publishing, Cham (2018). https://doi.org/10.1007/978-3-319-70749-5

Mutation and Prediction of COVID-19

Pei-Jun Zuo[1]([✉]), Long-Long Zuo[2], Zhi-Hong Li[1], and Li-Ping Li[1]

[1] Shantou University, Shantou, Guangdong, China
peijunzuo@hotmail.com, lpli@stu.edu.cn
[2] The Chinese University of Hong Kong, Hong Kong, China

Abstract. The spike protein of COVID-19 original strain D614 contains three antigenic long fragments with 37 amino acids or more, namely D614, N148 and I358. The antigen precision is 25.23, 26.92, and 31.66, respectively. The mutation time of these three long fragments was 3, 10 and 23.4 months after the COVID-19 outbreak. There is a correlation between antigen precision and mutation time, $R^2 = 0.996$. The precision of these long antigens determines the difficulty of mutation, which has statistical significance after statistical testing, $P = 0.03$. The mutations caused by long antigen mutations are the D614G strain, Delta strain, and Omicron strain that cause the world pandemic. This discovery will provide a simple tool for predicting mutations, which can be used not only to predict mutations of COVID-19, but also to predict mutations of pathogens of other infectious diseases and has reference value for mutation prediction of new infectious diseases in the future.

Keywords: Antigen · Evolution · D614

1 Background

As is well known, it is very difficult to predict the mutation of infectious disease pathogens two years in advance. As we are familiar with predicting weather forecasts. But the weather forecast is either one day or one week in advance. The weather forecast for two weeks in advance is still in the research stage [1]. It should be very difficult to predict a rainy day in 2 years. But can you predict whether it will be sunny one day in three months? The answer should be 'no'. There are too many influencing factors in weather forecasting.

How to predict earthquakes? In Japan, predicting earthquakes 5 min in advance may be a fact. The principle of predicting earthquakes is that the propagation speed of electromagnetic waves is greater than that of seismic waves. There is a report on predicting earthquakes 24 h in advance [2], However, there are still no reports of predicting earthquakes three months or two years in advance.

If we predict that there will be an outbreak of infectious diseases in 5 min or 24 h, all we can do is wear masks, which are useful for protecting vulnerable populations, but have limited effectiveness in controlling infectious disease pathogens.

Predicting the onset of infectious diseases three months in advance may be useful for preventing and controlling the harm of infectious diseases. We need 3 months to prepare

B.-Y. Cao et al. (Eds.): ICFIE 2022, LNDECT 207, pp. 422–433, 2024.
https://doi.org/10.1007/978-981-97-2891-6_32

personnel, instruments, reagents, funds, and even vaccines. In general, it takes 2 months to determine the reagent, so it is reasonable to respond to new infectious diseases within 3 months.

In order to predict infectious diseases, such as influenza, many factors are needed. If a multi-dimensional mathematical prediction model is built, information needs to be obtained from big data [3]. For influenza, mathematical models may be suitable because influenza outbreaks occur almost every year and during the same season. Despite this regularity, mathematical models are not helpful in predicting mutations. For irregular infectious diseases, it is even more difficult to predict. For example, SARS, which broke out in 2003, became the COVID-19 after mutation and came again in 2019. No one could predict that such frequent and numerous variations would occur in COVID-19 two years ago when COVID-19 just broke out.

On December 12, 2019, COVID-19 original strain D614 first broke out in Wuhan, China [4]. As of January 26, 2020, the Wuhan epidemic has confirmed 2794 infections and 80 deaths.

According to the sequencing data of 6244 global COVID-19 cases, the D614G mutant strain only accounted for 9.9% of the total number of COVID-19 cases before March 2020, but 10 days later, that is, on March 10, 2020, the D614G mutant strain soon accounted for 54% of the total cases, becoming the dominant strain [5]. The second generation COVID-19 mutant is D614G. The mutant strain caused the second wave of COVID-19 epidemic that began on March 10, 2020.

Based on D614G mutant, COVID-19 has 10 subtypes of mutations [6]. One of them is the Delta type, as shown in Table 1. In October 2020, the mutant strain of B.1.617.2 (Delta) COVID-19 was first prevalent in India [7]. The third generation COVID-19 should be Delta subtype.

Finally, by November 23, 2021, Omicron type began to become the dominant strain [8].

How did these four generations of COVID-19 evolve step by step?

Is it possible to predict the three major mutations that will occur on December 12, 2019, the initial outbreak of the COVID-19 epidemic?

The traditional predictive models for infectious diseases are S (susceptible), E (exposed), I (infected), and R (recovered) models. Abbreviated as the SEIR model [9]. But this prediction only predicts the number of cases in a few days. In fact, the number of future cases is only predicted based on the number of infectious sources and susceptible individuals. This model does not consider changes in pathogen variation or virulence at all.

The COVID-19 has been prevalent for more than three years, seriously threatening the life safety of the world. This epidemic not only endangers people's health, but also threatens the global economy and the development of the international community. The problem we are facing is how to quickly and effectively control infectious diseases in the event of a new outbreak. Our recent work is to analyze the antigenic variation of COVID-19 and its regularity. Our findings are that some characteristics of the virus can be used to predict the mutation of the spike antigen of COVID-19 and the date of the COVID-19 outbreak caused by the mutant strain, such as the sequence and date of the COVID-19 outbreak caused by D614G, Delta, and Omicron, respectively [10].

We analyzed four longest antigenic fragments without glycine from the original strain D614 of COVID-19, three of which contained tryptophan. The standard deviation of the molecular weight of amino acids contained in the fragment is calculated separately [11].

This paper attempts to find out why the mutation of COVID-19 first occurs in the location of D614, rather than other locations. Just like the first mutation, why did the Delta type of the second mutation occur at N148 and must be mutated after the D614G mutation, rather than reversing the order. Similarly, why did the third mutation Omcron type occur in the fragment of I358.

2   Material and Method

The protein access number of original strains D614 of COVID-19 in Wuhan is "QHD43416", including D614 antigen fragment. The protein access number of mutant D614G is "7KDK_A", including a point mutation D614G. The sequence was retrieved from the NCBI database of the National Institutes of Health in the United States.

The delta protein access number of COVID-19 is "7V8B_A", including the mutations of N148 (E156del, F157del, R158G), T19R, G142D, L452R, T478K, D614G, P681R, D950N. The sequence was retrieved from the NCBI database of the National Institutes of Health in the United States. Ten-point mutation are listed in Table 1.

The protein access number of COVID-19 Omicron subtype is "UGO97992" and "UGY75354". The included point mutations are I358 (S375F, S373P, S371L), N211del/L212I, Y145del, Y144del, V143del, G142D, T95I, V70del, H69del, A67V, Y505H, N501Y, Q498R, G496S, Q493R, E484A, T478K, S477N, G446S, N440K, K417N, G339D, D796Y, L981F, N969K, Q954H, D614K G. The 30-point mutations are also listed in Table 1.

Table 1. Dominant mutations of SARS-CoV-2.

Dominant date	Months	Sub Types	Spike mutations (number)
Dec. 12, 2019	0	D614	**D614** (0)
Mar. 10, 2020	3	D614G	**D614G** (1)
Oct. 2020	10	Delta	T19R, G142D, E156del, F157del, R158G, L452R, T478K, **D614G**, P681R, D950N (10)
Nov. 23, 2021	23.4	Omicron	N211del/L212I, Y145del, Y144del, V143del, G142D, T95I, V70del, H69del, A67V, Y505H, N501Y, Q498R, G496S, Q493R, E484A, T478K, S477N, G446S, N440K, K417N, S375F, S373P, S371L, G339D, D796Y, L981F, N969K, Q954H, **D614G** (30)

All mutant subtypes of COVID-19 occurred on the basis of D614G mutant. D614G is marked in bold.

The amino acid sequences of the D614 original strain, D614G mutant strain, Delta subtype, and Omicron subtype were compared using online software Cluster Omega.

In the comparison results of Cluster Omega, all glycine is labeled in yellow (as shown in Figs. 1, 2, and 3).

In the original strain D614, the four longest antigenic fragments without glycine were selected for further analysis. The amino acid molecular weights of these candidate fragments will be analyzed. The standard deviation of molecular weight was calculated using Excel.

Table 2. The molecular weights and SDs of D614, N148, I358 and candidate of potential mutant peptide.

Peptides	Amino acids	Mean of molecular weight	SD	Contains "W"
F718	43	125.3	20.69	N
I358	41	134.2 (Omicron)	31.66(3)	Y
N148	38	140.1 (Delta)	26.92(2)	Y
D614	37	129.3 (D614G)	25.23(1)	Y

Among the largest fragments with tryptophan "W", containing 27 amino acids or more between glycine "G", D614 has the smallest standard deviation of 25.23. Its meaning is that the precision is the "roughest" or the lowest. It is most likely that the mutation occurred first here, as confirmed by the D614G mutation. The evolutionary order is: first D614G, second Delta, and third Omicron.

The amino acid sequence of protein "QHD43416" includes D614, and "7KDK_A" includes D614G, which is derived from NCBI. The amino acid sequence of COVID-19 included D614 and D614G mutants, which were compared with the online software Cluster Omega. In the comparison results of Cluster Omega, all 'G' is marked in yellow. N148 contains the largest fragment of the Delta mutant strain, marked by an arrow.

In the comparison results of Cluster Omega, all 'G' is marked in yellow. D614 and I358 are marked by arrows. I358 contains the largest fragment of the Omicron mutant.

In the comparison results of Cluster Omega, all 'G' is marked in yellow. F718 is marked by an arrow. The F718 fragment does not include tryptophan W and has not mutated until this day.

3 Results

3.1 Original Strain D614, Confirmed Mutant D614G

As shown in Fig. 2, the D614 fragment contains 37 amino acids. Its amino acid sequence is TNTSNQVAVLYQDVNCTEVPIHADQLTPTWRVYST. As shown in Table 2, the average molecular weight of the amino acids it contains is 129.3, and the standard deviation between the molecular weights of the amino acids it contains is 25.23. This glycine free fragment contains the highest molecular weight amino acid tryptophan (W, molecular weight 204.2262). The amino acid at position 614 is aspartic acid, which has been confirmed to mutate into glycine after 3 months, namely the mutant D614G.

```
QHD43416    mfvflvllplvssqcvnlttrtqlppaytnsftrgvyypdkvfrssvlhstqdlflpffs    60
UG097992    mfvflvllplvssqcvnlttrtqlppaytnsftrgvyypdkvfrssvlhstqdlflpffs    60
UGY75354    mfvflvllplvssqcvnlttrtqlppaytnsftrgvyypdkvfrssvlhstqdlflpffs    60
7KDK_A      mfvflvllplvssqcvnlttrtqlppaytnsftrgvyypdkvfrssvlhstqdlflpffs    60
7V8B_A      mfvflvllplvssqcvnlrtrtqlppaytnsftrgvyypdkvfrssvlhstqdlflpffs    60
            *******************  ****************************************


QHD43416    nvtwfhaihvsgtngtkrfdnpvlpfndgvyfastek2sniirgwifgttldsktqslliv   120
UG097992    nvtwfhvi--sgtngtkrfdnpvlpfndgvyfasieksniirgwifgttldsktqslliv   118
UGY75354    nvtwfhvi--sgtngtkrfdnpvlpfndgvyfasieksniirgwifgttldsktqslliv   118
7KDK_A      nvtwfhaihvsgtngtkrfdnpvlpfndgvyfasteksniirgwifgttldsktqslliv   120
7V8B_A      nvtwfhaihvsgtngtkrfdnpvlpfndgvyfasteksniirgwifgttldsktqslliv   120
            ******. *  **************************  ***********************
                            N148→ → → → → → → → → → → → → → → → → → → → →
QHD43416    nnatnvvikvcefqfcndpflgvyyhknnkswmesefrvyssannctfeyvsqpflmdle   180
UG097992    nnatnvvikvcefqfcndpfld---hknnkswmesefrvyssannctfeyvsqpflmdle   175
UGY75354    nnatnvvikvcefqfcndpfld---hknnkswmesefrvyssannctfeyvsqpflmdle   175
7KDK_A      nnatnvvikvcefqfcndpflgvyyhknnkswmesefrvyssannctfeyvsqpflmdle   180
7V8B_A      nnatnvvikvcefqfcndpfldvyyhknnkswmes--gvyssannctfeyvsqpflmdle   178
            ********************.    *********    ********************


QHD43416    gkqgnfknlrefvfknidgyfkiyskhtpinlvrdlpqgfsaleplvdlpiginitrfqt   240
UG097992    gkqgnfknlrefvfknidgyfkiyskhtpi-ivrdlpqgfsaleplvdlpiginitrfqt   234
UGY75354    gkqgnfknlrefvfknidgyfkiyskhtpi-ivrdlpqgfsaleplvdlpiginitrfqt   234
7KDK_A      gkqgnfknlrefvfknidgyfkiyskhtpinlvrdlpqgfsaleplvdlpiginitrfqt   240
7V8B_A      gkqgnfknlrefvfknidgyfkiyskhtpinlvrdlpqgfsaleplvdlpiginitrfqt   238
            ***************************** :*****************************


QHD43416    llalhrsyltpgdsssgwtagaaayyvgylqprtfllkynengtitdavdcaldplsetk   300
UG097992    llalhrsyltpgdsssgwtagaaayyvgylqprtfllkynengtitdavdcaldplsetk   294
UGY75354    llalhrsyltpgdsssgwtagaaayyvgylqprtfllkynengtitdavdcaldplsetk   294
7KDK_A      llalhrsyltpgdsssgwtagaaayyvgylqprtfllkynengtitdavdcaldplsetk   300
7V8B_A      llalhrsyltpgdsssgwtagaaayyvgylqprtfllkynengtitdavdcaldplsetk   298
            ************************************************************
```

Fig. 1. Amino acid N148. (Color figure online)

3.2   Delta Subtype N148 Fragment

As shown in Fig. 1, the other longer segment without glycine is N148, which contains tryptophan and has 38 amino acids. Its amino acid sequence is VYYHKNNKSWME-SEFRVYSSANNCTFEYVSQPFLMDLE. As shown in Table 2, the average molecular weight of the amino acids it contains is 140.9, and the standard deviation between the molecular weights of the amino acids it contains is 26.92. N148 was also confirmed to mutate into a Delta subtype (E156del, F157del, R158G are mutations in Delta subtype) after 10 months.

```
                                                      I358→→→→→→→→→→→→
QHD43416    ctlksftvekgiyqtsnfrvqptesivrfpnitnlcpfgevfnatrfasvyawnrkrisn    360
UGO97992    ctlksftvekgiyqtsnfrvqptesivrfpnitnlcpfdevfnatrfasvyawnrkrisn    354
UGY75354    ctlksftvekgiyqtsnfrvqptesivrfpnitnlcpfdevfnatrfasvyawnrkrisn    354
7KDK_A      ctlksftvekgiyqtsnfrvqptesivrfpnitnlcpfgevfnatrfasvyawnrkrisn    360
7V8B_A      ctlksftvekgiyqtsnfrvqptesivrfpnitnlcpfgevfnatrfasvyawnrkrisn    358
            *******************************. ********************
         I358→→→→→→→→→→→→
QHD43416    cvadysvlynsasfstfkcygvsptklndlcftnvyadsfvirgdevrqiapgqtgkiad    420
UGO97992    cvadysvlynlapfftfkcygvsptklndlcftnvyadsfvirgdevrqiapgqtgniad    414
UGY75354    cvadysvlynlapfftfkcygvsptklxxxxxxnvyadsfvirgdevrqiapgqtgniad    414
7KDK_A      cvadysvlynsasfstfkcygvsptklndlcftnvyadsfvirgdevrqiapgqtgkiad    420
7V8B_A      cvadysvlynsasfstfkcygvsptklndlcftnvyadsfvirgdevrqiapgqtgkiad    418
            ********** * * ***********       **********************:***

QHD43416    ynyklpddftgcviawnsnnldskvggnynylyrlfrksnlkpferdisteiyqagstpc    480
UGO97992    ynyklpddftgcviawnsnkldskvsgnynylyrlfrksnlkpferdisteiyqagnkpc    474
UGY75354    ynyklpddftgcviawnsnkldskvsgnynylyrlfrksnlkpferdisteiyqagnkpc    474
7KDK_A      ynyklpddftgcviawnsnnldskvggnynylyrlfrksnlkpferdisteiyqagstpc    480
7V8B_A      ynyklpddftgcviawnsnnldskvggnynyryrlfrksnlkpferdisteiyqagskpc    478
            *******************:*****. ***** *********************. .**

QHD43416    ngvegfncyfplqsygfqptngvgyqpyrvvvlsfellhapatvcgpkkstnlvknkcvn    540
UGO97992    ngvagfncyfplrsysfrptygvghqpyrvvvlsfellhapatvcgpkkstnlvknkcvn    534
UGY75354    ngvagfncyfplrsysfrptygvghqpyrvvvlsfellhapatvcgpkkstnlvknkcvn    534
7KDK_A      ngvegfncyfplqsygfqptngvgyqpyrvvvlsfellhapatvcgpkkstnlvknkcvn    540
7V8B_A      ngvegfncyfplqsygfqptngvgyqpyrvvvlsfellhapatvcgpkkstnlvknkcvn    538
            *** ********:**. *:** ***:********************************

QHD43416    fnfngltgtgvltesnkkflpfqqfgrdiadttdavrdpqtleilditpcsfggvsvitp    600
UGO97992    fnfnglkgtgvltesnkkflpfqqfgrdiadttdavrdpqtleilditpcsfggvsvitp    594
UGY75354    fnfnglkgtgvltesnkkflpfqqfgrdiadttdavrdpqtleilditpcsfggvsvitp    594
7KDK_A      fnfngltgtgvltesnkkflpfqqfgrdiadttdavrdpqtleilditpcsfggvsvitp    600
7V8B_A      fnfngltgtgvltesnkkflpfqqfgrdiadttdavrdpqtleilditpcsfggvsvitp    598
            ******. *****************************************************
         D614→→→→→→→→→→→→→→→→→→→→
QHD43416    gtntsnqvavlyqdvnctevpvaihadqltptwrvystgsnvfqtragcligaehvnnsy    660
UGO97992    gtntsnqvavlyqgvnctevpvaihadqltptwrvystgsnvfqtragcligaeyvnnsy    654
UGY75354    gtntsnqvavlyqgvnctevpvaihadqltptwrvystgsnvfqtragcligaeyvnnsy    654
7KDK_A      gtntsnqvavlyqgvnctevpvaihadqltptwrvystgsnvfqtragcligaehvnnsy    660
7V8B_A      gtntsnqvavlyqgvnctevpvaihadqltptwrvystgsnvfqtragcligaehvnnsy    658
            *************. *******************************:*****
```

Fig. 2. Amino acids of D614 and I358. (Color figure online)

F718→→→→→→→→→→→→

QHD43416	ecdipigagicasyqtqtnsprrarsvasqsiiaytmslgaensvaysnnsiaiptnfti	720
UGO97992	ecdipigagicasyqtqtkshrrarsvasqsiiaytmslgaensvaysnnsiaiptnfti	714
UGY75354	ecdipigagicasyqtqtkshrrarsvasqsiiaytmslgaensvaysnnsiaiptnfti	714
7KDK_A	ecdipigagicasyqtqtnspgsassvasqsiiaytmslgaensvaysnnsiaiptnfti	720
7V8B_A	ecdipigagicasyqtqtnsrgsassvasqsiiaytmslgaensvaysnnsiaiptnfti	718

******************:*   * *********************************

F718→→→→→→→→→→→→→

QHD43416	svtteilpvsmtktsvdctmyicgdstecsnlllqygsfctqlnraltgiaveqdkntqe	780
UGO97992	svtteilpvsmtktsvdctmyicgdstecsnlllqygsfctqlkraltgiaveqdkntqe	774
UGY75354	svtteilpvsmtktsvdctmyicgdstecsnlllqygsfctqlkraltgiaveqdkntqe	774
7KDK_A	svtteilpvsmtktsvdctmyicgdstecsnlllqygsfctqlnraltgiaveqdkntqe	780
7V8B_A	svtteilpvsmtktsvdctmyicgdstecsnlllqygsfctqlnraltgiaveqdkntqe	778

:************

QHD43416	vfaqvkqiyktppikdfggfnfsqilpdpskpskrsfiedllfnkvtladagfikqygdc	840
UGO97992	vfaqvkqiyktppikyfggfnfsqilpdpskpskrsfiedllfnkvtladagfikqygdc	834
UGY75354	vfaqvkqiyktppikyfggfnfsqilpdpskpskrsfiedllfnkvtladagfikqygdc	834
7KDK_A	vfaqvkqiyktppikdfggfnfsqilpdpskpskrsfiedllfnkvtladagfikqygdc	840
7V8B_A	vfaqvkqiyktppikdfggfnfsqilpdpskpskrsfiedllfnkvtladagfikqygdc	838

**************** ***

QHD43416	lgdiaardlicaqkfngltvlpplltdemiaqytsallagtitsgwtfgagaalqipfam	900
UGO97992	lgdiaardlicaqkfkgltvlpplltdemiaqytsallagtitsgwtfgagaalqipfam	894
UGY75354	lgdiaardlicaqkfkgltvlpplltdemiaqytsallagtitsgwtfgagaalqipfam	894
7KDK_A	lgdiaardlicaqkfngltvlpplltdemiaqytsallagtitsgwtfgagaalqipfam	900
7V8B_A	lgdiaardlicaqkfngltvlpplltdemiaqytsallagtitsgwtfgagaalqipfam	898

**************:***

QHD43416	qmayrfngigvtqnvlyenqklianqfnsaigkiqdslsstasalgklqdvvnqnaqaln	960
UGO97992	qmayrfngigvtqnvlyenqklianqfnsaigkiqdslsstasalgklqdvvnhnaqaln	954
UGY75354	qmayrfngigvtqnvlyenqklianqfnsaigkiqdslsstasalgklqdvvnhnaqaln	954
7KDK_A	qmayrfngigvtqnvlyenqklianqfnsaigkiqdslsstasalgklqdvvnqnaqaln	960
7V8B_A	qmayrfngigvtqnvlyenqklianqfnsaigkiqdslsstasalgklqnvvnqnaqaln	958

:*:******

QHD43416	tlvkqlssnfgaissvlndilsrldkveaevqidrlitgrlqslqtyvtqqliraaeira	1020
UGO97992	tlvkqlsskfgaissvlndifsrldkveaevqidrlitgrlqslqtyvtqqliraaeira	1014
UGY75354	tlvkqlsskfgaissvlndifsrldkveaevqidrlitgrlqslqtyvtqqliraaeira	1014
7KDK_A	tlvkqlssnfgaissvlndilsrldkveaevqidrlitgrlqslqtyvtqqliraaeira	1020
7V8B_A	tlvkqlssnfgaissvlndilsrldppeaevqidrlitgrlqslqtyvtqqliraaeira	1018

********:***********:****   *****************************

Fig. 3. Amino acid of F718. (Color figure online)

3.3   Omicron Subtype I358 Fragment (S375F, S373P, S371L)

As shown in Fig. 2, other similar fragments include I358, which contains tryptophan and 41 amino acids. Its amino acid sequence is EVFNATRFASVYAWNRKRISNC-VADYSVLYNSASFSTFKCY. As shown in Table 2, the average molecular weight of the amino acids it contains is 134.2, and the standard deviation between the molecular weights of the amino acids it contains is 31.66. I358 was also confirmed to mutate into Omicron subtype (S375F, S373P, S371L are mutations in Omicron subtype) after 23.4 months.

3.4   There is a Dose-Response Relationship Between the Standard Deviation Between the Molecular Weights of Amino Acids in the Mutated Fragment and the Time Required for the Mutation of the Fragment

As shown in Fig. 4, for the convenience of expression, we refer to the standard deviation between the molecular weights of amino acids in the mutated fragment as the antigen standard deviation. The antigen standard deviation of D614 fragment is 25.23, and the time required for mutation to D614G is 3 months, which is March 10, 2020. The antigen standard deviation of N148 fragment is 26.92, and the time required for mutation to Delta type is 10 months, which is October 2020. Similarly, the antigen standard deviation of fragment I358 is 31.66, and the time required for mutation to Omicron type is 23.4 months, which is November 23, 2021. If x represents the standard deviation of antigen and y represents the time required for antigen mutation, the regression equation is $y = 88.467\ln(x) -282.05$. The correlation coefficient R is 0.998. After statistical testing, the probability of Class I errors occurring $P = 0.03$, < 0.05. There is a dose-response relationship between the standard deviation of antigen and the time required for the mutation, which is statistically significant.

The actual date of occurrence of D614G mutation is 3 months later, and the predicted date of mutation is 3.5 months. The actual date of the Delta mutation is 10 months later, and the predicted date of the mutation is 9.3 months. The actual date of Omicron mutation is 23.4 months later, and the predicted date of mutation is 23.6 months. The longest difference between a true mutation and a predicted mutation is 3 weeks.

There is a dose-dependent relationship between the latency required for antigen evolution and antigen precision. The probability of type A error occurring is $P = 0.03$, less than 0.05, is statistically significant. The predicted evolutionary order is consistent with the actual evolutionary history.

3.5   The Longest Glycine Free Fragment F718 Without Mutation

As shown in Fig. 3, the longest glycine free fragment is F718, which does not contain tryptophan and has 43 amino acids. Its amino acid sequence is AENSVAYSNNSIAIPT-NFTISVTTEILPVSMTKTSVDCTMYIC. No mutation occurred.

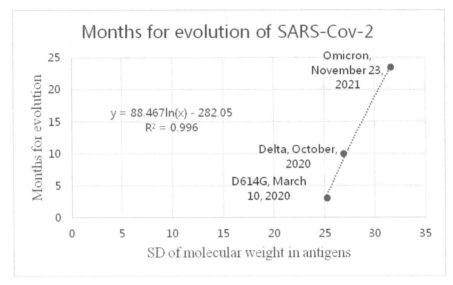

Fig. 4. Dose-dependent relationship between evolution time of COVID-19 and antigen precision.

4 Discussion

Question: Why does COVID-19 D614 mutate first to D614G, then to Delta type, and finally to Omicron type?

Discovery: It was found that the D614 mutation fragment had the smallest antigen standard deviation, the Delta mutation longest fragment N148 had the second smallest antigen standard deviation, and the Omicron longest mutation fragment I358 had the third smallest antigen standard deviation, showing significant differences.

Significance: During the COVID-19 pandemic, the antigenic standard deviation of mutant largest fragments can be used to predict the order and date of mutation.

For proteins or antigens, the standard deviation between the molecular weights of the amino acids they contain can be used as a marker of their "precision". The larger the standard deviation of antigens, the more "precise" they are. On the contrary, the smaller the standard deviation of the antigen, the "rougher" it is.

The standard deviation of D614 antigen is 25.23, and it has already mutated into D614G earliest.

Why did the D614 fragment mutate first, rather than the longer fragments F718, I 358, N148?

The antigen standard deviation of N148 fragment is 26.92, and the antigen standard deviation of I358 fragment is 31.66. Compared with the antigen standard deviation of 25.23 for the D614 fragment, neither has the smallest antigen standard deviation. There is no reason for these "precise" antigens to mutate in the presence of rough antigens. The crudest antigen first mutates around 3 months. After the rough antigen mutation, the second rough N148 upgraded to the roughest antigen, and the mutation also occurred

10 months after the outbreak of the epidemic. After the mutation of the second rough antigen, the third rough I358 fragment was upgraded to the roughest antigen, and the mutation also occurred 23.6 months after the outbreak of the epidemic. There is a dose-dependent relationship between the standard deviation of antigens and the time required for mutations to occur, and it is statistically significant. The order of predicted mutations is consistent with the actual order of occurrence. D614G is the first mutation, Delta subtype is the second order mutation, and Omicron is the third order mutation.

The antigen standard deviation of F718 fragment is only 20.69, even smaller than 25.23. Why is this 'rough' fragment not mutated? The possible reason is that the distribution of amino acids leads to some biases. For example, this fragment does not contain the largest amino acid, tryptophan (W). If a segment does not contain tryptophan, it may not mutate or mutate later than other segments.

Excluding fragments without "W" from potential candidates for evolution is due to the influence of statistical errors on the calculation of standard deviation.

Tryptophan also has its own biochemical functions, one of which is that it can be translated from a "terminator, TGA" into tryptophan, greatly extending the length of the peptide chain. There are reports that the TGA of spirochetes can be translated into tryptophan, while other organisms are signaling the termination of translation [12].

We all know that spirochetes can produce non-specific hypersensitivity reactions in the human body. Because of this, tryptophan may play a certain role in the evolution of viruses.

Tryptophan itself is also an inflammatory mediator. Potential evolution cannot occur in tryptophan free fragments.

To sum up, COVID-19 may start its evolutionary behavior from a segment containing tryptophan (W) between two glycine (G) and having "rough" characteristics, just as it mutates from D614 to D614G, then to Delta, and finally to Omicron.

This prediction only accurately matched the D614G, Delta, and Omicron mutant strains, but did not predict mutations in Alpha, Beta, and Gamma. It is precisely because mutations in Alpha, Beta, and Gamma were not predicted that this hypothesis is correct, as these three mutant strains have not become widespread in China. Without a large-scale epidemic in China, it cannot be considered a global pandemic. This hypothesis perfectly explains the four world COVID-19 pandemics caused by D614, D614G, Delta and Omicron types. This prediction can guide us to detect the pandemic waves caused by mutant strains in advance and filter out the small pandemic waves caused by mutant strains. This is what we need. Identify and resolve major conflicts, ignoring minor conflicts.

The mutation points of Delta and Omicron types do indeed include some amino acids that mutate from small molecular weights to large molecules. These mutations did not occur in the largest segment larger than or equal 27 amino acids. We found that mutations in the largest fragment greater than or equal to 27 amino acids determine the infectivity of the virus, so from D614 to D614G, to Delta type, and then to Omicron type, the infectivity becomes stronger and stronger. And all mutated fragments, including the largest fragment, determine the pathogenicity of the virus. Among all 10 mutation fragments of Delta type, compared with the original D614 strain in Wuhan, there were 9 mutation fragments with higher precision, while only one mutation fragment had the same precision. The overall trend was an increase in precision, so the pathogenicity also increased. Among the 43 mutated fragments in Omicron, over half of them showed a decrease in precision, and the overall trend was a decrease in precision. Therefore, Omicron has the weakest pathogenicity, even lower than the original strain. We will have specialized articles in the future to explain in detail the prediction of pathogenicity, and we will not elaborate on it here.

The calculation of antigen precision can help us detect virus mutations at least 3 generations or about 23 months in advance. This prediction not only correctly indicates that the order of virus evolution is D614, D614G, Delta, Omicron, but also accurately predicts that the time of evolution is 3 months for D614G, 10 months for Delta, and 23.4 months for Omicron type, respectively.

According to the calculation of antigen precision, it can also be applied to mutation prediction of other infectious diseases such as dengue fever, as well as to mutation prediction of tumors. The concept of antigen precision can unify the holographic theory and mutation theory of tumor mutations, hoping to make breakthroughs in predicting tumor mutations soon.

Acknowledgments. The project is supported by Research Starting Funding from Shantou University (Number NTF21021).

Recommender: LU Jiahai, Professor, Sun Yat-sen University in China.

References

1. Voosen, P.: A 2-week weather forecast may be as good as it gets. Science **363**(6429), 801 (2019)
2. Gerstenberger, M.C., et al.: Real-time forecasts of tomorrow's earthquakes in California. Nature **435**(7040), 328–331 (2005)
3. Gibson, G.C., et al.: Improving probabilistic infectious disease forecasting through coherence. PLoS Comput. Biol. **17**(1), e1007623 (2021)
4. Zhou, P., et al.: A pneumonia outbreak associated with a new Coronavirus of probable bat origin. Nature **579**(7798), 270–273 (2020)
5. Korber, B., et al.: Tracking changes in SARS-CoV-2 spike: evidence that D614G increases infectivity of the COVID-19 virus. Cell **182**(4), 812-827.e19 (2020). https://doi.org/10.1016/j.cell.2020.06.043
6. Jhun, H., Park, H.Y., Hisham, Y., Song, C.S., Kim, S.: SARS-CoV-2 delta (B.1.617.2) variant: a unique T478K mutation in receptor binding motif (RBM) of spike gene. Immune Netw. **21**(5), e32 (2021). https://doi.org/10.4110/in.2021.21.e32
7. Pascarella, S., et al.: SARS-CoV-2 B.1.617 Indian variants: are electrostatic potential changes responsible for a higher transmission rate? J. Med. Virol. **93**(12), 6551–6556 (2021)
8. Aleem, A., Akbar Samad, A.B., Slenker, A.K.: Emerging Variants of SARS-CoV-2 And Novel Therapeutics Against Coronavirus (COVID-19). StatPearls, Treasure Island (FL) (2022)
9. He, S., Peng, Y., Sun, K.: SEIR modeling of the COVID-19 and its dynamics. Nonlinear Dyn. **101**(3), 1667–1680 (2020)

10. Zuo, P., Li, L.: Antigen evolution from D614, to G614, and to Delta subtype of SARS-CoV-2 (2022). https://www.researchsquare.com/article/rs-1218337/v1
11. Zuo, P., et al.: Antigen evolution from D614, to G614, to delta, and to omicron subtype of SARS-CoV-2 (2022). https://www.researchsquare.com/article/rs-1859710/v1
12. Meng, Q., et al.: Spiralin-like protein SLP31 from spiroplasma eriocheiris as a potential antigen for immunodiagnostics of tremor disease in Chinese mitten crab Eriocheir sinensis. Folia Microbiol. (Praha) **55**(3), 245–250 (2010)

Author Index

B

Bai, Lanxi   189
Bin, Peng   292
Bing-yuan, Cao   292, 405

C

Cao, Jiekai   201
Cen, Yongru   189
Chai, Yaping   73
Chen, Jin-pei   64
Chen, Miaoxia   395
Chen, Qinqun   383
Chen, Weitong   96, 163
ChuXiong, Hu   343

D

Deng, Shengqi   201
Du, Wen Sheng   219
Duan, Yuhang   119

F

Fong, Yunshi   373
Fu, Xu   37

G

Gong, Zheyu   108
Guan, Jianbo   373
Guo, Ying   335
Guo, Yinhua   3
Guo, Yizhi   28

H

Hao, Yuexing   383
He, Guo-dong   64
He, Qiang   358
He, Qingping   201
He, Xiao   300
Huo, Jianhong   383

L

Li, Chongwen   383
Li, Dongxin   335
Li, Guohua   383
Li, Jialian   96, 163
Li, Jialu   383
Li, Li-Ping   422
Li, Qi   177
Li, Tai-Fu   300
Li, Taifu   358
Li, Xia   383
Li, Xianglong   151
Li, Yongtao   140
Li, Yu-Yan   300
Li, Yuyan   358
Li, Zhi-Hong   422
Li, Zhixuan   28
Li, Zhongping   232
Li, Ziqing   189
Liang, Sitong   279
Liang, Yiqi   28
Lin, Hongbin   96, 163
Lin, Li   108
Liu, Guiqing   383
Liu, Ning   279
Liu, Weiyu   189
Liu, Xi-yue   64
Liu, Zhihong   335
Luo, Haichang   73

M

Mai, Hong   177
Mao, Yang   119

N

Nan, Wu   343
Nie, Yulin   279

O

OuYang, YongBin   49, 319

P
Peng, Zhongyuan   140

Q
Qin, Fuhong   358
Qin, Zejian   37
Qing, Fu-Hong   300

S
Shi, Cunqin   3
Shibghatullah, Abdul Samad   395
Song, Bingying   73
Su, Peiwei   335

T
Tang, Hongwei   151
Tang, Qianyao   335
Tian-hui, Yin   405
Tong, Zhiyuan   232, 279

W
Wan, Jun   177
Wang, Jiayin   3
Wang, Shufeng   268
Wang, Shuyi   140
Wang, Wenzhu   201
Wang, Xianguo   358
Wei, Caimin   232, 279
Wei, Fuyi   335
Wei, Hang   383
Wu, Jiasheng   189
Wu, Qiwen   177
Wu, Shuitian   189
Wu, Wanying   151
Wu, Yu-cheng   64

X
Xiaoyu, Wang   343
Xie, Ying   177
Xu, Jian   251

Y
Yang, Jiexia   119
Yang, Xiaopeng   395
Yao, Anyi   28
Ying-bing, Chen   292
Yong-wen, Chen   292
Yu, Fusheng   3

Z
Zeng, Liting   373
Zeng, Qiao   300
Zhang, Congrun   73
Zhang, Hang   335
Zhang, Shao-Lin   300
Zhang, Weize   140
Zhang, Yifeng   119
Zhao, Hongchao   358
Zheng, Jiachong   119
Zhong, Wangwei   96
Zhong, Xingyi   129
Zhong, Yubin   28, 96, 108, 163
Zhong, Ziqi   140
Zhou, Xue-Gang   319
Zhu, Chang-xin   37
Zhu, ChangXin   49
Zhu, Guo-Cheng   251
Zhu, Wenya   177
Zhuang, Miaoxia   279
Zou, Zongbao   232
Zuo, Long-Long   422
Zuo, Pei-Jun   422

Printed in the United States
by Baker & Taylor Publisher Services